T0214237

Lecture Notes in Computer Science   12098

More information about this series at http://www.springer.com/series/7407

Marcella Anselmo · Gianluca Della Vedova ·
Florin Manea · Arno Pauly (Eds.)

Beyond the Horizon
of Computability

16th Conference on Computability in Europe, CiE 2020
Fisciano, Italy, June 29 – July 3, 2020
Proceedings

Editors
Marcella Anselmo ⓘ
University of Salerno
Fisciano, Italy

Gianluca Della Vedova ⓘ
University of Milano-Bicocca
Milan, Italy

Florin Manea ⓘ
University of Göttingen
Göttingen, Germany

Arno Pauly ⓘ
Swansea University
Swansea, UK

ISSN 0302-9743   ISSN 1611-3349   (electronic)
Lecture Notes in Computer Science
ISBN 978-3-030-51465-5   ISBN 978-3-030-51466-2   (eBook)
https://doi.org/10.1007/978-3-030-51466-2

LNCS Sublibrary: SL1 – Theoretical Computer Science and General Issues

This Springer imprint is published by the registered company Springer Nature Switzerland AG
The registered company address is: Gewerbestrasse 11, 6330 Cham, Switzerland

Preface

The conference Computability in Europe (CiE) is organized yearly under the auspices of the Association CiE, a European association of mathematicians, logicians, computer scientists, philosophers, physicists, biologists, historians, and others interested in new developments in computability and their underlying significance for the real world. CiE promotes the development of computability-related science, ranging over mathematics, computer science, and applications in various natural and engineering sciences, such as physics and biology, as well as related fields, such as philosophy and history of computing. CiE 2020 had as its motto *Beyond the Horizon of Computability*, reflecting the interest of CiE in research transgressing the traditional boundaries of computability theory.

CiE 2020 was the 16th conference in the series, and the first one to take place virtually. If not for the COVID-19 pandemic, CiE 2020 would have taken place in Salerno, Italy. We are grateful to the Computer Science Department of the University of Salerno for their willingness to host and support our conference. In the transition to a virtual conference, the patience and adaptability of the community was invaluable. We also would like to thank the trailblazers who shared their experience, in particular the organizers of BCTCS 2020 and EDBT/ICDT 2020.

The 15 previous CiE conferences were held in Amsterdam (The Netherlands) in 2005, Swansea (UK) in 2006, Siena (Italy) in 2007, Athens (Greece) in 2008, Heidelberg (Germany) in 2009, Ponta Delgada (Portugal) in 2010, Sofia (Bulgaria) in 2011, Cambridge (UK) in 2012, Milan (Italy) in 2013, Budapest (Hungary) in 2014, Bucharest (Romania) in 2015, Paris (France) in 2016, Turku (Finland) in 2017, Kiel (Germany) in 2018, and Durham (UK) in 2019. CiE 2021 will be held in Ghent (Belgium). Currently, the annual CiE conference is the largest international meeting focused on computability-theoretic issues. The proceedings containing the best submitted papers, as well as extended abstracts of invited, tutorial, and special session speakers, for all these meetings are published in the Springer series *Lecture Notes in Computer Science*.

The CiE conference series is coordinated by the CiE conference series Steering Committee consisting of Alessandra Carbone (Paris), Gianluca Della Vedova (Milan), Liesbeth De Mol (Lille), Mathieu Hoyrup (Nancy), Nataša Jonoska (Tampa FL),

Benedikt Löwe (Amsterdam/Cambridge/Hamburg), Florin Manea (Göttingen, chair), Klaus Meer (Cottbus), Russel Miller (New York), Mariya Soskova (Madison), and ex-officio members Paola Bonizzoni (Milan, President of the Association CiE) and Dag Normann (Oslo).

The Program Committee of CiE 2020 was chaired by Marcella Anselmo (University of Salerno, Italy) and Arno Pauly (Swansea University, UK). The committee, consisting of 27 members, selected the invited and tutorial speakers and the special session organizers, and coordinated the reviewing process of all submitted contributions.

The S. Barry Cooper Prize

S. Barry Cooper (1943–2015) was the founding President of the Association Computability in Europe; his vision on the fundamental concept of computability brought together several different research communities. In memory of Barry's visionary engagement, the Association Computability in Europe established the S. Barry Cooper Prize. The prize is awarded to a researcher who has contributed to a broad understanding and foundational study of computability by outstanding results, by seminal and lasting theory building, by exceptional service to the research communities involved, or by a combination of these.

The first S. Barry Cooper Prize was awarded as part of CiE 2020. Following nomination by the community, the Prize Committee (Anuj Dawar (chair), Peter Van Emde Boas, Yuri Gurevitch, Mariya Soskova, and Paola Bonizzoni) selected Bruno Courcelle for his work on the definability of graph properties in Monadic Second Order Logic, through a sequence of seminal papers and a book (joint with Joost Engelfriet). This forms an outstanding example of theory building, bringing together logic, computability, graph grammars, and various notions of graph width (tree-width, clique-width, and rank-width) and opening new avenues in our understanding of graph structure theory and the computability and complexity of graph algorithms. Besides its foundational character, the work has had great impact on a number of areas of computer science, including in parameterized algorithmics, verification and other areas, and has influenced a generation of researchers in this field. It has straddled the divide between the logical and algorithmic aspects of theoretical computer science.

Structure and Program of the Conference

The Program Committee invited six speakers to give plenary lectures at CiE 2020: Paolo Boldi (Milano, Italy), Véronique Bruyère (Mons, Belgium), Ekaterina Fokina (Vienna, Austria), Amoury Pouly (Oxford, UK), Antonio Restivo (Palermo, Italy), and Damien Woods (Maynooth, Ireland). The conference also had two plenary tutorials, presented by Virginia Vassilevska-Williams (MIT, USA) and Martin Ziegler (KAIST, South Korea).

In addition, the conference had six special sessions: Algorithmic Learning Theory, Combinatorial String Matching, Computable Topology, HAPOC session on Fairness in Algorithms, Large Scale Bioinformatics and Computational Sciences, and Modern Aspects of Formal Languages. Speakers in these special sessions were selected by the respective special session organizers and were invited to contribute a paper to this volume.

Algorithmic Learning Theory
Organizers: Frank Stephan and Lorenzo Carlucci
Speakers:
Ziyuan Gao
Luca San Mauro
Amir Yehudayoff
Thomas Zeugmann

Combinatorial String Matching
Organizers: Travis Gagie and Marinella Sciortino
Speakers:
Inge Li Gørtz
Markus Lohrey
Cinzia Pizzi
Przemek Uznanski

Computable Topology
Organizers: Matthew de Brecht and Mathieu Hoyrup
Speakers:
Takayuki Kihara
Alexander Melnikov
Matthias Schröder
Mariya Soskova

HAPOC session on Fairness in Algorithms
Organizers: Viola Schiaffonati and Teresa Scantamburlo
Speakers:
Krishna Gummadi
Christoph Heitz
Teresa Scantamburlo

Large Scale Bioinformatics and Computational Sciences
Organizers: Gianluca Della Vedova and Iman Hajirasouliha
Speakers:
Marco Aldinucci
Can Alkan
Valentina Boeva
Erik Garrison

Modern Aspects of Formal Languages
Organizers: Markus L. Schmid and Rosalba Zizza
Speakers:

Joel D. Day
Stavros Konstantinidis
Maria Madonia
Wim Martens

The members of the Program Committee of CiE 2020 selected for publication in this volume and for presentation at the conference 23 of the 49 non-invited submitted papers. Each paper received at least three reviews by the Program Committee and their subreviewers. In addition to the accepted contributed papers, this volume contains 21 invited papers. The production of the volume would have been impossible without the diligent work of our expert referees, both Program Committee members and subreviewers. We would like to thank all of them for their excellent work.

All authors who have contributed to this conference are encouraged to submit significantly extended versions of their papers, with additional unpublished research content, to *Computability. The Journal of the Association CiE*.

The Steering Committee of the conference series CiE is concerned about the representation of female researchers in the field of computability. In order to increase female participation, the series started the Women in Computability (WiC) program in 2007. In 2016, after the new constitution of the Association, CiE allowed for the possibility of creating special interest groups, a Special Interest Group named WiC was established. Also since 2016, the WiC program is sponsored by ACM's Women in Computing. This program includes a workshop, the annual WiC dinner, the mentorship program and a grant program for young female researchers. The WiC workshop continued in 2020, coordinated by Mariya Soskova. The speakers were Ekaterina Fokina, Marinella Sciortino, and Virginia Vassilevska-Williams.

The organizers of CiE 2020 would like to acknowledge and thank the following entities for their financial support (in alphabetical order): the Association for Symbolic Logic (ASL), the Commission for the History and Philosophy of Computing (HaPoC), the Computer Science Department of the University of Salerno, the *Gruppo Nazionale per il Calcolo Scientifico* (GNCS–INdAM), and Springer.

We gratefully thank all the members of the Organizing Committee of CiE 2020 for their work towards making the conference a successful event. We thank Andrej Voronkov for his EasyChair system which facilitated the work of the program Committee and the editors considerably.

May 2020

Marcella Anselmo
Gianluca Della Vedova
Florin Manea
Arno Pauly

Organization

Program Committee

Marcella Anselmo (Co-chair)	University of Salerno, Italy
Veronica Becher	Universidad de Buenos Aires, Argentina
Paola Bonizzoni	University of Milano-Bicocca, Italy
Laura Crosilla	University of Oslo, Norway
Liesbeth De Mol	University of Lille, France
Gianluca Della Vedova	University of Milano-Bicocca, Italy
Jerôme Durand-Lose	Université d'Orléans, France
Paweł Gawrychowski	University of Wroclaw, Poland
Mathieu Hoyrup	LORIA, France
Juliette Kennedy	University of Helsinki, Finland
Karoliina Lehtinen	The University of Liverpool, UK
Benedikt Löwe	University of Amsterdam, The Netherlands
Florin Manea	Universität Göttingen, Germany
Timothy McNicholl	Iowa State University, USA
Klaus Meer	BTU Cottbus-Senftenberg, Germany
Turlough Neary	University of Zurich, Switzerland
Daniel Paulusma	Durham University, UK
Arno Pauly (Co-chair)	Swansea University, UK
Karin Quaas	University of Leipzig, Germany
Viola Schiaffonati	Politecnico di Milano, Italy
Markus L. Schmid	Humboldt University Berlin, Germany
Thomas Schwentick	Technical University of Dortmund, Germany
Marinella Sciortino	University of Palermo, Italy
Victor Selivanov	Novosibirsk State University, Russia
Mariya Soskova	University of Wisconsin–Madison, USA
Peter Van Emde Boas	University of Amsterdam, The Netherlands
Linda Brown Westrick	Penn State University, USA

Organizing Committee

Marcella Anselmo (Co-chair)
Gianluca Della Vedova (Co-chair)
Manuela Flores
Rocco Zaccagnino
Rosalba Zizza

Additional Reviewers

Allouche, Jean-Paul
Anglès D'Auriac, Paul-Elliot
Avni, Guy
Bader, Michael
Bernardini, Giulia
Berwanger, Dieter
Biniaz, Ahmad
Carl, Merlin
Carton, Olivier
Casel, Katrin
Cenzer, Douglas
Choffrut, Christian
Crépeau, Claude
Damnjanovic, Zlatan
Day, Joel
De Naurois, Paulin Jacobé
Downey, Rod
Dudek, Bartlomiej
Egry-Nagy, Attila
Ehlers, Thorsten
Férée, Hugo
Ferreira, Fernando
Fiorentini, Camillo
Fiorino, Guido
Fleischmann, Pamela
Franklin, Johanna
Frid, Anna
Galeotti, Lorenzo
Georgiev, Ivan
Giammarresi, Dora
Habič, Miha
Haeusler, Edward Hermann
Hölzl, Rupert
Ishihara, Hajime
Janczewski, Wojciech
Joosten, Joost
Kawamura, Akitoshi
Kihara, Takayuki
Kötzing, Timo
Kristiansen, Lars

Kufleitner, Manfred
Linker, Sven
Lohrey, Markus
Lubarsky, Robert
Madelaine, Florent
Nartin, Barnaby
Melnikov, Alexander
Mignosi, Filippo
Miller, Russell
Monin, Benoît
Mummert, Carl
Nagy, Benedek
Neuen, Daniel
Neuwirth, Stefan
Onn, Shmuel
Palmiere, Ethan
Patey, Ludovic
Perifel, Sylvain
Piccinini, Gualtiero
Pinna, G. Michele
Pokorski, Karol
Rathjen, Michael
Rizzi, Raffaella
Romashchenko, Andrei
San Mauro, Luca
San Pietro, Pierluigi
Selivanova, Svetlana
Smith, Jonathan
Sorbi, Andrea
Totzke, Patrick
Towsner, Henry
Verlan, Sergey
Vidal, Amanda
Watrous, John
Welch, Philip
Yang, Fan
Yang, Yue
Yu, Liang
Ziegler, Martin
Zizza, Rosalba

Abstracts of Invited Talks

Acceleration of Read Mapping Through Hardware/Software Co-design

Can Alkan

Department of Computer Engineering, Bilkent University, Ankara, Turkey
calkan@cs.bilkent.edu.tr

Genome sequence analysis can enable significant advancements in areas such as personalized medicine, study of evolution, and forensics. However, effectively leveraging genome sequencing for advanced scientific and medical breakthroughs requires very high computational power. As prior works have shown, many of the core steps in genome sequence analysis are bottle necked by the current capabilities of computer systems, as these steps must process a large amount of data.

Read mapping process, which is identification of the most likely locations (under a sequence distance metric) of hundreds of millions to billions of short DNA segments within a long reference genome is a major computational bottleneck in genome analysis as the flood of sequencing data continues to overwhelm the processing capacity of existing algorithms and hardware. There is also an urgent need for rapidly incorporating clinical DNA sequencing and analysis into clinical practice for early diagnosis of genetic disorders or bacterial and viral infections. This makes the development of fundamentally new, fast, and efficient read mapper the utmost necessity.

This talk describes our ongoing journey in improving the performance of genome read mapping through hardware/software co-design. I will first introduce using parallelism in existing hardware such as single instruction multiple data (SIMD) instructions in commodity CPUs, and then describe use of massively parallel GPGPUs. I will then introduce our FPGA designs for the same purpose. Finally, I will discuss our latest acceleration efforts to map both short and long reads using the emerging processing-in-memory computation.

Centralities in Network Analysis

Paolo Boldi

Dipartimento di Informatica, Università degli Studi di Milano, via Celoria 18,
I-20133 Milan, Italy

Abstract. Graphs are one of the most basic representation tools employed in many fields of pure and applied science; they are the natural way to model a wide range of situations and phenomena. The emergence of social networking and social media has given a new strong impetus to the field of "social network analysis"; although the latter can be thought of as a special type of data mining, and can take into account different data related to the network under consideration, a particularly important type of study is what people usually call "link analysis". In a nutshell, link analysis tries to discover properties, hidden relations and typical patterns and trends from the study of the graph structure alone.

In other words, more formally, link analysis studies graph invariants: given a family $\mathcal{G}$ of graphs under consideration, a V-valued *graph invariant* [7] is any function $\pi : \mathcal{G} \to V$ that is invariant under graph isomorphisms (that is, $G \cong H$ implies $\pi(G) = \pi(H)$, where $\cong$ denotes isomorphism).

Binary invariants are those for which $V = \{true, false\}$: for instance, properties like "the graph is connected", "the graph is planar" and so on are all examples of binary graph invariants. Scalar invariants have values on a field (e.g., $V = \mathbb{R}$): for instance "the number of connected components", "the average length of a shortest path", "the maximum size of a connected component" are all examples of scalar graph invariants. Distribution invariants take as value a *distribution*: for instance "the degree distribution" or the "shortest-path–length distribution" are all examples of distribution invariants.

It is convenient to extend the definition of graph invariants to the case where the output is in fact a function assigning a value to each node. Formally, a V-valued *node-level graph invariant* is a function π that maps every $G \in \mathcal{G}$ to an element of V^{N_G} (i.e., to a function from the set of nodes N_G to V), such that for every graph isomorphism $f : G \to H$ one has $\pi(G)(x) = \pi(H)(f(x))$. Informally, π should have the same value on nodes that are exchanged by an isomorphism.

Introducing node-level graph invariants is crucial in order to be able to talk about graph centralities [8]. A node-level real-valued graph invariant that aims at estimating node importance is called a *centrality (index or measure)* [1]; given a centrality index $c : N_G \to \mathbb{R}$, we interpret $c(x) > c(y)$ as a sign of the fact that x is "more important" (more central) than y. Since the notion of being "important" can be declined to have different meanings, in different contexts, a wide range of indices were proposed in the literature.

The purpose of my talk is to present the idea of graph centrality in the context of Information Retrieval, and then to give a taxonomic and historically-aware description of the main centrality measures defined in the literature and commonly used in social-network analysis (see, e.g., [3–7]). I will then discuss how centrality measures can be compared with one another, and focus on the axiomatic approach [2], providing examples of how this approach can be used to highlight the features that a certain centrality does (or does not) possess.

References

1. Bavelas, A.: A mathematical model for group structures. Hum. Organ. **7**, 16–30 (1948)
2. Boldi, P., Vigna, S.: Axioms for centrality. Internet Math. **10**(3–4), 222–262 (2014)
3. Brin, S., Page, L.: The anatomy of a large-scale hypertextual web search engine. Comput. Netw. ISDN Syst. **30**(1), 107–117 (1998)
4. Craswell, N., Hawking, D., Upstill, T.: Predicting fame and fortune: PageRank or indegree? In: Proceedings of the Australasian Document Computing Symposium, ADCS2003, pp. 31–40 (2003)
5. Freeman, L.C.: A set of measures of centrality based on betweenness. Sociometry **40**(1), 35–41 (1977)
6. Katz, L.: A new status index derived from sociometric analysis. Psychometrika **18**(1), 39–43 (1953). https://doi.org/10.1007/bf02289026
7. Lovász, L.: Large Networks and Graph Limits., Colloquium Publications, vol. 60. American Mathematical Society (2012)
8. Newman, M.E.J.: Networks: An Introduction. Oxford University Press, Oxford, New York (2010)
9. Vigna, S.: Spectral ranking. Netw. Sci. **4**(4), 433–445 (2016)

A Game-Theoretic Approach for the Automated Synthesis of Complex Systems

Véronique Bruyère

UMONS - University of Mons, 20 Place du Parc, B-7000-Mons, Belgium
Veronique.Bruyere@umons.ac.be

Abstract. *Game theory* is a well-developed branch of mathematics that is applied to various domains like economics, biology, computer science, etc. It is the study of mathematical models of interaction and conflict between individuals and the understanding of their decisions assuming that they are rational [11, 13].

The last decades have seen a lot of research on the *automatic synthesis* of reliable and efficient systems by using the mathematical framework of game theory. One important line of research is concerned with *reactive systems* that must continuously react to the uncontrollable events produced by the environment in which they evolve. A *controller* of a reactive system indicates which actions it has to perform to satisfy a certain objective against any behavior of the environment. An example in air traffic management is the autopilot that controls the trajectory of the plane to guarantee a safe landing without any control on the weather conditions. Such a situation can be modeled by a *two-player game played on a finite directed graph*: the system and the environment are the two players, the vertices of the graph model the possible configurations of the system and the environment, and the infinite paths in the graph model the continuous interactions between them. As we cannot assume the cooperation of the environment, its objective is the negation of the objective of the system and we speak of *zero-sum* games. In this framework, checking whether the system is able to achieve its objective reduces to the existence of a *winning strategy* in the corresponding game, and building a controller reduces to computing such a strategy [8]. Whether such a controller can be automatically designed from the objective is known as the *synthesis problem*.

In this talk, we consider another, more recent, line of research concerned with the modelization and the study of *complex systems*. Instead of the simple situation of a system embedded in a hostile environment, we are now faced with systems/environments formed of several components each of them with their own objectives that are not necessarily conflicting. An example is a communication network composed of several nodes, each of them aiming at sending a message to some other nodes by using a certain frequency range. These objectives are conflicting or not, depending on the used frequencies. We model such complex systems by *multi-player non zero-sum games played on graphs*: the components of the complex system are the different players, each of them aiming at satisfying his own objective. In this context, the synthesis problem is a little different: winning strategies are no longer appropriate and are replaced by the concept of *equilibrium*, that is, a profile of strategies, one for each player, such that no player has an incentive to deviate from the play consistent with this profile [9]. Different kinds of relevant equilibria have been investigated among which the famous notions of *Nash equilibrium* [10] and *subgame perfect equilibrium* [12].

In the setting of multi-player non zero-sum games, *classical questions* are the following ones. What are the objectives for which there always exists an equilibrium of a certain given type? When the existence is not guaranteed, can we decide whether such an equilibrium exists or not. When the latter problem is decidable, what is its complexity class? Many results were first obtained for *Boolean* objectives, and in particular for *ω-regular* objectives like avoiding a deadlock, always granting a request, etc [8, 9]. In this context, an infinite path in the game graph is either winning or losing w.r.t. a player depending on whether his objective is satisfied or not. More recent results were then obtained for *quantitative* objectives such as minimizing the energy consumption or guaranteeing a limited response time to a request [6]. To allow such richer objectives, the game graph is augmented with weights and a payoff (or a cost) is then associated with each of its infinite paths [7]. When solving the above mentioned questions, for practical applicability of the studied models, it is also important to know how *complex* the strategies composing the equilibrium are. Given past interactions between the players, a strategy for a player indicates the next action he has to perform. The amount of memory on those past interactions is one of the ways to express the complexity of the strategy, the simplest strategies being those requiring no memory at all [8].

In this talk, we focus on *reachability objectives*: each player wants to reach a given subset of vertices (qualitative objective) or to reach it as soon as possible (quantitative objective). We also restrict the discussion to *turn-based* games (the players choose their actions in a turn-based way, and not concurrently) and to *pure* strategies (the next action is chosen in a deterministic way, and not according to a probability distribution). For those games, we explain the different techniques developed and the results obtained for both notions of Nash equilibrium and subgame perfect equilibrium [1–5]. The talk is made accessible to a large audience through illustrative examples.

References

1. Brihaye, T., Bruyère, V., De Pril, J., Gimbert, H.: On subgame perfection in quantitative reachability games. Log. Methods Comput. Sci. **9** (2012)
2. Brihaye, T., Bruyère, V, Goeminne, A., Raskin, J.-F., van den Bogaard, V.: The complexity of subgame perfect equilibria in quantitative reachability games. In: CONCUR Proceedings, vol. 140. LIPIcs, pp. 13:1–13:16. Schloss Dagstuhl - Leibniz-Zentrum für Informatik (2019)
3. Brihaye, T., Bruyère, V., Goeminne, A., Thomasset, N.: On relevant equilibria in reachability games. In: Filiot, E., Jungers, R., Potapov, I. (eds.) Reachability Problems, RP 2019. LNCS, vol. 11674, 48–62. Springer, Cham (2019). https://doi.org/10.1007/978-3-030-30806-3_5
4. Brihaye, T., Bruyère, V., Meunier, N., Raskin, J.-F.: Weak subgame perfect equilibria and their application to quantitative reachability. In: CSL Proceedings, vol. 41. LIPIcs, pp. 504–518. Schloss Dagstuhl - Leibniz-Zentrum für Informatik (2015)

5. Brihaye, T., De Pril, J., Schewe, S.: Multiplayer cost games with simple nash equilibria. In: Artemov, S., Nerode, A. (eds.) Logical Foundations of Computer Science, LFCS 2013. LNCS, vol. 7734, pp. 59–73. Springer, Heidelberg (2013). https://doi.org/10.1007/978-3-642-35722-0_5

6. Bruyère, V.: Computer aided synthesis: a game-theoretic approach. In: Charlier, É., Leroy, J., Rigo, M. (eds.) Developments in Language Theory, DLT 2017. LNCS, vol. 10396, pp. 3–35. Springer, Cham (2017). https://doi.org/10.1007/978-3-319-62809-7_1

7. Chatterjee, K., Doyen, L., Henzinger, T.A.: Quantitative languages. ACM Trans. Comput. Log. **11** (2010)

8. Grädel, E., Thomas, W., Wilke, T. (eds.): Automata, Logics, and Infinite Games: A Guide to Current Research. LNCS, vol. 2500. Springer, Heidelberg (2002). https://doi.org/10.1007/3-540-36387-4

9. Grädel, E., Ummels, M.: Solution concepts and algorithms for infinite multiplayer games. In: New Perspectives on Games and Interaction, vol. 4, pp. 151–178. Amsterdam University Press (2008)

10. Nash, J.F.: Equilibrium points in n-person games. In: PNAS, vol. 36, pp. 48–49. National Academy of Sciences (1950)

11. Osborne, M.J., Rubinstein, A.: A Course in Game Theory. MIT Press, Cambridge (1994)

12. Selten, R.: Spieltheoretische Behandlung eines Oligopolmodells mit Nachfrageträgheit. Zeitschrift für die gesamte Staatswissenschaft, 121:301–324 and 667–689 (1965)

13. von Neumann, J., Morgenstern, O.: Theory of Games and Economic Behavior. Princeton University Press (1944)

Hard Problems for Simple Word Equations: Understanding the Structure of Solution Sets in Restricted Cases

Joel D. Day

Loughborough University, UK
J.Day@lboro.ac.uk

A *word equation* is a tuple (α, β), usually written $\alpha = \beta$, such that α and β are words over a combined alphabet $X \cup \Sigma$ consisting of *variables* $X = \{x, y, z, \ldots\}$ and *terminal symbols* $\Sigma = \{\mathsf{a}, \mathsf{b}, \ldots\}$. A solution to a word equation is a substitution of the variables for words in Σ^* unifying the two terms α and β. In other words, a solution is a (homo)morphism $h : (X \cup \Sigma)^* \to \Sigma^*$ satisfying $h(\mathsf{a}) = \mathsf{a}$ for all $\mathsf{a} \in \Sigma$ such that $h(\alpha) = h(\beta)$. For example, one solution h to the word equation $x\mathsf{ab}y = y\mathsf{ba}x$ is given by $h(x) = \mathsf{b}$ and $h(y) = \mathsf{bab}$.

The task of deciding, for a given word equation E, whether a solution h to E exists is called the *satisfiability problem*. There are various reductions showing that the satisfiability problem for word equations is NP-hard. On the other hand, the satisfiability problem was shown to be decidable by Makanin who gave the first general algorithm [11]. Since then, improvements on the complexity upper bounds, first to PSPACE, and later to non-deterministic linear space, have resulted from further algorithms due to Plandowski [14] and Jeż [7, 8] respectively. Additionally, Plandowksi and Rytter [15] showed that solutions to word equations are highly compressible, and their compressed forms can be verified efficiently, leading to a (nondeterministic) algorithm running in time polynomial in $\log(N)$, where N is an upper bound on the length of the shortest solution. It is thought that N is at most exponential in the length of the equation, resulting in the conjecture that the satisfiability problem for word equations is NP-complete.

Determining the precise complexity of the satisfiability problem for word equations, in particular its inclusion in NP, remains one of the most important open problems on the topic. However it is often the case for practical applications such as in software verification, that more general versions of the satisfiability problem need to be solved. Specifically, word equations constitute one type of relation on words, and will typically occur in combination with various other conditions. Examples include membership of regular languages, numerical comparisons on e.g. word-lengths, and rewriting operators such as transducers and the ReplaceAll function (see e.g. [1]). In this context, it is not sufficient to know that an arbitrary solution exists, since it might violate the other conditions. Instead, we are interested in a class of satisfiability problems *modulo constraints* of a given type or combination of types.

The case of satisfiability modulo *regular constraints*, where for each variable x, a solution h must satisfy conditions of the form $h(x) \in \mathcal{L}_x$ where $\mathcal{L}_x$ is a regular language, is known to be **PSPACE**-complete [5, 14, 16]. Many other types of constraints lead to undecidable variants of the satisfiability problem, as is the case with e.g. abelian equivalence, letter-counting functions, and the ReplaceAll operator [2, 9]. The case of satisfiability modulo length constraints, which asks for the existence of a solution h to the equation such that the lengths of the words $h(x)$ for $x \in X$ additionally satisfy a given system of linear diophantine equations, remains a long standing open problem.

One of the main obstacles to solving word equations modulo constraints is that the set of solutions to a word equation can have a rather complex structure. One natural way to try to represent potentially-infinite sets of solutions to a given equation is using parametric words, consisting of word-parameters and numerical-parameters. For example, solutions h for the equation $xy = yx$ are described by the parametric words $h(x) = u^p, h(y) = u^q$, to be interpreted as "there exists a word u and numbers p, q such that $h(x)$ is p consecutive repetitions of u while $h(y)$ is q consecutive repetitions of u". Given such a representation, it is not difficult to imagine how we might go about deciding whether a solution exists which satisfies various kinds of additional constraints, including length and regular constraints. Unfortunately, a (finite) representation using parametric words is typically not possible, even for simple equations such as $xuy = yvx$ (see [13]). Equations whose solution-sets do not permit such a representation are called non-parametrisable, and for these instances we must consider other, less explicit representations.

For quadratic word equations – those for which each variable may occur at most twice – an algorithm exists based on rewriting operations called Nielsen transformations which, given an equation E, produces a finite directed graph $\mathcal{G}_E$ whose vertices are word equations (including one corresponding to E) with the property that solutions to E correspond exactly to walks (i.e. paths which may visit vertices and edges more than once) in the graph from E to the trivial equation $\varepsilon = \varepsilon$, where ε denotes the empty word. As such, properties of the graph $\mathcal{G}_E$ will often also say something about the set of solutions to E, and it is natural to draw comparisons between the structure of $\mathcal{G}_E$ and the structure of the solution-sets. A connection is made in [13] between the existence of certain structures in the graph and the non-parametric stability of E, while [10] demonstrates how restrictions on the structure of $\mathcal{G}_E$ lead to a decidable satisfiability problem modulo length constraints. Moreover, if, for a class of word equations $\mathcal{C}$, the diameter of $\mathcal{G}_E$ is bounded by a polynomial in the length of E for all $E \in \mathcal{C}$, then the satisfiability problem (without constraints) for $\mathcal{C}$ is in **NP**.

In general, the structure of the graphs $\mathcal{G}_E$ is not well understood. It follows from the fact that the satisfiability problem remains **NP**-hard even for very restricted subclasses of quadratic word equations [4, 6], that simply deciding whether a given equation E' is a vertex in $\mathcal{G}_E$ is **NP**-hard. Nevertheless, there are regularities and patterns in these graphs which allow us to begin to describe their structure. In the case of *regular* equations – the subclass of quadratic equations introduced in [12] for which each variable occurs at most once on each side – one can derive results describing these structures, and, as a consequence, bounds on parameters of the graphs such as diameter, number of vertices, and connectivity measures (see [3]), all of which are central to obtaining a better understanding of variants of the satisfiability problem.

References

1. Amadini, R.: A Survey on String Constraint Solving. arXiv e-prints arXiv:2002.02376, January 2020
2. Day, J.D., Ganesh, V., He, P., Manea, F., Nowotka, D.: The satisfiability of word equations: decidable and undecidable theories. In: Potapov, I., Reynier, P.A. (eds.) RP 2018. LNCS, vol. 11123, pp. 15–29. Springer, Cham (2018). https://doi.org/10.1007/978-3-030-00250-3_2
3. Day, J.D., Manea, F.: On the structure of solution sets to regular word equations. In: Proceedings 47th International Colloquium on Automata, Languages and Programming, ICALP 2020. LIPIcs (2020, to appear)
4. Day, J.D., Manea, F., Nowotka, D.: The hardness of solving simple word equations. In: Proceedings of the 42nd International Symposium on Mathematical Foundations of Computer Science, MFCS 2017. LIPIcs, vol. 83, pp. 18:1–18:14 (2017)
5. Diekert, V., Jeż, A., Plandowski, W.: Finding all solutions of equations in free groups and monoids with involution. Inf. Comput. **251**, 263–286 (2016)
6. Robson, J.M., Diekert, V.: On quadratic word equations. In: Meinel, C., Tison, S. (eds.) STACS 1999. LNCS, vol. 1563, pp. 217–226. Springer, Heidelberg (1999). https://doi.org/10.1007/3-540-49116-3_20
7. Jeż, A.: Recompression: a simple and powerful technique for word equations. J. ACM **63** (2016)
8. Jeż, A.: Word equations in nondeterministic linear space. In: Proceedings of the 44th International Colloquium on Automata, Languages, and Programming, ICALP 2017. LIPIcs, vol. 80, pp. 95:1–95:13 (2017)
9. Lin, A.W., Barceló, P.: String solving with word equations and transducers: towards a logic for analysing mutation XSS. In: ACM SIGPLAN Notices. vol. 51, pp. 123–136. ACM (2016)
10. Lin, A.W., Majumdar, R.: Quadratic word equations with length constraints, counter systems, and presburger arithmetic with divisibility. In: Lahiri, S., Wang, C. (eds.) ATVA 2018. LNCS, vol. 11138, pp. 352–369. Springer, Cham (2018). https://doi.org/10.1007/978-3-030-01090-4_21
11. Makanin, G.S.: The problem of solvability of equations in a free semigroup. Sbornik Math. **32**(2), 129–198 (1977)
12. Manea, F., Nowotka, D., Schmid, M.L.: On the complexity of solving restricted word equations. Int. J. Found. Comput. Sci. **29**(5), 893–909 (2018)
13. Petre, E.: An Elementary proof for the non-parametrizability of the Equation $xyz=zvx$. In: Fiala, J., Koubek, V., Kratochvíl, J. (eds.) MFCS 2004. LNCS, vol. 3153, pp. 807—817. Springer, Heidelberg (2004). https://doi.org/10.1007/978-3-540-28629-5_63
14. Plandowski, W.: Satisfiability of word equations with constants is in PSPACE. In: Proceedings of the 40th Annual Symposium on Foundations of Computer Science, FOCS 1999, pp. 495–500. IEEE (1999)
15. Plandowski, W., Rytter, W.: Application of Lempel-Ziv encodings to the solution of word equations. In: Larsen, K.G., Skyum, S., Winskel, G. (eds.) ICALP 1998. LNCS, vol. 1443, pp. 731–742. Springer, Heidelberg (1998)
16. Schulz, K.U.: Makanin's algorithm for word equations-two improvements and a generalization. In: Schulz, K.U. (ed.) IWWERT 1990. LNCS, vol. 572, pp. 85–150. Springer, Heidelberg (1992). https://doi.org/10.1007/3-540-55124-7_4

A Survey of Recent Results on Algorithmic Teaching

Ziyuan Gao

Department of Mathematics, National University of Singapore,
10 Lower Kent Ridge Road, Singapore 119076, Republic of Singapore
matgaoz@nus.edu.sg

Abstract. Algorithmic teaching studies the problem of determining an optimal training set that can guide a learning algorithm to identify a target concept as efficiently as possible. Various models of cooperative teaching and learning have been proposed in the literature, and they may be broadly classified into (1) batch models, where the teacher designs a training sample and presents it all at once to the learner, which then outputs a hypothesis, and (2) sequential or online models, where the teacher presents examples in a sequential manner and the learner updates its hypothesis (or retains its original hypothesis) each time a new datum is received.

The study of teaching-learning models not only has potential real-world applications, such as in modelling human-computer interaction and in the design of educational programmes, but also sheds new light on well-established notions in statistical learning theory. In particular, several teaching complexity measures such as the Recursive Teaching Dimension, the No-Clash Teaching Dimension and a recently introduced sequential model teaching dimension have been shown to be mathematically related to the VC-dimension and/or sample compression schemes. We will give an overview of recently introduced teaching models as well as results on the connections between various teaching complexity measures and the VC-dimension.

We conclude with a brief discussion of some possible notions of a 'helpful' distribution in the Probably Approximately Correct learning model as well as in probabilistic learning settings.

Tree Compression with Top Trees

Inge Li Gørtz

Technical University of Denmark, DTU Compute
inge@dtu.dk

Top tree compression is an elegant and powerful compression scheme for ordered and labeled trees [3]. Top tree compression is based on transforming the input tree T into another tree $\mathcal{T}$, called the *top tree*, that represents a balanced hierarchical clustering of T, and then DAG compressing $\mathcal{T}$. Surprisingly, the resulting *top DAG* achieves optimal compression compared to the information-theoretic lower bound, can compress exponentially better than DAG compression, is never much worse than DAG compression, and supports navigational queries in logarithmic time [1, 3–5]. In this talk we give an overview of the key ideas and techniques in top tree compression and highlight recent results on top tree compression and optimal string dictionaries [2].

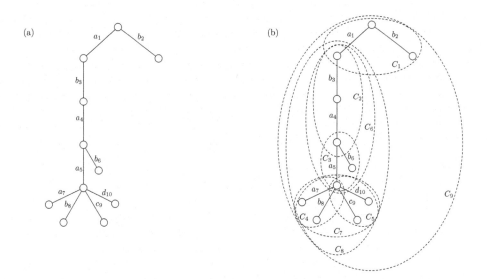

Fig. 1 A trie compressed with top tree compression

References

1. Bille, P., Fernstrøm, F., Gørtz, I.L.: Tight bounds for top tree compression. In: Proceedings of the 24th SPIRE, pp. 97–102 (2017)
2. Bille, P., Gawrychowski, P., Gørtz, I.L., Landau, G.M., Weimann, O.: Top tree compression of tries. In: Proceedings of the Symposium on Algorithms and Computation, ISAAC 2019. LIPIcs, vol. 149, pp. 4:1–4:18 (2019)
3. Bille, P., Gørtz, I.L., Weimann, O., Landau, G.M.: Tree compression with top trees. Inf. Comput. **243**, 166–177 (2015). announced at ICALP 2013
4. Dudek, B., Gawrychowski, P.: Slowing down top trees for better worst-case compression. In: Proceedings of the 29th CPM, pp. 16:1–16:8 (2018)
5. Hübschle-Schneider, L., Raman, R.: Tree compression with top trees revisited. In: Proceedings of the 14th SEA, pp. 15–27 (2015)

Efficient Counting of k-mers and Spaced Seeds to Speed-Up Alignment-Free Methods

Cinzia Pizzi ⓘ

Department of Information Engineering, University of Padova, Italy
cinzia.pizzi@dei.unipd.it

The comparison of sequences is both a core problem itself, and a preliminary step of more complex types of analysis. While many approaches have been developed during the past decades, the problem of sequence comparison is still a computational challenge in many application fields that are big data generators. For example, in Bioinformatics, the development of high-throughput technologies has made the computation of similarity measures based on sequence alignment unpractical in many contexts.

Today, an important portion of state-of-the-art tools for biosequence analysis are based on alignment-free approaches [1], where sequences are described in terms of content descriptors such as k-mers, or spaced seeds (i.e. gapped patterns). Thus, the efficient computation of words statistics, such as presence or frequency, is crucial for all those tools that are based on alignment-free techniques.

Word counting is a largely studied problem, with apparent relatively simple solutions. However, when dealing with tens or hundreds of Giga-bases, performing this task in a reasonable amount of time, given limited resources, becomes a challenge.

This lead to the development of tools that try to exploit system resources in different ways to achieve this goal. These tools are based on approaches that can be memory-based or disk-based, rely on data structures, statistical filters or sorting, and exploit parallelism by processing chunks of data and then merging the results (e.g. [2–4]).

In this talk I will discuss state-of-the-art k-mer counting, with a focus on recently proposed approaches to further speed-up sequence scanning when collecting statistics on k-mers [5] and spaced seeds [6].

References

1. Zielezinski, A., Vinga, S., Almeida, J., Karlowski, W.M.: Alignment-free sequence comparison: benefits, applications, and tools. Genome Biol. **18**, 186 (20170. https://doi.org/10.1186/s13059-017-1319-7
2. Marçais, G., Kingsford, C.: A fast, lock-free approach for efficient parallel counting of occurrences of k-mers. Bioinformatics **27**(6), 764–770 (2011). https://doi.org/10.1093/bioinformatics/btr011
3. Rizk, G., Lavenier, D., Chikhi, R.: DSK: k-mer counting with very low memory usage. Bioinformatics **29**(5), 652–653 (2013). https://doi.org/10.1093/bioinformatics/btt020

4. Kokot, M., Długosz, M., Deorowicz, S.: KMC 3: counting and manipulating k-mer statistics. Bioinformatics **33**(17), 2759–2761 (2017). https://doi.org/10.1093/bioinformatics/btx304
5. Pellegrina, L., Pizzi, C., Vandin, F.: Fast approximation of frequent k-mers and applications to metagenomics. J. Comput. Biol. **27**(4), 534–549 (2020) http://doi.org/10.1089/cmb.2019.0314
6. Petrucci, E., Noè, L., Pizzi, C., Comin, M.: Iterative spaced seed hashing: closing the gap between spaced seed hashing and k-mer hashing. J. Comput. Biol. **27**(2), 223–233 (2020). http://doi.org/10.1089/cmb.2019.0298

On the Repetitive Structure of Words

Antonio Restivo

DMI, Università di Palermo, via Archirafi, 34 - 90123 Palermo, Italy
antonio.restivo@unipa.it

Abstract. The investigations on repetitions in words began with the work of Axel Thue at the dawn of last century (cf. [3]). Nowadays they constitute one of the most fundamental areas of combinatorics on words and are important both in theory and practice (cf. [18, 19]). In this presentation we report some recent results related to repetitions in words, focusing on the combinatorial aspects. Our contribution is not to be intended as a complete survey on the repetitive structure of words: we have put aside some important aspects as, for instance, the researches around the "runs" theorem (cf. 1, 7, 17]).

A basic definition in this context is that of *power*. Recall that a k-power is a word of the form w^k, for some non-empty word w and a positive integer k. Recently, the notion of *antipower* has been introduced in [12] to contrast that of power. If r is a positive integer, an r-*antipower* is a word of the form $w_1 w_2 \ldots w_r$, where the w_i are words such that $|w_i| = |w_j|$ and $w_i \neq w_j$ for every pair (i, j) with $i \neq j$. Problems related to avoidability of antipowers have been recently approached in several papers (cf. [2, 4, 10, 14, 21]). These studies could be considered as the first contributions to an anti-Ramsey theory (cf. [11]) in the context of words. A significant result in this area concerns the simultaneous avoidance of powers and antipowers. In [12] it is proved that, for any pair (k, r) of positive integers, there is an integer $N(k, r)$ such that any word having length greater than or equal to $N(k, r)$ either contains a k-power or an r-antipower. The exact asymptotic behavior of the function $N(k, r)$ remains open. The most recent results on upper and lower bounds on $N(k, r)$ have been obtained in [13].

The notions of power and antipower correspond to the opposite extreme cases in an analysis of the repetitive structure of a word. Indeed, one could consider, as a measure of repetitiveness of a word w, the number of *distinct* factors in a factorization of w: if this number is small (in comparison to the length of w), then many factors are repeated (cf. [6, 22]). In this approach, powers, i.e. words with factorizations having only one (repeated) factor, are at the first level of the hierarchy, and correspond to maximal repetitiveness. Closely related to the notion of power are the notions of *primitive* word and *root*.

We then investigate the second level of this hierarchy, i.e. words that can be factorized using (an arbitrary number of) copies of just two words. If these two words are related by some morphism or anti-morphism, we have *pseudo-powers*. The concept of pseudo-power (or pseudo-repetition) has been introduced in [8] and draws its original motivations from computational biology. The notions of primitivity and root are extended to pseudo-powers and, in particular, it is proved that any word has a unique (pseudo-)root. For recent developments on pseudo-powers, and some extensions, the reader could refer to [9, 15, 16, 20].

In the last part of the presentation the approach to pseudo-powers is further generalized by dropping any relation between the two words. We introduce a notion of primitivity for a set of words and, given a word w, we say that a pair $\{u, v\}$ of words is a *bi-root* of w if w can be written as a concatenation of copies of u and v, and $\{u, v\}$ is a primitive set. In particular, we prove (cf. [5]) that the bi-root of a primitive word is unique provided the length of the word is sufficiently large with respect the size of the bi-root. The tight bound on this length remains an open problem. We also present some combinatorial properties of bi-roots proved in [5], and suggestions for further work.

References

1. Bannai, H., Tomohiro, I, Inenaga, S. Yuto, N., Takeda, M., Tsuruta, K. The "Runs" theorem. SIAM J. Comput. **46**(5), 1501–1514 (2017)
2. Berger, A., Defant, C.: On anti-powers in aperiodic recurrent words. preprint arXiv:1902. 01291 (2019)
3. Berstel, J.: Axel Thue's papers on repetitions in words: a Translation, Number **20** in Publications du Laboratoire de Combinatoire et Informatique mathematique, Universitè du Quebec a Montreal (1995)
4. Burcroft, A.: (k, λ)-anti-powers and other patterns, in words. Electron. J. Comb. 25, 4–41 (2018)
5. Castiglione, G., Fici, G., Restivo, A.: On sets of words of rank two. In: Mercaş, R., Reidenbach, D. (eds.) WORDS 2019. LNCS, vol. 11682, pp. 46–59. Springer, Cham (2019). https://doi.org/10.1007/978-3-030-28796-2_3
6. Condon, A., Manuch, J., Thachuk, C.: The complexity of string partitioning. J. Discrete Algorithms, **32**, 24–43 (2015)
7. Crochemore, M., Ilie, L., Rytter, W.: Repetitions in strings. Theoret. Comput. Sci. **410**, 5227–5235 (2009)
8. Czeizler, E., Kari, L., Seki, S.: On a special class of primitive words, Theoret, Comput. Sci. **411**, 617–630 (2010).
9. Czeizler, E., Czeizler, E., Kari, L., Seki, S.: An extension of the Lyndon-Schützenberger result to pseudoperiodic words **209**, 717–730 (2011)
10. Defant, C.: Anti-power prefixes of the Thue-Morse word. Electron. J. Comb. **24**, 1–32 (2017)
11. Erdös, P., Simonovits, M., Sós, V.T.: Anti-Ramsey theorems, infinite and finite sets. Colloq. Math. Soc. János Bolyai 633–643 (1975). (Colloq., Keszthely, 1973; dedicated to P. Erdös on his 60th birthday)
12. Fici, G., Restivo, A., Silva, M., Zamboni, L.: Anti-powers in infinite words. J. Comb. Theory Ser. A, **157**, 109–119 (2018)
13. Fleischer, L., Riasat, S., Shallit, J.: New bounds on antipowers in binary words. preprint arXiv: arXiv:1912.08147 (2019)
14. Gaetz, M.: Anti-Power Prefixes of the Thue-Morse Word. arXiv Preprint (2019)
15. Gawrychowski, P., Manea, F., Mercas, R., Nowotka, D.: Hide and seek with repetitions. J. Comput. Syst. Sci. **101**, 42–67 (2019)

16. Kari, L., Masson, B., Seki, S.: Properties of Pseudo-primitive words and their applications. Int. J. Found. Comput. Sci. **22**(2), 447–471 (2011).
17. Kolpakov, R.M., Kucherov, G.: Finding maximal repetitions in a word in linear time. In: 40th Annual Symposium on Foundations of Computer Science, New York City, NY, USA, pp. 596–604 (1999)
18. Lothaire, M.: Algebraic Combinatorics on Words. Cambridge University Press (2002)
19. Lothaire, M.: Applied Combinatorocs on Words. Cambridge University Press (2005)
20. Manea, F., Mercaş, R., Nowotka, D.: Fine and Wilf's theorem and pseudo-repetitions. In: Rovan, B., Sassone, V., Widmayer, P. (eds.) MFCS 2012. LNCS, vol. 7464, pp. 668–680. Springer, Heidelberg (2012). https://doi.org/10.1007/978-3-642-32589-2_58
21. Narayanan, S.: Functions on antipower prefix lengths of the Thue-Morse word. Discrete Math., to appear, preprint arXiv:1705.06310 (2017)
22. Schmid, M.L.: Computing equality-free and repetitive string factorizations. Theoret. Comput. Sci. **618**, 42–51 (2016)

Learning Algebraic Structures

Luca San Mauro

Vienna University of Technology, Austria

Algorithmic learning theory (**ALT**), initiated by Gold and Putnam in the 1960's, deals with the question of how a *learner*, provided with more and more data about some *environment*, is eventually able to achieve systematic knowledge about it. Classical paradigms of learning concern either learning of formal languages or learning of total functions. This is a convenient abstraction for representing the learning of a given flow of data, but it fails to portrait cases in which an agent deals with data embodying a structure.

So, in this work we want to make sense of the following question: *what does it mean to learn a structure?* To do so, we combine the technology of **ALT** with notions coming from computable structure theory, and develop a formal framework for learning structures in the limit.

In this framework, the learner receives larger and larger pieces of an arbitrary copy of an algebraic structure and, at each stage, is required to output a conjecture about the isomorphism type of such a structure. The learning is successful if the conjectures eventually stabilise to a correct guess. We use the infinitary logic $\mathcal{L}_{\omega_1,\omega}$ to provide a syntactic characterization of which families of structures can be learned. We apply this characterization to familiar cases and we show the following: there is an infinite learnable family of distributive lattices; no pair of Boolean algebras is learnable; no infinite family of linear orders is learnable.

In the last part of the talk, we compare alternative learning criteria and we report on recent progress about learning structures lying at various levels of the arithmetical hierarchy.

This is joint work with Nikolay Bazhenov, Ekaterina Fokina, and Timo Kötzing [1, 2].

References

1. Fokina, E., Kötzing, T., Mauro, L.S.: Limit learning equivalence structures. In: Proceedings of Machine Learning Research,vol. 98, pp. 383–403 (2019)
2. Bazhenov, N., Fokina, N., Mauro, L.S.: Learning families of algebraic structures from informant, submitted

Topological Spaces in Computable Analysis

Matthias Schröder[1,2]

[1] TU Darmstadt
[2] The University of Birmingham
matthias.schroeder@cca-net.de

Abstract. Computable Analysis investigates computability on the real numbers and related spaces. We discuss an approach to define computability on topological spaces and classify the class of topological spaces which can be dealt within this model. Moreover, we investigate the class of Co-Polish spaces. Co-Polish spaces play a big role in Type Two Complexity Theory.

Keywords: Computable analysis · Topological spaces · QCB-spaces · Co-polish spaces

1 Computability on QCB-Spaces

Computable Analysis investigates computability and complexity on spaces occurring in functional analysis like the real numbers, vector spaces and spaces of measures.

A well-established computational model to define computability on topological spaces is provided by K. Weihrauch's Type Two Theory of Effectivity (TTE) [15]. The main idea of TTE is to represent the objects of a given space by infinite sequences of natural numbers, i.e. by elements of the Baire space $\mathbb{N}^{\mathbb{N}}$. Such a naming function is a called a *representation* of that space. The actual computation is performed by a digital computer on these names. This means that a function f is computable if, whenever p is a name of an argument x, every finite prefix of some name of $f(x)$ can be computed from some finite prefix of the input name p. Examples of efficient implementations of exact real arithmetic are N. Müller's iRRAM [8], P. Collins' Ariadne [2], B. Lambov's RealLib [7], and M. Konečný's AERN [5].

It is obvious that one should restrict oneself to representations which reflect the mathematical properties of a given space appropriately. *Effective admissibility* is a notion of well-behavedness of representations [11, 13]. The class of topological spaces that can be endowed with an effectively admissible representation turn out to be the T_0-quotients of countably-based topological spaces [3, 11]. The common acronym is QCB_0-spaces. Examples of QCB-spaces are the Euclidean space $\mathbb{R}$, separable metrisable spaces and the space of distributions.

The ensuing category $\mathbf{QCB}_0$ of QCB_0-spaces as objects and continuous functions as morphisms has a remarkably rich structure. In particular it is cartesian closed, in contrast to its supercategory **Top** of topological spaces and its subcategory ω**Top** of topological spaces with a countable base. Cartesian closedness means that finite products $X \times Y$ and function spaces Y^X can be formed inside the category with reasonable properties, e.g., there exists an evaluation function $apply : Y^X \times X \to Y$ and for every morphism $f : Z \times X \to Y$ there exists exactly one morphism $curry(f) : Z \to Y^X$ that satisfies $f(z, x) = apply(curry(f)(z), x)$ for all $z \in Z$ and $x \in X$. For more details of how to construct finite products and function spaces inside $\mathbf{QCB}_0$ we refer to [10, 11, 13]. Moreover, $\mathbf{QCB}_0$ is countably complete and countably co-complete, meaning that countable products, subspaces, countable co-products and quotient spaces can be constructed. These excellent closure properties allow us to canonically deal with product spaces, function spaces and hyperspaces like the space $\mathcal{O}(X)$ of open subsets and the space $\mathcal{K}(X)$ of compact subspaces of a given admissibly represented topological space X, cf. [9, 13].

2 Co-polish Spaces

A nice subcategory of QCB-spaces is formed by the class of *Co-Polish spaces* [1, 12]. This class plays a big role in Complexity Theory. The name originates from the characterisation of Co-Polish spaces as being those $\mathbf{QCB}_0$-spaces X for which the function space $\mathbb{R}^X$ (formed in $\mathbf{QCB}_0$) is a Polish space (i.e., a separable and completely metrisable space).

The pivotal property which exhibits Co-Polish spaces as suitable for complexity theory is their characterisation as an inductive limit of an increasing sequence of compact metrisable spaces. This fact allows us to measure time complexity of functions between Co-Polish spaces in terms of two *discrete* parameters: on the one hand in terms of the output precision, as it is typical for complexity theory on the real numbers, and on the other hand, like in classical complexity theory, in terms of a discrete parameter on the input, e.g., the index of a compact metric subspace in which the input x lies.

So Co-Polish spaces enjoy the property that one can assign to each element a natural number as its size, like in discrete complexity theory. An upper bound of this size can be computed from any name of x. By contrast, for assigning a size to the elements of general spaces (like the function space $\mathbb{R}^{\mathbb{R}}$) one has to resort to elements of the Baire space. This idea has been developed by A. Kawamura and S. Cook in their approach to complexity in Computable Analysis [4].

Examples of Co-Polish spaces are all locally compact separable Hausdorff spaces (e.g. the reals), Silva spaces [6, 14] known from the theory of locally convex vector spaces (the space of distributions with compact support is an example of a Silva space) and, more generally, vector space duals of separable Banach spaces formed in the category of sequentially locally convex QCB-spaces.

References

1. de Brecht, M., Pauly, A., Schröder, M.: Overt choice. To appear in Journal of Computability. https://doi.org/10.3233/com-190253. Available as arXiv: arXiv:1902.05926v1 (2019)
2. Collins, P.J.: Ariadne: A tool for rigorous numerics and verification of nonlinear and hybrid systems. http://www.ariadne-cps.org (2017)
3. Escardó, M.H., Lawson, J.D., Simpson, A.K.: Comparing cartesian-closed categories of core compactly generated spaces. Topol. Its Appl. **143**, 105–145 (2004)
4. Kawamura, A., Cook, S.: Complexity Theory for Operators in Analysis. arXiv: arXiv:1305.0453v1 (2013)
5. Konečný, M.: AERN-Net: Exact real networks. http://hackage.haskell.org/cgi-bin/hackage-scripts/package/AERN-Net (2008)
6. Kunkle, D., Schröder, M.: Some examples of non-metrizable spaces allowing a simple type-2 complexity theory. Electron Notes Theor. Comput. Sci. **120**, 111–123 (2005)
7. Lambov, B.: RealLib: an efficient implementation of exact real arithmetic. Math. Struct. Comput. Sci. **17**(1), 81–98 (2007)
8. Müller, N.T.: The iRRAM: exact arithmetic in C++. In: Blanck, J., Brattka V., Hertling, P. (eds.) CCA 2000. LNCS, vol. 2064, pp. 222—252.Springer, Heidelberg (2001). https://doi.org/10.1007/3-540-45335-0_14
9. Pauly, A.: On the topological aspects of the theory of represented spaces. Computability **5**(2), 159–180 (2016)
10. Schröder, M.: Extended admissibility. Theor. Comput. Sci. **284**(2), 519–538 (2002)
11. Schröder, M.: Admissible representations for continuous computations. FernUniversität Hagen (2002)
12. Schröder, M.: Spaces allowing type-2 complexity theory revisited. Math. Logic Q. **50**(4/5), 443–459 (2004)
13. Schröder, M.: Admissibly Represented Spaces and QCB-Spaces. To appear as a chapter in the Handbook of Computability and Complexity in Analysis, Theory and Applications of Computability, Springer-Verlag. Available as arXiv: arXiv: 2004.09450v1 (2020)
14. Silva, J.S.: Su certi classi di spazi localmente convessi importanti per le applicazioni. Universita di Roma, Rendiconti di Matematico, 14, pp. 388–410 (1955)
15. Weihrauch, K.: Computable Analysis. Springer, Berlin (2000). https://doi.org/10.1007/978-3-642-56999-9

Molecular Algorithms Using Reprogrammable DNA Self-assembly

Damien Woods

Hamilton Institute and Department of Computer Science,
Maynooth University, Ireland
damien.woods@mu.ie
https://dna.hamilton.ie/

Abstract. The history of computing tells us that computers can be made of almost anything: silicon, gears and levers, neurons, flowing water, interacting particles or even light. Although lithographically patterned silicon surfaces have been by far the most successful of these, they give us a limited view of what computation is capable of. Algorithms typically control the flow of information by controlling the flow of electrons through digital-electronic devices, but in the field of molecular computing we imagine algorithms that control matter itself.

This talk will be about DNA molecules that interact with each other in a test tube to execute algorithms. We will show how DNA can be re-engineered to act not only as an information encoding polymer (as it is in biology) but also as a computational primitive for executing somewhat soggy computer programs. The talk will showcase some of our wet-lab results on implementing 21 different algorithms using self-assembling DNA strands [1]. We will also see how tools from the theory of computation can help us understand what kinds of computations molecules are capable of.

[1] Woods*, Doty*, Myhrvold, Hui, Zhou, Yin, Winfree. Diverse and robust molecular algorithms using reprogrammable DNA self-assembly. Nature 567:366-372. 2019. *Joint lead co-authors.

Keywords: DNA computing · Self-assembly · Theory of computation

D. Woods—Supported by European Research Council (ERC) award number 772766 and Science foundation Ireland (SFI) grant 18/ERCS/5746 (this abstract reflects only the authors' view and the ERC is not responsible for any use that may be made of the information it contains).

Contents

A Note on Computable Embeddings
for Ordinals and Their Reverses

Nikolay Bazhenov[1,2] and Stefan Vatev[3(⊠)]

[1] Sobolev Institute of Mathematics, Novosibirsk, Russia
[2] Novosibirsk State University, Novosibirsk, Russia
bazhenov@math.nsc.ru
[3] Faculty of Mathematics and Informatics, Sofia University, 5 James Bourchier blvd.,
1164 Sofia, Bulgaria
stefanv@fmi.uni-sofia.bg

Abstract. We continue the study of computable embeddings for pairs of structures, i.e. for classes containing precisely two non-isomorphic structures. Surprisingly, even for some pairs of simple linear orders, computable embeddings induce a non-trivial degree structure. Our main result shows that although $\{\omega \cdot 2, \omega^\star \cdot 2\}$ is computably embeddable in $\{\omega^2, (\omega^2)^\star\}$, the class $\{\omega \cdot k, \omega^\star \cdot k\}$ is *not* computably embeddable in $\{\omega^2, (\omega^2)^\star\}$ for any natural number $k \geq 3$.

Keywords: Computable embedding · Enumeration operator · Linear order

1 Introduction

The paper studies computability-theoretic complexity for classes of countable structures. A standard method of investigating this problem is to fix a particular notion of *reduction* $\leq_r$ between classes, and then to gauge the complexity of classes via the degrees induced by $\leq_r$.

One of the first examples of such reductions comes from descriptive set theory: Friedman and Stanley [9] introduced the notion of *Borel embedding*. Informally speaking, a *Borel embedding* Φ from a class $\mathcal{K}$ into a class $\mathcal{K}'$ is a Borel measurable function, which acts as follows. Given the atomic diagram of an arbitrary structure $A \in \mathcal{K}$ as an input, Φ outputs the atomic diagram of some structure $\Phi(A)$ belonging to $\mathcal{K}'$. The key property of Φ is that Φ is injective on isomorphism types, i.e. $\mathcal{A} \cong \mathcal{B}$ if and only if $\Phi(\mathcal{A}) \cong \Phi(\mathcal{B})$.

Calvert, Cummins, Knight, and Miller [4] (see also [15]) developed two different *effective versions* of Borel embeddings. Roughly speaking, a *Turing computable embedding* (or *tc-embedding*, for short) is a Borel embedding, which is

The first author was supported by RFBR, project number 20-31-70006. The second author was supported by BNSF DN 02/16/19.12.2016 and NSF grant DMS 1600625/2016. The authors also would like to thank Manat Mustafa for his hospitality during their visit of Nazarbayev University.

M. Anselmo et al. (Eds.): CiE 2020, LNCS 12098, pp. 1–13, 2020.
https://doi.org/10.1007/978-3-030-51466-2_1

realized by a Turing functional Φ. A *computable embedding* is realized by an enumeration operator. It turned out that one of these notions is strictly stronger than the other: If there is a computable embedding from $\mathcal{K}$ into $\mathcal{K}'$, then there is also a *tc*-embedding from $\mathcal{K}$ into $\mathcal{K}'$. The converse is not true, see Sect. 2 for formal details.

A powerful tool, which helps to work with Turing computable embeddings, is provided by the Pullback Theorem of Knight, Miller, and Vanden Boom [15]. Informally, this theorem says that *tc*-embeddings behave well, when working with syntactic properties: one can "pull back" computable infinitary sentences from the output class $\mathcal{K}'$ to the input class $\mathcal{K}$, while *preserving* the complexity of these sentences.

Nevertheless, Pullback Theorem and its consequences show that sometimes *tc*-embeddings are too coarse: they *cannot see* finer structural distinctions between classes. One of the first examples of this phenomenon was provided by Chisholm, Knight, and Miller [5]: Let VS be the class of infinite $\mathbb{Q}$-vector spaces, and let ZS be the class of models of the theory $\mathrm{Th}(\mathbb{Z}, S)$, where $(\mathbb{Z}, S)$ is the integers with successor. Then VS and ZS are equivalent with respect to *tc*-embeddings, but there is no computable embedding from VS to ZS.

Another example of this intriguing phenomenon can be found in the simpler setting of classes generated by pairs of linear orderings, closed under isomorphism. Recall that by ω one usually denotes the standard ordering of natural numbers. For a linear order L, by L^* we denote the reverse ordering, i.e. $a \leq_{L^*} b$ iff $b \leq_L a$.

Ganchev, Kalimullin and Vatev [10] gave one such example. For a structure $\mathcal{A}$, let $\tilde{\mathcal{A}}$ be the enrichment of $\mathcal{A}$ with a congruence relation $\sim$ such that every congruence class in $\tilde{\mathcal{A}}$ is infinite and $\tilde{\mathcal{A}}/_\sim \cong \mathcal{A}$. Then they showed that the class $\{\omega_S, \omega_S^*\}$ is *tc*-equivalent to the class $\{\tilde{\omega}_S, \tilde{\omega}_S^*\}$, whereas $\{\tilde{\omega}_S, \tilde{\omega}_S^*\}$ is not computably embeddable into $\{\omega_S, \omega_S^*\}$. Here ω_S and ω_S^* are linear orderings of type ω and ω^*, respectively, together with the successor relation.

One can prove (see, e.g., Theorem 3.1 in [3]) the following: Let L be a computable infinite linear order with a least, but no greatest element. Then the pair $\{L, L^*\}$ is equivalent to $\{\omega, \omega^*\}$ with respect to *tc*-embeddings. This result gives further evidence that, in a sense, *tc*-embeddings cannot work with finer algebraic properties: Here a *tc*-embedding Φ can only employ the existence (or non-existence) of the least and the greatest elements. If one considers, say, the pair $\{\omega^\omega, (\omega^\omega)^*\}$, then our Φ is not able to "catch" limit points, limits of limit points, etc. Section 2.1 gives a further discussion of interesting peculiarities of the pair $\{\omega, \omega^*\}$.

On the other hand, when one deals with *computable embeddings*, even finite sums of ω (together with their reverse orders) already exhibit a quite complicated structure: Let k and ℓ be non-zero natural numbers. Then there is a computable embedding from $\{\omega \cdot k, \omega^* \cdot k\}$ into $\{\omega \cdot \ell, \omega^* \cdot \ell\}$ if and only if k divides ℓ (Theorem 5.2 of [3]). In other words, in this particular setting the only possible computable embeddings are the simplest ones—by appending a fixed number of copies of an input order together. We note that it is quite non-trivial to prove that all other embeddings Ψ (e.g., a computable embedding from $\{\omega \cdot 3, \omega^\star \cdot 3\}$ to $\{\omega \cdot 4, \omega^\star \cdot 4\}$) are not possible—our proofs fully employ the peculiarities inherent to enumeration operators. These peculiarities have topological nature: indeed, one can establish the lack of continuous operators Ψ (in the Scott topology).

The current paper continues the investigations of [3]. We show that even adding the finite sums of ω^2 (and their inverses) to the mix makes the resulting picture more combinatorially involved (compare with Theorem 5.2 mentioned above).

2 Preliminaries

We will slightly abuse the notations: both the set of natural numbers and the standard ordering of this set will be denoted by ω. The precise meaning of the symbol ω will be clear from the context. We consider only computable languages, and structures with domain contained in ω. We assume that any considered class of structures $\mathcal{K}$ is closed under isomorphism, modulo the restriction on domains. For a structure $\mathcal{S}$, $D(\mathcal{S})$ denotes the atomic diagram of $\mathcal{S}$. We will often identify a structure and its atomic diagram. We refer to atomic formulas and their negations as basic.

Let $\mathcal{K}_0$ be a class of L_0-structures, and $\mathcal{K}_1$ be a class of L_1-structures. In the definition below, we use the following convention: An *enumeration operator* Γ is treated as a computably enumerable set of pairs (α, φ), where α is a finite set of basic $(L_0 \cup \omega)$-sentences, and φ is a basic $(L_1 \cup \omega)$-sentence. As usual, for a set X, we have $\Gamma(X) = \{\varphi : (\alpha, \varphi) \in \Gamma, \ \alpha \subseteq X\}$.

Definition 1 ([4,15])**.** *An enumeration operator Γ is a* computable embedding *of $\mathcal{K}_0$ into $\mathcal{K}_1$, denoted by $\Gamma \colon \mathcal{K}_0 \leq_c \mathcal{K}_1$, if Γ satisfies the following:*

1. *For any $\mathcal{A} \in \mathcal{K}_0$, $\Gamma(\mathcal{A})$ is the atomic diagram of a structure from $\mathcal{K}_1$.*
2. *For any $\mathcal{A}, \mathcal{B} \in \mathcal{K}_0$, we have $\mathcal{A} \cong \mathcal{B}$ if and only if $\Gamma(\mathcal{A}) \cong \Gamma(\mathcal{B})$.*

Any computable embedding has an important property of *monotonicity*: If $\Gamma \colon \mathcal{K}_0 \leq_c \mathcal{K}_1$ and $\mathcal{A} \subseteq \mathcal{B}$ are structures from $\mathcal{K}_0$, then we have $\Gamma(\mathcal{A}) \subseteq \Gamma(\mathcal{B})$ [4, Proposition 1.1].

Definition 2 ([4,15])**.** *A* Turing operator *$\Phi = \varphi_e$ is a* Turing computable embedding *of $\mathcal{K}_0$ into $\mathcal{K}_1$, denoted by $\Phi \colon \mathcal{K}_0 \leq_{tc} \mathcal{K}_1$, if Φ satisfies the following:*

1. *For any $\mathcal{A} \in \mathcal{K}_0$, the function $\varphi_e^{D(\mathcal{A})}$ is the characteristic function of the atomic diagram of a structure from $\mathcal{K}_1$. This structure is denoted by $\Phi(\mathcal{A})$.*
2. *For any $\mathcal{A}, \mathcal{B} \in \mathcal{K}_0$, we have $\mathcal{A} \cong \mathcal{B}$ if and only if $\Phi(\mathcal{A}) \cong \Phi(\mathcal{B})$.*

Proposition (Greenberg and, independently, Kalimullin; see [14,15]). *If $\mathcal{K}_0 \leq_c \mathcal{K}_1$, then $\mathcal{K}_0 \leq_{tc} \mathcal{K}_1$. The converse is not true.*

Both relations $\leq_c$ and $\leq_{tc}$ are preorders. If $\mathcal{K}_0 \leq_{tc} \mathcal{K}_1$ and $\mathcal{K}_1 \leq_{tc} \mathcal{K}_0$, then we say that $\mathcal{K}_0$ and $\mathcal{K}_1$ are *tc-equivalent*, denoted by $\mathcal{K}_0 \equiv_{tc} \mathcal{K}_1$. For a class $\mathcal{K}$, by $\deg_{tc}(\mathcal{K})$ we denote the family of all classes which are tc-equivalent to $\mathcal{K}$. Similar notations can be introduced for the c-reducibility.

We note that except the reductions $\leq_c$ and $\leq_{tc}$, there are many other approaches to comparing computability-theoretic complexity of classes of structures. These approaches include: transferring degree spectra and other algorithmic properties [13], Σ-reducibility [8,17], computable functors [11,16], Borel functors [12], primitive recursive functors [1,7], etc.

For two ω-chains $\overline{x} = (x_i)_{i=0}^{\infty}$ and $\overline{y} = (y_j)_{j=0}^{\infty}$, analogous to the relation $\subseteq^{\star}$ between sets, let us denote by $\overline{x} <^{\star} \overline{y}$ the following infinitary sentence

$$\bigvee_{q \in \omega} \bigwedge_{i,j > q} x_i < y_j.$$

The following proposition is essential for our results. It is a slight reformulation of Proposition 5.7 from [3].

Proposition 1. *Suppose* $\{\omega \cdot 2, \omega^{\star} \cdot 2\} \leq_c \{\mathcal{C}, \mathcal{D}\}$ *via* Γ, *where* $\mathcal{C}$ *is a linear order without infinite descending chains and* $\mathcal{D}$ *is an infinite order without infinite ascending chains. Let* $\mathcal{A}$ *and* $\mathcal{B}$ *be copies of* ω *with mutually disjoint domains. Then for any* ω-chains $(x_i)_{i=0}^{\infty}$ *and* $(y_i)_{i=0}^{\infty}$ *such that* $\Gamma(\mathcal{A}) \models \bigwedge_{i \in \omega} x_i < x_{i+1}$ *and* $\Gamma(\mathcal{B}) \models \bigwedge_{i \in \omega} y_i < y_{i+1}$ *such that*

$$\Gamma(\mathcal{A} + \mathcal{B}) \models \overline{x} <^{\star} \overline{y} \lor \overline{y} <^{\star} \overline{x}.$$

2.1 Further Background

This paper is focused on the degree $\deg_{tc}(\{\omega, \omega^{\star}\})$. Historically speaking, the choice of this particular degree was motivated by the following open question:

Problem (Kalimullin). It is easy to show that the pairs $\{\omega, \omega^{\star}\}$ and $\{\tilde{\omega}, \tilde{\omega}^{\star}\}$ are tc-equivalent. Moreover, $\{\omega, \omega^{\star}\} \leq_c \{\tilde{\omega}, \tilde{\omega}^{\star}\}$. Is there a computable embedding from $\{\tilde{\omega}, \tilde{\omega}^{\star}\}$ to $\{\omega, \omega^{\star}\}$?

This problem was a starting point of investigations of [3] and the current paper. One can attack the problem via employing model-theoretic properties of the structures (in a way similar to [5]). In particular, a naive way to distinguish these pairs would be the following. Each of the orders ω and $\omega^{\star}$ is rigid, while both $\tilde{\omega}$ and $\tilde{\omega}^{\star}$ have continuum many automorphisms. Maybe, this fact can help us to prove that $\{\tilde{\omega}, \tilde{\omega}^{\star}\} \not\leq_c \{\omega, \omega^{\star}\}$? Nevertheless, this is *not* the case—one can show that $\{\tilde{\omega}, \tilde{\omega}^{\star}\} \equiv_c \{(\omega^2, B), (\omega \cdot \omega^{\star}, B)\}$, where B is the standard block relation on a linear order. Since the structures (ω^2, B) and $(\omega \cdot \omega^{\star}, B)$ are both rigid, it seems that studying automorphism groups does not help in this setting.

We note that quite unexpectedly (at least for us), the theory of Turing computable embeddings found applications in algorithmic learning theory. Section 3.2 of [2] establishes connections between tc-embeddings and a particular paradigm of learnability for classes of countable structures. Informally speaking, this paradigm employs a learner whose goal is, given the atomic diagram of a structure $\mathcal{A}$, to learn the isomorphism type of $\mathcal{A}$. The learner is allowed to use both positive and negative data provided by the atomic diagram. Remarkably, the family $\{\omega, \omega^{\star}\}$ is learnable by a computable learner. We conjecture that our results can be also connected to learnability, specifically to its topological aspects (see, e.g., [6]). The reader is referred to [3] for more results on $\deg_{tc}(\{\omega, \omega^{\star}\})$.

3 Positive Results

Let $\mathcal{A}$ be a linear ordering and let us have, for all $a \in \mathcal{A}$, the linear orderings $\mathcal{B}_a$ with mutually disjoint domains. Following Rosenstein [18], we define the generalized sum $\mathcal{C} = \sum_{a \in \mathcal{A}} \mathcal{B}_a$ as the linear ordering such that $\dom(\mathcal{C}) = \bigcup_{a \in \mathcal{A}} \dom(\mathcal{B}_a)$ and for any $x, y \in \mathcal{C}$, we define $x <_{\mathcal{C}} y$ iff $x, y \in \mathcal{B}_a$ for some $a \in \mathcal{A}$ and $x <_{\mathcal{B}_a} y$, or $x \in \mathcal{B}_a$, $y \in \mathcal{B}_{a'}$ and $a <_{\mathcal{A}} a'$.

Theorem 1. *For any natural number $n \geq 1$,*

$$\{\omega \cdot n, \omega^{\star} \cdot n\} \leq_c \{\omega^2 \cdot n, (\omega^2)^{\star} \cdot n\}.$$

Proof. The same enumeration operator Γ works for all $n \geq 1$. For a linear ordering $\mathcal{L}$ and $a \in \mathcal{L}$, let $\mathcal{L}_a$ be the linear ordering consisting of pairs (a, b), where $b \in \mathcal{L}$, and ordered by the second component as in $\mathcal{L}$. Informally, for each element a in the input linear order $\mathcal{L}$, the enumeration operator outputs $\mathcal{L}_a$. Moreover, all pairs in $\Gamma(\mathcal{L})$ are ordered lexicographically by the order induced by $\mathcal{L}$. In other words, $\Gamma(\mathcal{L}) = \sum_{a \in \mathcal{L}} \mathcal{L}_a$.

- If $\mathcal{L} \cong \omega \cdot n$, then $\Gamma(\mathcal{L}) \cong \sum_{j \in n} \sum_{i \in \omega} \omega \cdot n = \sum_{j \in n} \omega^2 = \omega^2 \cdot n$.
- If $\mathcal{L} \cong \omega^{\star} \cdot n$, then $\Gamma(\mathcal{L}) \cong \sum_{j \in n} \sum_{i \in \omega^{\star}} \omega^{\star} \cdot n = \sum_{j \in n} (\omega^2)^{\star} = (\omega^2)^{\star} \cdot n$.

$\square$

For the next result, we need the following notation. For a linear ordering $\mathcal{L}$ and an element a in $\mathcal{L}$, we define

$$\mathtt{left}_{\mathcal{L}}(a) = |\{b \in \dom(\mathcal{L}) \mid b \leq_{\mathcal{L}} a\}|$$
$$\mathtt{right}_{\mathcal{L}}(a) = |\{b \in \dom(\mathcal{L}) \mid b \geq_{\mathcal{L}} a\}|$$
$$\mathtt{rad}_{\mathcal{L}}(a) = \min\{\mathtt{left}_{\mathcal{L}}(a), \mathtt{right}_{\mathcal{L}}(a)\}.$$

Informally, we will show that there exists an enumeration operator Γ which can "guess" whether an element a in the input linear ordering $\mathcal{L}$ has finite or infinite radius, denoted $\mathtt{rad}_{\mathcal{L}}(a)$.

Theorem 2. $\{\omega \cdot 2, \omega^{\star} \cdot 2\} \leq_c \{\omega^2, (\omega^2)^{\star}\}.$

Proof. We informally describe the work of the enumeration operator Γ. Suppose we have as input the finite linear ordering $\mathcal{L} = a_0 < a_1 < a_2 < \cdots < a_n$. For each a_i in $\mathcal{L}$, Γ outputs the pairs of the form (a_i, a_j), where $a_j \leq_{\mathbb{N}} \mathtt{rad}_{\mathcal{L}}(a_i)$, where $\leq_{\mathbb{N}}$ is the standard ordering of natural numbers. All pairs in the output structure are ordered in lexicographic order. This concludes the description of how Γ operates. Now we have two cases to consider for the input structure $\mathcal{A}$.

Suppose that $\mathcal{A} = \mathcal{A}_1 + \mathcal{A}_2$, where $\mathcal{A}_1$ and $\mathcal{A}_2$ are copies of ω. If $a \in \mathcal{A}_1$ is its k-th least element, then $\mathrm{rad}_{\mathcal{A}}(a) = k$ and hence a contributes at most k pairs to $\Gamma(\mathcal{A})$. If $a \in \mathcal{A}_2$, then clearly $\mathrm{rad}_{\mathcal{A}}(a) = \aleph_0$ and hence a contributes infinitely many pairs to $\Gamma(\mathcal{A})$, forming a linear ordering of type $\omega \cdot 2$.

We conclude that in this case

$$\Gamma(\mathcal{A}) \cong \sum_{i \in \omega} i + \sum_{i \in \omega} \omega \cdot 2 \cong \omega + \omega^2 = \omega^2.$$

Suppose that $\mathcal{A} = \mathcal{A}_1 + \mathcal{A}_2$, where $\mathcal{A}_1$ and $\mathcal{A}_2$ are copies of $\omega^\star$. If $a \in \mathcal{A}_1$, then a contributes infinitely many elements of type $\omega^\star \cdot 2$ in $\Gamma(\mathcal{A})$. If $a \in \mathcal{A}_2$ is its k-th greatest element, then a contributes at most k pairs in $\Gamma(\mathcal{A})$. We conclude that in this case

$$\Gamma(\mathcal{A}) \cong \sum_{i \in \omega^\star} \omega^\star \cdot 2 + \sum_{i \in \omega^\star} i \cong (\omega^2)^\star + \omega^\star = (\omega^2)^\star.$$

$\square$

Corollary 1. *For any natural number $n \geq 1$,*

$$\{\omega \cdot (n+1), \omega^\star \cdot (n+1)\} \leq_c \{\omega^2 \cdot n, (\omega^2)^\star \cdot n\}.$$

Proof. We use the same enumeration operator Γ as in Theorem 2. Suppose $\mathcal{A} = \mathcal{A}_0 + \mathcal{A}_1 + \cdots + \mathcal{A}_n$, where each $\mathcal{A}_i$ is a copy of ω. Then if $a \in \mathcal{A}_0$ is the k-th least element, then a contributes at most k pairs in $\Gamma(\mathcal{A})$. If $a \in \mathcal{A}_i$, where $i > 0$, then a contributes infinitely many pairs of the type of $\mathcal{A}$ to $\Gamma(\mathcal{A})$. It follows that

$$\Gamma(\mathcal{A}) \cong \sum_{i \in \omega} i + \underbrace{\sum_{i \in \omega} \omega \cdot (n+1) + \cdots + \sum_{i \in \omega} \omega \cdot (n+1)}_{n}$$

$$= \omega + \omega^2 \cdot n = \omega^2 \cdot n.$$

Suppose $\mathcal{A} = \mathcal{A}_0 + \mathcal{A}_1 + \cdots + \mathcal{A}_n$, where each $\mathcal{A}_i$ is a copy of $\omega^\star$. Then if $a \in A_n$ is the k-th greatest element, then a contributes at most k pairs in $\Gamma(\mathcal{A})$. If $a \in \mathcal{A}_i$, where $0 \leq i < n$, then a contributes infinitely many pairs of the type of $\mathcal{A}$ to $\Gamma(\mathcal{A})$. It follows that

$$\Gamma(\mathcal{A}) \cong \underbrace{\sum_{i \in \omega^\star} \omega^\star \cdot (n+1) + \cdots + \sum_{i \in \omega^\star} \omega^\star \cdot (n+1)}_{n} + \sum_{i \in \omega^\star} i$$

$$= (\omega^2)^\star \cdot n + \omega^\star = (\omega^2)^\star \cdot n.$$

$\square$

Corollary 2. *For any natural number $n \geq 1$,*

$$\{\omega \cdot 2, \omega^\star \cdot 2\} \leq_c \{\omega^2 \cdot n, (\omega^2)^\star \cdot n\}.$$

Proof. This is straightforward. Let Γ be the enumeration operator from Theorem 2. Then for a natural number $n \geq 1$, the embedding will be obtained by the enumeration operator, which, for linear ordering $\mathcal{A}$, simply copies n number of times the linear ordering $\Gamma(\mathcal{A})$. $\square$

4 The Case $\{\omega \cdot 3, \omega^\star \cdot 3\} \not\leq_c \{\omega^2, (\omega^2)^\star\}$

In this section, towards a contradiction, assume $\Gamma : \{\omega \cdot 3, \omega^\star \cdot 3\} \leq_c \{\omega^2, (\omega^2)^\star\}$. Let $\mathcal{B}$ be a copy of $\omega \cdot 3$ (or the reverse ordinal). In general, for a subordering $\mathcal{A}$ of $\mathcal{B}$, we may have that $\Gamma(\mathcal{A})$ is not a linear ordering. For example, we may have $x, y \in \Gamma(\mathcal{A})$, but none of the sentences $x < y$ or $y < x$ are in $\Gamma(\mathcal{A})$. Suppose $\Gamma(\mathcal{B}) \models x < y$. Then we claim that there is no extension $\mathcal{C}$ of $\mathcal{A}$ such that $\Gamma(\mathcal{C}) \models y < x$. In other words, although $\Gamma(\mathcal{A})$ does not "know" the relation between x and y, this relation is already fixed. Assume there is such an extension $\mathcal{C}$ for which $\Gamma(\mathcal{C}) \vdash y < x$. By compactness of enumeration operators, we may suppose that $\mathcal{C}$ extends $\mathcal{A}$ by only finitely many elements. We can find another finite extension $\mathcal{D}$ of $\mathcal{A}$ with $\mathrm{dom}(\mathcal{D}) \cap \mathrm{dom}(\mathcal{B}) = \mathrm{dom}(\mathcal{A})$ and $\mathrm{dom}(\mathcal{D}) \cap \mathrm{dom}(\mathcal{C}) = \mathrm{dom}(\mathcal{A})$ such that $\Gamma(\mathcal{D}) \models x < y \vee y < x$. Now we use monotonicity. If $\Gamma(\mathcal{D}) \models x < y$, then we must have $\Gamma(\mathcal{C} \cup \mathcal{D}) \models x < y \; \& \; y < x$. If $\Gamma(\mathcal{D}) \models y < x$, then we must have $\Gamma(\mathcal{B} \cup \mathcal{D}) \models x < y \; \& \; y < x$. In both cases we reach a contradiction.

Remark 1. It is safe to always suppose that if $\mathcal{A}$ is a linear ordering (or its corresponding reverse linear ordering), then $\Gamma(\mathcal{A})$ is also a linear ordering.

Let us denote by $a <^\infty b$ the computable infinitary sentence saying that there are infinitely many elements between a and b.

Proposition 2. *For any infinite and coinfinite set A, if there is a copy $\mathcal{A}$ of ω with $\mathrm{dom}(\mathcal{A}) = A$ such that $\Gamma(\mathcal{A}) \cong \omega^2$, then there is no copy $\mathcal{B}$ of ω with $\mathrm{dom}(\mathcal{B}) \subseteq \mathbb{N} \setminus A$ such that $\Gamma(\mathcal{B}) \cong \omega^2$.*

Proof. Assume that there are at least two copies $\mathcal{A}$ and $\mathcal{B}$ of ω, with mutually disjoint domains, such that $\Gamma(\mathcal{A}) \cong \omega^2$ and $\Gamma(\mathcal{B}) \cong \omega^2$. Then we can fix the infinite sequences $\overline{a} = (a_i)_{i=0}^\infty$ and $\overline{b} = (b_i)_{i=0}^\infty$ such that

$$\Gamma(\mathcal{A}) \models a_0 <^\infty a_1 <^\infty a_2 <^\infty \cdots$$
$$\Gamma(\mathcal{B}) \models b_0 <^\infty b_1 <^\infty b_2 <^\infty \cdots$$

Then by Proposition 1, we have $\Gamma(\mathcal{A} + \mathcal{B}) \models \overline{a} <^\star \overline{b} \vee \overline{b} <^\star \overline{a}$. It follows that $\Gamma(\mathcal{A} + \mathcal{B})$ extends a copy of $\omega^2 \cdot 2$, which is a contradiction because by monotonicity of enumeration operators this would mean that there is a copy $\mathcal{C}$ of $\omega \cdot 3$ extending $\mathcal{A} + \mathcal{B}$ such that $\Gamma(\mathcal{C})$ extends $\omega^2 \cdot 2$. $\square$

From now on, in this section, we suppose that we work with copies $\mathcal{A}$ of ω such that $\Gamma(\mathcal{A})$ has type *strictly less* than ω^2, i.e. there exist natural numbers n and ℓ such that $\Gamma(\mathcal{A}) \cong \omega \cdot n + \ell$.

Proposition 3. *There exists an infinite subset D of natural numbers and a number n such that any copy $\mathcal{A}$ of ω with $\mathrm{dom}(\mathcal{A}) \subseteq D$ is such that $\Gamma(\mathcal{A})$ has type at most $\omega \cdot n$.*

Proof. Towards a contradiction, assume that for any infinite subset D of natural numbers, for any n, there exists a copy $\mathcal{A}$ of ω with $\mathrm{dom}(\mathcal{A}) \subseteq D$ such that $\Gamma(\mathcal{A})$ is *at least* $\omega \cdot n$. This means that we can consider a sequence $\mathcal{A}_n$ of copies of ω, with mutually disjoint domains, such that $\Gamma(\mathcal{A}_n)$ has type *at least* $\omega \cdot n$. Now we can partition each copy $\mathcal{A}_n$ into an infinite sum of finite parts $(\alpha_{n,i})_{i=0}^{\infty}$ such that $\mathcal{A}_n = \sum_{i \in \omega} \alpha_{n,i}$. Then we can form a new copy $\mathcal{B}$ of ω in the following way: $\mathcal{B} = \sum_{i \in \omega} \sum_{n=0}^{i} \alpha_{n,i-n}$. In other words, $\mathcal{B} = \alpha_{0,0} + \alpha_{0,1} + \alpha_{1,0} + \alpha_{0,2} + \alpha_{1,1} + \alpha_{2,0} + \alpha_{0,3} + \cdots$ Then $\mathcal{B}$ contains $\mathcal{A}_n$ for all n and by monotonicity, $\Gamma(\mathcal{B})$ has type greater than $\omega \cdot n$ for all n. We conclude that $\Gamma(\mathcal{B})$ has type at least ω^2, which is a contradiction. $\qquad\square$

Remark 2. Proposition 3 allows us to proceed as in Section 7 of [3] and suppose that we have fixed an infinite set D and a number n such that any copy $\mathcal{A}$ of ω with $\mathrm{dom}(\mathcal{A}) \subseteq D$ is such that $\Gamma(\mathcal{A}) \cong \omega \cdot n$. From here on, all copies of ω that we consider will have as domains coinfinite subsets of D.

Proposition 4. *Let $\mathcal{A}$ and $\mathcal{B}$ be two such copies of ω, with mutually disjoint domains, such that for the ω-chains $\bar{a}_i = (a_{i,j})_{j=0}^{\infty}$ and $\bar{b}_i = (b_{i,j})_{j=0}^{\infty}$, where $i = 1, \ldots, n$, we have $\Gamma(\mathcal{A}) \models \bar{a}_1 < \bar{a}_2 < \cdots < \bar{a}_n$ and $\Gamma(\mathcal{B}) \models \bar{b}_1 < \bar{b}_2 < \cdots < \bar{b}_n$. Then*

$$\Gamma(\mathcal{A} + \mathcal{B}) \models \bar{a}_n <^* \bar{b}_n.$$

Proof. Assume not. By Proposition 1 we would have $\Gamma(\mathcal{A} + \mathcal{B}) \models \bar{b}_n <^* \bar{a}_n$. Let $a_{n,0} \in \Gamma(\alpha)$ for some finite part α of $\mathcal{A}$. Then $\mathcal{C} = \alpha + \mathcal{B}$ is a copy of ω such that $\Gamma(\mathcal{C}) \models \bar{b}_1 < \bar{b}_2 < \cdots < \bar{b}_n < a_{n,0}$. It follows that $\Gamma(\mathcal{C})$ extends a copy of $\omega \cdot n + 1$, which is a contradiction with Remark 2. $\qquad\square$

Proposition 5. *Let $\mathcal{A}$, $\mathcal{B}$, and $\mathcal{C}$ be copies of ω. Suppose that*

$$\Gamma(\mathcal{C}) \models \bar{c}_1 < \bar{c}_2 < \cdots < \bar{c}_n,$$

where $\bar{c}_i = (c_{i,j})_{j=0}^{\infty}$ are ω-chains. Then there exists an infinite subsequence $(i_s)_{s=0}^{\infty}$ such that

$$\Gamma(\mathcal{A} + \mathcal{B} + \mathcal{C}) \models \bigwedge_{s \in \omega} c_{n,i_s} <^\infty c_{n,i_{s+1}}.$$

Proof. Assume not. Then $\Gamma(\mathcal{A} + \mathcal{B} + \mathcal{C}) \models \bar{c}_1 < \cdots < \bar{c}_n + \mathcal{D}$, where $\mathcal{D}$ has the type of ω^2. Let $d \in \mathcal{D}$ be such that $d \in \Gamma(\alpha + \beta + \mathcal{C})$, where α and β are finite parts of $\mathcal{A}$ and $\mathcal{B}$ respectively. Then $\alpha + \beta + \mathcal{C}$ is a copy of ω, but $\Gamma(\alpha + \beta + \mathcal{C})$ extends a copy of $\omega \cdot n + 1$, which is a contradiction with Remark 2. $\qquad\square$

Proposition 6. *For any linear ordering $\mathcal{L}$ of type $\omega \cdot 3$, there is a linear ordering $\mathcal{M}$ of type $\omega \cdot 2$ with $dom(\mathcal{L}) = dom(\mathcal{M})$ and $\Gamma(\mathcal{M}) \cong \omega^2$.*

Proof. Let $\mathcal{L} = \mathcal{A} + \mathcal{B} + \mathcal{C}$, where $\mathcal{A}$, $\mathcal{B}$, and $\mathcal{C}$ are copies of ω. By Proposition 5, consider the infinite sequence $\bar{c} \in \Gamma(\mathcal{C})$ such that

$$\Gamma(\mathcal{A} + \mathcal{B} + \mathcal{C}) \models \bigwedge_{i \in \omega} c_i <^\infty c_{i+1}.$$

Assume that for some finite parts α and β of $\mathcal{A}$ and $\mathcal{B}$ respectively, for some i, $\Gamma(\alpha + \beta + \mathcal{C}) \models c_i <^\infty c_{i+1}$. But since $\alpha + \beta + \mathcal{C}$ is a copy of ω, and $\Gamma(\mathcal{C}) \cong \omega \cdot n$, then $\Gamma(\alpha + \beta + \mathcal{C})$ would extend a copy of $\omega \cdot (n+1)$, which is a contradiction with Remark 2. It follows that any such finite parts α and β contribute finitely many elements to any interval of the form (c_i, c_{i+1}).

Let $\bar{u}_i = (u_{i,j})_{j=0}^\infty$ be ω-chains such that we can partition $\mathcal{A}$ and $\mathcal{B}$ into finite parts such that $\mathcal{A} = \sum_{i \in \omega} \alpha_i$ and $\mathcal{B} = \sum_{i \in \omega} \beta_i$ and for all i,

$$\Gamma(\alpha_i + \beta_i + \mathcal{C}) \models \bigwedge_{j=0}^{i} c_j < u_{j,i-j} < c_{j+1}.$$

Then, by monotonicity, we obtain the following:

$$\Gamma(\sum_{i \in \omega}(\alpha_i + \beta_i) + \mathcal{C}) \models \bigwedge_{i \in \omega} \bigwedge_{j \in \omega} c_i < u_{i,j} < c_{i+1}.$$

It follows that $\mathcal{M} = \sum_{i \in \omega}(\alpha_i + \beta_i) + \mathcal{C}$ is a copy of $\omega \cdot 2$ with $dom(\mathcal{M}) = dom(\mathcal{L})$ which produces a copy of ω^2. □

Proposition 7. *Let $\mathcal{L}$ and $\mathcal{M}$ be disjoint copies of $\omega \cdot 2$ such that $\Gamma(\mathcal{L}) \cong \omega^2$ and $\Gamma(\mathcal{M}) \cong \omega^2$. Then there is a copy $\mathcal{N}$ of $\omega \cdot 3$ such that $\Gamma(\mathcal{N})$ extends a copy of $\omega^2 \cdot 2$.*

Proof. Let $\mathcal{L} = \mathcal{A} + \mathcal{B}$ and $\mathcal{M} = \mathcal{C} + \mathcal{D}$, where $\mathcal{A}$, $\mathcal{B}$, $\mathcal{C}$ and $\mathcal{D}$ are copies of ω. Let us fix the ω-chains $\bar{b}_i = (b_{i,j})_{j=0}^\infty$ and $\bar{d}_i = (d_{i,j})_{j=0}^\infty$ where $i = 1, \ldots, n$ such that $\Gamma(\mathcal{B}) \models \bar{b}_1 < \bar{b}_2 < \cdots < \bar{b}_n$ and $\Gamma(\mathcal{D}) \models \bar{d}_1 < \bar{d}_2 < \cdots < \bar{d}_n$. By Proposition 5, we can suppose that the ω-chains $\bar{b}_n$ and $\bar{d}_n$ are such that

$$\Gamma(\mathcal{A} + \mathcal{B}) \models \bigwedge_{i \in \omega} b_{n,i} <^\infty b_{n,i+1} \tag{1}$$

$$\Gamma(\mathcal{C} + \mathcal{D}) \models \bigwedge_{i \in \omega} d_{n,i} <^\infty d_{n,i+1}, \tag{2}$$

Now by Proposition 4 we have that

$$\Gamma(\mathcal{B} + \mathcal{D}) \models \bar{b}_n <^* \bar{d}_n. \tag{3}$$

For an arbitrary partition of $\mathcal{B}$ and $\mathcal{C}$ into finite parts such that $\mathcal{B} = \sum_{i \in \omega} \beta_i$ and $\mathcal{C} = \sum_{i \in \omega} \gamma_i$, let us consider the copy $\mathcal{N}$ of $\omega \cdot 3$, where

$$\mathcal{N} = \mathcal{A} + \sum_{i \in \omega}(\beta_i + \gamma_i) + \mathcal{D}.$$

By monotonicity, (3) implies that $\Gamma(\mathcal{N}) \models \overline{b}_n <^* \overline{d}_n$. Now, again by monotonicity, (1) and (2) imply that $\Gamma(\mathcal{N})$ extends a copy of $\omega^2 \cdot 2$.   □

Now we are ready to finish the proof. Consider two disjoint copies $\mathcal{L}$ and $\mathcal{M}$ of $\omega \cdot 3$ such that $\Gamma(\mathcal{L}) \cong \omega^2$ and $\Gamma(\mathcal{M}) \cong \omega^2$. By Proposition 6, we obtain two disjoint copies $\mathcal{L}_1$ and $\mathcal{M}_1$ of $\omega \cdot 2$ such that $\Gamma(\mathcal{L}_1) \cong \omega^2$ and $\Gamma(\mathcal{M}_1) \cong \omega^2$. Then by Proposition 7, from $\mathcal{L}_1$ and $\mathcal{M}_1$ we can construct a copy $\mathcal{N}$ of $\omega \cdot 3$ such that $\Gamma(\mathcal{N}) \not\cong \omega^2$. Thus, we have proven the following theorem.

Theorem 3. $\{\omega \cdot 3, \omega^* \cdot 3\} \not\leq_c \{\omega^2, (\omega^2)^*\}$.

Corollary 3. *For any non-zero natural number n,*

$$n \geq 2 \iff \{\omega \cdot 3, \omega^* \cdot 3\} \leq_c \{\omega^2 \cdot n, (\omega^2)^* \cdot n\}.$$

Proof. First consider the direction $(\Rightarrow)$. For each $n \geq 2$ we will show how to build an enumeration operator Γ_n. Notice that by Corollary 1 we have an enumeration operator $\Gamma_2 : \{\omega \cdot 3, \omega^* \cdot 3\} \leq_c \{\omega^2 \cdot 2, (\omega^*)^2 \cdot 2\}$. Moreover, by Theorem 1, we have an enumeration operator $\Gamma_3 : \{\omega \cdot 3, \omega^* \cdot 3\} \leq_c \{\omega^2 \cdot 3, (\omega^*)^2 \cdot 3\}$.

Let $n = 2k$ for some $k \geq 1$. Then Γ_n works so that, for any input $\mathcal{A}$, it outputs k disjoint copies of $\Gamma_2(\mathcal{A})$.

Let $n = 2k + 3$ for some $k \geq 0$. Then Γ_n works so that, for any input $\mathcal{A}$, it outputs k disjoint copies of $\Gamma_2(\mathcal{A})$ together with a copy of $\Gamma_3(\mathcal{A})$. The direction $(\Leftarrow)$ is exactly Theorem 3.   □

5   The General Case

Here, using the same techniques as in Sect. 4, we will obtain the following theorem.

Theorem 4. *For any $k \geq 3$, $\{\omega \cdot k, \omega^* \cdot k\} \not\leq_c \{\omega^2, (\omega^2)^*\}$.*

Again towards a contradiction, assume that we have fixed a number $k \geq 3$ and an enumeration operator $\Gamma : \{\omega \cdot k, \omega^* \cdot k\} \leq_c \{\omega^2, (\omega^2)^*\}$. Since Proposition 2 and Proposition 3 still apply in this more general case, we can use Remark 2 and suppose we have fixed a number n such that we always work with copies $\mathcal{A}$ of ω such that $\Gamma(\mathcal{A}) \cong \omega \cdot n$. By essentially repeating the proof of Proposition 5, we obtain the following proposition.

Proposition 8. *Let $\mathcal{A}_1, \mathcal{A}_2, \ldots, \mathcal{A}_k$ be copies of ω. Suppose that*

$$\Gamma(\mathcal{A}_k) \models \overline{c}_1 < \overline{c}_2 < \cdots < \overline{c}_n,$$

where $\overline{c}_i = (c_{i,j})_{j=0}^{\infty}$ are ω-chains. Then there exists an infinite subsequence $(i_s)_{s=0}^{\infty}$ such that

$$\Gamma\left(\sum_{j=1}^{k} \mathcal{A}_j\right) \models \bigwedge_{s \in \omega} c_{n,i_s} <^{\infty} c_{n,i_{s+1}}.$$

The next proposition is a generalization of Proposition 6.

Proposition 9. *For any linear ordering $\mathcal{L}$ of type $\omega \cdot k$, there is a linear ordering $\mathcal{M}$ of type $\omega \cdot 2$ with $dom(\mathcal{L}) = dom(\mathcal{M})$ and $\Gamma(\mathcal{M}) \cong \omega^2$.*

Proof. Let $\mathcal{L} = \sum_{i=1}^{k} \mathcal{A}_i$, where $\mathcal{A}_i$ are copies of ω. By Proposition 8, consider the ω-chain $(c_i)_{i=0}^{\infty}$ in $\Gamma(\mathcal{A}_k)$ such that $\Gamma(\sum_{j=1}^{k} \mathcal{A}_j) \models \bigwedge_{i \in \omega} c_i <^{\infty} c_{i+1}$.

As in the proof of Proposition 6, for any ℓ, let $\overline{u}_\ell = (u_{\ell,j})_{j=0}^{\infty}$ be an ω-chain such that we can partition $\mathcal{A}_i$ into finite parts with $\mathcal{A}_i = \sum_{j \in \omega} \alpha_{i,j}$, where $i = 1, 2, \ldots, k-1$, where for all j,

$$\Gamma(\sum_{i=1}^{k-1} \alpha_{i,j} + \mathcal{A}_k) \models \bigwedge_{\ell=0}^{j} c_\ell < u_{\ell,j-\ell} < c_{\ell+1}.$$

Then, by monotonicity, we obtain the following:

$$\Gamma(\sum_{j \in \omega} \sum_{i=1}^{k-1} \alpha_{i,j} + \mathcal{A}_k) \models \bigwedge_{\ell \in \omega} \bigwedge_{j \in \omega} c_\ell < u_{\ell,j} < c_{\ell+1}.$$

It follows that $\mathcal{M} = \sum_{j \in \omega} \sum_{i=1}^{k-1} \alpha_{i,j} + \mathcal{A}_k$ is a copy of $\omega \cdot 2$ with $dom(\mathcal{M}) = dom(\mathcal{L})$ which produces a copy of ω^2. □

Let us take two disjoint copies $\mathcal{L}$ and $\mathcal{M}$ of $\omega \cdot k$ such that $\Gamma(\mathcal{L}) \cong \omega^2$ and $\Gamma(\mathcal{M}) \cong \omega^2$. By Proposition 9, we obtain two disjoint copies $\mathcal{L}_1$ and $\mathcal{M}_1$ of $\omega \cdot 2$ such that $\Gamma(\mathcal{L}_1) \cong \omega^2$ and $\Gamma(\mathcal{M}_1) \cong \omega^2$. Then by Proposition 7, from $\mathcal{L}_1$ and $\mathcal{M}_1$ we can construct a copy $\mathcal{N}$ of $\omega \cdot 3$ such that $\Gamma(\mathcal{N})$ extends a copy of $\omega^2 \cdot 2$. By monotonicity, any copy $\hat{\mathcal{N}}$ of $\omega \cdot k$ extending $\mathcal{N}$ will be such that $\Gamma(\hat{\mathcal{N}}) \not\cong \omega^2$. We conclude that $\{\omega \cdot k, \omega^\star \cdot k\} \not\leq_c \{\omega^2, (\omega^2)^\star\}$.

6   Positive Results for Powers of ω

Proposition 10. *For any $n \geq 1$, $\{\omega^n, (\omega^n)^\star\} \leq_c \{\omega^{2n}, (\omega^{2n})^\star\}$.*

Proof. Standard cartesian product construction as in [18, Definition 1.40]. □

Theorem 5. $\{\omega^2, (\omega^2)^\star\} \leq_c \{\omega^3, (\omega^3)^\star\}$.

Proof. The idea here is to replace each point by an interval of the form $[a, b]$, which means that this interval will have type $\omega \cdot k + \ell$ in the first case and $\ell + \omega^\star \cdot k$ in the second case.

We informally describe the work of the enumeration operator Γ. Let us consider some finite diagram $\delta(\overline{a})$ of the input structure $\mathcal{A}$. For each a in $\delta(\overline{a})$, Γ executes the following steps: Find elements b and c such that $b \leq_{\mathcal{A}} a \leq_{\mathcal{A}} c$, where $b, c \leq_{\mathbb{N}} a$, such that b is the $\leq_{\mathcal{A}}$-least such element and c is the $\leq_{\mathcal{A}}$-greatest such element in $\delta(\overline{a})$. For all elements d in $\delta(\overline{a})$ such that $b \leq_{\mathcal{A}} d \leq_{\mathcal{A}} c$, Γ enumerates in the output structure the pair (a, d). All pairs are ordered lexicographically. This concludes the description of Γ. Now we have two cases to consider.

Suppose that $\mathcal{A} = \sum_{i \in \omega} \mathcal{A}_i$, where $\mathcal{A}_i$ are copies of ω. It is easy to see that for each i, there are only finitely many elements in $\mathcal{A}_i$, which contribute finitely many pairs in $\Gamma(\mathcal{A})$. For instance, let a be the $<_{\mathbb{N}}$-least element in $\mathcal{A} \setminus \mathcal{A}_0$. It follows that in $\mathcal{A}_0$ only the elements which are $<_{\mathbb{N}}$-less than a contribute finitely many pairs in $\Gamma(\mathcal{A})$. We have

$$\Gamma(\mathcal{A}) = \sum_{a \in \mathcal{A}_0} (\omega \cdot k_{a,0} + \ell_{a,0}) + \cdots + \sum_{a \in \mathcal{A}_i} (\omega \cdot k_{a,i} + \ell_{a,i}) + \cdots$$
$$= \omega^2 + \cdots + \omega^2 + \cdots = \omega^3.$$

For the second case, suppose that $\mathcal{A} = \sum_{i \in \omega^*} \mathcal{A}_i$, where $\mathcal{A}_i$ are copies of ω^*. Again, for each i, there are only finitely many elements in $\mathcal{A}_i$, which contribute finitely many pairs in $\Gamma(\mathcal{A})$. It follows that

$$\Gamma(\mathcal{A}) = \cdots + \sum_{a \in \mathcal{A}_i} (\ell_{a,i} + \omega^* \cdot k_{a,i}) + \cdots + \sum_{a \in \mathcal{A}_0} (\ell_{a,0} + \omega^* \cdot k_{a,0})$$
$$= \cdots + (\omega^2)^* + \cdots + (\omega^2)^* = (\omega^3)^*.$$

$\square$

Using the same enumeration operator Γ as in Theorem 5, we obtain the following corollary.

Corollary 4. *For any $n \geq 1$, $\{\omega^n, (\omega^n)^*\} \leq_c \{\omega^{2n-1}, (\omega^{2n-1})^*\}$.*

Corollary 5. *For any $n \geq 2$, $\{\omega^2, (\omega^2)^*\} \leq_c \{\omega^n, (\omega^n)^*\}$.*

Proof. For any natural number $n \geq 2$, we briefly describe the enumeration operator $\Gamma_n : \{\omega^2, (\omega^2)^*\} \leq_c \{\omega^n, (\omega^n)^*\}$.

- If $n = 2k$, where $k \geq 1$, then for any input $\mathcal{A}$, Γ_n outputs $\mathcal{A}^k$.
- If $n = 3$, then Γ_3 is the enumeration operator from Theorem 5.
- If $n = 2k + 3$, where $k \geq 1$, then for any input $\mathcal{A}$, Γ_n outputs $\Gamma_3(\mathcal{A}) \cdot \mathcal{A}^k$.

$\square$

7   Future Work

We strongly conjecture that by employing the methods of this paper, one can prove that $\{\omega \cdot 3, \omega^* \cdot 3\} \nleq_c \{\omega^3, (\omega^3)^*\}$. Furthermore, it would be interesting to consider pairs of structures $\{\mathcal{A}, \mathcal{B}\}$ such that $\mathcal{A}$ and $\mathcal{B}$ are *not* linear orders, but still $\{\mathcal{A}, \mathcal{B}\} \equiv_{tc} \{\omega, \omega^*\}$. We note that in this case, $\mathcal{A}$ and $\mathcal{B}$ *cannot* be Boolean algebras (see Proposition 4.6 of [2]).

References

1. Bazhenov, N., Downey, R., Kalimullin, I., Melnikov, A.: Foundations of online structure theory. Bull. Symb. Log. **25**(2), 141–181 (2019). https://doi.org/10.1017/bsl.2019.20
2. Bazhenov, N., Fokina, E., San Mauro, L.: Learning families of algebraic structures from informant. Preprint, arXiv:1905.01601 (2019)
3. Bazhenov, N., Ganchev, H., Vatev, S.: Computable embeddings for pairs of linear orders. Preprint, arXiv:1901.01933 (2019)
4. Calvert, W., Cummins, D., Knight, J.F., Miller, S.: Comparing classes of finite structures. Algebra Log. **43**(6), 374–392 (2004). https://doi.org/10.1023/B:ALLO.0000048827.30718.2c
5. Chisholm, J., Knight, J.F., Miller, S.: Computable embeddings and strongly minimal theories. J. Symb. Log. **72**(3), 1031–1040 (2007). https://doi.org/10.2178/jsl/1191333854
6. de Brecht, M., Yamamoto, A.: Topological properties of concept spaces (full version). Inf. Comput. **208**(4), 327–340 (2010). https://doi.org/10.1016/j.ic.2009.08.001
7. Downey, R., Harrison-Trainor, M., Kalimullin, I., Melnikov, A., Turetsky, D.: Graphs are not universal for online computability. J. Comput. Syst. Sci. **112**, 1–12 (2020). https://doi.org/10.1016/j.jcss.2020.02.004
8. Ershov, Y.L., Puzarenko, V.G., Stukachev, A.I.: HF-computability. In: Cooper, S.B., Sorbi, A. (eds.) Computability in Context, pp. 169–242. Imperial College Press, London (2011). https://doi.org/10.1142/9781848162778_0006
9. Friedman, H., Stanley, L.: A Borel reducibility theory for classes of countable structures. J. Symb. Log. **54**(3), 894–914 (1989). https://doi.org/10.2307/2274750
10. Ganchev, H., Kalimullin, I., Vatev, S.: Computable embedding of classes of algebraic structures with congruence relation. Uchenye Zapiski Kazanskogo Universiteta. Seriya Fiziko-Matematicheskie Nauki **160**(4), 731–737 (2018). in Russian
11. Harrison-Trainor, M., Melnikov, A., Miller, R., Montalbán, A.: Computable functors and effective interpretability. J. Symb. Log. **82**(1), 77–97 (2017). https://doi.org/10.1017/jsl.2016.12
12. Harrison-Trainor, M., Miller, R., Montalbán, A.: Borel functors and infinitary interpretations. J. Symb. Log. **83**(4), 1434–1456 (2018). https://doi.org/10.1017/jsl.2017.81
13. Hirschfeldt, D.R., Khoussainov, B., Shore, R.A., Slinko, A.M.: Degree spectra and computable dimensions in algebraic structures. Ann. Pure Appl. Logic **115**(1–3), 71–113 (2002). https://doi.org/10.1016/S0168-0072(01)00087-2
14. Kalimullin, I.S.: Computable embeddings of classes of structures under enumeration and turing operators. Lobachevskii J. Math. **39**(1), 84–88 (2018). https://doi.org/10.1134/S1995080218010146
15. Knight, J.F., Miller, S., Vanden Boom, M.: Turing computable embeddings. J. Symb. Log. **72**(3), 901–918 (2007). https://doi.org/10.2178/jsl/1191333847
16. Miller, R., Poonen, B., Schoutens, H., Shlapentokh, A.: A computable functor from graphs to fields. J. Symb. Log. **83**(1), 326–348 (2018). https://doi.org/10.1017/jsl.2017.50
17. Puzarenko, V.G.: A certain reducibility on admissible sets. Sib. Math. J. **50**(2), 330–340 (2009). https://doi.org/10.1007/s11202-009-0038-z
18. Rosenstein, J.G.: Linear Orderings. Pure and Applied Mathematics, vol. 98. Academic Press, New York (1982)

Clockability for Ordinal Turing Machines

Merlin Carl[(✉)]

Institut für mathematische, naturwissenschaftliche und technische Bildung, Abteilung
für Mathematik und ihre Didaktik. Europa-Universität Flensburg,
Flensburg, Germany
`merlin.carl@uni-konstanz.de`

Abstract. We study clockability for Ordinal Turing Machines (OTMs).
In particular, we show that, in contrast to the situation for ITTMs,
admissible ordinals can be OTM-clockable, that Σ_2-admissible ordinals
are never OTM-clockable and that gaps in the OTM-clockable ordinals
are always started by admissible limits of admissible ordinals. This partially answers two questions in [3].

1 Introduction

In ordinal computability, "clockability" denotes the property of an ordinal that
it is the halting time of some program. The term was introduced in [9], which was
the paper that triggered the bulk of research in the area of ordinal computability
by introducing Infinite Time Turing Machines (ITTMs).[1] By now, a lot is known
about clockability for ITTMs. To give a few examples: In [9], it was proved that
there are gaps in the ITTM-clockable ordinals, i.e., there are ordinals $\alpha < \beta < \gamma$
such that α and γ are ITTM-clockable, but β is not. Moreover, it is known
that no admissible ordinal is ITTM-clockable (Hamkins and Lewis, [9]), that the
first ordinal in a gap is always admissible (Welch, [14]), that the supremum λ
of the ITTM-writable ordinals (i.e. ordinals coded by a real number that is the
output of some halting ITTM-computation) equals the supremum of the ITTM-
clockable ordinals (Welch, [14]), that an ITTM-clockable γ has a code that is
ITTM-writable in γ many steps (Welch, [14]) and that ITTM-writable ordinals
have real codes that are ITTM-writable at the point the next clockable appears.
Moreover, it is known that not every admissible below λ starts a gap, there are
admissibles properly inside gaps, and occasionally many of them (Carl, Durand,
Lafitte, Ouazzani, [6]). And indeed, clockability turned out to be a central topic
in ordinal computability; it was, for example, crucial for Welch's analysis of the
computational strength of ITTMs.

Besides ITTMs, clockability was also considered for Infinite Time Register
Machines (ITRMs), where the picture turned out to be quite different: In particular, there are no gaps in the ITRM-clockable ordinals (see [5]), and in fact,

---

[1] As one of our referees pointed out, there are earlier considerations of machine models
computing along an ordinal time axis; however, none of them was studied in the detail
that ITTMs were.

© Springer Nature Switzerland AG 2020
M. Anselmo et al. (Eds.): CiE 2020, LNCS 12098, pp. 14–25, 2020.
https://doi.org/10.1007/978-3-030-51466-2_2

the ITRM-clockable ordinals are exactly those below $\omega_\omega^{\mathrm{CK}}$, which thus includes ω_n^{CK} for every $n \in \omega$, i.e. the first ω many admissible ordinals.

For other models, clockability received comparably little attention. This work arose out of a question of T. Kihara during the CTFM[2] conference in 2019 in Wuhan who, after hearing that admissible ordinals are never ITTM-clockable, asked whether the same holds for OTMs. After most of the results of this paper had been proved, we found two questions in the report of the 2007 BIWOC (Bonn International Workshop on Ordinal Computability) [3] concering this topic: the first (p. 42, question 9), due to J. Reitz, was whether ω_1^{CK} was OTM-clockable, the second, due to J. Hamkins, whether gap-starting ordinals for OTMs can be characterized as "something stronger" than being admissible. In [3], both are considered to be answered by the claim that no admissible ordinal is OTM-clockable, which is attributed to J. Reitz and S. Warner. Upon personal inquiry, Reitz told us that they had a sketch of a proof which, however, did not entirely work; what it does show with a few modifications, though, is that Σ_2-admissible ordinals are not OTM-clockable, and the argument that Reitz sketched in personal correspondence to us in fact resembles the one of Theorem 6 below. We thus regard Reitz and Warner as the first discoverers of this theorem. Both the argument of Reitz and Warner from 2007 and the one we found during the CTFM in 2019 are adaptations of Welch's argument that admissible ordinals are not ITTM-clockable.

The statement actually made in [3], is, however, false: As we will show below, ω_n^{CK} is OTM-clockable for any $n \in \omega$. Thus, there are plenty of admissible ordinals that are OTM-clockable, and the answer to the first question is positive. The idea is to use the ITRM-clockability of these ordinals, which follows from Lemma 3 in [5], together with a slightly modified version of the obvious procedure for simulating ITRMs on OTMs. This actually shows that ω_n^{CK} is clockable on an ITTM with tape length α as soon as $\alpha > \omega$. Thus, the strong connection between admissibility and clockability seems to depend rather strongly on the details of the ITTM-architecture. We remark that this is a good example of how the studies of different models of infinitary computability can fruitfully interact: At least for us, it would not have been possible to find this result while only focusing on OTMs.

Moreover, we will answer the second question in the positive as well by showing that, if α starts a gap in the OTM-clockable ordinals, then α is an admissible limit of admissible ordinals.[3]

Of course, the gap between "admissible limit of admissible ordinals" and "Σ_2-admissible" is quite wide. In particular, we do not know whether every gap starting ordinal for OTMs is Σ_2-admissible, though we conjecture this to be false.

---

[2] International Conference on Computability Theory and Foundations of Mathematics.

[3] The notion of admissibility will play a prominent role in this paper. Readers unfamiliar with it are referred to Barwise [1].

2   Ordinal Turing Machines

Ordinal Turing Machines (OTMs) were introduced by Koepke in [10] as a kind of "symmetrization" of ITTMs: Instead of having a tape of length ω and the whole class of ordinals as their working time, OTMs have a tape of proper class length On while retaining On as their "working time" structure. We refer to [10] for details.

In contrast to Koepke's definition but in closer analogy with the setup of ITTMs, we allow finitely many tapes instead of a single one. Each tape has a head, and the heads move independently of each other; the program for such an OTM is simply a program for a (finite) multihead Turing machine. At limite times, the inner state (which is coded by a natural number), the cell contents and the head positions are all determined as the inferior limits of the sequences of the respective earlier values. At successor steps, an OTM-program is carried out as if on a finite Turing machine with the addition that, when a head is moved to the left from a limit posistion, it is reset to the start of the tape. Though models of ordinal computability generally enjoy a good degree of stability under such variations as far as computational strength is concerned, this often makes a difference when it comes to clockability. Intuitively, simulating several tapes with separate read-write-heads on a single tape requires one to check the various head positions to determine whether the simulated machine has halted, which leads to a delay in halting. For ITTMs, this is e.g. demonstrated in [13]. For OTMs, insisting on a single tape would lead to a theory that is "morally" the same as the one described here, but make the results much less compelling and the proofs more technically involved and harder to follow.[4] Thus, allowing multiple tapes seems to be a good idea.

An important property of OTMs that will be used below is the existence of an OTM-program P that 'enumerates L'; in particular, P will write (a code for) the constructible level L_α on the tape in $< \alpha'$ many steps, where α' is the smallest exponentially closed ordinal $> \alpha$ (this notation will be used throughout the paper).

The following picture of OTM-computations may be useful to some readers: Let us imagine the tape split into ω-blocks. Then an OTM-computation proceeds like this: The head works for a bit in one ω-block, then leaves it to the right, works for a bit in the new ω-portion, again leaves it to the right and so on, until eventually the computation either halts or the head is moved back from a limit position, i.e., goes back to 0 and starts over. Thus, if one imagines an ω-portion as single point, then the head moves from left to right, jumps back to 0, moves right again etc. Moreover, in each ω-portion, we have a classical ITTM-computation (up to the limit rules for the head position and the inner state, which make little difference).

We fix some terminology for the rest of this paper.

---

[4] For example, by simulating multitape machines on a single-type machine in a rather straightforward way, one can see that the following holds: If α is exponentially closed and clockable by an OTM, then $\alpha \cdot 2$ is clockable by an OTM using only one tape.

Definition 1. *If M is one of ITRM, ITTM or OTM and α is an ordinal, then α is called M-clockable if and only if there is an M-program that halts at time $\alpha + 1$.[5] α is called M-writable if and only if there is a real number coding α that is M-computable. An M-clockable gap is an interval $[\alpha, \beta)$ of ordinals such that $\alpha < \beta$, no element of $[\alpha, \beta)$ is M-clockable and $[\alpha, \beta)$ is maximal in the sense that there are cofinally many M-clockable ordinals below α and β is M-clockable. In this case, we say that α "starts" the gap and call α a "gap starting ordinal" or "gap starter" for M.*

3   Basic Observations

We start with some useful observations that can mostly be obtained by easy adaptations of the corresponding results about ITTM-clockability.

We start by noting that the analogue of the speedup-theorem for ITTMs from [9] holds for multitape-OTMs. This is proved by an adaptation of the argument for the speedup-theorems for ITTMs. The main difference is that, in contrast to ITTMs, OTMs do not have their head on position 0 at every limit time and that the head may make long "jumps" when moved to the left from a limit position. This generates a few extra complications.

To simplify the proof, we start by building up a few preliminaries.

For the ITTM-speedup, the following compactnes property is used: If P halts in $\delta + n$ many steps and the head is located at position k at time δ, then only the n cells contents before and after the kth one at time δ are relevant for this. Now, this is a fixed string s of $2n$ bits. In [9], a construction is described that achieves that the information whether these $2n$ cells currently contain s at a limit time γ is coded on some extra tapes at time γ. Due to the special limit rules for ITTMs that set the head back to position 0 at every limit time, the Hamkins-Lewis-proof has this information stored at the initial tape cells, but the construction is easily modified to store the respective information on any other tape position.

We will use it in the following way: Suppose that P is an OTM-program that halts at time $\delta + n$, where δ is a limit ordinal and $n \in \omega$. We want to "speed up" P by n steps, i.e. to come up with a program Q that halts in δ many steps. Suppose that P halts with the head on position $\gamma + k$, where γ is a limit ordinal and $k \in \omega$. Let m be $k - n$ if $k - n \geq 0$ and 0, otherwise, and let s be the bit string present on positions $\gamma + m$ until $\gamma + k + n$ at time δ. Then we use the Hamkins-Lewis-construction to ensure that the information whether the bit string present on positions $\eta + m$ until $\eta + k + n$ is equal to s on the $(\eta + k)$th cells of three extra tapes, for each limit ordinal η.

An extra complication arises from the possibility of a "setback": Within the n steps from time δ to time $\delta + n$, it may happen that the head is moved left from position δ, thus ending up at the start of the tape. Clearly, it will then take $< n$ many further steps at the start of the tape and only consider the first n

---

[5] The +1 allows limit ordinals to appear as halting times and thus simplifies the theory.

bits during this time. However, we need to know what these bits are - or rather, whether they are the "right ones", i.e., the ones present at time δ - while our head is located at position $\delta + k$. The idea is then to store this information in the inner state of the sped-up program. We thus create extra states: The new state $2i$ will represent the old state i together with the information that the first n bits were the "right ones" (i.e. the same ones as at time δ) and $2i + 1$ will represent the old state i together with the information that some of these bits deviated from those at time δ. To achieve this, we use an extra tape T_4. At the start of Q, a 1 is written to each of the first n cells of T_4; after that, the head on T_4 is set back to position 0 and then moved along with the head of P. In this way, we will always know whether the head of P is currently located at one of the first n cells. Whenever this is the case, we insert some intermediate steps to read out the first n bits, update the inner state and move the head back to its original position. (This requires some additional states, but we skip the details). Note that, if η is a limit time and the first n bits have been changed unboundedly often before η, then the head will be located at one of these positions at time η by the liminf-rule and thus, a further update will take place so that the state will correctly represent the configuration afterwards. On the other hand, if the first n bits were only changed boundedly often before time η, then let $\bar{\eta}$ be the supremum of these times. We just saw that the state will represent the configuration correctly finitely many steps after time $\bar{\eta}$, after which the first n cell contents remain unchanged, so that the state is still correct at time η. In each case, updating this information and returning to the original configuration will take only finitely many extra steps and thus not cause a delay at limit times.[6]

In the following construction, we will need to know whether the head is currently located at a cell the index of which is of the form $\delta + k$, where δ is a limit ordinal and k is a fixed natural number. To achieve this, we add three tapes T_0, T_1 and T_2 to P. The tape T_0 serves as a flag: By having two cells with alternating contents 01 and 10, we can detect a limit time as a time at which both cells contain 0. On T_2, we move the head along with the head on P and place a 1 on a cell whenever we encounter a cell on which a 0 is written. Thus, the head occupies a certain limit position for the first time if and only if the head on T_1 reads a 0 at a limit time. Finally, on T_2, we move the head along with the heads on T_1 and the main tape. Whenever the head on T_1 reads a 0 at a limit time, we interrupt the computation, move the head on T_2 for k many steps to the right, write a 1, move the head k many places to the left, and continue. In this way, the head on T_2 will read a 1 if and only if the head on the main tape is at a position of the desired form. As this merely inserts finitely many steps occasionally, running this procedure along with an OTM-program P will still carry out δ many steps of P at time δ whenever δ is a limit ordinal. We will say that the head is "at a $\delta + k$-position" if the index of the cell where it is

---

[6] This leaves us with the case that the head occupies one of the first n tape positions at time δ, in which case even a finite delay would increase our running time. However, in this special case, no setback will take place during the last n steps of the computation, so the construction described in this paragraph can simply be skipped.

currently located is of this form with δ a limit ordinal and, by the construction just described, we can use formulations like "if the head is currently at a $\delta + k$-position" in describing OTM-programs without affecting the running time at limit ordinals.

Lemma 1. *If $\alpha + n$ is OTM-clockable and $n \in \omega$, then α is OTM-clockable.*

Proof. It is clear that finite ordinals are OTM-clockable and that OTM-clockable ordinals are closed under addition (by simply running one program after the other).[7] Thus, it suffices to consider the case that α is a limit ordinal. Moreover, we assume for simplicity that P uses only one tape.[8]

Let P be an OTM-program that runs for $\alpha + n$ many steps, where α is a limit ordinal. We want to construct a program Q that runs for α many steps. Let the head position at time α be equal to $\delta + k$, where δ is a limit ordinal and $k \in \omega$. As above, let m be $k - n$ if $k - n \geq 0$ and otherwise let $m = 0$. Let s be the bit string present on the positions $\delta + m$ until $\delta + k + n$ at time α, and let t be the string present on the first n positions.

Using the constructions explained above, Q now works as follows: Run P. At each step, determine whether the head is currently at a location of the form $\eta + k$ with η a limit ordinal and whether one of the two following conditions holds:

1. The head is currently at one of the first n positions and the bit string currently present on the positions $\eta + m$ up to $\eta + k + n$ is equal to s.
2. The head is currently not on one of the first n positions, the bit string currently present on the positions $\eta + m$ up to $\eta + k + n$ is equal to s and whether the bit string currently present on the first n positions is equal to t.

If not, continue with P. Otherwise, halt. As described above, the necessary information can be read off from the various extra tapes and the inner state simultaneously. Now it is clear that, if Q halts at time β, then P will halt at time $\beta + n$. Thus, Q halts at time α, as desired.

Definition 2. *Let σ be the minimal ordinal such that $L_\sigma \prec_{\Sigma_1} L$, i.e. such that L_σ is a Σ_1-submodel of L.*

Proposition 3. *Every OTM-clockable ordinals is $< \sigma$, and their supremum is σ.*

Proof. The statement 'The program P halts' is Σ_1. Moreover, any halting OTM-computation is contained in L. Consequently, if P halts, its computation is contained in L, and hence in L_σ, and thus, the halting time of P, if it exists, is $< \sigma$.

On the other hand, every real number in L_σ is OTM-computable (see, e.g., [12], proof of Corollary 3), including codes for all ordinals $< \sigma$, and thus we can

---

[7] It is folklore (and easy to see) that, for any reasonable model of computation, clockable ordinals are closed under ordinal arithmetic, i.e. under addition, multiplication and exponentiation, see e.g. [9] or [5]. This also holds true for OTMs.

[8] If P uses several tapes, the construction below is carried out for each of these.

write such a code for any ordinal $\alpha < \sigma$ and then run through this code, which takes at least α many steps. Thus, there is an OTM-clockable ordinal above α for every $\alpha < \sigma$.

Proposition 4. *There are gaps in the OTM-clockable ordinals. That is, there are ordinals $\alpha < \beta < \gamma$ such that α and γ are OTM-clockable, but β is not.*

Proof. This works like the argument in Hamkins and Lewis ([9], Theorem 3.4) for the existence of gaps in the ITTM-clockable ordinals: Take the OTM-program that simultaneously simulates all OTM-programs and halts as soon as it arrives at a level at which no OTM-program halts. If there were no gap, then this program would halt after all OTM-halting times, which is a contradiction.

The following is an OTM-version of Welch's "quick writing theorem" (see [14], Lemma 48) for ITTMs.

Lemma 2. *If an ordinal α is OTM-clockable, then a real number coding α is OTM-writable in $< \alpha'$ many steps, where α' denotes the next exponentially closed ordinal after α.*

Proof. If α is clocked by some OTM-program P, then $L_{\alpha+\omega}$ believes that P halts. Thus, there is a Σ_1-statement that becomes true between L_α and $L_{\alpha+\omega}$ for the first time and hence, by finestructure (see [2], Lemma 1), a real number coding $\alpha + 1$ is contained in $L_{\alpha+\omega}$. But the OTM-program Q that enumerates L will have (a code for) $L_{\alpha+\omega}$ on the tape in $< \alpha'$ many steps. So we can simply run this program until we arrive at a code c for a limit L-level that believes that P halts for the first time. Now, we can easily find out the desired real code for α in the code for $L_{\alpha+\omega}$ (by searching the coded structure for an element which it believes to be the halting time of P).

Proposition 5. *If $\beta < \alpha$ is exponentially closed and OTM-clockable and there is a total $\Sigma_1(L_\alpha)$-function $f : \beta \rightarrow \alpha$ such that f is cofinal in α, then α is OTM-clockable.*

Proof. This works by the same argument as the "only admissibles start gaps"-theorem for ITTMs, see Welch [14]: Suppose for a contradiction that α starts an OTM-gap, but is not admissible.

Pick $\beta < \alpha$ OTM-clockable and $f : \beta \rightarrow \alpha$ such that f is $\Sigma_1(L_\alpha)$ and cofinal in α. Let B be an OTM-program that clocks β. By the last lemma, we can compute a real code for β in $< \beta' \leq \alpha$ many steps. Run the OTM that enumerates L. If β is exponentially closed, then we will have a code for L_β on the tape at time β. In addition, for each new L-level, check which ordinals recieve f-images when evaluating the definition of f in that level. Determine the largest ordinal γ such that f is defined on γ. Whenever γ increases, say from γ_0 to γ_1, let δ be such that $\gamma_0 + \delta = \gamma_1$ and run B for δ many steps. When B halts, all elements of β have images, so we have arrived at time α.

This suffices for an OTM-analogue of Welch's theorem [14], Theorem 50:

Corollary 1. *If α starts a gap in the OTM-clockable ordinals, then α is admissible.*

Proof. As α starts an OTM-gap, it is exponentially closed.

If α is not admissible, there is a total cofinal $\Sigma_1(L_\alpha)$-function $f : \beta \to \alpha$ with $\beta < \alpha$. Pick $\gamma \in (\beta, \alpha)$ OTM-clockable and large enough so that all parameters used in the definition of f are contained in L_γ. By Lemma 2, we can write a real code for L_γ, and thus for all of its elements in time $< \gamma' \le \alpha$. We can now clock α as in Proposition 5, a contradiction.

4 Σ_2-admissible Ordinals Are Not OTM-clockable

We now show that no Σ_2-admissible ordinal α can be the halting time of a parameter-free OTM-computation. The proof is mostly an adapatation of argument in Hamkins and Lewis [9] for the non-clockability of admissible ordinals by ITTMs to the extra subtleties of OTMs.

Theorem 6. *No Σ_2-admissible ordinal is OTM-clockable.*

Proof. We will show this for the case of a single-tape OTM for the sake of simplicity.

Let α be Σ_2-admissible and assume for a contradiction that α is the halting time of the parameter-free OTM-program P. At time α, suppose that the read-write-head is at position ρ, the program is in state $s \in \omega$ and the head reads the symbol $z \in \{0,1\}$. As one cannot move the head more than α many places to the right in α many steps, we have $\rho \le \alpha$.

By the limit rules, z must have been the symbol on cell ρ cofinally often before time α and similarly, s must have been the program state cofinally often before time α. By recursively building an increasing 'interleaving' sequence of ordinals of both kinds, we see that the set S of times at which the program state was s and the symbol on ρ was z, we see that S is closed and unbounded in α.

We now distinguish three cases.

Case 1: $\rho < \alpha$ and the head position ρ was assumed cofinally often before time α.

Let β be the order type of the set of times at which ρ was the head position in the computation of P. We show that $\beta = \alpha$. If not, then $\beta < \alpha$; let $f : \beta \to \alpha$ be the function sending each $\iota < \beta$ to the ιth time at which ρ was the head position. Then f is Σ_1 over L_α and thus, by admissibility of α, $f[\beta]$ is bounded in α, contradicting the case assumption.

Let T be the set of times at which ρ was the head position. Then, by the limit rules and the case assumption, T is closed and unbounded in α.

As S and T are both Σ_1 over L_α and α is admissible, it follows that $S \cap T$ is also closed and unbounded in α. In particular, there is an element $\gamma < \alpha$ in $S \cap T$, i.e. there is a time $< \alpha$ at which the head was on position ρ, the cell ρ contained the symbol z and the inner state was s. But then, the situation that prompted P to halt at time α was already given at time $\gamma < \alpha$, so P cannot have run up to time α, a contradiction.

Case 2: $\rho < \alpha$ and the head position ρ was assumed boundedly often before time α.

By the liminf rule for the determination of the head position at time α, this implies that, for every $\iota < \rho$, there is a time $\tau_\iota < \alpha$ such that, from time τ_ι on, the head never occupied a position $< \iota$. The function $f : \iota \mapsto \tau_\iota$ is Π_1 over L_α (we have $f(\iota) = \tau$ if and only if, for all $\beta > \tau$ and all partial P-computations of length β, the head position in the final state of the partial computation was $\geq \iota$) and thus in particular Σ_2 over L_α. By Σ_2-admissibility of α and the case assumption $\rho < \alpha$, the set $f[\rho]$ must be bounded in α, say by $\gamma < \alpha$. But this implies that, after time γ, all head positions were $\geq \rho$. As ρ was assumed only boundedly often as the head position, this means that, from some time $< \alpha$ on, all head positions were actually $> \rho$. But then, ρ cannot be the inferior limit of the sequence of earlier head positions at time α, contradicting the case assumption that the head is on position ρ at time α.

Case 3: $\rho = \alpha$.

This implies that the head is on position ρ for the first time at time α, so that we must have $z = 0$, as there was no chance to write on the ρth cell before time α.

Let S be the set of times $< \alpha$ at which some head position was assumed for the first time during the computation of P. By the same reason as above, this newly reached cell will contain 0 at that time. If we can show that there is such a time $< \alpha$ at which the inner state is also s, we are done, because that would mean that the halting situation at time α was already given at an earlier time, contradicting the assumption that P halts at time α.

As $\rho > 0$, there must be an ordinal $\tau < \alpha$ such that the head was never on position 0 after time τ (otherwise, the liminf rule would force the head to be on position 0 at time α). This means that the head was never moved to the left from a limit position after time τ. This further implies that, after time τ, for any position β that the head occupied, all later positions were at most finitely many positions to the left of β and hence that, if β is a limit ordinal, then it never occupied a position $< \beta$ afterwards. In particular, the sequence of limit positions that the head occupied after time τ is increasing. Note that the set of head positions occupied before time τ is bounded in α, say by ξ. Let S' be the set of elements $\iota > \tau$ of S such that, at time ι, the head occupied a limit position $> \xi$ for the first time. Then S' is a closed and unbounded subset of S.

As s is the program state at the limit time α, there must be $\gamma < \alpha$ such that, after time γ, the program state was never $< s$ and moreover, the program state s itself must have occured cofinally often in α after that time.

But now, building an increasing ω-sequence of times starting with γ that alternately belong to S' and have the program state s, we see that its limit δ is $< \alpha$ and is a time at which the head was reading z and the state was s, we have the desired contradiction.

Since each case leads to a contradiction, our assumption on P must be false; as P was arbitrary, α is not a parameter-free OTM-halting time.

To see now that the theorem holds for any finite number of tapes, consider the argument below for each tape separately, note that we showed above that case 2 cannot occur while cases 1 and 3 both imply that, as far as the tape under consideration is concerned, the halting configuration occurred on a closed unbounded set of times before time α. Thus, one can again build an increasing 'interleaving' sequence of times at which each head read the same symbol as in the halting configuration and the inner state was the one in the halting configuration. The supremum of this sequence will be $< \alpha$, leading again to the contradiction that the program must have halted before α.

5   Existence of Admissible OTM-clockable Ordinals

We will now show that at least the first ω many admissible ordinals are OTM-clockable, thus answering the first question mentioned in the introduction positively. To this end, we need some preliminaries about Infinite Time Register Machines (ITRMs). ITRMs were introduced by Koepke in [11]; we sketch their architecture and refer to [11] for further information. An ITRM has finitely many registers, each of which stores one natural number. ITRM-programs are just programs for (classical) register machines. At successor times, an ITRM proceeds like a classical register machine. At limit levels, the active program line index and the register contents are defined to be the inferior limits of the sequences of earlier program line indices and respective register contents. When that limit is not finite in the case of a register content, the new content is defined to be 0, and one speaks of an 'overflow' of the respective register.

We recall Lemma 3 from [5]:

Theorem 7. *There are no gaps in the ITRM-clockable ordinals. That is, if $\alpha < \beta$ and β is ITRM-clockable, then α is ITRM-clockable.*

Combining this result with the main result of [11] on the computational strength of ITRMs, we obtain:

Lemma 3. *The ITRM-clockable ordinals are exactly those below ω_ω^{CK}. In particular, ω_n^{CK} is ITRM-clockable for all $n \in \omega$.*

Lemma 4. *Let α be ITRM-clockable. Then α is OTM-clockable.*

Proof. If $\alpha < \omega^2$, this is straightforward. Now let $\alpha \geq \omega^2$.

Let P be an ITRM-program that clocks α. We simulate P by an OTM-program that takes the same running time.

The simulation of ITRMs by OTMs here works like this: Use a tape for each register, have i many 1s, followed by 0s, on a tape to represent that the respective register contains $i \in \omega$; in addition, after a simulation step is finished, the head position on this tape represents the register content, i.e. it is at the first 0 on the tape.

For an ITTM, the simulation takes an extra ω many steps to halt because it takes time to detect an overflow. For an OTM, one can simply use one extra tape for each register, write 1 to their ωth positions at the start of the computation, move their heads along with the heads on the register simulating tapes and know that there is an overflow as soon as one of the heads on the extra tapes reads a $1.$[9] Since $\alpha \geq \omega^2$, the initial placement of 1s on the ωth tape positions does not affect the running time.

Corollary 2. *For every $n \in \omega$, ω_n^{CK} is OTM-clockable.*

This answers the first question mentioned above in the positive. By a relativization of the above argument, we can achieve the same for the second (i.e. whether gap starters for OTMs are something "better" than admissible):

Theorem 8. *Let $\alpha = \beta^+$ be a successor admissible. Then α does not start an OTM-clockable gap.*

Proof. Suppose for a contradiction that $\alpha = \beta^+$ starts an OTM-clockable gap. Then there is an OTM-clockable ordinal $\gamma \in (\beta, \alpha)$; pick one. By Lemma 2 above, a real code c for γ is OTM-writable in $< \alpha$ many steps. Suppose c has been written. Then $\omega_1^{CK,c} \geq \alpha$. Thus, α is ITRM-clockable in the oracle c. But now, α is OTM-clockable by first writing c and then ITRM-clocking α relative to c, contradicting the assumption that α starts a gap.

Corollary 3. *Every gap-starting ordinal for OTMs is an admissible limit of admissible ordinals.*

This allows a considerable strengthening of Corollary 2:

Corollary 4. *Every admissible ordinal up to the first admissible limit of admissible ordinals is OTM-clockable.*

6   Conclusion and Further Work

We showed that OTM-gaps are always started by limits of admissible ordinals and that, while admissible ordinals can be OTM-clockable, Σ_2-admissible ordinals cannot. This provokes the following questions:

Question: Is every gap-starting ordinal for OTMs Σ_2-admissible?

---

[9] The fact that more tapes are needed the more registers P uses may be seen as a little defect. (Note that, by the results of [11], the halting times of ITRM-programs using n registers are bounded by ω_{n+1}^{CK} so that indeed arbitrarily large numbers of registers - and thus of tapes - are required to make the above construction work for all α_n^{CK} with $n \in \omega$.) It would certainly be nicer to have a uniform bound on the number of required tapes. And indeed, by a slightly refined argument using that only two of the used registers are ultimately relevant for the halting of an ITRM, such a bound can be obtained.

Question: What is the minimal gap-starting ordinal for OTMs? Does it coincide with first Σ_2-admissible ordinal?

Further worthile topics include clockability for OTMs with a fixed ordinal parameter α and for other models of computability, like the "hypermachines" of Friedman and Welch (see [8]), α-ITTMs (see [7]) or α-ITRMs (see [4]), where the main question left open in [4] is to determine the supremum of the α-ITRM-clockable ordinals.

References

1. Barwise, J.: Admissible Sets and Structures: An Approach to Definability Theory. Springer, Berlin (1975)
2. Boolos, G., Putnam, H.: Degrees of unsolvability of constructible sets of integers. J. Symb. Log. **33**, 497–513 (1968)
3. Dimitriou, I. (ed.): Bonn International Workshop on Ordinal Computability. Report Hausdorff Centre for Mathematics, Bonn (2007). http://www.math.uni-bonn.de/ag/logik/events/biwoc/index.html
4. Carl, M.: Taming Koepkes Zoo II: Register Machines. Preprint (2020). arXiv:1907.09513v4
5. Carl, M., Fischbach, T., Koepke, P., Miller, R., Nasfi, M., Weckbecker, G.: The basic theory of infinite time register machines. Arch. Math. Logic **49**(2), 249–273 (2010)
6. Carl, M., Durand, B., Lafitte, G., Ouazzani, S.: Admissibles in gaps. In: Kari, J., Manea, F., Petre, I. (eds.) CiE 2017. LNCS, vol. 10307, pp. 175–186. Springer, Cham (2017). https://doi.org/10.1007/978-3-319-58741-7_18
7. Carl, M., Ouazzani, S., Welch, P.: Taming Koepke's Zoo. In: Manea, F., Miller, R.G., Nowotka, D. (eds.) CiE 2018. LNCS, vol. 10936, pp. 126–135. Springer, Cham (2018). https://doi.org/10.1007/978-3-319-94418-0_13
8. Friedman, S., Welch, P.: Hypermachines. J. Symb. Log. **76**(2), 620–636 (2011)
9. Hamkins, L.: Infinite time turing machines. J. Symb. Log. **65**(2), 567–604 (2000)
10. Koepke, P.: Turing computations on ordinals. Bull. Symb. Log. **11**, 377–397 (2005)
11. Koepke, P.: Ordinal computability. In: Ambos-Spies, K., Löwe, B., Merkle, W. (eds.) CiE 2009. LNCS, vol. 5635, pp. 280–289. Springer, Heidelberg (2009). https://doi.org/10.1007/978-3-642-03073-4_29
12. Seyfferth, B., Schlicht, P.: Tree representations via ordinal machines. Computablility **1**(1), 45–57 (2012)
13. Seabold, D., Hamkins, J.: Infinite time turing machines with only one tape. Math. Log. Quart. **47**(2), 271–287 (1999)
14. Welch, P.: Characteristics of discrete transfinite time turing machine models: halting times, stabilization times, and normal form theorems. Theor. Comput. Sci. **410**, 426–442 (2009)

Some Notes on Spaces of Ideals and Computable Topology

Matthew de Brecht[(✉)]

Graduate School of Human and Environmental Studies, Kyoto University,
Kyoto, Japan
matthew@i.h.kyoto-u.ac.jp

1 Introduction

It was shown in [4] that the *quasi-Polish spaces* introduced in [2] can be equivalently characterized as spaces of ideals in the following sense.

Definition 1 (see [4]). *Let $\prec$ be a transitive relation on $\mathbb{N}$. A subset $I \subseteq \mathbb{N}$ is an* ideal *(with respect to $\prec$) if and only if:*

1. $I \neq \emptyset$,   *(I is non-empty)*
2. $(\forall a \in I)(\forall b \in \mathbb{N})\,(b \prec a \Rightarrow b \in I)$,   *(I is a lower set)*
3. $(\forall a, b \in I)(\exists c \in I)\,(a \prec c \,\&\, b \prec c)$.   *(I is directed)*

The collection $\mathbf{I}(\prec)$ of all ideals has the topology generated by basic open sets of the form $[n]_\prec = \{I \in \mathbf{I}(\prec) \mid n \in I\}$. □

We often apply the above definition to other countable sets with the implicit assumption that it has been suitably encoded as a subset of $\mathbb{N}$. If $\prec$ is actually a partial order, then the definition of ideal above agrees with the usual definition of an ideal from order theory. Note that $\mathbf{I}(\prec) \subseteq \bigcup_{n \in \mathbb{N}} [n]_\prec$ and if $I \in [a]_\prec \cap [b]_\prec$ then there is $c \in \mathbb{N}$ with $I \in [c]_\prec \subseteq [a]_\prec \cap [b]_\prec$, so $\{[n]_\prec \mid n \in \mathbb{N}\}$ really is a basis for $\mathbf{I}(\prec)$ and not just a subbasis. Also note that proving the claim in the previous sentence requires all three of the axioms that define ideals.

We first give some basic examples. If $=$ is the equality relation on $\mathbb{N}$, then $\mathbf{I}(=)$ is homeomorphic to $\mathbb{N}$ with the discrete topology. If $\prec$ is the strict prefix relation on the set $\mathbb{N}^{<\mathbb{N}}$ of finite sequences of natural numbers, then $\mathbf{I}(\prec)$ is homeomorphic to the Baire space $\mathbb{N}^{\mathbb{N}}$. If $\subseteq$ is the usual subset relation on the set $\mathcal{P}_{\mathrm{fin}}(\mathbb{N})$ of finite subsets of $\mathbb{N}$, then $\mathbf{I}(\subseteq)$ is homeomorphic to $\mathcal{P}(\mathbb{N})$, the powerset of the natural numbers with the Scott-topology.

Spaces of the form $\mathbf{I}(\prec)$ for some transitive computably enumerable (c.e.) relation on $\mathbb{N}$ provide an effective interpretation of quasi-Polish spaces. This effective interpretation as spaces of ideals was first investigated in [4], where

This work was supported by JSPS Core-to-Core Program, A. Advanced Research Networks and by JSPS KAKENHI Grant Number 18K11166. The author thanks Tatsuji Kawai, Takayuki Kihara, Arno Pauly, Matthias Schröder, Victor Selivanov, and Hideki Tsuiki for helpful discussions.

M. Anselmo et al. (Eds.): CiE 2020, LNCS 12098, pp. 26–37, 2020.
https://doi.org/10.1007/978-3-030-51466-2_3

they are called *precomputable quasi-Polish spaces*, but they are equivalent to the *computable quasi-Polish spaces* in [10], and they naturally correspond to c.e. propositional geometric theories via the duality in [7] (see [1] for extending this duality beyond propositional logic). In many applications it is useful to assume that $\mathbf{I}(\prec)$ also comes with a c.e. set $E_{\prec} = \{n \in \mathbb{N} | [n]_{\prec} \neq \emptyset\}$, which provides an effective interpretation of *overt quasi-Polish spaces*. These are called *computable quasi-Polish spaces* in [4], and are equivalent to the *effective quasi-Polish spaces* in [8], and correspond to *effectively enumerable computable quasi-Polish spaces* in the terminology of [10]. Dually, they correspond to c.e. propositional geometric theories where satisfiability is semidecidable.

In this paper, we show some basic results on spaces of ideals, with an emphasis on the connections with computable topology. We also hope that our approach will help clarify the relationship between quasi-Polish spaces and domain theory (see *abstract basis* in [5] or [6]), and implicitly demonstrate how the theory of quasi-Polish spaces can be developed within relatively weak subsystems of second-order arithmetic (see the work on *poset spaces* in [11]).

2   Computable Functions

Computability of functions between spaces of ideals can be defined in a way that is compatible with the TTE framework [16]. We briefly review the TTE approach to computability on countably based T_0-spaces, but see [13] for the extension to the cartesian closed category of admissibly represented spaces and [12] for more general represented spaces.

Given a countably based T_0-space X with fixed basis $(B_i)_{i \in \mathbb{N}}$, the *standard (admissible) representation* of X is the partial function $\delta_X :\subseteq \mathbb{N}^{\mathbb{N}} \to X$ defined as $\delta_X(p) = x \iff range(p) = \{i \in \mathbb{N} | x \in B_i\}$. A function $f \colon X \to Y$ between spaces with standard admissible representations is computable if and only if there is a computable (partial) function $F :\subseteq \mathbb{N}^{\mathbb{N}} \to \mathbb{N}^{\mathbb{N}}$ such that $f \circ \delta_X = \delta_Y \circ F$. It follows that a function $f \colon \mathbf{I}(\prec_1) \to \mathbf{I}(\prec_2)$ is computable if and only if there is an algorithm which transforms any enumeration of the elements of any $I \in \mathbf{I}(\prec_1)$ into an enumeration of the elements of $f(I) \in \mathbf{I}(\prec_2)$.

We define a code for a partial function to be any subset $R \subseteq \mathbb{N} \times \mathbb{N}$. Each code R encodes the partial function $\ulcorner R \urcorner :\subseteq \mathbf{I}(\prec_1) \to \mathbf{I}(\prec_2)$ defined as

$$\ulcorner R \urcorner(I) = \{n \in \mathbb{N} | (\exists m \in I) \langle m, n \rangle \in R\},$$
$$dom(\ulcorner R \urcorner) = \{I \in \mathbf{I}(\prec_1) | \ulcorner R \urcorner(I) \in \mathbf{I}(\prec_2)\}.$$

Theorem 2. *Let $\prec_1$ and $\prec_2$ be transitive relations on $\mathbb{N}$. A total function $f \colon \mathbf{I}(\prec_1) \to \mathbf{I}(\prec_2)$ is computable if and only if there is a c.e. code $R \subseteq \mathbb{N} \times \mathbb{N}$ such that $f = \ulcorner R \urcorner$.*

Proof. It is clear that if $f = \ulcorner R \urcorner$ for some c.e. code R, then there is an algorithm which transforms any enumeration of the elements of any $I \in \mathbf{I}(\prec_1)$ into an enumeration of the elements of $f(I) \in \mathbf{I}(\prec_2)$. Therefore, f is computable.

For the other direction, assume $f\colon \mathbf{I}(\prec_1) \to \mathbf{I}(\prec_2)$ is computable. It is a standard result that there is a computable enumeration $(U_n)_{n\in\mathbb{N}}$ of c.e. subsets of $\mathbb{N}$ such that $f^{-1}([n]_{\prec_2}) = \bigcup_{m\in U_n}[m]_{\prec_1}$. Let $R = \{\langle m,n\rangle | m \in U_n\}$. Given $I \in \mathbf{I}(\prec_1)$, if $n \in \ulcorner R \urcorner(I)$, then there is some $m \in I$ with $\langle m,n\rangle \in R$, hence $m \in U_n$. Thus $I \in [m]_{\prec_1} \subseteq f^{-1}([n]_{\prec_2})$ which implies $n \in f(I)$. Conversely, if $n \in f(I)$ then $I \in f^{-1}([n]_{\prec_2})$, so there must be $m \in U_n$ with $I \in [m]_{\prec_1}$. It follows that $\langle m,n\rangle \in R$ and that $n \in \ulcorner R \urcorner(I)$. Therefore, R is a c.e. code satisfying $f = \ulcorner R \urcorner$. $\qquad\square$

3   Basic Constructions

3.1   Products

Given relations $\prec_1$ and $\prec_2$ on $\mathbb{N}$, define the relation $\prec_{1,2}^{\times}$ on $\mathbb{N}$ as

$$\langle a,b\rangle \prec_{1,2}^{\times} \langle a',b'\rangle \iff a \prec_1 a' \ \& \ b \prec_2 b',$$

where $\langle \cdot,\cdot\rangle\colon \mathbb{N} \times \mathbb{N} \to \mathbb{N}$ is a computable bijection. Then $\mathbf{I}(\prec_{1,2}^{\times})$ is computably homeomorphic to the product $\mathbf{I}(\prec_1) \times \mathbf{I}(\prec_2)$ via the pairing function $\langle \cdot,\cdot\rangle\colon \mathbf{I}(\prec_1) \times \mathbf{I}(\prec_2) \to \mathbf{I}(\prec_{1,2}^{\times})$

$$\langle I_1, I_2\rangle = \{\langle a,b\rangle | a \in I_1 \ \& \ b \in I_2\}$$

and the projections $\pi_i\colon \mathbf{I}(\prec_{1,2}^{\times}) \to \mathbf{I}(\prec_i)$ $(i \in \{1,2\})$

$$\pi_1(I) = \{a \in \mathbb{N} | (\exists b \in \mathbb{N})\langle a,b\rangle \in I\},$$
$$\pi_2(I) = \{b \in \mathbb{N} | (\exists a \in \mathbb{N})\langle a,b\rangle \in I\}.$$

We leave most of the proof to the reader as an exercise, but we will show that $\pi_1(I)$ really is a lower set because it is a nice example of how directedness and transitivity often compensate for the lack of reflexivity of the relations. Assume $I \in \mathbf{I}(\prec_{1,2}^{\times})$ and $a \in \pi_1(I)$ and $a_0 \prec_1 a$. Then there is $b \in \mathbb{N}$ with $\langle a,b\rangle \in I$. If $\prec_2$ was reflexive, then we would have $\langle a_0,b\rangle \prec_{1,2}^{\times} \langle a,b\rangle \in I$, and since I is a lower set we would immediately conclude $\langle a_0,b\rangle \in I$. But without reflexivity, we must instead use the directedness of I to first obtain $\langle a',b'\rangle \in I$ with $\langle a,b\rangle \prec_{1,2}^{\times} \langle a',b'\rangle$, and then we have $\langle a_0,b\rangle \prec_{1,2}^{\times} \langle a',b'\rangle \in I$ by the transitivity of $\prec_1$ and $\prec_2$. We still get the desired conclusion $\langle a_0,b\rangle \in I$ (hence $a_0 \in \pi_1(I)$), albeit with a slight detour that required directedness and transitivity.

A simple modification of Definition 2.3.13 in [11] can be used to construct countable products from an enumeration $(\prec_i)_{i\in\mathbb{N}}$ of transitive relations.

3.2   Co-products

We get co-products (i.e., disjoint unions) by defining the relation $\prec_{1,2}^{+}$ on $\mathbb{N}$ as

$$\langle a,i\rangle \prec_{1,2}^{+} \langle a',j\rangle \iff i = j \in \{1,2\} \ \& \ a \prec_i a'.$$

Then it is an easy exercise to show that $\mathbf{I}(\prec_{1,2}^{+})$ is computably homeomorphic to the co-product $\mathbf{I}(\prec_1) + \mathbf{I}(\prec_2)$. It should be clear to the reader how to extend this to countable co-products.

3.3 Π_2^0-subspaces and Equalizers

Let $\prec$ be a transitive relation on $\mathbb{N}$. The Σ_1^0-subsets (or c.e. open subsets) of $\mathbf{I}(\prec)$ are encoded by c.e. subsets $U \subseteq \mathbb{N}$ by defining

$$\ulcorner U \urcorner = \bigcup_{n \in U} [n]_\prec.$$

The Π_2^0-subsets of $\mathbf{I}(\prec)$ are encoded by computable enumerations $(U_i, V_i)_{i \in \mathbb{N}}$ of c.e. subsets of $\mathbb{N}$ by defining

$$\ulcorner (\forall i) U_i \Rightarrow V_i \urcorner = \{ I \in \mathbf{I}(\prec) | (\forall i \in \mathbb{N}) [I \in \ulcorner U_i \urcorner \Rightarrow I \in \ulcorner V_i \urcorner] \}.$$

The next theorem is part of the characterization of precomputable quasi-Polish spaces from [4]. We provide a direct proof for convenience.

Theorem 3. (see [4]). *Let $\prec$ be a transitive c.e. relation on $\mathbb{N}$. Given a code of a Π_2^0-subset A of $\mathbf{I}(\prec)$, one can computably obtain a transitive c.e. relation $\sqsubset$ on $\mathbb{N}$ such that $\mathbf{I}(\sqsubset)$ is computably homeomorphic to A.*

Proof. Assume $A = \ulcorner (\forall i) U_i \Rightarrow V_i \urcorner$ for some computable enumeration $(U_i, V_i)_{i \in \mathbb{N}}$ of c.e. subsets of $\mathbb{N}$. Let $\prec^{(\cdot)}$ be a decidable relation such that

$$m \prec n \iff (\exists k \in \mathbb{N}) \, m \prec^{(k)} n, \text{ and}$$
$$k \le k' \,\&\, m \prec^{(k)} n \implies m \prec^{(k')} n.$$

Let $(U_i^{(k)})_{i,k \in \mathbb{N}}$ be a double enumeration of decidable subsets of $\mathbb{N}$ such that

$$U_i = \bigcup_{k \in \mathbb{N}} U_i^{(k)} \text{ and } k \le k' \Rightarrow U_i^{(k)} \subseteq U_i^{(k')}.$$

For $F_1, F_2 \in \mathcal{P}_{\text{fin}}(\mathbb{N})$ and $k_1, k_2 \in \mathbb{N}$, define $\langle F_1, k_1 \rangle \sqsubset \langle F_2, k_2 \rangle$ if and only if the following all hold:

1. $k_1 < k_2$
2. $F_1 \subseteq F_2$
3. $F_2 \neq \emptyset$
4. $(\forall m \le k_1) \left[[(\exists n \in F_1) \, m \prec^{(k_1)} n] \Rightarrow m \in F_2 \right]$
5. $(\forall a, b \in F_1)(\exists c \in F_2)[a \prec c \,\&\, b \prec c]$
6. $(\forall i \le k_1)[F_1 \cap U_i^{(k_1)} \neq \emptyset \Rightarrow F_2 \cap V_i \neq \emptyset]$.

It is clear that $\sqsubset$ is c.e., and the monotonicity assumptions on $\prec^{(\cdot)}$ and $U_i^{(k)}$ imply that if $\langle F_1, k_1 \rangle \sqsubset \langle F_2, k_2 \rangle$ and $F \subseteq F_1$ and $k \le k_1$ then $\langle F, k \rangle \sqsubset \langle F_2, k_2 \rangle$, hence $\sqsubset$ is transitive.

Define $f : \mathbf{I}(\sqsubset) \to \ulcorner (\forall i) U_i \Rightarrow V_i \urcorner$ by $f(I) = \bigcup_{\langle F,k \rangle \in I} F$. Given $I \in \mathbf{I}(\sqsubset)$, the directedness of I implies any $\langle F_0, k_0 \rangle \in I$ can be extended to a finite $\sqsubset$-chain in I of arbitrary length, hence for any $k \in \mathbb{N}$ there are $\langle F_1, k_1 \rangle, \langle F_2, k_2 \rangle \in I$ with $\langle F_0, k_0 \rangle \sqsubset \langle F_1, k_1 \rangle \sqsubset \langle F_2, k_2 \rangle$ and $k < k_1$. Thus $\langle F_0, k \rangle \sqsubset \langle F_2, k_2 \rangle$, which implies $\langle F_0, k \rangle \in I$. It follows that $n \in f(I)$ if and only if $\langle \{n\}, k \rangle \in I$ for some (equivalently, every) $k \in \mathbb{N}$. By using the directedness of I this observation generalizes from singletons to all finite sets, and so for any $F \in \mathcal{P}_{\text{fin}}(\mathbb{N})$ we have $F \subseteq f(I)$ if and only if $\langle F, k \rangle \in I$ for some (equivalently, every) $k \in \mathbb{N}$. Then conditions 3, 4, and 5 in the definition of $\sqsubset$ imply that $f(I)$ is indeed an ideal of $\prec$, and condition 6 implies $f(I) \in \ulcorner (\forall i)U_i \Rightarrow V_i \urcorner$. Thus f is well-defined, and it is clearly a computable injection.

A computable inverse of f is given by $g \colon \ulcorner (\forall i)U_i \Rightarrow V_i \urcorner \rightarrow \mathbf{I}(\sqsubset)$ defined as $g(I) = \{\langle F, k \rangle \mid k \in \mathbb{N} \ \& \ F \subseteq I \text{ is finite}\}$. The only part of the proof that g is well-defined which requires a little thought is showing that $g(I)$ is directed for each $I \in \ulcorner (\forall i)U_i \Rightarrow V_i \urcorner$, but it is not difficult to see that if $\langle F_1, k_1 \rangle, \langle F_2, k_2 \rangle \in g(I)$, then one can find a finite $G \subseteq I$ which contains $F_1 \cup F_2$ and enough of I to satisfy conditions 3 through 6 and obtain $\langle F_1, k_1 \rangle, \langle F_2, k_2 \rangle \sqsubset \langle G, k_1 + k_2 + 1 \rangle \in g(I)$. The claim that g is an inverse to f follows from the observations in the previous paragraph. □

If R and S are codes for total functions $\ulcorner R \urcorner, \ulcorner S \urcorner \colon \mathbf{I}(\prec_1) \rightarrow \mathbf{I}(\prec_2)$, then for any $I \in \mathbf{I}(\prec_1)$ we have $\ulcorner R \urcorner(I) = \ulcorner S \urcorner(I)$ if and only if

$$(\forall n \in \mathbb{N}) \left[n \in \ulcorner R \urcorner(I) \iff n \in \ulcorner S \urcorner(I) \right].$$

This is a Π_2^0-subset of $\mathbf{I}(\prec_1)$ whenever $\prec_1$, $\prec_2$, R, and S are c.e. It follows that we can computably obtain a c.e. relation $\sqsubset$ such that $\mathbf{I}(\sqsubset)$ is an equalizer of $\ulcorner R \urcorner$ and $\ulcorner S \urcorner$.

Note that Theorem 3 is the best result possible because if $\prec$, $\sqsubset$, R, and S are c.e. such that $\ulcorner R \urcorner \colon \mathbf{I}(\sqsubset) \rightarrow \mathbf{I}(\prec)$ is total with partial inverse $\ulcorner S \urcorner \colon \mathbf{I}(\prec) \rightarrow \mathbf{I}(\sqsubset)$ (meaning $(\forall J \in \mathbf{I}(\sqsubset))[J = \ulcorner S \urcorner(\ulcorner R \urcorner(J))]$) then $I \in range(\ulcorner R \urcorner)$ if and only if $I \in dom(\ulcorner S \urcorner)$ and $I = \ulcorner R \urcorner(\ulcorner S \urcorner(I))$. Since $dom(\ulcorner S \urcorner)$ is Π_2^0, it follows that $\mathbf{I}(\sqsubset)$ is computably homeomorphic to the Π_2^0-subset $range(\ulcorner R \urcorner)$ of $\mathbf{I}(\prec)$.

4 Examples from Computable Topology

4.1 Completion of (Computable) Separable Metric Spaces

Let (X, d) be a separable metric space. Fix a countable dense subset $D \subseteq X$, and define a transitive relation $\prec$ on $P = D \times \mathbb{N}$ as

$$\langle x, n \rangle \prec \langle y, m \rangle \iff d(x, y) < 2^{-n} - 2^{-m}.$$

This definition guarantees that the open ball with center x and radius 2^{-n} contains the closed ball with center y and radius 2^{-m}. The pair $(P, \prec)$ is a countable substructure of the *formal balls* of (X, d), a well-known construction in domain theory (see Section V-6 of [5] and Section 7.3 of [6]). It is straightforward to see that $\prec$ is transitive by using the triangular inequality for d.

If $I \in \mathbf{I}(\prec)$ then it contains a cofinal infinite ascending $\prec$-chain $(\langle x_i, n_i \rangle)_{i \in \mathbb{N}}$, which means that $\langle x_i, n_i \rangle \prec \langle x_{i+1}, n_{i+1} \rangle$ for all $i \in \mathbb{N}$ and that for any $\langle x, n \rangle \in I$ there is $i \in \mathbb{N}$ with $\langle x, n \rangle \prec \langle x_i, n_i \rangle$. Note that $(n_i)_{i \in \mathbb{N}}$ is strictly increasing because $0 \leq d(x_i, x_{i+1}) < 2^{-n_i} - 2^{-n_{i+1}}$, and therefore $(x_i)_{i \in \mathbb{N}}$ is a Cauchy sequence. It follows that $\lim_{i \to \infty} d(x, x_i)$ is well-defined for all $x \in D$.

Next we show that $\langle x, n \rangle \in I$ if and only if $\lim_{i \to \infty} d(x, x_i) < 2^{-n}$. For any $\langle x, n \rangle \in I$ the cofinality of $(\langle x_i, n_i \rangle)_{i \in \mathbb{N}}$ implies there is $i_0 \in \mathbb{N}$ with $\langle x, n \rangle \prec \langle x_i, n_i \rangle$ for all $i \geq i_0$. Let $\varepsilon > 0$ be such that $d(x, x_{i_0}) = 2^{-n} - 2^{-n_{i_0}} - \varepsilon$. Then for $i \geq i_0$ we have

$$d(x, x_i) \leq d(x, x_{i_0}) + d(x_{i_0}, x_i) < (2^{-n} - 2^{-n_{i_0}} - \varepsilon) + (2^{-n_{i_0}} - 2^{-n_i}) < 2^{-n} - \varepsilon,$$

hence $\lim_{i \to \infty} d(x, x_i) \leq 2^{-n} - \varepsilon < 2^{-n}$. Conversely, assume $x \in D$ and there is $\varepsilon > 0$ such that $\lim_{i \to \infty} d(x, x_i) < 2^{-n} - \varepsilon$. Fix $i \in \mathbb{N}$ such that $d(x, x_i) + \varepsilon < 2^{-n}$ and $2^{-n_i} < \varepsilon$. Then $d(x, x_i) < d(x, x_i) + \varepsilon - 2^{-n_i} < 2^{-n} - 2^{-n_i}$, hence $\langle x, n \rangle \prec \langle x_i, n_i \rangle$, which implies $\langle x, n \rangle \in I$.

It is now easy to see that $\mathbf{I}(\prec)$ is homeomorphic to the completion $(\widehat{X}, \widehat{d})$ of (X, d). The usual admissible representation for $\widehat{X}$ is to represent each $x \in \widehat{X}$ by the fast Cauchy sequences $(x_i)_{i \in \mathbb{N}}$ in D that converge to x (by *fast Cauchy* we mean $d(x_i, x_{i+1}) < 2^{-(i+1)}$ for each $i \in \mathbb{N}$). From an enumeration of $I \in \mathbf{I}(\prec)$ we can extract a cofinal infinite ascending $\prec$-chain $(\langle x_i, n_i \rangle)_{i \in \mathbb{N}}$ in I so that $(x_i)_{i \in \mathbb{N}}$ is a fast Cauchy sequence determining a point in $\widehat{X}$. In the other direction, given a fast Cauchy sequence $(x_i)_{i \in \mathbb{N}}$ in D, we have $d(x_i, x_{i+1}) < 2^{-(i+1)} = 2^{-i} - 2^{-(i+1)}$ for each $i \in \mathbb{N}$, hence $(\langle x_i, i \rangle)_{i \in \mathbb{N}}$ is an infinite ascending $\prec$-chain which generates an ideal $I \in \mathbf{I}(\prec)$. This determines a homeomorphism between $\mathbf{I}(\prec)$ and $\widehat{X}$.

A computable metric space (X, d) comes with an indexing $\alpha \colon \mathbb{N} \to D$ for some dense $D \subseteq X$ in such a way that $\{(q, r, i, j) \in \mathbb{Q}^2 \times \mathbb{N}^2 \mid q < d(\alpha(i), \alpha(j)) < r\}$ is computably enumerable. Defining $\langle i, n \rangle \prec \langle j, m \rangle$ if and only if $d(\alpha(i), \alpha(j)) < 2^{-n} - 2^{-m}$ determines a transitive c.e. relation $\prec$ such that $\mathbf{I}(\prec)$ and $\widehat{X}$ are computably homeomorphic.

4.2 Completion of Computable Topological Spaces

A (countably based) *computable topological space* (also called an *effective topological space*; see [8–10, 14, 17]) is a tuple (X, φ, S) where:

1. X is a T_0-space (we write $\mathbf{O}(X)$ for its topology),
2. $\varphi \colon \mathbb{N} \to \mathbf{O}(X)$ is an enumeration of a basis for X,
3. $S \subseteq \mathbb{N}^3$ is a c.e. set satisfying $\varphi(n) \cap \varphi(m) = \bigcup \{\varphi(k) \mid \langle n, m, k \rangle \in S\}$ for each $n, m \in \mathbb{N}$.

Note that the only effective aspect of this definition is the c.e. set S, and there are no specifications as to how the space X and the enumeration φ should be defined. As a result, if (X, φ, S) is a computable topological space, then for any subspace $Y \subseteq X$ we can restrict φ in the obvious way to obtain a map $\varphi' \colon \mathbb{N} \to \mathbf{O}(Y)$ such that (Y, φ', S) is also a computable topological space. A common extension of the above definition additionally requires that $\{n \in \mathbb{N} \mid \varphi(n) \neq \emptyset\}$ is a c.e. set, but even in this case one can define highly non-constructive dense subspaces of a computable topological space which are still computable topological spaces.

Since the effective part of the above definition is compatible with infinitely many computable topological spaces, a natural question to ask is whether there is any *canonical* computable topological space associated to a given c.e. set S. This question leads to Definition 4 below. In the following, for any continuous function $f\colon X \to Y$, the function $\mathbf{O}(f)\colon \mathbf{O}(Y) \to \mathbf{O}(X)$ is defined as $\mathbf{O}(f)(U) = f^{-1}(U)$.

Definition 4. *Let* $S \subseteq \mathbb{N}^3$ *be a c.e. set. A computable topological space* (X, φ, S) *is* complete *if and only if for any computable topological space* (Y, ψ, S) *there is a unique computable embedding* $e\colon Y \to X$ *satisfying* $\psi = \mathbf{O}(e) \circ \varphi$.

Intuitively, (X, φ, S) is a complete computable topological space if and only if all other computable topological spaces associated to S are essentially just restrictions of the kind (Y, φ', S) we saw earlier. Also note that any complete computable topological space associated to S is unique up to computable homeomorphism. The next lemma shows that every c.e. subset $S \subseteq \mathbb{N}^3$ determines a complete computable topological space.

Lemma 5. *For any c.e. subset* $S \subseteq \mathbb{N}^3$, *there is a* Π^0_2-*subspace* $X \subseteq \mathcal{P}(\mathbb{N})$ *such that* $\varphi\colon \mathbb{N} \to \mathbf{O}(X)$ *defined as* $\varphi(n) = \{x \in X | n \in x\}$ *is an enumeration of a basis for* X *and* (X, φ, S) *is a complete computable topological space.*

Proof. Let $S \subseteq \mathbb{N}^3$ be a c.e. subset. Define $X \subseteq \mathcal{P}(\mathbb{N})$ so that $x \in X$ if and only if the following conditions are all satisfied:

(i) $x \neq \emptyset$,
(ii) $(\forall \langle n, m, k \rangle \in S)\,[k \in x \Rightarrow \{n, m\} \subseteq x]$, and
(iii) $(\forall n, m \in \mathbb{N})\,[\{n, m\} \subseteq x \Rightarrow (\exists k \in x)\,\langle n, m, k \rangle \in S]$.

It is clear that X is a Π^0_2-subspace of $\mathcal{P}(\mathbb{N})$. We first show that φ is an enumeration of a basis for X. It is clear that each $\varphi(n)$ is an open subset of X, and that $\{\varphi(n) | n \in \mathbb{N}\}$ covers X because each $x \in X$ is non-empty. Next, note that if $\langle n, m, k \rangle \in S$, then condition (ii) implies $\varphi(k) \subseteq \varphi(n) \cap \varphi(m)$. So for any $x \in \varphi(n) \cap \varphi(m)$, by using condition (iii) it follows that there is $k \in \mathbb{N}$ with $x \in \varphi(k) \subseteq \varphi(n) \cap \varphi(m)$. Therefore, φ is an enumeration of a basis for X. It is then easy to see (using condition (iii) again), that (X, φ, S) is a computable topological space.

Given another computable topological space (Y, ψ, S), define $e\colon Y \to X$ as $e(y) = \{n \in \mathbb{N} | y \in \psi(n)\}$. We first show that $e(y) \in X$ for each $y \in Y$. Clearly, $e(y)$ is non-empty because the basis enumerated by ψ must cover Y. Next, if $\langle n, m, k \rangle \in S$ then $\psi(k)$ must be a subset of $\psi(n) \cap \psi(m)$ by condition (3) of the definition of a computable topological space, hence $e(y)$ satisfies condition (ii). Finally, condition (iii) is satisfied because if $\{n, m\} \subseteq e(y)$ then $y \in \psi(n) \cap \psi(m)$ hence there must be $\langle n, m, k \rangle \in S$ with $y \in \psi(k)$ which implies $k \in e(y)$. Therefore, e is well-defined.

Using the fact that ψ enumerates a basis for Y, it is easy to see that e is a computable topological embedding. Furthermore, $y \in \psi(n)$ if and only if $n \in e(y)$ if and only if $e(y) \in \varphi(n)$, hence $\psi(n) = e^{-1}(\varphi(n))$. This proves that $\psi = \mathbf{O}(e) \circ \varphi$, and it is clear that e is the only possible embedding of Y into X that satisfies this property. $\qquad\square$

We take a brief moment to consider the extension where a computable topological space comes with an additional c.e. set $E = \{n \in \mathbb{N} | \varphi(n) \neq \emptyset\}$. Completeness is defined as in Definition 4, but with quantification over spaces of the form (Y, ψ, S, E). There is no guarantee that arbitrarily chosen S and E will be compatible, but if they are compatible with at least one computable topological space, then a complete space can be obtained by adding a fourth (Π_2^0) axiom "$(\forall n \in \mathbb{N}) [n \in x \Rightarrow n \in E]$" to the construction in the proof of Lemma 5. These modifications could also be made to the following theorem, which shows that complete computable topological spaces provide an effective interpretation of quasi-Polish spaces that is equivalent to the approach using spaces of ideals.

Theorem 6. *Every complete computable topological space is computably homeomorphic to $\mathbf{I}(\prec)$ for some transitive c.e. relation $\prec$ on $\mathbb{N}$. Conversely, given a transitive c.e. relation $\prec$ on $\mathbb{N}$ one can computably obtain a c.e. subset $S \subseteq \mathbb{N}^3$ such that $(\mathbf{I}(\prec), \varphi_\prec, S)$ is a complete computable topological space, where $\varphi_\prec \colon \mathbb{N} \to \mathbf{O}(\mathbf{I}(\prec))$ is the standard enumeration of a basis for $\mathbf{I}(\prec)$ given by $\varphi_\prec(n) = [n]_\prec$.*

Proof. The first claim follows from Lemma 5 and Theorem 3.

For the converse, let $\prec$ be a transitive c.e. relation on $\mathbb{N}$. Define

$$S = \{\langle n, m, k \rangle \in \mathbb{N}^3 | n \prec k \text{ and } m \prec k\},$$

and let (X, φ, S) be the complete computable topological space for S as constructed in the proof of Lemma 5. The proof will be completed by showing that $X = \mathbf{I}(\prec)$ as subsets of $\mathcal{P}(\mathbb{N})$.

First we show $\mathbf{I}(\prec) \subseteq X$. Fix $I \in \mathbf{I}(\prec)$. It is clear that I satisfies condition (i) of the definition of X. Next, condition (ii) is satisfied because if $\langle n, m, k \rangle \in S$ and $k \in I$, then $n, m \prec k$ by the definition of S, hence $\{n, m\} \subseteq I$ because I is a lower set. Finally, condition (iii) is satisfied because if $\{n, m\} \subseteq I$ the directedness of I implies there is $k \in I$ with $n, m \prec k$, hence $\langle n, m, k \rangle \in S$. Therefore, $I \in X$.

To show $X \subseteq \mathbf{I}(\prec)$, fix any $x \in X$. Clearly x is non-empty. Next, assume $k \in x$ and $n \prec k$. Then $\langle n, n, k \rangle \in S$, hence condition (ii) on X implies $n \in x$, so x is a lower set. Finally, if $n, m \in x$ then condition (iii) on X implies there is $\langle n, m, k \rangle \in S$ with $k \in x$. By definition of S we have $n \prec k$ and $m \prec k$, which shows that x is directed. Therefore, $x \in \mathbf{I}(\prec)$. $\square$

The above theorem shows that we get a computably equivalent definition of computable topological space if we simply define them to be a pair $(\prec, X)$, where $\prec$ is a transitive c.e. relation and $X \subseteq \mathbf{I}(\prec)$. A more rigorous approach would also require a precise definition of the set X, for example by defining a (countably based) "computable topological space" to be a pair $(\prec, \Phi_X)$ that contains an explicit (finite) formula Φ_X with a single free variable I that defines the set $X = \{I \in \mathbf{I}(\prec) | \Phi_X(I)\}$ within some fixed formal system. This would lead us more into the realm of effective descriptive set theory, but adopting such a definition would guarantee that *computable* topological spaces are unambiguously defined by a finite amount of information.

5 Powerspaces

Given a topological space X, we write $\mathbf{A}(X)$ for the lower powerspace of X (the closed subsets of X with the lower Vietoris topology), and $\mathbf{K}(X)$ for the upper powerspace of X (the saturated compact subsets of X with the upper Vietoris topology). Our notation follows that of [3], where other basic results on quasi-Polish powerspaces can be found. For countably based spaces, the lower powerspace defined here is equivalent to the space of (closed) overt sets in [4,12]. In this section, we show how to represent powerspaces as spaces of ideals using the construction introduced in [15] for ω-algebraic domains (which is equivalent to the case that $\prec$ is a partial order within our framework). We fix a transitive relation $\prec$ on $\mathbb{N}$ for the rest of this section.

5.1 Lower Powerspace

A basis for the lower Vietoris topology on $\mathbf{A}(\mathbf{I}(\prec))$ is given by sets of the form

$$\bigcap_{n \in F} \lozenge[n]_\prec = \{A \in \mathbf{A}(\mathbf{I}(\prec)) \mid (\forall n \in F)(\exists I \in A)\, n \in I\}$$

for $F \in \mathcal{P}_{\mathrm{fin}}(\mathbb{N})$. Define the transitive relation $\prec_L$ on $\mathcal{P}_{\mathrm{fin}}(\mathbb{N})$ as

$$F \prec_L G \text{ if and only if } (\forall m \in F)\,(\exists n \in G)\, m \prec n.$$

Transitivity of $\prec_L$ easily follows from the transitivity of $\prec$, and it is clear that $\prec_L$ is c.e. whenever $\prec$ is. Next, define $f_L \colon \mathbf{A}(\mathbf{I}(\prec)) \to \mathbf{I}(\prec_L)$ as

$$f_L(A) = \{F \in \mathcal{P}_{\mathrm{fin}}(\mathbb{N}) \mid (\forall m \in F)(\exists I \in A)\, m \in I\}$$

and $g_L \colon \mathbf{I}(\prec_L) \to \mathbf{A}(\mathbf{I}(\prec))$ as

$$g_L(J) = \{I \in \mathbf{I}(\prec) \mid (\forall m \in I)(\exists F \in J)\, m \in F\}.$$

We will need the following lemma when we prove that these two functions are well-defined computable homeomorphisms.

Lemma 7. *If $J \in \mathbf{I}(\prec_L)$ and $F \in \mathcal{P}_{\mathrm{fin}}(\mathbb{N})$, then $(\forall m \in F)\, g_L(J) \cap [m]_\prec \neq \emptyset$ if and only if $F \in J$.*

Proof. First assume $(\forall m \in F)\, g_L(J) \cap [m]_\prec \neq \emptyset$. For each $m \in F$, there is $I \in g_L(J)$ with $m \in I$, and as I is directed, there is $n \in I$ with $m \prec n$, but since $I \in g_L(J)$ there must be $G \in J$ with $n \in G$, and therefore $\{m\} \prec_L G \in J$. This shows that $\{m\} \in J$ for each $m \in F$. Since F is finite and J is directed, there is $H \in J$ such that $(\forall m \in F)\, \{m\} \prec_L H$. It follows that $F \prec_L H$, and therefore $F \in J$.

For the converse, assume $F \in J$, and fix any $m \in F$. Since J is directed there exists an infinite sequence $F = F_0 \prec_L F_1 \prec_L F_2 \prec_L \cdots$ with $F_i \in J$ for each $i \in \mathbb{N}$. From the definition of $\prec_L$, there exists an infinite sequence $m = m_0 \prec m_1 \prec m_2 \prec \cdots$ with $m_i \in F_i$ for each $i \in \mathbb{N}$. Then $I = \{n \in \mathbb{N} \mid (\exists i \in \mathbb{N})\, n \prec m_i\}$ is in $\mathbf{I}(\prec)$ and $m \in I$. For any $n \in I$ there is $i \in \mathbb{N}$ with $n \prec m_i \in F_i$, thus $\{n\} \prec_L F_i \in J$ which implies $\{n\} \in J$. Therefore, $I \in g_L(J) \cap [m]_\prec$. $\square$

Theorem 8. $\mathbf{A}(\mathbf{I}(\prec))$ *and* $\mathbf{I}(\prec_L)$ *are computably homeomorphic.*

Proof. We will prove that f_L and g_L are well-defined computable inverses of each other in several steps.

- f_L *is well-defined:* We must show that $f_L(A)$ is an ideal.

  1. *($f_L(A)$ is non-empty).* $f_L(A) \neq \emptyset$ because $\emptyset \in f_L(A)$.
  2. *($f_L(A)$ is a lower set).* If $G \in f_L(A)$ and $F \prec_L G$, then for any $m \in F$ there is $n \in G$ with $m \prec n$. There is some $I \in A$ with $n \in I$, and also $m \in I$ because I is a lower set. Therefore, $F \in f_L(A)$.
  3. *($f_L(A)$ is directed).* Assume $F, G \in f_L(A)$. For each $m \in F \cup G$ there is some $I \in A$ with $m \in I$, and by directedness of I we can choose some $n_m \in I$ with $m \prec n_m$. Combine these choices into a single (finite) set $H = \{n_m | m \in F \cup G\}$. Then $H \in f_L(A)$ and $F, G \prec_L H$.

- g_L *is well-defined:* We must show that $g_L(J)$ is a closed subset of $\mathbf{I}(\prec)$. If $I \notin g_L(J)$, then by definition of $g_L(J)$ there must be $m \in I$ such that $(\forall F \in J)\, m \notin F$. Then $[m]_{\prec}$ is an open neighborhood of I that does not intersect $g_L(J)$, hence $g_L(J)$ is closed.
- f_L *is computable:* Clearly, $f_L(A) \in [F]_{\prec_L}$ if and only if $A \in \bigcap_{m \in F} \Diamond[m]_{\prec}$.
- g_L *is computable:* Lemma 7 is the statement $g_L(J) \in \bigcap_{m \in F} \Diamond[m]_{\prec}$ if and only if $J \in [F]_{\prec_L}$.
- $f_L(g_L(J)) = J$: The above proofs that f_L and g_L are computable imply that $F \in f_L(g_L(J))$ if and only if $g_L(J) \in \bigcap_{m \in F} \Diamond[m]_{\prec}$ if and only if $F \in J$.
- $g_L(f_L(A)) = A$: The above proofs that g_L and f_L are computable imply that $g_L(f_L(A)) \in \bigcap_{m \in F} \Diamond[m]_{\prec}$ if and only if $F \in f_L(A)$ if and only if $A \in \bigcap_{m \in F} \Diamond[m]_{\prec}$. $\qquad\square$

5.2   Upper Powerspace

A basis for the upper Vietoris topology on $\mathbf{K}(\mathbf{I}(\prec))$ is given by sets of the form

$$\Box \bigcup_{n \in F} [n]_{\prec} = \{K \in \mathbf{K}(\mathbf{I}(\prec)) | (\forall I \in K)(\exists n \in F)\, n \in I\}$$

for $F \in \mathcal{P}_{\mathrm{fin}}(\mathbb{N})$. Define the transitive relation $\prec_U$ on $\mathcal{P}_{\mathrm{fin}}(\mathbb{N})$ as

$$F \prec_U G \text{ if and only if } (\forall n \in G)(\exists m \in F)\, m \prec n.$$

Transitivity of $\prec_U$ easily follows from the transitivity of $\prec$, and it is clear that $\prec_U$ is c.e. whenever $\prec$ is. Next, define $f_U \colon \mathbf{K}(\mathbf{I}(\prec)) \to \mathbf{I}(\prec_U)$ as

$$f_U(K) = \{F \in \mathcal{P}_{\mathrm{fin}}(\mathbb{N}) | (\forall I \in K)(\exists m \in F)\, m \in I\}$$

and $g_U \colon \mathbf{I}(\prec_U) \to \mathbf{K}(\mathbf{I}(\prec))$ as

$$g_U(J) = \{I \in \mathbf{I}(\prec) | (\forall F \in J)(\exists m \in I)\, m \in F\}.$$

We will need the following lemma when we prove that these two functions are well-defined computable homeomorphisms.

Lemma 9. *If $J \in \mathbf{I}(\prec_U)$ and $S \subseteq \mathbb{N}$, then $g_U(J) \subseteq \bigcup_{m \in S}[m]_\prec$ if and only if there is finite $F \subseteq S$ with $F \in J$.*

Proof. For the easy direction, assume $F \subseteq S$ is finite and $F \in J$. Then every $I \in g_U(J)$ intersects F, which implies $g_U(J) \subseteq \bigcup_{m \in F}[m]_\prec \subseteq \bigcup_{m \in S}[m]_\prec$.

Conversely, assume $g_U(J) \subseteq \bigcup_{m \in S}[m]_\prec$. Since J is an ideal and countable, there is a sequence $(F_i)_{i \in \mathbb{N}}$ in J satisfying $(\forall i \in \mathbb{N})\, F_i \prec_U F_{i+1}$ and $(\forall F \in J)(\exists i \in \mathbb{N})\, F \prec_U F_i$. It is straightforward to see that $I \in g_U(J)$ if and only if $(\forall i \in \mathbb{N})\, F_i \cap I \neq \emptyset$. Define T to be the set of all $\sigma \in \mathbb{N}^{<\mathbb{N}}$ satisfying:

1. $(\forall i < len(\sigma) - 1)\, \sigma(i) \prec \sigma(i+1)$,
2. $(\forall i < len(\sigma))\, \sigma(i) \in F_i$,
3. $(\forall i < len(\sigma))(\forall m \in S)\, m \nprec \sigma(i)$.

Clearly T is closed under subsequences, hence T is a finitely branching tree because of item 2. If T contained an infinite path p then the ideal $I = \{n \in \mathbb{N} \mid (\exists i \in \mathbb{N})\, n \prec p(i)\}$ would be in $g_U(J)$ even though item 3 prevents I from being in $\bigcup_{m \in S}[m]_\prec$, which would be a contradiction. It follows from König's lemma that T is finite. Let $k \in \mathbb{N}$ be an upper bound for $\{len(\sigma) \mid \sigma \in T\}$.

Assume for a contradiction that there is $n_k \in F_k$ such that $(\forall m \in S)\, m \nprec n_k$. If $k > 0$, then $F_{k-1} \prec_U F_k$, hence there is $n_{k-1} \in F_{k-1}$ with $n_{k-1} \prec n_k$, and transitivity of $\prec$ implies $(\forall m \in S)\, m \nprec n_{k-1}$. Continuing in this way, we can construct a finite sequence $\sigma \in \mathbb{N}^{<\mathbb{N}}$ as $\sigma(k) = n_k$, $\sigma(k-1) = n_{k-1}$, and so on, in such a way that $\sigma \in T$ but $len(\sigma) = k+1$, which contradicts the choice of k.

Therefore, for each $n \in F_k$ there is $m_n \in S$ with $m_n \prec n$. Then $F = \{m_n \mid n \in F_k\}$ is a finite subset of S satisfying $F \prec_U F_k$, hence $F \in J$. $\square$

Theorem 10. $\mathbf{K}(\mathbf{I}(\prec))$ *and* $\mathbf{I}(\prec_U)$ *are computably homeomorphic.*

Proof. We will prove that f_U and g_U are well-defined computable inverses of each other in several steps.

- f_U *is well-defined:* We must show that $f_U(K)$ is an ideal.

  1. ($f_U(K)$ *is non-empty*). Ideals are non-empty, so we can fix some $m_I \in I$ for each $I \in K$. By compactness of K there is a finite subset F of $\{m_I \mid I \in K\}$ such that $K \subseteq \bigcup_{m_I \in F}[m_I]_\prec$. Then $F \in f_U(K)$, hence $f_U(K) \neq \emptyset$.
  2. ($f_U(K)$ *is a lower set*). Assume $G \in f_U(K)$ and $F \prec_U G$. For any $I \in K$ there exists $n \in G \cap I$, and since $F \prec_U G$ there is $m \in F$ with $m \prec n$. Then $m \in F \cap I$ because I is a lower set, and it follows that $F \in f_U(K)$.
  3. ($f_U(K)$ *is directed*). Assume $F, G \in f_U(A)$. For each $I \in K$ there exist $m_I \in F \cap I$ and $n_I \in G \cap I$. Since I is an ideal, there is $p_I \in I$ with $m_I \prec p_I$ and $n_I \prec p_I$. By compactness of K there is a finite subset H of $\{p_I \mid I \in K\}$ such that $K \subseteq \bigcup_{p_I \in H}[p_I]_\prec$. Then $H \in f_U(K)$ and $F, G \prec_U H$.

- $g_U(J)$ *is well-defined:* We must show that $g_U(J)$ is a saturated compact subset of $\mathbf{I}(\prec)$. It is clear that $g_U(J)$ is saturated, because the specialization order on $\mathbf{I}(\prec)$ is subset inclusion, and if I intersects each $F \in J$ then so does any superset I' of I. To show compactness, assume $S \subseteq \mathbb{N}$ is such that $g_U(J) \subseteq \bigcup_{m \in S}[m]_\prec$. Using Lemma 9, there is finite $F \subseteq S$ with $F \in J$, hence $g_U(J) \subseteq \bigcup_{m \in F}[m]_\prec$.

- f_U *is computable:* Clearly, $f_U(K) \in [F]_{\prec_U}$ if and only if $K \in \Box \bigcup_{m \in F} [m]_{\prec}$.
- g_U *is computable:* Lemma 9 implies $g_U(J) \in \Box \bigcup_{m \in F} [m]_{\prec}$ if and only if $J \in [F]_{\prec_U}$.
- $f_U(g_U(J)) = J$: The above proofs that f_U and g_U are computable imply that $F \in f_U(g_U(J))$ if and only if $g_U(J) \in \Box \bigcup_{m \in F} [m]_{\prec}$ if and only if $F \in J$.
- $g_U(f_U(K)) = K$: The above proofs that g_U and f_U are computable imply that $g_U(f_U(K)) \in \Box \bigcup_{m \in F} [m]_{\prec}$ if and only if $F \in f_U(K)$ if and only if $K \in \Box \bigcup_{m \in F} [m]_{\prec}$.

$\Box$

References

1. Chen, R.: Borel functors, interpretations, and strong conceptual completeness for $\mathcal{L}_{\omega_1 \omega}$. Trans. Am. Math. Soc. **372**, 8955–8983 (2019)
2. de Brecht, M.: Quasi-polish spaces. Ann. Pure Appl. Log. **164**, 356–381 (2013)
3. de Brecht, M., Kawai, T.: On the commutativity of the powerspace constructions. Log. Methods Comput. Sci. **15**, 1–25 (2019)
4. de Brecht, M., Pauly, A., Schröder, M.: Overt choice, to appear in the journal Computability, preprint arXiv: 1902.05926) (2019)
5. Gierz, G., Hofmann, K.H., Keimel, K., Lawson, J.D., Mislove, M.W., Scott, D.S.: Continuous Lattices and Domains. Cambridge University Press, Cambridge (2003)
6. Goubault-Larrecq, J.: Non-Hausdorff Topology and Domain Theory. Cambridge University Press, Cambridge (2013)
7. Heckmann, R.: Spatiality of countably presentable locales (proved with the Baire category theorem). Math. Struct. Comp. Sci. **25**, 1607–1625 (2015)
8. Hoyrup, M., Rojas, C., Selivanov, V., Stull, D.M.: Computability on quasi-polish spaces. In: Hospodár, M., Jirásková, G., Konstantinidis, S. (eds.) DCFS 2019. LNCS, vol. 11612, pp. 171–183. Springer, Cham (2019). https://doi.org/10.1007/978-3-030-23247-4_13
9. Korovina, M., Kudinov, O.: Towards computability over effectively enumerable topological spaces. Electron. Notes Theor. Comput. Sci. **221**, 115–125 (2008)
10. Korovina, M., Kudinov, O.: On higher effective descriptive set theory. In: Kari, J., Manea, F., Petre, I. (eds.) CiE 2017. LNCS, vol. 10307, pp. 282–291. Springer, Cham (2017). https://doi.org/10.1007/978-3-319-58741-7_27
11. Mummert, C.: On the reverse mathematics of general topology, Ph.D. thesis, Pennsylvania State University (2005)
12. Pauly, A.: On the topological aspects of the theory of represented spaces. Computability **5**(2), 159–180 (2016)
13. Schröder, M.: Extended admissibility. Theor. Comput. Sci. **284**(2), 519–538 (2002)
14. Selivanov, V.: On the difference hierarchy in countably based T_0-spaces. Electron. Notes Theor. Comput. Sci. **221**, 257–269 (2008)
15. Smyth, M.B.: Power domains and predicate transformers: a topological view. In: Diaz, J. (ed.) ICALP 1983. LNCS, vol. 154, pp. 662–675. Springer, Heidelberg (1983). https://doi.org/10.1007/BFb0036946
16. Weihrauch, K.: Computable Analysis. Springer, Heidelberg (2000)
17. Weihrauch, K., Grubba, T.: Elementary computable topology. J. Univ. Comput. Sci. **15**(6), 1381–1422 (2009)

Parallelizations in Weihrauch Reducibility and Constructive Reverse Mathematics

Makoto Fujiwara[(✉)] [iD]

School of Science and Technology, Meiji University,
1-1-1 Higashi-Mita, Tama-ku, Kawasaki-shi, Kanagawa 214-8571, Japan
makotofujiwara@meiji.ac.jp

Abstract. In the framework of finite-type arithmetic, we characterize the notion that an existence statement is primitive recursive Weihrauch reducible to the parallelization of another existence statement by a standard derivability notion in constructive reverse mathematics.

Keywords: Weihrauch reducibility · Parallelization · Constructive reverse mathematics

1 Introduction

A large amount of mathematical statements are of the logical form

$$\forall f(A(f) \rightarrow \exists g B(f,g)), \tag{1}$$

where f and g may be tuples. Such statements are often called **existence statements** since they argue the existence of some objects. One can see existence statements represented as sentences of the form (1) as problems to be solved. In such a context, any f such that $A(f)$ holds is called an **instance** of the problem and g is called a **solution** to the instance. Uniform relationships between existence statements have been investigated extensively in computable analysis and classical reverse mathematics ([4–8,15] etc.). The investigation usually employs the following reduction: a Π_2^1 sentence P of the form (1) is reducible to another Π_2^1 sentence Q of the form (1) if there exist Turing functionals Φ and Ψ such that whenever f_1 is an instance of P, then $\Phi(f_1)$ is an instance of Q, and whenever g_2 is a solution to $\Phi(f_1)$, then $g_1 := \Psi(f_1, g_2)$ is a solution to f_1. This is a particular case of **Weihrauch reducibility** for Π_2^1 sentences with Baire space as their represented spaces (see [8, Appendix]). For a detailed account of uniformity, the **parallelizations**:

$$\forall \langle f_n \rangle_{n \in \mathbb{N}} (\forall n^{\mathbb{N}} A(f_n) \rightarrow \exists \langle g_n \rangle_{n \in \mathbb{N}} \forall n^{\mathbb{N}} B(f_n, g_n)),$$

have been studied in computable analysis ([4,5]) and also in classical reverse mathematics under the name of sequential versions ([8,11,12]). In particular, the weak König lemma WKL, is not Weihrauch reducible to the intermediate value theorem IVT, but is so to the parallelization of IVT (see [4]).

© Springer Nature Switzerland AG 2020
M. Anselmo et al. (Eds.): CiE 2020, LNCS 12098, pp. 38–49, 2020.
https://doi.org/10.1007/978-3-030-51466-2_4

On the other hand, there is another development on the relation between existence statements from constructive mathematics [3], where every existence is shown by giving a construction of the witness entirely in the proofs. Ishihara and others have developed reverse mathematics first informally in Bishop's constructive mathematics, and later formally in two-sorted intuitionistic arithmetic ([1,9,14] etc.). In particular, the intermediate value theorem IVT is known to be equivalent to the weak König lemma WKL over the constructive base theory containing a countable choice principle (see [1, Section 1]).

Interestingly, there are several corresponding results between constructive reverse mathematics and computability-theoretic investigations on parallelizations or sequential versions, including the above mentioned facts on WKL and IVT. In particular, it seems to be believed in the community of computable analysis that constructive equivalences of two existence statements in Bishop's constructive mathematics (which accepts the use of a countable choice principle) correspond to Weihrauch equivalences of their parallelizations (see [6, Footnote c]). In this paper, we verify that this experimental belief is plausible by showing some meta-theorems in the framework of finite-type arithmetic together with observations in some concrete examples. This sheds light on the correspondence between computable analysis and constructive reverse mathematics which have been developed independently until recently. The investigation in this paper is based on the previous work [10] of the author himself.

As a framework for our investigation, we employ an extensional variant $\mathsf{E\text{-}HA}^\omega$ of intuitionistic arithmetic in all finite types and its fragment $\widehat{\mathsf{E\text{-}HA}}^\omega{\restriction}$ from [17, Section 3.3]. Recall that finite types are defined as follows: $\mathbb{N}$ is a type; if σ and τ are types, then so is $\sigma \to \tau$. Note that $\widehat{\mathsf{E\text{-}HA}}^\omega{\restriction}$ has a recursor only of type $\mathbb{N}$ and its induction schema is restricted to quantifier-free formulas. The λ-abstraction is officially defined by using the combinators. The set of the closed terms of $\mathsf{E\text{-}HA}^\omega$ and that for $\widehat{\mathsf{E\text{-}HA}}^\omega{\restriction}$ are denoted by $\mathbf{T}$ and $\mathbf{T_0}$ respectively. The set-theoretic functionals definable in $\mathbf{T}$ (resp. $\mathbf{T_0}$) are called Gödel (resp. Kleene) primitive recursive functionals of finite type. A classical variant $\mathsf{E\text{-}PA}^\omega$ (resp. $\widehat{\mathsf{E\text{-}PA}}^\omega{\restriction}$) is obtained from $\mathsf{E\text{-}HA}^\omega$ (resp. $\widehat{\mathsf{E\text{-}HA}}^\omega{\restriction}$) by adding the axiom scheme of excluded middle $A \vee \neg A$. The language of our systems contains a binary predicate symbol $=_\mathbb{N}$ for equality between objects of type $\mathbb{N}$ only. Throughout this paper, we employ the same notations as in [10]. Note that the type superscripts (for terms) and subscripts (for equality) are omitted when they are clear from the context. A tuple of terms is denoted with a underline as $\underline{t}$. In addition, $\dot{-}$ denotes the primitive recursive cut-off subtraction, and $\{0,1\}^m$ denotes the set of all binary sequences of length m. Recall that an $\exists$-free formula is a formula which does not contain $\vee$ and $\exists$. A countable choice principle $\mathsf{AC}^{0,\omega}$ is the following schema:

$$(\mathsf{AC}^{0,\omega})\forall x^\mathbb{N}\exists f^\tau A(x,f) \to \exists F^{\mathbb{N}\to\tau}\forall x^\mathbb{N} A(x, Fx),$$

where τ is any type. This principle is crucial for our meta-theorems (see Remark 8). For the other principles in this paper, we refer the reader to [10, Section 1.1].

2   Meta-theorems

In this decade, there are several attempts to reveal the proper relation between constructive reverse mathematics and Weihrauch reducibility in classical reverse mathematics and computable analysis ([10,12,13,18,20] etc.). In particular, the author formalized in [10, Definition 2.5] the primitive recursive variants of Weihrauch reduction between existence statements P and Q formalized as $\widehat{\mathsf{E\text{-}HA}}^{\omega}\!\upharpoonright$-sentences $\forall \underline{f}(A_1(\underline{f}) \to \exists \underline{g} B_1(\underline{f}, g))$ and $\forall \underline{f}(A_2(\underline{f}) \to \exists \underline{g} B_2(\underline{f}, g))$ in the context of finite-type arithmetic as follows:

- For a finite-type arithmetic S^{ω} containing $\mathsf{E\text{-}HA}^{\omega}$, P is **Gödel-primitive-recursive Weihrauch reducible** to Q in S^{ω} if there exist closed terms $\underline{s}$ and $\underline{t}$ (of suitable types) in $\mathbf{T}$ such that S^{ω} proves

$$\forall \underline{f}(A_1(\underline{f}) \to A_2(\underline{s}\underline{f})) \wedge \forall \underline{f}, g'\left(B_2(\underline{s}\underline{f}, g') \wedge A_1(\underline{f}) \to B_1(\underline{f}, t\underline{f}\ g')\right). \qquad (2)$$

- For a finite-type arithmetic S^{ω} containing $\widehat{\mathsf{E\text{-}HA}}^{\omega}\!\upharpoonright$, P is **Kleene-primitive-recursive Weihrauch reducible** to Q in S^{ω} if there exist closed terms $\underline{s}$ and $\underline{t}$ (of suitable types) in $\mathbf{T_0}$ such that S^{ω} proves (2).

In addition, P is **normally reducible** to Q in S^{ω} if S^{ω} proves

$$\forall \underline{f}\left(A_1(\underline{f}) \to \exists \underline{f}'\left(A_2(\underline{f}') \wedge \forall g'\left(B_2(\underline{f}', g') \to \exists \underline{g} B_1(\underline{f}, g)\right)\right)\right).$$

The notions of Gödel/Kleene-primitive-recursive Weihrauch reducibility is a natural restriction of formalized Weihrauch reducibility e.g. in [8,18] where Turing functionals for the reduction are replaced by primitive recursive functionals in the sense of Gödel/Kleene. The normal reducibility, which requires a proof of $Q \to P$ with a specific form, is a stronger notion than just proving $Q \to P$ (see [10, Remark 2.9]). Since intuitionistic finite-type arithmetic with a choice principle roughly corresponds to Bishop's constructive mathematics, one may regard the normal reducibility in a nearly intuitionistic finite-type arithmetic as a sort of constructive reducibility. In [10, Theorem 2.10], the author showed a meta-theorem stating that the primitive-recursive Weihrauch reducibility verifiably in a fragment of classical finite-type arithmetic is equivalent to the normal reducibility in the corresponding (nearly) intuitionistic finite-type arithmetic for all existence statements formalized with $\exists$-free formulas. Thus constructive reducibility can be captured by the primitive-recursive variant of Weihrauch reducibility with an additional restriction on the verification theory (which has not been taken into account in computable analysis). Of course, the Weihrauch reductions between concrete existence statements are not always primitive recursive (in the sense of Gödel/Kleene). In addition, there are many existence statements which are not formalized with $\exists$-free formulas. Nonetheless, there seem to be plenty of examples to which the meta-theorem is applicable. In fact, the Weihrauch reductions between concrete existence statements can be verified usually in a weak theory (cf. [10, Section 3]). On the other hand, any characterization on parallelizations was still missing.

We call sentences of the form $\forall \underline{f}(A(\underline{f}) \to \exists \underline{g}B(\underline{f},\underline{g}))$, where $\underline{f} :\equiv f_1^{\sigma_1}, \ldots, f_k^{\sigma_k}$ and $\underline{g} :\equiv g_1^{\tau_1}, \ldots, g_l^{\tau_l}$ are finite tuples of variables, **normal existential sentences**. In the language of finite-type arithmetic, the parallelization of this sentence is formalized as

$$\forall f_1^{\mathbb{N}\to\sigma_1}, \ldots, f_k^{\mathbb{N}\to\sigma_k} \left(\begin{array}{l} \forall n^{\mathbb{N}} A\left(f_1 n, \ldots, f_k n\right) \to \\ \exists g_1^{\mathbb{N}\to\tau_1}, \ldots, g_l^{\mathbb{N}\to\tau_l} \forall n^{\mathbb{N}} B\left(f_1 n, \ldots, f_k n, g_1 n, \ldots, g_l n\right) \end{array} \right),$$

which is again a normal existential sentence. For each normal existential sentence P, we write its parallelization as $\widehat{P}$. Throughout this paper, we sometimes identify a normal existential sentence with the indicated existence statement. The following proposition is shown straightforwardly by the soundness of modified realizability [17, Theorem 5.8].

Proposition 1. *For normal existential sentences* P *and* Q *with* $\exists$-*free formulas of* $\widehat{\text{E-HA}}^{\omega} \upharpoonright$ *(in the sense of Theorem 6), if* P *is normally reducible to* Q *in* E-HA$^{\omega}$+ $AC^{\omega} + IP^{\omega}_{\text{ef}}$, *then* $\widehat{P}$ *is normally reducible to* $\widehat{Q}$ *in* E-HA$^{\omega}$. *The analogous result also holds where* E-HA$^{\omega}$ *is replaced by* E-HA$^{\omega} + QF\text{-}AC^{0,0} + \Sigma_1^0\text{-}DNS^0$, E-HA$^{\omega} +$ $\Pi_1^0\text{-}AC^{0,0} + \Sigma_2^0\text{-}DNS^0$, E-HA$^{\omega} + AC^{\omega} + mr\text{-}DNS^{\omega}$, $\widehat{\text{E-HA}}^{\omega} \upharpoonright, \widehat{\text{E-HA}}^{\omega} \upharpoonright + QF\text{-}AC^{0,0} +$ $\Sigma_1^0\text{-}DNS^0$, $\widehat{\text{E-HA}}^{\omega} \upharpoonright + \Pi_1^0\text{-}AC^{0,0} + \Sigma_2^0\text{-}DNS^0$, *or* $\widehat{\text{E-HA}}^{\omega} \upharpoonright + AC^{\omega} + mr\text{-}DNS^{\omega}$.

The following corollary is obtained immediately from [10, Theorem 2.10] and Proposition 1 together with the general fact that P is normally reducible to $\widehat{P}$.

Corollary 2. *Let* S^{ω} *be one of the systems* E-PA$^{\omega}$, E-PA$^{\omega} + QF\text{-}AC^{0,0}$, E-PA$^{\omega} +$ $\Pi_1^0\text{-}AC^{0,0}$, E-PA$^{\omega} + AC^{\omega}$, $\widehat{\text{E-PA}}^{\omega} \upharpoonright$, $\widehat{\text{E-PA}}^{\omega} \upharpoonright + QF\text{-}AC^{0,0}$, $\widehat{\text{E-PA}}^{\omega} \upharpoonright + \Pi_1^0\text{-}AC^{0,0}$, *and* $\widehat{\text{E-PA}}^{\omega} \upharpoonright + AC^{\omega}$. *For* P *and* Q *as in Proposition 1, if* P *is Gödel/Kleene-primitive-recursive Weihrauch reducible to* Q *in* S^{ω}, *then* P *is so to* $\widehat{Q}$ *in* S^{ω}.

Remark 3. The relation between WKL and IVT in computable analysis, which is mentioned in Sect. 1, shows that the converse of Corollary 2 does not hold.

From the perspective of Corollary 2 and Remark 3, it is worthwhile to characterize the property that P is Gödel/Kleene-primitive-recursive Weihrauch reducible to $\widehat{Q}$ in S^{ω} in some natural context of constructive reverse mathematics. For this purpose, we introduce the following notions:

Definition 4. Let P and Q be existence statements formalized as normal existential sentences $\forall \underline{f}(A_1(\underline{f}) \to \exists \underline{g}B_1(\underline{f},\underline{g}))$ and $\forall f_1, \ldots, f_k(A_2(f_1, \ldots, f_k) \to \exists \underline{g}B_2(f_1, \ldots, f_k, \underline{g}))$ of $\widehat{\text{E-HA}}^{\omega} \upharpoonright$ respectively. For a finite-type arithmetic S^{ω} containing $\widehat{\text{E-HA}}^{\omega} \upharpoonright$, P **is normally T-derivable from** Q **in** S^{ω} if there exist closed terms $s_1, \ldots, s_k$ (of suitable types) in **T** such that S^{ω} proves

$$\forall \underline{f} \left(A_1(\underline{f}) \to \forall m^{\mathbb{N}} A_2(s_1 m \underline{f}, \ldots, s_k m \underline{f}) \right) \tag{3}$$

and

$$\forall \underline{f} \left(\begin{array}{l} A_1(\underline{f}) \wedge \forall m \left(A_2(s_1 m \underline{f}, \ldots, s_k m \underline{f}) \to \exists \underline{g}' B_2\left(s_1 m \underline{f}, \ldots, s_k m \underline{f}, \underline{g}'\right)\right) \\ \to \exists \underline{g} B_1(\underline{f}, \underline{g}) \end{array} \right). \tag{4}$$

The notion that P **is normally T_0-derivable from** Q in S^ω is defined in the same manner with using $\mathbf{T_0}$ instead of $\mathbf{T}$. In fact, "$A_2(s_1 m\underline{f}, \ldots, s_k m\underline{f}) \to$" in (4) is redundant in the presence of (3).

Remark 5. For existence statements P : $\forall \underline{f}(A_1(\underline{f}) \to \exists \underline{g} B_1(\underline{f}, \underline{g}))$ and Q, the fact that P is normally derivable from Q demands some proof of Q $\to$ P with the following structure:

1. Fix $\underline{f}$ such that $A_1(\underline{f})$;
2. Assuming $A_1(\underline{f})$, derive $\exists \underline{g} B_1(\underline{f}, \underline{g})$ by using Q for the countably many instances which are given primitive recursively in $\underline{f}$.

For existence statements P and Q, it is quite common in practical mathematics (not only in constructive mathematics) to show Q $\to$ P in this manner.

On the other hand, the normal derivability is properly weaker than the normal reducibility in the context of (nearly) intuitionistic systems. Note also that the normal derivability relation is reflexive and transitive.

Theorem 6. *Let* P *and* Q *be existence statements formalized as normal existential sentences* $\forall \underline{f}(A_1(\underline{f}) \to \exists \underline{g} B_1(\underline{f}, \underline{g}))$ *and* $\forall f_1, \ldots, f_k(A_2(f_1, \ldots, f_k) \to \exists g' B_2(f_1, \ldots, f_k, g'))$ *of* $\widehat{\mathsf{E\text{-}HA}}^\omega \!\restriction$ *respectively with $\exists$-free formulas* $A_1, A_2, B_1,$ *and* B_2. *Then the following hold:*

1. P *is Gödel-primitive-recursive Weihrauch reducible to* $\widehat{\mathsf{Q}}$ *in* $\mathsf{E\text{-}PA}^\omega$ *if and only if* P *is normally* **T**-*derivable from* Q *in* $\mathsf{E\text{-}HA}^\omega + \mathsf{AC}^{0,\omega}$.
2. P *is Kleene-primitive-recursive Weihrauch reducible to* $\widehat{\mathsf{Q}}$ *in* $\widehat{\mathsf{E\text{-}PA}}^\omega \!\restriction$ *if and only if* P *is normally* $\mathbf{T_0}$-*derivable from* Q *in* $\widehat{\mathsf{E\text{-}HA}}^\omega \!\restriction + \mathsf{AC}^{0,\omega}$.

Proof. (1) Assume that P is Gödel-primitive-recursive Weihrauch reducible to $\widehat{\mathsf{Q}}$ in $\mathsf{E\text{-}PA}^\omega$. As in the proof of [10, Theorem 2.10], by the negative translation (see [17, Section 10.1]), we have that $\mathsf{E\text{-}HA}^\omega$ proves $\forall \underline{f}(A_1(\underline{f}) \to \forall n^\mathbb{N} A_2(s_1 \underline{f} n, \ldots, s_k \underline{f} n))$ and $\forall \underline{f}, \underline{G'}(\forall n^\mathbb{N} B_2(s_1 \underline{f} n, \ldots, s_k \underline{f} n, \underline{G'} n) \wedge A_1(\underline{f}) \to B_1(\underline{f}, \underline{t} \underline{f} \underline{G'}))$ for some closed terms $s_1, \ldots, s_k$ and $\underline{t}$ in $\mathbf{T}$. Put a closed term $\tilde{s}_i$ as $\lambda n, \underline{f}. s_i \underline{f} n$ for each $i \in \{1, \ldots, k\}$. Then we have that $\mathsf{E\text{-}HA}^\omega$ proves

$$\forall \underline{f}(A_1(\underline{f}) \to \forall n^\mathbb{N} A_2(\tilde{s}_1 n \underline{f}, \ldots, \tilde{s}_k n \underline{f})) \tag{5}$$

and

$$\forall \underline{f}, \underline{G'}\left(\forall n^\mathbb{N} B_2(\tilde{s}_1 n \underline{f}, \ldots, \tilde{s}_k n \underline{f}), \underline{G'} n) \wedge A_1(\underline{f}) \to \exists \underline{g} B_1(\underline{f}, \underline{g})\right). \tag{6}$$

Applying $\mathsf{AC}^{0,\omega}$ to (6), we have

$$\forall \underline{f}\left(\forall n^\mathbb{N} \exists g' B_2(\tilde{s}_1 n \underline{f}, \ldots, \tilde{s}_k n \underline{f}), g') \wedge A_1(\underline{f}) \to \exists \underline{g} B_1(\underline{f}, \underline{g})\right). \tag{7}$$

Then it follows from (5) and (7) that

$$\forall \underline{f}\left(\begin{array}{l} A_1(\underline{f}) \wedge \forall n^\mathbb{N}\left(A_2(\tilde{s}_1 n \underline{f}, \ldots, \tilde{s}_k n \underline{f}) \to \exists g' B_2\left(\tilde{s}_1 n \underline{f}, \ldots, \tilde{s}_k n \underline{f}, g'\right)\right) \\ \to \exists \underline{g} B_1(\underline{f}, \underline{g}) \end{array}\right). \tag{8}$$

Thus P is normally **T**-derivable from Q in $\mathsf{E\text{-}HA}^\omega + \mathrm{AC}^{0,\omega}$.

For the converse direction, assume that $\mathsf{E\text{-}HA}^\omega + \mathrm{AC}^{0,\omega}$ proves (5) and (8) for some closed terms $\tilde{s}_1, \ldots, \tilde{s}_k$ in **T**. By (8), we have that $\mathsf{E\text{-}HA}^\omega + \mathrm{AC}^{0,\omega}$ proves (7), and hence, (6).

As in the proof of [10, Theorem 2.10], by the soundness of modified realizability [17, Theorem 5.8], there exist closed terms $\underline{t}$ in **T** such that $\mathsf{E\text{-}HA}^\omega$ proves

$$\forall \underline{f}, \underline{G'} \left(A_1(\underline{f}) \wedge \forall n^{\mathbb{N}} B_2 \left(\tilde{s}_1 n \underline{f}, \ldots, \tilde{s}_k n \underline{f}, \underline{G'} n \right) \to B_1(\underline{f}, \underline{t}\underline{f}\,\underline{G'}) \right). \tag{9}$$

Put a closed term s_i as $\lambda \underline{f}, n. \tilde{s}_i n \underline{f}$ for each $i \in \{1, \ldots, k\}$. Then, by (5) and (9), we have that P is Gödel-primitive-recursive Weihrauch reducible to $\widehat{Q}$ in $\mathsf{E\text{-}HA}^\omega$ (and hence, so is in $\mathsf{E\text{-}PA}^\omega$) with $s_1, \ldots, s_k$ and $\underline{t}$ as the witnesses.

The same proof works also for (2). $\qquad\square$

Combining the proof of Theorem 6 with [10, Lemma 2.1] and [10, Lemma 2.2] as in [10, Theorem 2.10], we have the following:

Theorem 7. *Let* P *and* Q *be existence statements formalized as normal existential sentences* $\forall \underline{f}(A_1(\underline{f}) \to \exists \underline{g} B_1(\underline{f}, \underline{g}))$ *and* $\forall \underline{f'}(A_2(\underline{f'}) \to \exists \underline{g'} B_2(\underline{f'}, \underline{g'}))$ *of* $\widehat{\mathsf{E\text{-}HA}}^\omega \upharpoonright$ *respectively with* $\exists$-*free formulas* $A_1, A_2, B_1,$ *and* B_2. *Then*

1. P *is Gödel-primitive-recursive Weihrauch reducible to* $\widehat{Q}$ *in* $\mathsf{E\text{-}PA}^\omega + \mathrm{QF\text{-}AC}^{0,0}$ *(resp.* $\mathsf{E\text{-}PA}^\omega + \Pi_1^0\text{-}\mathrm{AC}^{0,0}$, $\mathsf{E\text{-}PA}^\omega + \mathrm{AC}^\omega$*) if and only if* P *is normally* **T**-*derivable from* Q *in* $\mathsf{E\text{-}HA}^\omega + \mathrm{AC}^{0,\omega} + \Sigma_1^0\text{-}\mathrm{DNS}^0$ *(resp.* $\mathsf{E\text{-}HA}^\omega + \mathrm{AC}^{0,\omega} + \Sigma_2^0\text{-}\mathrm{DNS}^0$, $\mathsf{E\text{-}HA}^\omega + \mathrm{AC}^\omega + \mathrm{mr\text{-}DNS}^\omega$*).*

2. P *is Kleene-primitive-recursive Weihrauch reducible to* $\widehat{Q}$ *in* $\widehat{\mathsf{E\text{-}PA}}^\omega \upharpoonright +$ $\mathrm{QF\text{-}AC}^{0,0}$ *(resp.* $\widehat{\mathsf{E\text{-}PA}}^\omega \upharpoonright + \Pi_1^0\text{-}\mathrm{AC}^{0,0}$, $\widehat{\mathsf{E\text{-}PA}}^\omega \upharpoonright + \mathrm{AC}^\omega$*) if and only if* P *is normally* **T₀**-*derivable from* Q *in* $\widehat{\mathsf{E\text{-}HA}}^\omega \upharpoonright + \mathrm{AC}^{0,\omega} + \Sigma_1^0\text{-}\mathrm{DNS}^0$ *(resp.* $\widehat{\mathsf{E\text{-}HA}}^\omega \upharpoonright + \mathrm{AC}^{0,\omega} + \Sigma_2^0\text{-}\mathrm{DNS}^0$, $\widehat{\mathsf{E\text{-}HA}}^\omega \upharpoonright + \mathrm{AC}^\omega + \mathrm{mr\text{-}DNS}^\omega$*).*

Remark 8. The meta-theorem where $\mathrm{AC}^{0,\omega}$ is replaced by $\mathrm{QF\text{-}AC}^{0,0}$ in Theorem 7.(2) does not hold: If it holds, by Proposition 10 below, we have that WKL is normally **T₀**-derivable from LLPO in $\widehat{\mathsf{E\text{-}HA}}^\omega \upharpoonright + \mathrm{QF\text{-}AC}^{0,0} + \Sigma_1^0\text{-}\mathrm{DNS}^0$, and hence, WKL is provable in $\widehat{\mathsf{E\text{-}PA}}^\omega \upharpoonright + \mathrm{QF\text{-}AC}^{0,0}$ (cf. Remark 5). This is a contradiction (see [16]). The same argument holds also for Theorem 7.(1).

3 Application

There seem to be many results (proofs) in computable analysis and constructive reverse mathematics to which our meta-theorems are applicable. For the purpose of demonstrating the availability of our meta-theorems, we inspect the existing proofs in both of the contexts on the relation between the weak König lemma WKL and the lesser limited principle of omniscience LLPO. This is the core for the relation between the intermediate value theorem IVT and WKL which is mentioned in Sect. 1.

Recall that WKL states that for any infinite binary tree, there exists an infinite path through the tree (see [10, Section 3.1] for the formal definition) and LLPO is formalized as

$$\forall f_0^{\mathbb{N} \to \mathbb{N}}, f_1^{\mathbb{N} \to \mathbb{N}} \left(\begin{array}{c} \neg \left(\exists x^{\mathbb{N}} f_0(x) = 0 \wedge \exists y^{\mathbb{N}} f_1(y) = 0 \right) \\ \to \exists k^{\mathbb{N}}((k = 0 \to \neg \exists x \, f_0(x) = 0) \wedge (k \neq 0 \to \neg \exists y \, f_1(y) = 0)) \end{array} \right)$$

in the language of $\widehat{\text{E-HA}}^\omega{\restriction}$. Both of them are normal existential sentences of the form to which our meta-theorems are applicable. We also recall the following disjunctive variant of $\Pi_1^0\text{-AC}^{0,0}$ from [2,14]:

$$\Pi_1^0\text{-AC}_\vee^{0,0} : \begin{array}{l} \forall n^{\mathbb{N}} \left(\forall x^{\mathbb{N}} A_{\text{qf}}(n, x) \vee \forall y^{\mathbb{N}} B_{\text{qf}}(n, y) \right) \\ \to \exists h \forall n \left((h(n) = 0 \to \forall x \, A_{\text{qf}}(n, x)) \wedge (h(n) \neq 0 \to \forall y \, B_{\text{qf}}(n, y)) \right), \end{array}$$

where A_{qf} and B_{qf} are quantifier-free. In the context of constructive reverse mathematics, Ishihara [14, Section 5] first showed that WKL is equivalent to LLPO plus $\Pi_1^0\text{-AC}_\vee^{0,0}$ over a weak intuitionistic arithmetic. More recently, Berger, Ishihara, and Schuster [2, Section 6] provided a simpler proof of the fact. On the other hand, in the context of computable analysis, Brattka and Gherardi [5, Theorem 8.2] showed that WKL is Weihrauch equivalent to the parallelization of LLPO while it is not so to LLPO.

Proposition 9. WKL *is normally* $\mathbf{T_0}$-*derivable from* LLPO *in* $\widehat{\text{E-HA}}^\omega{\restriction} +$ $\Pi_1^0\text{-AC}_\vee^{0,0} + \text{QF-AC}^{0,0} + \Sigma_1^0\text{-DNS}^0$.

Proof. The proof is basically the same as that for [2, Theorem 27]. We reason informally in $\widehat{\text{E-HA}}^\omega{\restriction} + \Pi_1^0\text{-AC}_\vee^{0,0} + \text{QF-AC}^{0,0} + \Sigma_1^0\text{-DNS}^0$ and let $T^{\mathbb{N} \to \mathbb{N}}$ be an infinite binary tree (officially given by its characteristic function as in [10, Section 3.1]).

For each $i \in \{0, 1\}$ and each code $u^{\mathbb{N}}$ of a finite sequence of natural numbers, define $C_i(u, T)$ as $\exists m^{\mathbb{N}} (\neg D(m, u * \langle i \rangle) \wedge D(m, u * \langle 1 - i \rangle))$, where $D(m, u)$ expresses that there exists a finite binary sequence $v^{\mathbb{N}}$ of length m such that $u * v$ is contained in T. Then we have $\forall u^{\mathbb{N}} \neg (C_0(u, T) \wedge C_1(u, T))$. Notice that there exist closed terms s_0 and s_1 of type $\mathbb{N} \to ((\mathbb{N} \to \mathbb{N}) \to (\mathbb{N} \to \mathbb{N}))$ in $\mathbf{T_0}$ such that

$$\exists m^{\mathbb{N}} \, s_i u T m = 0 \leftrightarrow C_i(u, T)$$

for each $i \in \{0, 1\}$. Thus we have $\forall u^{\mathbb{N}} \neg (\exists m^{\mathbb{N}} \, s_0 u T m = 0 \wedge \exists m^{\mathbb{N}} \, s_1 u T m = 0)$. This validates the first condition of normal derivability in Definition 4.

For the second condition, assume also that for each $u^{\mathbb{N}}$, there exists $k^{\mathbb{N}}$ such that $k = 0 \to \neg \exists m^{\mathbb{N}} \, s_0 u T m = 0$ and $k \neq 0 \to \neg \exists m^{\mathbb{N}} \, s_1 u T m = 0$. Using a dependence choice principle which is derived from $\Pi_1^0\text{-AC}_\vee^{0,0}$ (see [2,

Corollary 5]), we have a function $h^{\mathbb{N} \to \mathbb{N}}$ such that $\forall u^{\mathbb{N}} \neg \exists m^{\mathbb{N}} s_{h(u)} \left(\overline{h}u \right) Tm = 0$, equivalently, $\forall u^{\mathbb{N}} \neg C_{h(u)} \left(\overline{h}u, T \right)$. As in the proof of [2, Theorem 27], one can show that $\forall n^{\mathbb{N}}, m^{\mathbb{N}} D(m, \overline{h}n)$ by using Π_1^0 induction which is provable in $\widehat{\text{E-HA}}^\omega \restriction + \text{QF-AC}^{0,0} + \Sigma_1^0\text{-DNS}^0$ (see [9, Lemma 20]). Then it follows that h is an infinite path through T. $\qquad\Box$

Applying Theorem 7.(2) to Proposition 9, as a corollary, we obtain the following result in the context of computable analysis:

Proposition 10. WKL *is Kleene-primitive-recursive Weihrauch reducible to* $\widehat{\text{LLPO}}$ *in* $\widehat{\text{E-PA}}^\omega \restriction + \text{QF-AC}^{0,0}$.

On the other hand, by refining the proof of [5, Theorem 8.2] (where WKL itself is used for the verification as it is mentioned there), we can also have a direct proof of Proposition 10:

Proof. Recall that $\widehat{\text{LLPO}}$ states that for all f_0 and f_1 of type $\mathbb{N} \to (\mathbb{N} \to \mathbb{N})$, if $\forall n^{\mathbb{N}} \neg \left(\exists x^{\mathbb{N}} f_0 nx = 0 \wedge \exists x^{\mathbb{N}} f_1 nx = 0 \right)$, then there exists $h^{\mathbb{N} \to \mathbb{N}}$ such that

$$\forall n^{\mathbb{N}} \left(\left(hn = 0 \to \neg \exists x^{\mathbb{N}} f_0 nx = 0 \right) \wedge \left(hn \neq 0 \to \neg \exists x^{\mathbb{N}} f_1 nx = 0 \right) \right). \qquad (10)$$

We reason informally in $\widehat{\text{E-PA}}^\omega \restriction + \text{QF-AC}^{0,0}$ and let $T^{\mathbb{N} \to \mathbb{N}}$ be an infinite binary tree (officially given by its characteristic function).

As in the proof of [5, Theorem 8.2], for each $i \in \{0, 1\}$, let $P_{k,i}^T$ denote the set of finite binary sequences u such that $u * \langle i \rangle$ is incomparable with all branches in T of length k. We define f_0 and f_1 of type $\mathbb{N} \to (\mathbb{N} \to \mathbb{N})$ primitive recursively (in the sense of Kleene) in the given tree T as

$$f_i u x = \begin{cases} 0 & \text{if } x \text{ is the least } k \text{ such that } u \in P_{k,i}^T \setminus P_{k,1-i}^T, \\ 1 & \text{otherwise,} \end{cases}$$

for $i \in \{0, 1\}$. For each $u^{\mathbb{N}}$, we have $\neg \left(\exists x^{\mathbb{N}} f_0 ux = 0 \wedge \exists x^{\mathbb{N}} f_1 ux = 0 \right)$ straightforwardly by definition.

For the second condition, let h satisfy (10) for f_0 and f_1 defined above. Define $p^{\mathbb{N} \to \mathbb{N}}$ primitive recursively in h as

$$p(k) = \begin{cases} 0 & \text{if } h \left(\overline{p}k \right) = 0, \\ 1 & \text{otherwise,} \end{cases}$$

For verifying that p is an infinite path through T, it suffices to show that for all $n^{\mathbb{N}}$ and $m^{\mathbb{N}}$, there exists $u \in \{0, 1\}^m$ such that $\overline{p}n * u$ is in T. In the following, we show this assertion by Π_1^0 induction (which is provable in $\widehat{\text{E-PA}}^\omega \restriction + \text{QF-AC}^{0,0}$) on $n^{\mathbb{N}}$. When $n = 0$, we are done since T is infinite. Assume that for all $m^{\mathbb{N}}$, there exists $u \in \{0, 1\}^m$ such that $\overline{p}n * u$ is in T. Our goal is to show the corresponding assertion for $n + 1$. Based on classical logic, we consider the following 4 cases:

1. There exists $k^{\mathbb{N}}$ such that $\overline{p}n \in P_{k,1}^T \setminus P_{k,0}^T$;

2. There exists $k^{\mathbb{N}}$ such that $\bar{p}n \in P^T_{k,0} \setminus P^T_{k,1}$;
3. There is no $k^{\mathbb{N}}$ such that $\bar{p}n \in P^T_{k,0} \cup P^T_{k,1}$;
4. There exists $k^{\mathbb{N}}$ such that $\bar{p}n \in P^T_{k,0} \cap P^T_{k,1}$.

We first work in the first case. Using classical logic, it follows from our induction hypothesis that at least one of the following holds:

– For all $m^{\mathbb{N}}$, there exists $u \in \{0,1\}^m$ such that $\bar{p}n * \langle 0 \rangle * u$ is in T;
– For all $m^{\mathbb{N}}$, there exists $u \in \{0,1\}^m$ such that $\bar{p}n * \langle 1 \rangle * u$ is in T.

If the latter holds, then for $m_0 := k \dotminus (n+1)$, there exists $u \in \{0,1\}^{m_0}$ such that $\bar{p}n * \langle 1 \rangle * u$ is in T, which contradicts $\bar{p}n \in P^T_{k,1}$. Thus the former holds. Since $p(n)$ must be 0 by definition, we have our goal. In the second case, by mimicking the above argument, we have our goal as well. Next, we work in the third case. Fix $m^{\mathbb{N}}_1$ and assume that for all $u \in \{0,1\}^{m_1}$, $\bar{p}(n+1) * u$ is not in T. Then all branches in T of length $n + 1 + m_1$ (note that there is at least one branch of this length by induction hypothesis) are incomparable with $\bar{p}(n+1)$, and hence, $\bar{p}n \in P^T_{n+1+m_1,p(n)}$, which is a contradiction. Thus there exists $u \in \{0,1\}^{m_1}$ such that $\bar{p}(n+1) * u$ is in T. Finally, we show that the fourth case does not occur. If $\bar{p}n \in P^T_{k,0} \cap P^T_{k,1}$, then for any $s \in \{0,1\}^k$ in T, s is incomparable with both of $\bar{p}n * \langle 0 \rangle$ and $\bar{p}n * \langle 1 \rangle$. Note that k must be greater than n. By induction hypothesis, there exists $u \in \{0,1\}^{k-n}$ such that $\bar{p}n * u$ is in T. This is a contradiction. □

Remark 11. We obtain Proposition 9 by applying Theorem 7.(2) to Proposition 10 with the following observation: In particular cases where Q is LLPO in Theorem 7.(2), the proof shows that only $\Pi^0_1\text{-AC}^{0,0}_\vee$ rather than $\text{AC}^{0,\omega}$ is enough for the corresponding argument. On the other hand, one needs $\text{QF-AC}^{0,0}$ and $\Sigma^0_1\text{-DNS}^0$ for deriving the negative translation of $\text{QF-AC}^{0,0}$ (see [10, Lemma 2.1.(3)]).

Remark 12. One can notice that the proof of Proposition 9 and the proof of Proposition 10 are somewhat similar. Nevertheless, the primitive recursive witnesses for the latter is not obvious from the proof of Proposition 9. On the other hand, the proof of Proposition 10 heavily uses classical logic for the verification, and hence, the former is also not an immediate consequence from the latter. This observation illustrates that our meta-theorems should give rise to new results in one of the contexts from the results or the proofs in the other context.

At the end of this section, we briefly deal with the relation between the intermediate value theorem IVT (see [10, Section 3.2] for the formal definition) and WKL which is mentioned in Sect. 1.

Proposition 13. 1. $\text{DICH}_\mathbb{R}$ is normally reducible to IVT in $\widehat{\text{E-HA}}^\omega \restriction +$ $\text{QF-AC}^{0,0}$, where

$$\text{DICH}_\mathbb{R} : \forall f^{\mathbb{N} \to \mathbb{N}} \exists k^{\mathbb{N}} ((k = 0 \to f \geq_\mathbb{R} 0) \wedge (k \neq 0 \to f \leq_\mathbb{R} 0)).$$

2. LLPO *is normally reducible to* $\mathrm{DICH}_\mathbb{R}$ *in* $\widehat{\mathsf{E\text{-}HA}}^\omega{\restriction} + \mathrm{QF\text{-}AC}^{0,0}$.
3. $\widehat{\mathrm{WKL}}$ *is normally reducible to* WKL *in* $\widehat{\mathsf{E\text{-}HA}}^\omega{\restriction} + \mathrm{QF\text{-}AC}^{0,0}$.

Proof. (1): Inspect the argument in [19, 6.1.2]. (2): Inspect the argument in [19, 5.2.12]. (3): Inspect the proof of (1) $\rightarrow$ (3) of [11, Lemma 5]. □

By Propositions 1, 9, 10, and 13, together with [10, Proposition 3.8], we have the following:

Corollary 14. WKL *is Kleene-primitive-recursive Weihrauch reducible to* $\widehat{\mathrm{IVT}}$ *and vice versa in* $\widehat{\mathsf{E\text{-}PA}}^\omega{\restriction} + \mathrm{QF\text{-}AC}^{0,0}$.

Corollary 15. WKL *is normally* $\mathbf{T_0}$*-derivable from* IVT *and vice versa in* $\widehat{\mathsf{E\text{-}HA}}^\omega{\restriction} + \Pi_1^0\text{-}\mathrm{AC}_\vee^{0,0} + \mathrm{QF\text{-}AC}^{0,0} + \Sigma_1^0\text{-}\mathrm{DNS}^0$.

4 Another Possible Consequence from Constructive Reverse Mathematics

A lot of existing proofs in constructive reverse mathematics show not only provability but rather normal derivability (see Remark 5). However, this is not always the case. For example, in the proof of deriving the convex weak König lemma WKL^c from IVT [1, Theorem 3], for a given infinite convex tree T, IVT is first used to construct an infinite convex subtree T' having at most 2 branches for each height, and then it is used again for taking an infinite path through T'. Thus, while the first instance to which one applies IVT is provided primitive recursively in a given infinite convex tree T, the second instance is not so. In this section, we characterize this kind of proofs by the notion of Weihrauch reducibility to the consecutive composition of the finitely many copies of an existence statement, which has been studied recently in computable analysis (e.g. [7,15,18]).

Definition 16. Let P and Q be existence statements formalized as normal existential sentences $\forall \underline{f}(A_1(\underline{f}) \rightarrow \exists \underline{g} B_1(\underline{f}, \underline{g}))$ and $\forall \underline{f}'(A_2(\underline{f}') \rightarrow \exists \underline{g}' B_2(\underline{f}', \underline{g}'))$ of $\widehat{\mathsf{E\text{-}HA}}^\omega{\restriction}$ respectively.

- For a finite-type arithmetic S^ω containing $\mathsf{E\text{-}HA}^\omega$, P is Gödel-primitive-recursive Weihrauch reducible to the 2-copies of Q in S^ω if there exist closed terms $\underline{s}$, $\underline{t}$, and $\underline{u}$ (of suitable types) in $\mathbf{T}$ such that S^ω proves

$$
\begin{aligned}
&\forall \underline{f}(A_1(\underline{f}) \rightarrow A_2(\underline{s}\underline{f})) \wedge \\
&\forall \underline{f}, \underline{g}' \left(B_2(\underline{s}\underline{f}, \underline{g}') \wedge A_1(\underline{f}) \rightarrow A_2(\underline{t}\underline{f}\ \underline{g}') \right) \wedge \\
&\forall \underline{f}, \underline{g}', \underline{g}'' \left(B_2(\underline{t}\underline{f}\ \underline{g}', \underline{g}'') \wedge B_2(\underline{s}\underline{f}, \underline{g}') \wedge A_1(\underline{f}) \rightarrow B_1(\underline{f}, \underline{u}\underline{f}\ \underline{g}'\ \underline{g}'') \right).
\end{aligned}
\tag{11}
$$

- For a finite-type arithmetic S^ω containing $\widehat{\mathsf{E\text{-}HA}}^\omega{\restriction}$, P is Kleene-primitive-recursive Weihrauch reducible to the 2-copies of Q in S^ω if there exist closed terms $\underline{s}$, $\underline{t}$, and $\underline{u}$ (of suitable types) in $\mathbf{T_0}$ such that S^ω proves (11).

– For a finite-type arithmetic S^ω containing $\widehat{\text{E-HA}}^\omega{\upharpoonright}$, P is normally reducible to the 2-copies of Q in S^ω if S^ω proves

$$\forall \underline{f} \left(\begin{array}{l} A_1(\underline{f}) \to \\ \exists \underline{f'} \left(A_2(\underline{f'}) \wedge \forall \underline{g'} \left(\begin{array}{l} B_2(\underline{f'}, \underline{g'}) \to \\ \exists \underline{f''} \left(A_2(\underline{f''}) \wedge \forall \underline{g''} \left(\begin{array}{l} B_2(\underline{f''}, \underline{g''}) \to \\ \exists \underline{g} B_1(\underline{f}, \underline{g}) \end{array} \right) \right) \end{array} \right) \right) \end{array} \right).$$

The versions for k-copies ($k = 3, 4, 5, \dots$) are also defined in the same manner. Note that [10, Definition 2.5] is the case of $k = 1$.

The proof of [10, Theorem 2.10] allows us to generalize it as follows:

Theorem 17. *Let k be a fixed natural number. Let P and Q be existence statements formalized as normal existential sentences $\forall \underline{f}(A_1(\underline{f}) \to \exists \underline{g} B_1(\underline{f}, \underline{g}))$ and $\forall \underline{f'}(A_2(\underline{f'}) \to \exists \underline{g'} B_2(\underline{f'}, \underline{g'}))$ of $\widehat{\text{E-HA}}^\omega{\upharpoonright}$ respectively with $\exists$-free formulas A_1, A_2, B_1, and B_2. Then the following hold:*

1. *P is Gödel-primitive-recursive Weihrauch reducible to the k-copies of Q in E-PA^ω (resp. $\text{E-PA}^\omega + \text{QF-AC}^{0,0}$, $\text{E-PA}^\omega + \Pi_1^0\text{-AC}^{0,0}$, $\text{E-PA}^\omega + \text{AC}^\omega$) if and only if P is normally reducible to the k-copies of Q in E-HA^ω (resp. $\text{E-HA}^\omega + \text{QF-AC}^{0,0} + \Sigma_1^0\text{-DNS}^0$, $\text{E-HA}^\omega + \Pi_1^0\text{-AC}^{0,0} + \Sigma_2^0\text{-DNS}^0$, $\text{E-HA}^\omega + \text{AC}^\omega + \text{mr-DNS}^\omega$).*

2. *P is Kleene-primitive-recursive Weihrauch reducible to the k-copies of Q in $\widehat{\text{E-PA}}^\omega{\upharpoonright}$ (resp. $\widehat{\text{E-PA}}^\omega{\upharpoonright} + \text{QF-AC}^{0,0}$, $\widehat{\text{E-PA}}^\omega{\upharpoonright} + \Pi_1^0\text{-AC}^{0,0}$, $\widehat{\text{E-PA}}^\omega{\upharpoonright} + \text{AC}^\omega$) if and only if P is normally reducible to the k-copies of Q in $\widehat{\text{E-HA}}^\omega{\upharpoonright}$ (resp. $\widehat{\text{E-HA}}^\omega{\upharpoonright} + \text{QF-AC}^{0,0} + \Sigma_1^0\text{-DNS}^0$, $\widehat{\text{E-HA}}^\omega{\upharpoonright} + \Pi_1^0\text{-AC}^{0,0} + \Sigma_2^0\text{-DNS}^0$, $\widehat{\text{E-HA}}^\omega{\upharpoonright} + \text{AC}^\omega + \text{mr-DNS}^\omega$).*

Applying Theorem 17.(2) for $k = 2$ to the proof of deriving WKL^c from IVT in [1, Theorem 3], one can obtain a non-trivial result in computable analysis:

Proposition 18. WKL^c *is Kleene-primitive-recursive Weihrauch reducible to the 2-copies of* IVT *in* $\widehat{\text{E-PA}}^\omega{\upharpoonright} + \text{QF-AC}^{0,0}$.

Remark 19. The notion of normal derivability does not imply the notion of normal reducibility to k-copies (verifiably even in a nearly intuitionistic system containing a choice principle) for any natural number k: if so, by Corollary 15, we have that WKL is normally reducible to the k-copies of IVT, and hence, WKL is Kleene-primitive-recursive Weihrauch reducible to the k-copies of IVT by Theorem 17. However, this is not the case because any computable instance of IVT has a computable solution while WKL is not so.

Acknowledgement. The author thanks Vasco Brattka and Kenshi Miyabe for their suggestions which motivate this work and also for their valuable comments. He also thanks one of the anonymous referees, who provides the author with valuable information on the literature. This work is supported by JSPS KAKENHI Grant Numbers JP18K13450 and JP19J01239.

References

1. Berger, J., Ishihara, H., Kihara, T., Nemoto, T.: The binary expansion and the intermediate value theorem in constructive reverse mathematics. Arch. Math. Logic **58**, 203–217 (2018). https://doi.org/10.1007/s00153-018-0627-2
2. Berger, J., Ishihara, H., Schuster, P.: The weak König lemma, Brouwer's fan theorem, de Morgan's law, and dependent choice. Rep. Math. Logic **47**, 63 (2012)
3. Bishop, E.: Foundations of Constructive Analysis. McGraw-Hill Book Co., New York (1967)
4. Brattka, V., Gherardi, G.: Effective choice and boundedness principles in computable analysis. Bull. Symb. Logic **17**(1), 73–117 (2011)
5. Brattka, V., Gherardi, G.: Weihrauch degrees, omniscience principles and weak computability. J. Symb. Logic **76**(1), 143–176 (2011)
6. Brattka, V., Le Roux, S., Miller, J.S., Pauly, A.: Connected choice and the Brouwer fixed point theorem. J. Math. Logic **19**(01), 1950004 (2019)
7. Brattka, V., Pauly, A.: On the algebraic structure of Weihrauch degrees. Log. Methods Comput. Sci. **14**(4), 1–36 (2018)
8. Dorais, F.G., Dzhafarov, D.D., Hirst, J.L., Mileti, J.R., Shafer, P.: On uniform relationships between combinatorial problems. Trans. Am. Math. Soc. **368**(2), 1321–1359 (2016)
9. Fujiwara, M.: Bar induction and restricted classical logic. In: Iemhoff, R., Moortgat, M., de Queiroz, R. (eds.) WoLLIC 2019. LNCS, vol. 11541, pp. 236–247. Springer, Heidelberg (2019). https://doi.org/10.1007/978-3-662-59533-6_15
10. Fujiwara, M.: Weihrauch and constructive reducibility between existence statements. Computability (2020, to appear)
11. Hirst, J.L.: Representations of reals in reverse mathematics. Bull. Pol. Acad. Sci. Math. **55**(4), 303–316 (2007)
12. Hirst, J.L., Mummert, C.: Reverse mathematics and uniformity in proofs without excluded middle. Notre Dame J. Form. Log. **52**(2), 149–162 (2011)
13. Hirst, J.L., Mummert, C.: Using Ramsey's theorem once. Arch. Math. Logic **58**(7), 857–866 (2019). https://doi.org/10.1007/s00153-019-00664-z
14. Ishihara, H.: Constructive reverse mathematics: compactness properties. In: From Sets and Types to Topology and Analysis. Oxford Logic Guides, vol. 48, pp. 245–267. Oxford University Press, Oxford (2005)
15. Kihara, T., Pauly, A.: Finite choice, convex choice and sorting. In: Gopal, T.V., Watada, J. (eds.) TAMC 2019. LNCS, vol. 11436, pp. 378–393. Springer, Cham (2019). https://doi.org/10.1007/978-3-030-14812-6_23
16. Kohlenbach, U.: Higher order reverse mathematics. In: Reverse Mathematics 2001, pp. 281–295. Lecture Notes in Logic, Cambridge University Press (2005)
17. Kohlenbach, U.: Applied Proof Theory: Proof Interpretations and Their Use in Mathematics. Springer Monographs in Mathematics. Springer, Berlin (2008). https://doi.org/10.1007/978-3-540-77533-1
18. Kuyper, R.: On Weihrauch reducibility and intuitionistic reverse mathematics. J. Symb. Log. **82**(4), 1438–1458 (2017)
19. Troelstra, A.S., van Dalen, D.: Constructivism in Mathematics, An Introduction, Vol. I, Studies in Logic and the Foundations of Mathematics, vol. 121. North Holland, Amsterdam (1988)
20. Uftring, P.: The characterization of Weihrauch reducibility in systems containing E-PA$^\omega$ + QF-AC^{0,0}. Preprint arXiv:2003.13331 (2020)

Liouville Numbers and the Computational Complexity of Changing Bases

Sune Kristian Jakobsen and Jakob Grue Simonsen$^{(\boxtimes)}$

Department of Computer Science, University of Copenhagen (DIKU),
Universitetsparken 5, 2100 Copenhagen Ø, Denmark
simonsen@diku.dk

Abstract. We study the computational complexity of uniformly converting the base-a expansion of an irrational numbers to the base-b expansion. In particular, we are interested in subsets of the irrationals where such conversion can be performed with little overhead. We show that such conversion is possible, essentially with polynomial overhead, for the set of irrationals that are not Liouville numbers. Furthermore, it is known that there are irrational numbers x such that the expansion of x in one integer base is efficiently computable, but the expansion of x in certain other integer bases is not. We prove that any such number must be a Liouville number.

Keywords: Computability · Computational complexity · Diophantine approximation · Number theory

1 Introduction

Let $a, b \geq 1$ be integers, and let $\mathbf{prim}(a)$ and $\mathbf{prim}(b)$ be the sets of prime factors of a and b. If $\mathbf{prim}(b) \subseteq \mathbf{prim}(a)$ there is an easily computable constant k that only depends on the exponents in the prime decompositions of a and b such that for any irrational number α with $0 < \alpha < 1$ and any $n \in \mathbb{N}$, the first n digits of the base-b expansion of α can be obtained efficiently knowing only the first kn digits of the base-a expansion of α.

However, if $\mathbf{prim}(b) \nsubseteq \mathbf{prim}(a)$, there are irrational numbers whose base-a expansion can be computed efficiently (say, in polynomial time), but whose base-b expansion cannot. Early partial results go back to Specker [21] and Mostowski [16], while Lachlan [13] showed that the set of primitive recursive reals to base b are a subset of the primitive recursive reals to base a iff $\mathbf{prim}(b) \subseteq \mathbf{prim}(a)$, a result recently extended by Kristiansen [12] to show that for each sufficiently large subrecursive class $\mathcal{S}$, there are irrationals with base-a expansion computable in polynomial time, but whose base-b expansion is not $\mathcal{S}$-computable.

The above phenomena all concern *non-uniform* complexity in the sense that the complexity of expansions of single numbers are concerned. One can also study a *uniform* version where the complexity of Turing machines provided with a, b, and the base-a expansion of *any* irrational number from a well-behaved set

© Springer Nature Switzerland AG 2020
M. Anselmo et al. (Eds.): CiE 2020, LNCS 12098, pp. 50–62, 2020.
https://doi.org/10.1007/978-3-030-51466-2_5

must produce the digits of the base-b expansion without unbounded search. If good subrecursive bounds on such uniform conversion exists for some subset $\mathcal{T}$ of the irrational numbers, it follows that numbers exhibiting the "wild" behaviour of [12,13,16,21] cannot be elements of $\mathcal{T}$.

This paper is devoted to proving that set $\mathbb{R} \setminus (\mathbb{Q} \cup \mathcal{L})$ of irrational numbers that are *not* Liouville numbers is an example of such a set $\mathcal{T}$—and thus that any "wild" number must be Liouville. As almost all "naturally occurring" irrational numbers (algebraic numbers, π, e, numbers with very slow-growing partial quotients, etc.) are not Liouville, this shows that "wild" differences in the computational complexity across integer bases is a somewhat artificial property of irrationals. We believe that some of the machinery we introduce to prove this result may be useful to reveal the connections between the computational complexity of expansions to integer bases and traditional number theory beyond what is already done in the literature (see, e.g. [3,4,8]).

1.1 Some Intuition

Converting a number from base a to base b is easy if the number of digits that need to be examined is limited—colloquially, if the "lookahead" is small. Consider a Turing machine converting an irrational number from base a to base b using the standard schoolbook algorithm. Write $\langle x \rangle_a$ and $\langle x \rangle_b$ for the base-a and base-b expansions of the irrational number x. To find the initial n digits $v_1, \ldots, v_n \in \{0, \ldots, b-1\}$ of $\langle x \rangle_b$ amounts to finding a particular integer $k = \sum_{i=1}^{n} v_i b^{n-i}$ such that $kb^{-n} < x < (k+1)b^{-n}$; the standard schoolbook method of doing so is to write the rational number kb^{-n} in base a and compare the result to successive digits of the base-a expansion of x until a digit is found where the two sequences differ sufficiently. The number of digits that the machine needs to consider to find the nth digit of $\langle x \rangle_b$ is bounded above by some integer $s(n)$ where $|x - kb^{-n}| \geq a^{-s(n)}$—because, roughly, if the base-a expansions of x and kb^{-n} did not differ in the $s(n)$th digit, we would have $|x - kb^{-n}| < a^{-s(n)}$. Hence, for efficient conversion from base a to base b, it is natural to consider real numbers x where the "lookahead" function $s(n)$ does not grow too rapidly. Furthermore, as $a^{-s(n)} = b^{-(\log a / \log b)s(n)}$, we can rewrite the above inequality as $|x - kb^{-n}| \geq b^{-(\log a / \log b)s(n)}$, and it is thus natural to consider a subset of real numbers where the lookahead is not contingent on a, but efficient conversion *to* base b *from* any integer base is possible, in which case the criterion above naturally becomes $|x - kb^{-n}| \geq b^{-t(n)}$ where the function $t(n)$ should be independent of a (we call this criterion (b,t)-*sanity*, see Sect. 3).

The reader should by now appreciate that the "lookahead" is a special case of the more general phenomenon of the convergence speed of rapidly converging sequences of rational approximations to x, that is, finding rational numbers p/q with $|x - p/q| < g(q)$ where g is a rapidly decreasing function. Indeed, for base-a expansions q will always be a negative power of a, and we will always have $p/q < x$ if p/q is a truncation of the base-a expansion of x. Thus, the study of finite truncations of base-a expansions is a limited special case of *Diophantine approximation* (see, e.g. [7]). Consider creating an irrational number x such that its base-a expansion is not efficiently computable; a typical first attempt would be to take a very rapidly growing function $f : \mathbb{N} \longrightarrow \mathbb{N}$ such that $f(n)$ is known to *not* be computable within appropriate bounds, and consider the irrational number $x = \sum_{i=1}^{\infty} a^{-f(i)}$; this is essentially the same approach that Liouville used when defining Liouville's constant $\sum_{i=1}^{\infty} 10^{-i!}$ as the first irrational number explicitly proven to be transcendent [14]. It is thus not surprising that there should be strong connections between the wider class of Liouville numbers and the problem of converting between integer bases.

2   Preliminaries

We assume familiary with standard computability theory and basic complexity theory at the level of introductory textbooks (see, e.g. [2,9,20]). Familiarity with basic computable analysis (e.g., [11,22]) will make the paper easier to read, but is not needed.

Notation. We write $\mathbb{N}$ for the set of positive integers and $\mathbb{Q}$ for the set of rationals. If $f : \mathbb{N} \longrightarrow \mathbb{N}$ and $j \geq 1$ is an integer, we write $f^{\circ j}$ for the jth iterate of f, that is, $f^{\circ 1}(n) = f(n)$ and $f^{\circ j}(n) = f(f^{\circ(j-1)}(n))$ for all $n \in \mathbb{N}$. We write $\mathbf{poly}(n)$ for an unspecified polynomial in n. For $a \in \mathbb{N}$, we write $\Sigma_a = \{0, \ldots, a-1\}$. We usually view Σ_a as an alphabet of a symbols and denote by Σ^* the set of finite, possibly empty, strings over Σ, by Σ^+ the set of finite non-empty string over Σ, by Σ^{ω} the set of right-infinite strings over Σ, and set $\Sigma^{\leq \omega} = \Sigma^* \cup \Sigma^{\omega}$. The binary representation of a is denoted by $\mathbf{a}_{\mathrm{bin}}$. The open interval of all reals between 0 and 1 is denoted by $(0,1)$, and if $x \in (0,1)$ is a real number and $a > 1$ is an integer, we denote by $\langle x \rangle_a$ the greedy base-a expansion of x, that is $\langle x \rangle_a = (e_n)_{n \in \mathbb{N}}$ where $x = \sum_{i=1}^{\infty} e_i/a^i$ such that each $e_i \in \{0, \ldots, a-1\}$ and each successive e_i is chosen as large as possible. We write $\langle x \rangle_a|_{\leq n}$ for the length-n initial prefix of $\langle x \rangle_a$, $\langle x \rangle_a|_n$ for the nth element of $\langle x \rangle_a$, and define $k_{x,a,n} = \sum_{i=1}^{n} e_i a^{n-i}$, i.e. $k_{x,a,n} \cdot a^{-n} = \sum_{i=1}^{n} e_i a^{-i}$, so $k_{x,a,n} a^{-n}$ is the multiple of a^{-n} corresponding to the length-n prefix $\langle x \rangle_a|_{\leq n}$ of the base-a expansion of x.

Turing Machines and Conversion Between Bases. Let $a \geq 2$ be an integer and $f : \mathbb{N} \longrightarrow \{0, \ldots, a-1\}$ be a map. A Turing machine M with input and output alphabet $\{0,1\}$ is said to *compute* f if, for each positive integer n, M will, on input $\mathbf{n}_{\mathrm{bin}}$, output $\mathbf{f(n)}_{\mathrm{bin}}$. We assume that if x is a real number and is given as input to a Turing machine M, then (a binary encoding of the infinite sequence of elements of) $\langle x \rangle_a$ is supplied to M on a particular input tape. All Turing machines considered in this paper will thus be (type-2) Turing machines: Let Σ be a finite alphabet with $0,1 \in \Sigma$. A type-2 Turing machine M is a Turing machine with k read-only input tapes accepting infinite inputs (called ω-tapes), m read-only input tapes accepting finite inputs, one (write-only, one-way) output tape and finitely many additional work tapes. Such a machine computes a partial function $\phi_M : (\Sigma^{\omega})^k \times (\Sigma^+)^m \rightharpoonup \Sigma^*$ in the usual way (i.e., it has to reach a halting state in finite time, and the output is what is present on the output tape at that time). Time and space-complexity will throughout the paper be specified in terms of the content on the input tapes accepting finite inputs (if a Turing machine runs in time at most $T(n)$, then it can examine at most the initial $T(n)$ elements on the ω-tapes). Note also that we will typically give the desired number of output symbols (e.g., the first n elements in $\langle x \rangle_b$) as input in binary $\mathbf{n}_{\mathrm{bin}}$, hence using at most $1 + \lfloor \log n \rfloor$ bits, but time complexities will be specified as functions of n, not $\mathbf{n}_{\mathrm{bin}}$. We assume that both the input ω-tapes and the output tape has binary alphabet. Thus, $\langle x \rangle_a$ will be coded on the input as an infinite sequence of binary representations of elements of $\{0, 1 \ldots, a - 1\}$ (each taking space $1 + \lfloor \log a \rfloor$); similarly, output in base b will be encoded as elements of $\{0, \ldots, b - 1\}$ with each digit using $1 + \lfloor \log b \rfloor$ bits of space.

Liouville Numbers. The *irrationality measure* of a real number x, denoted $\mu(x)$, is the infimum of the set of positive reals μ such that the inequality $|x - p/q| < 1/q^\mu$ has only finitely many distinct solutions $(p, q) \in \mathbb{Z} \times \mathbb{N}$ (conversely, $\mu(x)$ is the supremum of the set of positive reals such that the inequality has infinitely many distinct solutions).

A real number x is a *Liouville number* [14] if it is irrational and for every integer $c \in \mathbb{N}$ there are integers p and q with $q \geq 2$ such that $|x - p/q| < q^{-c}$. We denote the set of Liouville numbers by $\mathcal{L}$. Hence, $\mathcal{L}$ is the set of reals having infinite (recall that $\inf \emptyset = \infty$ by convention) irrationality measure.

An Ancillary Result. The multiple of b^{-n} that best approximates x is either $k_{x,b,n}$ or $k_{x,b,n} + 1$. We state this straightforward result explicitly as we shall refer to it several times:

Proposition 1. *Let $x \in (0, 1)$ be irrational. Then, for all integers $b \geq 2$ and $n \geq 1$, we have $k_{x,b,n} b^{-n} < x < (k_{x,b,n} + 1) b^{-n}$, and either (i) $\forall k \in \mathbb{Z}. |x - k_{x,b,n} b^{-n}| \leq |x - k b^{-n}|$, or (ii) $\forall k \in \mathbb{Z}. |x - (k_{x,b,n} + 1) b^{-n}| \leq |x - k b^{-n}|$.*

3 Rational Approximations and Sanity

The discussion in the paper's introduction prompts the definition of *sane* numbers below.

Definition 1. *Let $x \in (0, 1)$ be a real number, and $b \geq 2$ an integer. Then, x is said to be: (b, t)-sane if there is a non-decreasing and unbounded map $t : \mathbb{N} \longrightarrow \mathbb{N}$ such that for all integers k, n with $n \geq 1$ we have $|x - k \cdot b^{-n}| \geq b^{-t(n)}$. The map t is said to be a witness of (b, t)-sanity of x. Furthermore, x is said to be uniformly sane if there exists a non-decreasing and unbounded map $t : \mathbb{N} \longrightarrow \mathbb{N}$ such that x is (b, t)-sane for all b. Again, t is said to be a witness of uniform sanity of x.*

Sane numbers are irrational: If p/q is rational, then it has a finite base-q expansion, whence it cannot be q-sane witnessed by any unbounded map t.

Observe that for every irrational x and every b, there is a non-decreasing and unbounded function t such that x is (b,t)-sane: for each $n \in \mathbb{N}$, there is a $k \in \mathbb{Z}$ such that $|x - kb^{-n}|$ is minimal. Let k' be such a k, define $d_n = \lceil -(\log |x - k'b^{-n}|)/\log b \rceil$, and define $t(1) = d_1$, and $t(n) = \max\{d_n, t(n-1)\}$ for $n > 1$. Note that this t is unbounded as x is irrational.

If there is no *slow*-growing function t such that x is (b,t)-sane, then–intuitively–bounding x away from very good rational approximations is difficult, and hence converting between bases may require large lookahead. As every irrational x is (b,t_b)-sane for at least one function t_b depending on b (and x), it is natural to consider the function $t' : \mathbb{N} \times \mathbb{N} \longrightarrow \mathbb{N}$ such that $t'(b,n) = t_b(n)$. Having small lookahead in all bases b then intuitively corresponds to the function t' not growing too fast in either of its arguments.

Proposition 2. *Let $x \in (0,1)$ be irrational. If $b \geq 2$ is an integer and $t : \mathbb{N} \longrightarrow \mathbb{N}$ is a non-decreasing and unbounded map such that for all but finitely many integer pairs (k,n) with $n \geq 1$ we have $|x - k \cdot b^{-n}| \geq b^{-t(n)}$, then there exists an e_x (which may depend on x) such that x is $(b, n \mapsto e_x t(n))$-sane.*

To show that the growth rate of the functions witnessing sanity actually matters, we now prove existence of a real number the sanity of which can be witnessed by a fast-growing function, but not witnessed by functions that grow slightly more slowly.

Proposition 3. *Let $b \geq 2$ be an integer. Furthermore, let $t : \mathbb{N} \longrightarrow \mathbb{N}$ be a map such that $t(1) \geq 3$ and $t(n) - t(n-1) \geq n+1$ for all $n > 1$. Then, for any non-decreasing unbounded function $s : \mathbb{N} \longrightarrow \mathbb{N}$ such that $s(n) > t(n)$, the number $x = \sum_{j=1}^{\infty} b^{-t^{\circ j}(1)}$ is (b,s)-sane. However, if $u : \mathbb{N} \longrightarrow \mathbb{N}$ is any function such that $u(n) < t(n)$ for all sufficiently large n, then x is not (b,u)-sane.*

Proof. First observe that x is irrational as $\langle x \rangle_b$ is not finite. Observe also that the requirement $t(n) - t(n-1) \geq n+1$ for all $n > 1$ entails that t is non-decreasing (in fact, strictly increasing) and unbounded. Furthermore, note that we must have $t(t(n)) > t(n+1)$ for all $n \geq 1$, because the fact that t is strictly increasing implies that $t(t(n)) > t(t(n) - t(n-1)) > t(n+1)$. Finally, note that as $t(1) \geq 3$ and $t(n) \geq t(n-1) + n + 1$ by assumption, it follows that $t(n) \geq n+1$ for all $n \in \mathbb{N}$.

For ease of notation, write $z_1 = t(1), z_2 = t(t(1)), \ldots, z_j = t^{\circ j}(1), \ldots$. As t is strictly increasing, $\langle x \rangle_b$ has ones at positions $z_1, z_2, \ldots, z_j, \ldots$, and zeros at all other positions. Observe, for $j \geq 1$, that the number of zeros following the occurrence of 1 at position z_j is at least

$$z_{j+1} - z_j - 1 = t(z_j) - z_j - 1 \geq z_j + 1 + t(z_j - 1) - z_j - 1 = t(z_j - 1) > z_j$$

For each n, let j be the largest integer such that $z_j \leq n < z_{j+1}$ (such a j exists because t is strictly increasing, so $z_j < z_{j+1} < n$ is only possible for finitely many j). Then $\langle x \rangle_b$ has zeros at all positions $z_j + 1, \ldots, z_{j+1} - 1$, and we thus have $|x - k_{x,b,n}b^{-n}| = |x - k_{x,b,z_j}b^{-z_j}|$. But as $\langle x \rangle_b$ also contains a 1 at position z_{j+1}, we have

$$k_{x,n,z_j}b^{-z_j} + b^{-z_{j+1}} = k_{x,n,z_{j+1}}b^{-z_{j+1}} \tag{1}$$

Now, by the above, and by Proposition 1, we have $k_{x,n,z_j}b^{-z_j} < k_{x,n,z_{j+1}}b^{-z_{j+1}} < x$, and thus $|x - k_{x,b,z_j}b^{-z_j}| \geq b^{-z_{j+1}}$. Therefore:

$$|x - k_{x,b,n}b^{-n}| = |x - k_{x,b,z_j}b^{-z_j}| \geq b^{-z_{j+1}} = b^{-t(z_j)} \geq b^{-t(n)} > b^{-s(n)}$$

where the last inequality follows from the fact that t is strictly increasing and $z_j \leq n$. Hence, $|x - k_{x,b,n}b^{-n}| \geq b^{-s(n)}$.

By Proposition 1, we have $0 < x - k_{x,b,z_{j+1}}b^{-z_{j+1}}$, and thus:

$$0 < x - k_{x,b,n}b^{-n} = x - k_{x,n,z_j}b^{-z_j} = x - (k_{x,b,z_{j+1}}b^{-z_{j+1}} - b^{-z_{j+1}}) =$$
$$x - k_{x,b,z_{j+1}}b^{-z_{j+1}} + b^{-z_{j+1}} = x - (k_{x,b,z_{j+2}}b^{-z_{j+2}} - b^{-z_{j+2}}) + b^{-z_{j+1}} =$$
$$x - k_{x,b,z_{j+2}}b^{-z_{j+2}} + b^{-z_{j+2}} + b^{-z_{j+1}} < 2b^{-z_{j+2}} + b^{-z_{j+1}}$$

Now, $z_{j+1} \geq t(t(1)) \geq t(3) \geq 4$, and we thus have:

$$z_{j+2} = t(z_{j+1}) \geq z_{j+1} + 1 + t(z_{j+1} - 1) \geq z_{j+1} + 1 + t(3) \geq z_{j+1} + 5$$

Hence, $b^{-z_{j+2}} \leq b^{-z_{j+1}-5}$, and thus:

$$|x - (k_{x,b,n} + 1)b^{-n})| = b^{-n} - (x - k_{x,b,n}b^{-n}) \geq b^{-n} - (2b^{-z_{j+2}} + b^{-z_{j+1}})$$
$$\geq b^{-z_{j+1}} - 2b^{-z_{j+2}} \qquad \geq b^{-z_{j+1}} - 2b^{-z_{j+1}-5}$$
$$= (1 - 2/b^5)b^{-z_{j+1}} \qquad \geq (15/16)b^{-z_{j+1}}$$
$$\geq b^{-1} \cdot b^{-z_{j+1}} \qquad = b^{-(1+z_{j+1})}$$
$$= b^{-(1+t(z_j))} \qquad \geq b^{-(1+t(n))}$$
$$\geq b^{-s(n)}$$

where the first inequality in the second line above follows from the fact that $n < z_{j+1}$ and thus $b^{-n} \geq b \cdot b^{-z_{j+1}} \geq 2b^{-z_{j+1}}$. By Proposition 1, either $k_{x,b,n}$ or $k_{x,b,n} + 1$ is an integer that minimizes $|x - db^{-n}|$ among all $d \in \mathbb{Z}$. Hence, for all $k \in \mathbb{Z}$, we have $|x - kb^{-n}| \geq b^{-s(n)}$, showing that x is (b, s)-sane, as desired.

Now, pick any $j \geq 1$ and set $n = z_j$. Then by (1) above we have:

$$|x - k_{x,b,n}b^{-n}| = x - (k_{x,b,z_{j+1}}b^{-z_{j+1}} - b^{-z_{j+1}})$$
$$= |x - k_{x,b,z_{j+1}}b^{-z_{j+1}}| + b^{-z_{j+1}}$$
$$< 2b^{-z_{j+1}} = 2b^{-t(z_j)} = 2b^{-t(n)}$$
$$\leq b^{-t(n)+1}$$

But by assumption we have $u(n) < t(n)$ for all sufficiently large n, whence $b^{-t(n)+1} \leq b^{-u(n)}$, and thus $|x - k_{x,b,n}b^{-n}| < b^{-u(n)}$ for infinitely many n (because $n = z_j$, and $j \in \mathbb{N}$ was chosen arbitrary). Hence, x is not (b, u)-sane. $\square$

Functions t satisfying the assumptions of Proposition 3 are not hard to devise. For example, any polynomial $n \mapsto 2 + n^q$ (for $q \geq 2$) satisfies the requirements.

Proposition 3 shows that, for each b, there is a hierarchy of (b, t)-sane numbers for successively faster-growing functions t, and there are numbers that require arbitrarily fast-growing witnesses for (b, t)-sanity. It is natural to conjecture that the same phenomena hold for uniform sanity, but–surprisingly–it turns out (see Lemma 2) not to be the case.

We have the following key lemma:

Lemma 1. *A number $x \in (0, 1)$ is uniformly sane iff it is irrational and is not a Liouville number.*

Proof. Proceed as follows:

- Let x be uniformly sane witnessed by the map t. Then, for all integers b, k, n with $b \geq 2$ and $n \geq 1$, we have $|x - kb^{-n}| \geq b^{-t(n)}$, in particular $|x - k/b| \geq b^{-t(1)}$. By the comments after Definition 1, x is irrational. Assume, for contradiction, that $x \in \mathcal{L}$. As $t(1)$ is a positive integer, there are integers b, k with $b \geq 2$ such that $|x - k/b| < b^{-t(1)}$, and we obtain the contradiction. Hence, $x \notin \mathcal{L}$.
- Let $x \notin \mathcal{L} \cup \mathbb{Q}$. Assume, for contradiction, that x is not uniformly sane. Then for each $c \in \mathbb{N}$, the map $n \mapsto cn$ does not witness uniform sanity, whence there are integers b_c, k_c, n_c with $b_c \geq 2$ and $n_c \geq 1$ such that $|x - k_c b_c^{-n_c}| < b_c^{-cn_c}$. Setting $q = b_c^{n_c}$ and $p = k_c$ we obtain $|x - p/q| < q^{-c}$. As $c \geq 1$ was arbitrary, $x \in \mathcal{L}$, a contradiction. Hence, x is uniformly sane.

$\square$

Let $\mathcal{E}^2$ be the set of total functions on the naturals in the second level of the Grzegorczyk hierarchy. A real number x is said to be $\mathcal{E}^2$-irrational if there is $f \in \mathcal{E}^2$ such that for all integers p, q with $q > 0$ we have $|x - p/q| > 1/f(q)$. By a result of Georgiev, a real number is $\mathcal{E}^2$-irrational iff it is irrational and not a Liouville number [10]; thus, by Lemma 1, the set of uniformly sane numbers is exactly the set of $\mathcal{E}^2$-irrational numbers.

Lemma 1 has the surprising consequence that fast-growing functions $t : \mathbb{N} \longrightarrow \mathbb{N}$ are *never* needed as witnesses for uniform sanity—uniform sanity can always be witnessed by a linear map:

Lemma 2. *A number $x \in (0, 1)$ is uniformly sane iff uniform sanity can be witnessed by a map of the form $t(n) = cn$, for some $c \in \mathbb{N}$ such that $c \geq \mu(x)$.*

Furthermore, no function $t(n) = cn$ with $c < \mu(x)$ witnesses uniform sanity of x.

Proof. If uniform sanity is witnessed by a linear map, then obviously x is uniformly sane.

Conversely, assume that x is uniformly sane and, for contradiction, that there is no map t of the form $t(n) = cn$ that witnesses this. Then, for every $c \in \mathbb{N}$ there are integers k, b, m with $b \geq 2$ and $m \geq 1$ such that $|x - kb^{-m}| < b^{-cm} = (b^m)^{-c}$.

Set $p = k$ and $q = b^m$; thus, for each integer $c \geq 1$, there are p, q with $q \geq 2$ such that $|x - p/q| < q^{-c}$, whence $x \in \mathcal{L}$, contradicting Lemma 1; hence, some map of the form $t(n) = cn$ witnesses uniform sanity of x. Further assume, for contradiction, that $c < \mu(x)$. Then, by uniform sanity, we have for all integers b, k, n with $b \geq 2$ and $n \geq 1$ that $|x - kb^{-n}| \geq b^{-cn}$. Setting $n = 1$, we thus have for any real number d with $c \leq d \leq \mu(x)$ and any integers p, q with $q \geq 2$ that $|x - p/q| \geq q^{-c} \geq q^{-d} \geq q^{-\mu(x)}$. Thus, the inequality $|x - p/q| < q^{-d}$ has only finitely many solutions in integers p, q (namely the case $q = 1$ where $|x - p| < 1^{-c} = 1$ might have the solutions $p = 0$ or $p = 1$ as $x \in (0, 1)$). Thus, as there are infinitely many d with $c \leq d < \mu(x)$, the number $\mu(x)$ cannot be the supremum of the set of numbers d such that $|x - p/q| < q^{-d}$ has infinitely many solutions in integers p, q, a contradiction.   □

We do not know whether it is always possible to choose $n \mapsto \lceil \mu(x) \rceil n$ as a witness of uniform sanity.

Lemma 1 furnishes a method for proving that concrete real numbers are (uniformly) sane: the set of Liouville numbers is exactly the set of real numbers having irrationality measure infinity. Thus: if a real number has finite irrationality measure it is uniformly sane. By the Thue-Siegel-Roth theorem [18], all algebraic irrational numbers have irrationality measure 2 and are thus uniformly sane, as are numbers with continued fractions whose partial quotients grow very slowly as $o(n)$, for example e. Further examples of specific uniformly sane numbers can be found where finite upper bounds on their irrationality measure have been proven. For example, π (the first bound by Mahler, $\mu(\pi) \leq 30$ [15], has been improved on many occasions; at the time of writing, the best known bound is $\mu(\pi) \leq 7.60630852 \cdots$ [19]), and Apéry's constant [1]. Similarly, all Martin-Löf random reals are not Liouville [8], hence are uniformly sane. Lemma 1 also implies that almost all real numbers are uniformly sane: By standard results, the set of Liouville numbers has Lebesgue measure zero [17] (and has Hausdorff dimension zero, hence d-dimensional Hausdorff measure zero for all positive integers $d > 0$ [17]). Likewise, the set of uniformly sane numbers is a G_δ-set, hence co-meagre.

3.1   A Digression: Normal Numbers

Recall that a real number x is b-*normal* [5] if every string of symbols $s \in \{0, \ldots, b-1\}^+$ occurs in $\langle x \rangle_b$ with limiting frequency $b^{-|s|}$, and b-*simply normal* if every element of $\{0, \ldots, b-1\}$ occurs in $\langle x \rangle_b$ with limiting frequency $1/b$. Clearly, a b-normal number is b-simply normal. We have:

Proposition 4. *Let $x \in (0, 1)$ be irrational and let $b \geq 2$ be an integer. If x is b-simply normal, then there is $c_x \in \mathbb{N}$ such that x is $(b, n \mapsto c_x n)$-sane.*

Recall also that x is said to be *absolutely normal* if it is b-normal for every $b \geq 2$. Proposition 4 yields the following corollary:

Corollary 1. *For any absolutely normal number $x \in (0, 1)$ and any integer $b \geq 2$, there is $c_{b,x} \in \mathbb{N}$ such that x is $(b, n \mapsto c_{b,x} n)$-sane.*

By a result of Bugeaud [6] there are uncountably many absolutely normal Liouville numbers, hence, by Lemma 1 there are uncountably many absolutely normal numbers that are not uniformly sane. By Corollary 1 each of these normal numbers are $(b, n \mapsto c_{b,x}n)$-sane for each b with some constant $c_{b,x}$ dependent on b and x. However, for each absolutely normal Liouville number, the sequence $c_{2,x}, c_{3,x}, \ldots$ must grow unboundedly as otherwise the numbers would be uniformly sane. Examples of computable absolutely normal Liouville numbers can be found in [4]. Another consequence of Proposition 4 is that every b-normal number has a very tame witness for sanity—and that numbers *requiring* fast-growing witnesses (such as the ones constructed in Proposition 3) cannot be b-normal for any b.

4   Uniform Conversion with Subrecursive Overhead Between Arbitrary Integer Bases

The following theorem shows that sanity implies that changing bases can be done without using unbounded search, indeed polynomial-time overhead is sufficient:

Theorem 1. *There is a (type-2) Turing machine M and a polynomial P with positive integer coefficients satisfying the following:*

For any integers $a, b \geq 2$, any non-decreasing and unbounded $t : \mathbb{N} \longrightarrow \mathbb{N}$, any (b, t)-sane number $x \in (0, 1)$, and any $n \in \mathbb{N}$, M will on input a_{bin}, b_{bin}, n_{bin}, and $\langle x \rangle_a$ (on an ω-tape) output $\langle x \rangle_b|_{\leq n}$ in time $T(a, b, x)(n) \leq P(t(n) \log(\max\{a, b\}))$.

Proof. The proof is essentially just an application of the schoolbook algorithm for changing the base of an irrational number. We first describe M and subsequently bound its running time.

(Start of Description of M)
M works in n stages, with each stage outputting the next digit of $\langle x \rangle_b$ until the entire sequence $\langle x \rangle_b|_{\leq n}$ has been output. For $1 \leq i \leq n$, at the beginning of the ith stage, M has on its work tapes the (binary representations of) (i) i, (ii) the string $\langle x \rangle_b|_{\leq i-1} = s_1 \cdots s_{i-1}$ of the first $i-1$ digits of $\langle x \rangle_b$, and (iii) the number $z_{i-1} = s_1 b^{-1} + \cdots s_{i-1} b^{-(i-1)} = k_{x,b,i-1} b^{-(i-1)}$ in base b. Initially, $i = 0$, and $\langle x \rangle_b|_{\leq 0} = \epsilon$, and $z_0 = 0$.

In stage i, M finds the ith digit, $s_i = \langle x \rangle_b|_i$, as follows: M uses binary search in $\{0, \ldots, b-1\}$ to find the largest $s \in \{0, \ldots, b-1\}$ such that

$$z_{i-1} + sb^{-i} = s_1 b^{-1} + \cdots + s_{i-1} b^{-(i-1)} + sb^{-i} < x$$

By definition, the largest such s is $s_i = \langle x \rangle_b|_i$, and M increments i and sets $\langle x \rangle_b|_{\leq i} = \langle x \rangle_b|_{\leq i-1} s$ and $z_i = z_{i-1} + sb^{-i}$.

For each s, checking whether $z_{i-1} + sb^{-i} < x$ is done by inspecting sufficiently many digits of $\langle x \rangle_a$. For ease of notation, define $r = \lceil (\log b)/(\log a) \rceil t(i)$. As x

is (b, t)-sane, we have $|x - kb^{-i}| \geq b^{-t(i)} = a^{-\frac{t(i) \log b}{\log a}}$ for all integers k, so in particular for the number

$$z_i + sb^{-i} = k_{x,b,i-1}b^{-(i-1)} + sb^{-i} = (bk_{x,b,i-1} + s)b^{-i}$$

we have:

$$|x - (z_i + sb^{-i})| = |x - (bk_{x,b,i-1} + s)b^{-i}| \geq a^{-\frac{t(i) \log b}{\log a}} \geq a^{-r}$$

By Proposition 1 we have:

$$x - a^{-r} < k_{x,a,r}a^{-r} < x < (k_{x,a,r} + 1)a^{-r} < x + a^{-r}$$

and thus we have *either*

$$z_i + sb^{-i} < k_{x,a,r}a^{-r} \tag{2}$$

or

$$z_i + sb^{-i} > (k_{x,a,r} + 1)a^{-r} \tag{3}$$

Clearly, if (2) holds, $s_i = \langle x \rangle_b|_i \geq s$, and if (3) holds, $s_i = \langle x \rangle_b|_i < s$. Thus, M needs only scan the initial r base-a elements of $\langle x \rangle_a$ (and compute the rational numbers above) to compute $\langle x \rangle_b|_i$; each of these elements are representable in at most $(1 + \lfloor a \rfloor)$ bits, hence can be read in time $O(r \log a) = \mathbf{poly}(t(i) \log(\max\{a, b\}))$. Observe that M does not need to know t or even compute r: it can simply compute $z_o + sb^{-i}$ and then brute-force scan enough digits of $\langle x \rangle_a$ until either (2) or (3) holds.

(End of Description of M)
By the above it is clear that M does not use unbounded search to output $\langle x \rangle_b|_{\leq n}$, and indeed the search in $\langle x \rangle_a$ is limited to the r initial symbols of it, whence subrecursive conversion is obviously possible. We now show that the conversion is indeed efficient by establishing the existence of the polynomial P.

Time Use of M: In stage i, all computations are performed on integer arguments smaller than $\max\{b^i, a^r\}$, or on rational numbers p/q where $p, q \leq \max\{b^i, a^r\}$. All exponentiation involves only computing powers of a and b, respectively, and all exponents involve negative powers of magnitude at most $\max\{i, r\}$; by repeated squaring each power can be computed in time at most:

$$\mathbf{poly}(\max\{i, r\} \log(\max\{a, b\})) = \mathbf{poly}(t(i) \log(\max\{a, b\}))$$

Apart from squaring, all multiplications and divisions involve at most two numbers, and by schoolbook arithmetic are thus computable in time polynomial in the logarithm of the largest integer involved, hence in time at most:

$$\mathbf{poly}(\log(\max\{b^i, a^r\})) = \mathbf{poly}(t(i) \log(\max\{a, b\}))$$

The remaining arithmetical operations are sums of at most $\max\{i, r\}$ rational numbers with numerators and denominators that are positive integers bounded above by $\max\{b^i, r\}$. By schoolbook arithmetic, this can be done in time at most:

$$\max\{i, r\} \cdot \mathbf{poly}(\log(\max\{b^i, a^r\})) = \mathbf{poly}(t(i) \log(\max\{a, b\}))$$

As there are n stages $0, \ldots, i, \ldots, n$, and each stage–by the above analysis–takes time at most $\mathbf{poly}(t(i) \log(\max\{a, b\}))$, the total time use of M to print $\langle x \rangle_b|_{\leq n}$ is at most $O(n\mathbf{poly}(t(n) \log(\max\{a, b\}))) = \mathbf{poly}(t(n) \log(\max\{a, b\}))$, as desired.
□

Observe in Theorem 1 that the polynomial P is independent of x, a, and b. *However* $t(n)$ will in general depend on both b and x. If we consider a class of reals for which $t(n)$ is bounded above by a slow-growing function, we can obtain stronger results, to wit the following theorem for numbers that are not Liouville:

Theorem 2. *There is a Turing machine M and a polynomial P with the following property: For any $x \in (0, 1) \setminus (\mathbb{Q} \cup \mathcal{L})$, there exists $m_x \in \mathbb{N}$ such that for all integers $a, b \geq 2$ and $n \in \mathbb{N}$, M will on input a_{bin}, b_{bin}, n_{bin}, and $\langle x \rangle_a$ (on an ω-tape) output $\langle x \rangle_b|_{\leq n}$ in time $T(a, b, x)(n) \leq m_x \cdot P(n \log(\max\{a, b\}))$.*

Proof. Let M and P be the Turing machine and polynomial of Theorem 1. By Lemma 1, every $x \in (0, 1) \setminus (\mathbb{Q} \cup \mathcal{L})$ is uniformly sane, and by Lemma 2, uniform sanity is witnessed by some function $t(n) = c_x n$ for some $c_x \in \mathbb{N}$ that depends on x. By Theorem 1, M outputs $\langle x \rangle_b|_{\leq n}$ in time at most $P(c_x n \log(\max\{a, b\})) \leq c_x^d \cdot P(n \log(\max\{a, b\}))$ (for some $d \in \mathbb{N}$). Setting $m_x = c_x^d$ now furnishes the result.
□

Inspection of the proof of Theorem 1 reveals that the Turing machine M reads at most $r = \lceil (\log b)/(\log a) \rceil t(n)$ base-a symbols of $\langle x \rangle_a$, and thus for a non-Liouville number x at most $\lceil (\log b)/(\log a) \rceil c_x n$ base-a symbols for some $c_x \geq \mu(x)$. If we fix x, we can absorb the constant m_x in Theorem 2 into the polynomial characterising the running time, and we obtain the following:

Corollary 2. *Let $x \in (0, 1) \setminus (\mathbb{Q} \cup \mathcal{L})$. There is a polynomial R such that if $a, b \geq 2$, and $\langle x \rangle_a|_{\leq n}$ is computable in time at most $T(n)$ for all n, then there is a constant c_x such that $\langle x \rangle_b|_{\leq n}$ is computable in time at most $R(T(c_x n))$ for all n.*

Thus in particular, every real number x where finite prefixes of $\langle x \rangle_a$ *can* be computed in polynomial time in n, but the finite prefixes of $\langle x \rangle_b$ *cannot*, must be a Liouville number.

5   Future Work

This paper has concerned the connection between Liouville numbers and the construction of real numbers with great disparity in the computational complexity needed to compute their expansion in different bases, and more generally the computational complexity needed to obtain "good" rational approximations to irrationals—indeed, the notion of expansion to integer bases is an example of a very particular kind of approximation, but we expect the results of this paper to hold, *mutatis mutandis* for other approximations with rationals as well. Furthermore, the set of Liouville numbers is almost certainly an *over*-approximation of the set of "poorly behaved" irrationals where conversion with polynomial overhead is not in general possible; it is interesting to pinpoint a proper subset of the Liouville numbers—hopefully connected to existing areas of number theory—that precisely contain those numbers having egregious differences in the complexity of their various integer base expansions.

Acknowledgments. We are grateful to Siddharth Bhaskar and the referees for useful feedback, and to one referee in particular for pointing out the connection between uniform sanity and $\mathcal{E}^2$-irrationality.

References

1. Apéry, R.: Irrationalité de $\zeta(2)$ et $\zeta(3)$. Astérisque **61**, 11–13 (1979)
2. Arora, S., Barak, B.: Computational Complexity: A Modern Approach. Cambridge University Press, Cambridge (2009)
3. Becher, V., Bugeaud, Y., Slaman, T.: The irrationality exponents of computable numbers. Proc. Am. Math. Soc. **144**, 1509–1521 (2014)
4. Becher, V., Heiber, P.A., Slaman, T.A.: A computable absolutely normal Liouville number. Math. Comput. **84**(296), 2939–2952 (2015)
5. Borel, É.: Les probabilités dénombrables et leurs applications arithmétiques. Supplemento di Rendiconti del circolo matematico di Palermo **27**, 247–271 (1909)
6. Bugeaud, Y.: Nombres de Liouville et nombres normaux. Comptes Rendus de l'Académie des Sciences Paris **335**(2), 117–120 (2002)
7. Bugeaud, Y.: Distribution Modulo One and Diophantine Approximation. Cambridge Tracts in Mathematics, vol. 193. Cambridge University Press, Cambridge (2012)
8. Calude, C.S., Staiger, L.: Liouville, computable, Borel normal and Martin-Löf random numbers. Theory Comput. Syst. **62**(7), 1573–1585 (2018)
9. Du, D.Z., Ko, K.I.: Theory of Computational Complexity. Wiley Interscience, Hoboken (2000)
10. Georgiev, I.: Continued fractions of primitive recursive real numbers. Math. Logic Q. **61**(4–5), 288–306 (2015)
11. Ko, K.I.: Complexity Theory of Real Functions. Progress in Theoretical Computer Science. Birkhäuser, Boston (1991). https://doi.org/10.1007/978-1-4684-6802-1
12. Kristiansen, L.: On subrecursive representability of irrational numbers, part II. Computability **8**, 43–65 (2018)
13. Lachlan, A.: Recursive real numbers. J. Symb. Logic **28**, 1–16 (1963)
14. Liouville, J.: Remarques relatives à des classes très-étendues de quantités dont la valeur n'est ni algébriques, ni même réductible à des irrationelles algébriques. Comptes rendus de l'académie des Sciences de Paris **18**, 883–885 (1844)
15. Mahler, K.: On the approximation of π. Nederl. Akad. Wetensc. Proc. Ser. A **56**, 30–42 (1953)
16. Mostowski, A.: On computable sequences. Fund. Math. **44**(1), 37–51 (1957)
17. Oxtoby, J.: Measure and Category. Graduate Texts in Mathematics, vol. 2. Springer, New York (1980). https://doi.org/10.1007/978-1-4615-9964-7
18. Roth, K.: Rational approximations to algebraic numbers. Mathematika **2**, 1–20 (1955)
19. Salikhov, V.: On the measure of irrationality of the number π. Math. Notes **88**(3–4), 563–573 (2010)

20. Sipser, M.: Introduction to the Theory of Computation. PWS Publishing Company, New Orleans (1997)
21. Specker, E.: Nicht konstruktiv beweisbare Sätze der Analysis. J. Symb. Log. **14**(3), 145–158 (1949)
22. Weihrauch, K.: Computable Analysis: An Introduction, 1st edn. Springer, Heidelberg (2013)

On Interpretability Between Some Weak Essentially Undecidable Theories

Lars Kristiansen[1,2(✉)] and Juvenal Murwanashyaka[1]

[1] Department of Mathematics, University of Oslo, Oslo, Norway
{larsk,juvenalm}@math.uio.no
[2] Department of Informatics, University of Oslo, Oslo, Norway

Abstract. We introduce two essentially undecidable first-order theories WT and T. The intended model for the theories is a term model. We prove that WT is mutually interpretable with Robinson's R. Moreover, we prove that Robinson's Q is interpretable in T.

1 Introduction

A first-order theory T is *undecidable* if there is no algorithm for deciding if $T \vdash \phi$. If every consistent extension of an undecidable theory T also is undecidable, then T is *essentially undecidable*.

We introduce two first-order theories, WT and T, over the language $\mathcal{L}_T = \{\bot, \langle \cdot, \cdot \rangle, \sqsubseteq\}$ where $\bot$ is a constant symbol, $\langle \cdot, \cdot \rangle$ is a binary function symbol and $\sqsubseteq$ is a binary relation symbol. The intended model for these theories is a term model: The universe is the set of all variable-free $\mathcal{L}_T$-terms. Each term is interpreted as itself, and $\sqsubseteq$ is interpreted as the subterm relation (s is a subterm of t iff $s = t$ or $t = \langle t_1, t_2 \rangle$ and s is a subterm of t_1 or t_2).

The non-logical axioms of WT are given by the two axiom schemes:

(WT$_1$) $$s \neq t$$

where s and t are distinct variable-free terms.

(WT$_2$) $$\forall x [\, x \sqsubseteq t \;\leftrightarrow\; \bigvee_{s \in \mathcal{S}(t)} x = s \,]$$

where t is a variable-free term and $\mathcal{S}(t)$ is the set of all subterms of t. There are no other non-logical axioms except those given by these two simple schemes, and at a first glance WT seems to be a very weak theory. Still it turns out that Robinson's essentially undecidable theory R is interpretable in WT, and thus it follows that also WT is essentially undecidable. The theory T is given by the four axioms:

T$_1$ $\forall xy [\, \langle x, y \rangle \neq \bot \,]$
T$_2$ $\forall x_1 x_2 y_1 y_2 [\, \langle x_1, x_2 \rangle = \langle y_1, y_2 \rangle \;\rightarrow\; (\, x_1 = y_1 \wedge x_2 = y_2 \,) \,]$
T$_3$ $\forall x [\, x \sqsubseteq \bot \;\leftrightarrow\; x = \bot \,]$
T$_4$ $\forall xyz [\, x \sqsubseteq \langle y, z \rangle \;\leftrightarrow\; (\, x = \langle y, z \rangle \vee x \sqsubseteq y \vee x \sqsubseteq z \,) \,]$.

© Springer Nature Switzerland AG 2020
M. Anselmo et al. (Eds.): CiE 2020, LNCS 12098, pp. 63–74, 2020.
https://doi.org/10.1007/978-3-030-51466-2_6

It is not difficult to see that T is a consistent extension of WT. Thus, since WT is essentially undecidable, we can conclude right away that also T is essentially undecidable. Furthermore, since every model of the finitely axiomatizable theory T is infinite, T cannot be interpretable in WT, and the obvious conjecture would be that T is mutually interpretable with Robinson's Q.

The Axioms of R

$$R_1 \; \bar{n} + \bar{m} = \overline{n+m} \; ; \quad R_2 \; \bar{n} \times \bar{m} = \overline{nm} \; ; \quad R_3 \; \bar{n} \neq \bar{m} \text{ for } n \neq m \; ;$$
$$R_4 \; \forall x [\, x \leq \bar{n} \rightarrow x = 0 \vee \ldots \vee x = \bar{n} \,] \; ; \quad R_5 \; \forall x [\, x \leq \bar{n} \vee \bar{n} \leq x \,]$$

The Axioms of Q

$$Q_1 \; \forall xy [\, Sx = Sy \rightarrow x = y \,] \; ; \quad Q_2 \; \forall x [\, Sx \neq 0 \,] \; ; \quad Q_3 \; \forall x [\, x \neq 0 \rightarrow \exists y [\, x = Sy \,] \,] \; ;$$
$$Q_4 \; \forall x [\, x + 0 = x \,] \; ; \quad Q_5 \; \forall xy [\, x + Sy = S(x+y) \,] \; ; \quad Q_6 \; \forall x [\, x \times 0 = 0 \,] \; ;$$
$$Q_7 \; \forall xy [\, x \times Sy = (x \times y) + x \,] \; ; \quad Q_8 \; \forall xy [\, x \leq y \leftrightarrow \exists z [\, x + z = y \,] \,]$$

Fig. 1. The axioms of R are given by axiom schemes where $n, m \in \mathbb{N}$ and $\bar{n}$ denotes the nth numeral, that is, $\bar{0} \equiv 0$ and $\overline{n+1} \equiv S\bar{n}$.

The seminal theories R and Q are theories of arithmetic. The theory R is given by axiom schemes, and Q is a finitely axiomatizable extension of R, see Fig. 1 (Q is also known as Robinson arithmetic and is more or less Peano arithmetic without the induction scheme). It was proved in Tarski et al. [9] that R and Q are essentially undecidable. Another seminal essentially undecidable first-order theory is Grzegorcyk's TC. This is a theory of concatenation. The language is $\{*, \alpha, \beta\}$ where α and β are constant symbols and $*$ is a binary function symbol. The standard TC model is the structure where the universe is $\{a, b\}^+$ (all finite nonempty strings over the alphabet $\{a, b\}$), $*$ is concatenation, α is the string a and β is the string b. It was proved in Grzegorzyk and Zdanowski [3] that TC is essentially undecidable. It was later proved that TC is mutually interpretable with Q, see Visser [10] for further references. The theory $\mathsf{WTC}^{-\epsilon}$ is a weaker variant of TC that has been shown to be mutually interpretable with R, see Higuchi and Horihata [4] for more details and further references. The axioms of TC and $\mathsf{WTC}^{-\epsilon}$ can be found in Fig. 2.

The overall picture shows three finitely axiomatizable and essentially undecidable first-order theories of different character and nature: Q is a theory of arithmetic, TC is a theory of concatenation, and T is a theory of terms (it may also be viewed as a theory of binary trees). All three theories are mutually interpretable with each other, and each of them come with a weaker variant given by axiom schemes. These weaker variants are also essentially undecidable and mutually interpretable with each other.

The theory T has, in contrast to Q and TC, a purely universal axiomatization, that is, there are no occurrences of existential quantifiers in the axioms. Moreover, its weaker variant WT has a neat and very compact axiomatization compared to R and $\mathsf{WTC}^{-\epsilon}$.

The Axioms of $\mathsf{WTC}^{-\epsilon}$

$\mathsf{WTC}_1^{-\epsilon}$ $\forall xyz\big[\ \big(\ x*(y*z)\sqsubseteq \underline{t}\vee (x*y)*z\sqsubseteq \underline{t}\ \big)\ \rightarrow\ x*(y*z)=(x*y)*z\ \big]$;

$\mathsf{WTC}_2^{-\epsilon}$ $\forall xyzu\big[\ x*y=z*u\ \wedge\ x*y\sqsubseteq \underline{t}\ \rightarrow\ \big(\ (x=z\wedge y=u)\ \vee$
$\qquad\qquad\qquad \exists w[\ (x*w=z\wedge w*u=y)\vee (z*w=x\wedge w*y=u]\ )\ \big]$;

$\mathsf{WTC}_3^{-\epsilon}$ $\forall xy[\ \alpha\neq x*y\ ]$; $\mathsf{WTC}_4^{-\epsilon}$ $\forall xy[\ \beta\neq x*y\ ]$; $\mathsf{WTC}_5^{-\epsilon}$ $\alpha\neq\beta$

where $x\sqsubseteq y$ is defined by

$$x=y\ \vee\ \exists z_1 z_2[\ z_1*x=y\ \vee\ x*z_2=y\ \vee\ (z_1*x)*z_2=y\ \vee\ z_1*(x*z_2)=y\ ]\ .$$

The Axioms of TC

TC_1 $\forall xyz[\ x*(y*z)=(x*y)*z\ ]$;

TC_2 $\forall xyzu\big[\ x*y=z*u\rightarrow\big(\ (x=z\wedge y=u)\ \vee$
$\qquad\qquad\qquad \exists w[\ (x*w=z\wedge w*u=y)\vee (z*w=x\wedge w*y=u]\ )\ \big]$;

TC_3 $\forall xy[\ \alpha\neq x*y\ ]$; TC_4 $\forall xy[\ \beta\neq x*y\ ]$; TC_5 $\alpha\neq\beta$

Fig. 2. $\mathsf{WTC}_1^{-\epsilon}$ and $\mathsf{WTC}_2^{-\epsilon}$ are axiom schemes where $t\in\{a,b\}^+$ and $\underline{t}$ is a term inductively defined by: $\underline{a}\equiv\alpha$, $\underline{b}\equiv\beta$, $\underline{au}\equiv\alpha*\underline{u}$ and $\underline{bu}\equiv\beta*\underline{u}$.

Another interesting theory which is known to be mutually interpretable with Q, and thus also with TC and T, is the adjunctive set theory AST. More on AST and adjunctive set theory can found in Damnjanovic [2]. For recent results related to the work in the present paper, we refer the reader to Jerabek [5], Cheng [1] and Kristiansen and Murwanashyaka [7].

The rest of this paper is fairly technical, and we will assume that the reader is familiar with first-order theories and the interpretation techniques introduced in Tarski et al. [9]. In Sect. 2 we prove that R and WT are mutually interpretable. In Sect. 3 we prove that Q is interpretable in T. We expect that T can be interpreted in Q by standard techniques available in the literature.

2   R and WT Are Mutually Interpretable

The theory R^- over the language of Robinson arithmetic is given by the axiom schemes

R_1^- $\overline{n}+\overline{m}=\overline{n+m}$; R_2^- $\overline{n}\times\overline{m}=\overline{nm}$; R_3^- $\overline{n}\neq\overline{m}$ for $n\neq m$;
R_4^- $\forall x[\ x\leq\overline{n}\ \leftrightarrow\ x=0\vee\ \dots\ \vee x=\overline{n}\ ]$

where $n,m\in\mathbb{N}$. Recall that $\overline{n}$ denotes the nth numeral, that is, $\overline{0}\equiv 0$ and $\overline{n+1}\equiv S\overline{n}$.

We now proceed to interpret R^- in WT. We choose the domain $I(x)\equiv x=x$ (thus we can just ignore the domain). Furthermore, we translate the successor function $S(x)$ as the function given by $\lambda x.\langle x,\perp\rangle$, and we translate the constant 0 as $\langle\perp,\perp\rangle$. Let $\overline{n}^\star$ denote the translation of the numeral $\overline{n}$. Then we have $\overline{n+1}^\star\equiv\langle\overline{n}^\star,\perp\rangle$. It follows from WT_1 that the translation of each instance of R_3^- is a theorem of WT since $\overline{m}^\star$ and $\overline{n}^\star$ are different terms whenever $m\neq n$.

We translate $x \leq y$ as $x \sqsubseteq y \wedge x \neq \bot$. It is easy to see that

$$\mathsf{WT} \vdash \forall x [\, x \sqsubseteq \overline{n}^{\star} \wedge x \neq \bot \quad \leftrightarrow \quad \bigvee_{s \in \mathcal{T}(n)} x = s \,] \tag{1}$$

where $\mathcal{T}(n) = \mathcal{S}(\overline{n}^{\star}) \setminus \{\bot\}$ and $\mathcal{S}(\overline{n}^{\star})$ denotes the set of all subterms of $\overline{n}^{\star}$. We observe that $\mathcal{T}(n) = \{\overline{k}^{\star} \mid k \leq n\}$ and that (1) indeed is the translation of the axiom scheme R_4^-. Hence we conclude that the translation of each instance of R_4^- is a theorem of WT.

Next we discuss the translation of $+$. The idea is to obtain $n + i$ through a formation sequence of length i. Such a sequence will be represented by a term of the form

$$\langle \ldots \langle \langle \langle \overline{n}^{\star}, \overline{0}^{\star} \rangle, \langle \overline{n+1}^{\star}, \overline{1}^{\star} \rangle \rangle, \langle \overline{n+2}^{\star}, \overline{2}^{\star} \rangle \rangle \ldots, \langle \overline{n+i}^{\star}, \overline{i}^{\star} \rangle \rangle. \tag{2}$$

Accordingly we translate $x + y = z$ by the predicate $add(x, y, z)$ given by the formula

$$(y = \overline{0}^{\star} \wedge z = x) \vee \Big\{ y \neq \overline{0}^{\star} \wedge \exists W [\, \langle x, \overline{0}^{\star} \rangle \sqsubseteq W \wedge$$

$$\forall X \, \forall Y \sqsubseteq y [\, \langle X, Y \rangle \sqsubseteq W \wedge Y \neq y \wedge Y \neq \bot \quad \rightarrow$$

$$(\, \langle \langle X, \bot \rangle, \langle Y, \bot \rangle \rangle \sqsubseteq W \wedge (\, \langle Y, \bot \rangle = y \rightarrow \langle X, \bot \rangle = z \,) \,) \,] \,] \Big\}.$$

Lemma 1. *For any $m, n \in \mathbb{N}$, we have*

$$\mathsf{WT} \vdash \forall z [\, add(\overline{n}^{\star}, \overline{m}^{\star}, z) \leftrightarrow z = \overline{n+m}^{\star} \,].$$

Proof. First we prove that $\mathsf{WT} \vdash add(\overline{n}^{\star}, \overline{m}^{\star}, \overline{n+m}^{\star})$. This is obvious if $m = 0$. Assume $m > 0$. Let

$$S_0^n \equiv \langle \overline{n}^{\star}, \overline{0}^{\star} \rangle \quad \text{and} \quad S_{i+1}^n \equiv \langle S_i^n, \langle \overline{n+i+1}^{\star}, \overline{i+1}^{\star} \rangle \rangle$$

and observe that S_i^n is of the form (2). We will argue that we can choose the W in the definition of $add(x, y, z)$ to be the term S_m^n.

So let $W = S_m^n$. By the axioms of WT, we have $\langle \overline{n}^{\star}, \overline{0}^{\star} \rangle \sqsubseteq W$. Assume

$$\langle X, Y \rangle \sqsubseteq W \text{ and } Y \neq y = \overline{m}^{\star} \text{ and } Y \sqsubseteq y = \overline{m}^{\star} \text{ and } Y \neq \bot.$$

By the axioms of WT, we have that $Y \sqsubseteq \overline{m}^{\star}$, $Y \neq \overline{m}^{\star}$ and $Y \neq \bot$ imply $Y = \overline{k}^{\star}$ for some $k < m$. Since $\langle X, Y \rangle \sqsubseteq W$, we know by WT_2 that $\langle X, Y \rangle$ is one of the subterms of W. By WT_1 and the form of S_m^n, we conclude that $X = \overline{n+k}^{\star}$. Furthermore, the form of S_m^n and WT_2 then ensures that $\langle \langle X, \bot \rangle, \langle Y, \bot \rangle \rangle \sqsubseteq W = S_m^n$. Moreover, if $\langle Y, \bot \rangle = \overline{m}^{\star}$, then by WT_1, we must have $k = m - 1$, and thus, $\langle X, \bot \rangle = \langle \overline{n+(m-1)}^{\star}, \bot \rangle = \overline{n+m}^{\star}$. This proves that we can deduce $add(\overline{n}^{\star}, \overline{m}^{\star}, \overline{n+m}^{\star})$ from the axioms of WT, and thus we also have

$$\mathsf{WT} \vdash \forall z [\, z = \overline{n+m}^{\star} \rightarrow add(\overline{n}^{\star}, \overline{m}^{\star}, z)].$$

Next we prove that the converse implication $add(\overline{n}^\star, \overline{m}^\star, z) \rightarrow z = \overline{n+m}^\star$ follows from the axioms of WT (and thus the lemma follows). This is obvious when $m = 0$. Assume $m \neq 0$ and $add(\overline{n}^\star, \overline{m}^\star, z)$. Then we have W such that $\langle \overline{n}^\star, \overline{0}^\star \rangle \sqsubseteq W$ and

$$\forall X \, \forall Y \sqsubseteq \overline{m}^\star [\, \langle X, Y \rangle \sqsubseteq W \, \wedge \, Y \neq \overline{m}^\star \, \wedge \, Y \neq \perp \quad \rightarrow$$
$$(\, \langle \langle X, \perp \rangle, \langle Y, \perp \rangle \rangle \sqsubseteq W \, \wedge \, (\, \langle Y, \perp \rangle = \overline{m}^\star \rightarrow \langle X, \perp \rangle = z \,) \,) \,]. \quad (3)$$

Since $\langle n, \overline{0}^\star \rangle \sqsubseteq W$ and (3) hold, we have $\langle \overline{n+k+1}^\star, \overline{k+1}^\star \rangle \sqsubseteq W$ for any $k < m$. It also follows from (3) that $z = \overline{n+k+1}^\star$ when $m = k+1$.   □

It follows from the preceding lemma that there for any $n, m \in \mathbb{N}$ exists a unique $k \in \mathbb{N}$ such that $WT \vdash add(\overline{n}^\star, \overline{m}^\star, \overline{k}^\star)$. We translate $x + y = z$ by the predicate ϕ_+ where $\phi_+(x, y, z)$ is the formula

$$(\, \exists!u[add(x, y, u)] \, \wedge \, add(x, y, z) \,) \, \vee \, (\, \neg\exists!u[add(x, y, u)] \, \wedge \, z = \perp \,). \quad (4)$$

The second disjunct of (4) ensures the functionality of our translation, that is, it ensures that $WT \vdash \forall xy \exists! x \phi_+(x, y, z)$ (the same technique is used in [6]). By Lemma 1, we have $WT \vdash \phi_+(\overline{n}^\star, \overline{m}^\star, \overline{n+m}^\star)$. This shows that the translation of any instance of the axiom scheme R_1^- can be deduced from the axioms of WT.

We can also achieve a translation of $x \times y = z$ such that the translation of each instance of R_2^- can be deduced from the axioms of WT. Such a translation claims the existence of a term S_m^n where

$$S_1^n \equiv \langle \overline{n}^\star, \overline{1}^\star \rangle \quad \text{and} \quad S_{i+1}^n \equiv \langle S_i^n, \langle \overline{(i+1)n}^\star, \overline{i+1}^\star \rangle \rangle$$

and will more or less be based on the same ideas as our translation of $x + y = z$. We omit the details.

Theorem 2. R *and* WT *are mutually interpretable.*

Proof. We have seen how to interpret R^- in WT. It follows straightforwardly from results proved in Jones and Shepherdson [6] that R^- and R are mutually interpretable. Thus R is interpretable in WT. A result of Visser [11] states that a theory is interpretable in R if and only if it is locally finitely satisfiable, that is, each finite subset of the non-logical axioms has a finite model. Since WT clearly is locally finitely satisfiable, WT is interpretable in R.   □

3   Q is Interpretable in T

The language of the arithmetical theory Q^- is $\{0, S, M, A\}$ where 0 is a constant symbol, S is a unary function symbol, and A and M are ternary predicate symbols. The non-logical axioms of the first-order theory Q^- are the the following:

A $\forall xyz_1z_2[\, A(x, y, z_1) \wedge A(x, y, z_2) \rightarrow z_1 = z_2 \,]$;
M $\forall xyz_1z_2[\, M(x, y, z_1) \wedge M(x, y, z_2) \rightarrow z_1 = z_2 \,]$;
Q_1 $\forall xy[\, x \neq y \rightarrow Sx \neq Sy \,]$; Q_2 $\forall x[\, Sx \neq 0 \,]$; Q_3 $\forall x[\, x = 0 \vee \exists y[\, x = Sy \,] \,]$;
G_4 $\forall x[\, A(x, 0, x) \,]$; G_5 $\forall xyu[\, \exists z[\, A(x, y, z) \wedge u = Sz \,] \rightarrow A(x, Sy, u) \,]$;
G_6 $\forall x[\, M(x, 0, 0) \,]$; G_7 $\forall xyu[\, \exists z[\, M(x, y, z) \wedge A(z, x, u) \,] \rightarrow M(x, Sy, u) \,]$.

Svejdar [8] proved that Q^- and Q are mutually interpretable. We will prove that Q^- is interpretable in T.

The first-order theory T^+ is T extended by the two non-logical axioms

$$T_5 \ \forall x[\ x \sqsubseteq x\ ] \quad \text{and} \quad T_6 \ \forall xyz[\ x \sqsubseteq y \wedge y \sqsubseteq z \ \rightarrow \ x \sqsubseteq z\ ].$$

Lemma 3. T^+ *is interpretable in* T.

Proof. We simply relativize quantification to the domain

$$I \ = \ \{\ x \mid \ x \sqsubseteq x \wedge \forall uv[\ u \sqsubseteq v \wedge v \sqsubseteq x \ \rightarrow \ u \sqsubseteq x\ ]\ \}\ .$$

Suppose $x_1, x_2 \in I$. We show that $\langle x_1, x_2 \rangle \in I$. Since $\langle x_1, x_2 \rangle = \langle x_1, x_2 \rangle$, we have $\langle x_1, x_2 \rangle \sqsubseteq \langle x_1, x_2 \rangle$ by T_4. Suppose now that $u \sqsubseteq v \wedge v \sqsubseteq \langle x_1, x_2 \rangle$. We need to show that $u \sqsubseteq \langle x_1, x_2 \rangle$. By T_4 and $v \sqsubseteq \langle x_1, x_2 \rangle$, at least one of the following three cases holds: (a) $v = \langle x_1, x_2 \rangle$, (b) $v \sqsubseteq x_1$, (c) $v \sqsubseteq x_2$. *Case (a):* Since $u \sqsubseteq v$ and $v = \langle x_1, x_2 \rangle$, we have $u \sqsubseteq \langle x_1, x_2 \rangle$ by our logical axioms. *Case (b):* $u \sqsubseteq v \wedge v \sqsubseteq x_1$ implies $u \sqsubseteq x_1$ since $x_1 \in I$. By T_4, we have $u \sqsubseteq \langle x_1, x_2 \rangle$. *Case (c):* We have $u \sqsubseteq \langle x_1, x_2 \rangle$ by an argument symmetric to the one used in Case (b). Hence, $\forall uv[\ u \sqsubseteq v \wedge v \sqsubseteq \langle x_1, x_2 \rangle \rightarrow u \sqsubseteq \langle x_1, x_2 \rangle\ ]$.

This proves that I is closed under $\langle \cdot, \cdot \rangle$. It follows from T_3 that $\bot \in I$, and thus I satisfies the domain condition. Clearly, the translation of each non-logical axiom of T^+ is a theorem of T. □

We now proceed to interpret Q^- in T^+. We choose the domain N given by

$$N(x) \ \equiv \ x \neq \bot \ \wedge \ \forall y \sqsubseteq x[\ y = \bot \ \vee \ \exists z[\ y = \langle z, \bot \rangle\ ]\ ]\ .$$

Lemma 4. *We have (i)* $T^+ \vdash N(\langle \bot, \bot \rangle)$, *(ii)* $T^+ \vdash \forall x[N(x) \rightarrow N(\langle x, \bot \rangle)]$ *and (iii)* $T^+ \vdash \forall yz[\ N(y) \wedge z \sqsubseteq y \ \rightarrow \ (\ z = \bot \ \vee \ N(z)\ )\ ]$.

Proof. It follows from T_1, T_3 and T_4 that (i) holds. In order to see that (ii) holds, assume $N(x)$ (we will argue that $N(\langle x, \bot \rangle)$ holds). Suppose $y \sqsubseteq \langle x, \bot \rangle$. Now, $N(\langle x, \bot \rangle)$ follows from

$$y = \bot \ \vee \ \exists z[\ y = \langle z, \bot \rangle\ ]. \tag{5}$$

Thus it is sufficient to argue that (5) holds. By T_4, we know that $y \sqsubseteq \langle x, \bot \rangle$ implies $y = \langle x, \bot \rangle \vee y \sqsubseteq x \vee y \sqsubseteq \bot$. The case $y = \langle x, \bot \rangle$: We obviously have $\exists z[\ y = \langle z, \bot \rangle\ ]$ and thus (5) holds. The case $y \sqsubseteq x$: (5) holds since $N(x)$ holds. The case $y \sqsubseteq \bot$: We have $y = \bot$ by T_3, and thus (5) holds. This proves (ii).

We turn to the proof of (iii). Suppose $N(y) \wedge z \sqsubseteq y$ (we show $z = \bot \vee N(z)$). Assume $w \sqsubseteq z$. By T_6, we have $w \sqsubseteq y$, moreover, since $N(y)$ holds, we have $w = \bot \vee \exists u[w = \langle u, \bot \rangle]$. Thus, we conclude that

$$\forall w \sqsubseteq z[\ w = \bot \ \vee \ \exists u[\ w = \langle u, \bot \rangle\ ]\ ]. \tag{6}$$

Now

$$z = \bot \ \vee \ \underbrace{(\ z \neq \bot \ \wedge \ \forall w \sqsubseteq z[\ w = \bot \ \vee \ \exists u[\ w = \langle u, \bot \rangle\ ]\ ]\ )}_{N(z)}$$

follows tautologically from (6). □

We interpret 0 as $\langle \perp, \perp \rangle$. We interpret the successor function Sx as $\lambda x.\langle x, \perp \rangle$. To improve the readability we will occasionally write $\dot{0}$ in place of $\langle \perp, \perp \rangle$, $\dot{S}t$ in place of $\langle t, \perp \rangle$ and $t \in N$ in place of $N(t)$. We will also write $\exists x \in N[\ \eta\ ]$ and $\forall x \in N[\ \eta\ ]$ in place of, respectively, $\exists x[\ N(x) \wedge \eta\ ]$ and $\forall x[\ N(x) \rightarrow \eta\ ]$. Furthermore, $\mathbf{Q}x_1, \ldots, x_n \in N$ is shorthand for $\mathbf{Q}x_1 \in N \ldots \mathbf{Q}x_n \in N$ where $\mathbf{Q}$ is either $\forall$ or $\exists$.

Lemma 5. *The translations of Q_1, Q_2 and Q_3 are theorems of T^+.*

Proof. The translation of Q_1 is $\forall x, y \in N[\ x \neq y \rightarrow \dot{S}x \neq \dot{S}y\ ]$. By T_2, we have $x \neq y \rightarrow \dot{S}x \neq \dot{S}y$ for any x, y, and thus, the translation of Q_1 is a theorem of T^+.

The translation of Q_2 is $\forall x \in N[\ \dot{S}x \neq \dot{0}\ ]$. Assume $x \in N$. Then we have $x \neq \perp$, and by T_2, we have $\dot{S}n \equiv \langle x, \perp \rangle \neq \langle \perp, \perp \rangle \equiv \dot{0}$.

The translation of Q_3 is $\forall x \in N[\ x = \dot{0} \vee \exists y \in N[\ x = \dot{S}y\ ]\ ]$. Assume $x \in N$, that is, assume

$$x \neq \perp \wedge \forall y \sqsubseteq x[\ y = \perp \vee \exists z[\ y = \langle z, \perp \rangle\ ]\ ]\ . \tag{7}$$

By T_5, we have $x \sqsubseteq x$. By (7) and $x \sqsubseteq x$, we have

$$x \neq \perp \wedge (\ x = \perp \vee \exists z[\ x = \langle z, \perp \rangle\ ]\ )$$

and then, by a tautological inference, we also have $\exists z[x = \langle z, \perp \rangle]$. Thus, we have z such that $\langle z, \perp \rangle \equiv \dot{S}z = x \in N$. By Lemma 4 (iii), we have $z = \perp \vee z \in N$. If $z = \perp$, we have $x = \langle \perp, \perp \rangle \equiv \dot{0}$. If $z \in N$, we have $z \in N$ such that $x = \dot{S}z$. Thus, $T^+ \vdash \forall x \in N[x = \dot{0} \vee \exists y \in N[x = \dot{S}y]]$. $\square$

Before we give the translation of A, we will provide some intuition. The predicate $A(a, b, c)$ holds in the standard model for Q^- iff $a + b = c$. Let $\widetilde{0} \equiv \dot{0}$ and $\widetilde{n+1} \equiv \dot{S}\tilde{n}$, and observe that $a + b = c$ iff there exists an $\mathcal{L}_T$-term of the form

$$\langle \ldots \langle \langle \langle \perp, \langle \tilde{a}, \widetilde{0} \rangle \rangle, \langle \widetilde{a+1}, \widetilde{1} \rangle \rangle, \langle \widetilde{a+2}, \widetilde{2} \rangle \rangle \ldots, \langle \widetilde{a+b}, \tilde{b} \rangle \rangle \tag{8}$$

where $c = a + b$. We will give a predicate ϕ_A such that $\phi_A(\tilde{a}, \tilde{b}, w)$ holds in T^+ iff w is of the form (8). Thereafter we will use ϕ_A to give the translation Ψ_A of A.

Let $\phi_A(x, y, w) \equiv$

$$(\ y = \dot{0} \rightarrow w = \langle \perp, \langle x, \dot{0} \rangle \rangle\ ) \wedge \exists w' \exists z \in N[\ w = \langle w', \langle z, y \rangle \rangle\ ] \wedge$$
$$\forall u \forall Y, Z \in N[\ \theta_A(u, w, Y, Z)\ ]$$

where $\theta_A(u, w, Y, Z) \equiv$

$$\langle u, \langle Z, Y \rangle \rangle \sqsubseteq w \wedge Y \neq \dot{0} \rightarrow$$
$$\exists v\ \exists Y'Z' \in N[\ Z = \dot{S}Z' \wedge Y = \dot{S}Y' \wedge u = \langle v, \langle Z', Y' \rangle \rangle \wedge$$
$$(\ Y' = \dot{0} \rightarrow (\ Z' = x \wedge v = \perp\ )\ )\ ]\ .$$

The translation Ψ_A of A is $\Psi_A(x, y, z) \equiv$

$$\exists w \left[\phi_A(x, y, w) \wedge \exists w' \left[w = \langle w', \langle z, y \rangle \rangle \right] \wedge \forall u \left[\phi_A(x, y, u) \rightarrow u = w \right] \right].$$

Lemma 6.

$$\mathsf{T}^+ \vdash \forall x \in N \forall w \left[\phi_A(x, \dot{0}, w) \leftrightarrow w = \langle \perp, \langle x, \dot{0} \rangle \rangle \right].$$

Proof. We assume $x \in N$ and prove the equivalence

$$\phi_A(x, \dot{0}, w) \leftrightarrow w = \langle \perp, \langle x, \dot{0} \rangle \rangle \tag{9}$$

The left-right direction of (9) follows straightforwardly from the definition of ϕ_A. To prove the right-left implication of (9), we need to prove $\phi_A(x, \dot{0}, \langle \perp, \langle x, \dot{0} \rangle \rangle)$. It is easy to see that $\phi_A(x, \dot{0}, \langle \perp, \langle x, \dot{0} \rangle \rangle)$ holds if

$$\forall u \forall Y, Z \in N \left[\theta_A(u, \langle \perp, \langle x, \dot{0} \rangle \rangle, Y, Z) \right] \tag{10}$$

holds, and to show (10), it suffices to show that

$$x, Y, Z \in N \quad \text{and} \quad \langle u, \langle Z, Y \rangle \rangle \sqsubseteq \langle \perp, \langle x, \dot{0} \rangle \rangle \quad \text{and} \quad Y \neq \dot{0} \tag{11}$$

is a contradiction. (If (11) is a contradiction, then (10) will hold as the antecedent of θ_A will be false for all $x, Y, Z \in N$ and all u.)

By T_4 and $\langle u, \langle Z, Y \rangle \rangle \sqsubseteq \langle \perp, \langle x, \dot{0} \rangle \rangle$ we have to deal with the following three cases: (a) $\langle u, \langle Z, Y \rangle \rangle = \langle \perp, \langle x, \dot{0} \rangle \rangle$, (b) $\langle u, \langle Z, Y \rangle \rangle \sqsubseteq \perp$ and (c) $\langle u, \langle Z, Y \rangle \rangle \sqsubseteq \langle x, \dot{0} \rangle$. *Case: (a):* We have $Y = \dot{0}$ by T_2, but we have $Y \neq \dot{0}$ in (11). *Case (b):* We have $\langle u, \langle Z, Y \rangle \rangle = \perp$ by T_3, and this contradicts T_1. *Case (c):* By T_4, this case splits into the three subcases: (a') $\langle u, \langle Z, Y \rangle \rangle = \langle x, \dot{0} \rangle$, (b') $\langle u, \langle Z, Y \rangle \rangle \sqsubseteq x$ and (c') $\langle u, \langle Z, Y \rangle \rangle \sqsubseteq \dot{0}$. *Case (a'):* We have $\langle u, \langle Z, Y \rangle \rangle = \langle x, \langle \perp, \perp \rangle \rangle$ since $\dot{0}$ is shorthand for $\langle \perp, \perp \rangle$. Thus, by T_2, we have $Z = \perp$ and $Y = \perp$. This contradicts $Y, Z \in N$. *Case (b'):* We have $\langle u, \langle Z, Y \rangle \rangle \sqsubseteq x$ and $x \in N$. By Lemma 4 (iii), we have $\langle u, \langle Z, Y \rangle \rangle = \perp$ or $\langle u, \langle Z, Y \rangle \rangle \in N$. Now, $\langle u, \langle Z, Y \rangle \rangle = \perp$ contradicts T_1. Furthermore, by our definitions, $\langle u, \langle Z, Y \rangle \rangle \in N$ implies that

$$\forall y_0 \sqsubseteq \langle u, \langle Z, Y \rangle \rangle [y_0 = \perp \vee \exists z_0 [y_0 = \langle z_0, \perp \rangle]].$$

By T_5, we have $\langle u, \langle Z, Y \rangle \rangle = \perp \vee \exists z_0 [\langle u, \langle Z, Y \rangle \rangle = \langle z_0, \perp \rangle]$, and this yields a contradiction together with T_1 and T_2. Case (c') is similar to Case (a'), but a bit simpler. This completes the proof of the lemma. $\qquad \square$

Lemma 7.

$$\mathsf{T}^+ \vdash \forall x, y \in N \forall z w w' [w = \langle w', \langle z, y \rangle \rangle \wedge \phi_A(x, y, w) \rightarrow$$
$$\phi_A(x, \dot{S}y, \langle w, \langle \dot{S}z, \dot{S}y \rangle \rangle)].$$

Proof. We assume

$$x, y \in N \text{ and } w = \langle w', \langle z, y \rangle \rangle \text{ and } \phi_A(x, y, w). \tag{12}$$

We need to prove $\phi_A(x, \dot{S}y, \langle w, \langle \dot{S}z, \dot{S}y \rangle \rangle) \equiv$

$$(\dot{S}y = \dot{0} \to w = \langle \bot, \langle x, \dot{0} \rangle \rangle) \wedge$$
$$\exists w_0 \exists z_0 \in N[\langle w, \langle \dot{S}z, \dot{S}y \rangle \rangle = \langle w_0, \langle z_0, \dot{S}y \rangle \rangle] \wedge$$
$$\forall u \forall Y, Z \in N[\theta_A(u, \langle w, \langle \dot{S}z, \dot{S}y \rangle \rangle, Y, Z)] \tag{13}$$

First we prove

$$z \in N \quad \text{and} \quad \dot{S}z \in N \tag{14}$$

Since $\phi_A(x, y, w)$ holds by our assumptions (12), we have $z_1 \in N$ and w_1 such that $w = \langle w_1, \langle z_1, y \rangle \rangle$. We have also assumed $w = \langle w', \langle z, y \rangle \rangle$. By T_2, we have $z = z_1$, and thus $z \in N$. By Lemma 4 (ii), we have $\dot{S}z \in N$. This proves (14).

The second conjunct of (13) follows straightforwardly from (14). (simply let z_0 be $\dot{S}z$ and let w_0 be w). The first conjunct follows easily from T_2 and the assumption $y \in N$. Thus, we are left to prove the third conjunct of (13), namely

$$\forall u \forall Y, Z \in N\big[\langle u, \langle Z, Y \rangle \rangle \sqsubseteq \langle w, \langle \dot{S}z, \dot{S}y \rangle \rangle \wedge Y \neq \dot{0} \to$$
$$\exists v \, \exists Y' Z' \in N\big[Z = \dot{S}Z' \wedge Y = \dot{S}Y' \wedge u = \langle v, \langle Z', Y' \rangle \rangle \wedge$$
$$(Y' = \dot{0} \to (Z' = x \wedge v = \bot)) \big] \big] \tag{15}$$

In order to do so, we assume

$$Y, Z \in N \text{ and } \langle u, \langle Z, Y \rangle \rangle \sqsubseteq \langle w, \langle \dot{S}z, \dot{S}y \rangle \rangle \text{ and } Y \neq \dot{0} \tag{16}$$

and prove

$$\exists v \, \exists Y' Z' \in N\big[Z = \dot{S}Z' \wedge Y = \dot{S}Y' \wedge u = \langle v, \langle Z', Y' \rangle \rangle \wedge$$
$$(Y' = \dot{0} \to (Z' = x \wedge v = \bot)) \big]. \tag{17}$$

By our assumptions (16), we have $\langle u, \langle Z, Y \rangle \rangle \sqsubseteq \langle w, \langle \dot{S}z, \dot{S}y \rangle \rangle$, and then T_4 yields three cases: (a) $\langle u, \langle Z, Y \rangle \rangle = \langle w, \langle \dot{S}z, \dot{S}y \rangle \rangle$, (b) $\langle u, \langle Z, Y \rangle \rangle \sqsubseteq w$ and (c) $\langle u, \langle Z, Y \rangle \rangle \sqsubseteq \langle \dot{S}z, \dot{S}y \rangle$. We prove that that (17) holds in each of these three cases.

Case (a): By T_2, we have $u = w$, $Z = \dot{S}z$ and $Y = \dot{S}y$. By (14), we have $z \in N$. By (12), we have $y \in N$. Moreover, by (12), we also have $u = w = \langle w', \langle z, y \rangle \rangle$. Thus there exist v and $Y', Z' \in N$ such that

$$Z = \dot{S}Z' \wedge Y = \dot{S}Y' \wedge u = \langle v, \langle Z', Y' \rangle \rangle .$$

If $y = \dot{0}$, we must have $\langle v, \langle z, y \rangle \rangle = w = \langle \bot, \langle x, \dot{0} \rangle \rangle$ since $\phi_A(x, y, w)$ holds by our assumptions (12). By T_2, this implies $z = x$ and $v = \bot$. This proves that (17) holds in Case (a).

Case (b): By our assumptions (12), we have $\phi_A(x, y, w)$, and thus we also have $\theta_A(u, w, Y, Z) \equiv$

$$\langle u, \langle Z, Y \rangle \rangle \sqsubseteq w \wedge Y \neq \dot{0} \to$$
$$\exists v \, \exists Y' Z' \in N\big[Z = \dot{S}Z' \wedge Y = \dot{S}Y' \wedge u = \langle v, \langle Z', Y' \rangle \rangle) \wedge$$
$$(Y' = \dot{0} \to (Z' = x \wedge v = \bot)) \big]. \tag{18}$$

We are dealing with a case where the antecedent of (18) holds, and thus (17) holds.

Case (c): This case is not possible. By T_4, this case splits into the subcases: (a') $\langle u, \langle Z, Y \rangle \rangle = \langle \dot{S}z, \dot{S}y \rangle$, (b') $\langle u, \langle Z, Y \rangle \rangle \sqsubseteq \dot{S}z$ and (c') $\langle u, \langle Z, Y \rangle \rangle \sqsubseteq \dot{S}y$. We prove that each of these subcases contradicts our axioms. *Case (a'):* Recall that $\dot{S}y$ is shorthand for $\langle y, \bot \rangle$. Thus, by T_2, we have $Y = \bot$. This contradicts the assumption (12) that $Y \in N$. *Case (b'):* By Lemma 4 (iii), we have $\langle u, \langle Z, Y \rangle \rangle = \bot \lor N(\langle u, \langle Z, Y \rangle \rangle)$. Now, $\langle u, \langle Z, Y \rangle \rangle = \bot$ contradicts T_1. Furthermore, $N(\langle u, \langle Z, Y \rangle \rangle)$ implies that there is z_0 such that $\langle u, \langle Z, Y \rangle \rangle = \langle z_0, \bot \rangle$. By T_2, we have $\langle Z, Y \rangle = \bot$. This contradicts T_1. Case (c') is similar to Case (b'). This proves that (17) holds, and thus we conclude that the lemma holds. □

Lemma 8.

$$T^+ \vdash \forall xy \in N \forall w [\ \phi_A(x, \dot{S}y, w) \ \rightarrow$$
$$\exists u \in N \exists w' [\ w = \langle w', \langle u, \dot{S}y \rangle \rangle \ \land \ \phi_A(x, y, w') \ ] \ ].$$

Proof. Let $x, y \in N$ and assume $\phi_A(x, \dot{S}y, w)$. Thus, we have w' and $z \in N$ such that

$$w = \langle w', \langle z, \dot{S}y \rangle \rangle \quad \text{and} \quad \forall u \forall Y, Z \in N [\ \theta_A(u, w, Y, Z) \ ] \tag{19}$$

Use the assumptions (19) to prove that $\phi_A(x, y, w') \equiv$

$$(\ y = \dot{0} \rightarrow w' = \langle \bot, \langle x, \dot{0} \rangle \rangle \ ) \ \land \ \exists w'' \exists z \in N [\ w' = \langle w'', \langle z, y \rangle \rangle \ ] \ \land$$
$$\forall u \forall Y, Z \in N [\ \theta_A(u, w', Y, Z) \ ] \tag{20}$$

holds. We omit the details. □

Lemma 9. *The translations of* A, G_4 *and* G_5 *are theorems of* T^+.

Proof. The translation of the axiom A is

$$\forall x, y, z_1, z_2 \in N [\ \Psi_A(x, y, z_1) \land \Psi_A(x, y, z_2) \ \rightarrow \ z_1 = z_2 \ ].$$

Assume $\Psi_A(x, y, z_1)$ and $\Psi_A(x, y, z_2)$. Then it follows straightforwardly from the definition of Ψ_A and T_2 that $z_1 = z_2$. Hence the translation is a theorem of T^+.

The translation of G_4 is $\forall x \in N [\Psi_A(x, \dot{0}, x)]$, that is

$$\forall x \in N \exists w [\ \phi_A(x, \dot{0}, w) \ \land \ \exists w' [\ w = \langle w', \langle x, \dot{0} \rangle \rangle \ ] \ \land$$
$$\forall u [\ \phi_A(x, \dot{0}, u) \rightarrow u = w \ ] \ ].$$

We have

$$T^+ \vdash \phi_A(x, \dot{0}, \langle \bot, \langle x, \dot{0} \rangle \rangle) \text{ and } T^+ \vdash \forall u [\phi_A(x, \dot{0}, u) \rightarrow u = \langle \bot, \langle x, \dot{0} \rangle \rangle$$

by Lemma 6, and it easy to see that the translation of G_4 is a theorem of T^+.

The translation of G_5 is

$$\forall x, y, u \in N[\ \exists z \in N[\ \Psi_A(x, y, z) \wedge u = \dot{S}z\ ] \to \Psi_A(x, \dot{S}y, u)\ ]. \tag{21}$$

In order to prove that (21) can be deduced from the axioms of T^+, we assume $\Psi_A(x, y, z) \wedge u = \dot{S}z$. Then we need to prove $\Psi_A(x, \dot{S}y, \dot{S}z) \equiv$

$$\exists w[\ \phi_A(x, \dot{S}y, w)\ \wedge\ \exists w'[\ w = \langle w', \langle \dot{S}z, \dot{S}y \rangle \rangle\ ]\ \wedge$$
$$\forall u[\ \phi_A(x, \dot{S}y, u) \to u = w\ ]\ ]. \tag{22}$$

By our assumption $\Psi_A(x, y, z)$ there is a unique w_1 such that $\phi_A(x, y, w_1)$ and $w_1 = \langle w_0, \langle z, y \rangle \rangle$ for some w_0. By Lemma 7, we have $\phi_A(x, \dot{S}y, \langle w_1, \langle \dot{S}z, \dot{S}y \rangle \rangle)$. Thus, we have w_2 such that $\phi_A(x, \dot{S}y, w_2)$ and $w_2 = \langle w_1, \langle \dot{S}z, \dot{S}y \rangle \rangle$. It is easy to see that (22) holds if w_2 is unique. Thus we are left to prove the uniqueness of w_2, more precisely, we need to prove that

$$\forall W_2[\ \phi_A(x, \dot{S}y, W_2)\ \to\ W_2 = w_2\ ]. \tag{23}$$

In order to prove (23), we assume $\phi_A(x, \dot{S}y, W_2)$ (we will prove $W_2 = w_2 = \langle w_1, \langle \dot{S}z, \dot{S}y \rangle \rangle$). By our assumption $\phi_A(x, \dot{S}y, W_2)$ and Lemma 8, we have $u_0 \in N$ and W_1 such that $W_2 = \langle W_1, \langle u_0, \dot{S}y \rangle \rangle$ and $\phi_A(x, y, W_1)$. We have argued that there is a unique $w_1 = \langle w_0, \langle z, y \rangle \rangle$ such that $\phi_A(x, y, w_1)$ holds. By this uniqueness, we have $W_1 = w_1 = \langle w_0, \langle z, y \rangle \rangle$. So far we have proved

$$w_2 = \langle \overbrace{\langle w_0, \langle z, y \rangle \rangle}^{w_1}, \langle \dot{S}z, \dot{S}y \rangle \rangle \text{ and } W_2 = \langle \overbrace{\langle w_0, \langle z, y \rangle \rangle}^{W_1}, \langle u_0, \dot{S}y \rangle \rangle$$

and then we are left to prove that $u_0 = \dot{S}z$. By our assumption $\phi_A(x, \dot{S}y, W_2)$, we have v and $Z', Y' \in N$ such that $u_0 = \dot{S}Z'$, $\dot{S}y = \dot{S}Y'$ and $W_1 = \langle v, \langle Z', Y' \rangle \rangle$. Thus, $\langle v, \langle Z', Y' \rangle \rangle = \langle w_0, \langle z, y \rangle \rangle$. By T_2, we have $z = Z'$, and thus, $u_0 = \dot{S}Z' = \dot{S}z$. This proves that (23) holds. $\qquad\square$

We will now give the translation Ψ_M of M. Let $\phi_M(x, y, w) \equiv$

$$(\ y = \dot{0} \to w = \langle \bot, \langle \dot{0}, \dot{0} \rangle \rangle\ )\ \wedge\ \exists w' \exists z \in N[\ w = \langle w', \langle z, y \rangle \rangle\ ]\ \wedge$$
$$\forall u \forall Y, Z \in N\ \theta_M(u, w, Y, Z)$$

where $\theta_M(u, w, Y, Z) \equiv$

$$\langle u, \langle Z, Y \rangle \rangle \sqsubseteq w \wedge Y \neq \dot{0}\ \to\ \exists v\ \exists Y', Z' \in N[\ \Psi_A(Z', x, Z)\ \wedge$$
$$Y = \dot{S}Y'\ \wedge\ u = \langle v, \langle Z', Y' \rangle \rangle\ \wedge\ (\ Y' = \dot{0} \to Z' = \dot{0} \wedge v = \bot\ )\ ].$$

We let $\Psi_M(x, y, z) \equiv$

$$\exists w[\ \phi_M(x, y, w)\ \wedge\ \exists w'[\ w = \langle w', \langle z, y \rangle \rangle\ \wedge\ \forall u[\ \phi_M(x, y, u) \to u = w\ ]\ ].$$

The translations of M, G_6 and G_7 are

M $\quad \forall x, y, z_1, z_2 \in N[\ \Psi_M(x, y, z_1) \wedge \Psi_M(x, y, z_2) \rightarrow z_1 = z_2\ ]$
$G_6 \quad \forall x \in N[\ M(x, \dot{0}, \dot{0})\ ]$
$G_7 \quad \forall x, y, u \in N[\ \exists z \in N[\ \Psi_M(x, y, z) \wedge \Psi_A(z, x, u)\ ] \rightarrow \Psi_M(x, \dot{S}y, u)\ ]$.

The proof of the next lemma follows the lines of the proof of Lemma 9. We omit the details.

Lemma 10. *The translations of* M, G_6 *and* G_7 *are theorems of* T^+.

Theorem 11. Q *is interpretable in* T.

Proof. It is proved in Svejdar [8] that Q is interpretable in Q^-. It follows from the lemmas above that Q^- is interpretable in T^+ which again is interpretable in T. Hence the theorem holds. $\qquad \square$

References

1. Cheng, Y.: Finding the limit of incompleteness I. arXiv:1902.06658v2
2. Damnjanovic, Z.: Mutual interpretability of Robinson arithmetic and adjunctive set theory. Bull. Symb. Logic **23**, 381–404 (2017)
3. Grzegorczyk, A., Zdanowski, K.: Undecidability and concatenation. In: Ehrenfeucht, A., et al. (eds.) Andrzej Mostowski and Foundational Studies, pp. 72–91. IOS, Amsterdam (2008)
4. Higuchi, K., Horihata, Y.: Weak theories of concatenation and minimal essentially undecidable theories. Arch. Math. Logic **53**, 835–853 (2014). https://doi.org/10.1007/s00153-014-0391-x
5. Jerabek, E.: Recursive functions and existentially closed structures. J. Math. Logic (2019). https://doi.org/10.1142/S0219061320500026
6. Jones, J., Shepherdson, J.: Variants of Robinson's essentially undecidable theory R. Archiv für mathematische Logik und Grundlagenforschung **23**, 61–64 (1983)
7. Kristiansen, L., Murwanashyaka, J.: First-order concatenation theory with bounded quantifiers. Arch. Math. Logic (accepted)
8. Svejdar, V.: An interpretation of Robinson arithmetic in its Grzegorczyk's weaker variant. Fundamenta Informaticae **81**, 347–354 (2007)
9. Tarski, A., Mostowski, A., Robinson, R.M.: Undecidable Theories. North-Holland, Amsterdam (1953)
10. Visser, A.: Growing commas. A study of sequentiality and concatenation. Notre Dame J. Formal Logic **50**, 61–85 (2009)
11. Visser, A.: Why the theory R is special. In: Tennant, N. (ed.) Foundational Adventures. Essays in Honour of Harvey Friedman, pp. 7–23. College Publications, UK (2014)

On the Complexity of Conversion Between Classic Real Number Representations

Lars Kristiansen[1,2] and Jakob Grue Simonsen[3(⊠)]

[1] Department of Mathematics, University of Oslo, Oslo, Norway
[2] Department of Informatics, University of Oslo, Oslo, Norway
[3] Department of Computer Science, DIKU, University of Copenhagen,
Copenhagen, Denmark
simonsen@diku.dk

Abstract. It is known that while it is possible to convert between many different representations of irrational numbers (e.g., between Dedekind cuts and Cauchy sequences), it is in general not possible to do so subrecursively: conversions in general need to perform unbounded search. This raises the question of categorizing the pairs of representations between which either subrecursive conversion is possible, or is not possible.

The purpose of this paper is to prove the following positive result: for a number of well-known representations (Beatty sequences, Dedekind cuts, General base expansions, Hurwitz characteristics, and Locators) conversion between the representations can be performed effectively and with good subrecursive bounds.

Keywords: Computable analysis · Computational complexity · Subrecursion · Representation of irrational numbers

The benefits of various representations of real numbers by computable functions is well-studied [10, 16–18, 21–23], and it is a standard result that the set of "computable reals" is the same in most representations, but that uniformly computable conversion between different representations is not always possible [18, 22]. When computable conversion *is* possible, it is in general necessary to perform unbounded search, and efficient, or *subrecursive*, conversion cannot be done. For example, for any sufficiently large subrecursive class S of functions satisfying mild conditions (e.g., the set of primitive recursive functions or the set of Kalmár elementary functions), write S_F, S_D, and S_C for the sets of irrational numbers representable by continued fractions, Dedekind cuts, and rapidly converging Cauchy sequences computable by functions in S. Then it is known that $S_F \subsetneq S_D \subsetneq S_C$ [13, 21], and thus *a fortiori* there can in general be no S-computable uniform conversion from the Cauchy representation to the Dedekind cut representation, or from the Dedekind cut representation to the continued fraction representation (for, if S is closed under composition, such uniform conversions would imply $S_F = S_D = S_C$).

© Springer Nature Switzerland AG 2020
M. Anselmo et al. (Eds.): CiE 2020, LNCS 12098, pp. 75–86, 2020.
https://doi.org/10.1007/978-3-030-51466-2_7

In this paper we derive upper bounds on the computational complexity of conversion between various representations of irrational numbers where subrecursive conversion *is* possible. In general, an irrational α in some representation R_1 will be computable by some function f (e.g., for the continued fraction representation $[3; 7, 15, 1, 292, \ldots]$ of π, $f(0) = 3, f(1) = 7, \ldots$), and in some other representation R_2 by some function g (e.g., for the decimal expansion of π, $g(0) = 3, g(1) = 1, g(2) = 4$); we are interested in the computational resources required to compute $g(n)$ when given access to the function f—in general this will require both querying the function f a number of times and performing a number of other operations, which we collectively call the "overhead" of conversion. In addition to the intrinsic value of this is the consequence that, roughly, if the overhead is a function in S, and S satisfies natural closure properties, it follows that $\mathcal{S}_{R_1} \subseteq \mathcal{S}_{R_2}$.

The results of the paper are shown in the diagram below; all results in the *left-hand side* of the diagram are proved explicitly in the paper. The arrows in the *right-hand side* of the diagram are known from the literature [13–15]; we defer more precise bounds on these (to wit, the existence of primitive recursive bounds) to future research.

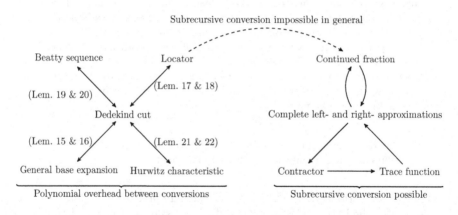

For the purposes of the present paper, we are only interested in *upper bounds* on conversion overhead. Our results show that conversion between Dedekind cuts and other classic representations can be done with polynomial overhead (and, by composition, conversion between any two of the considered representations in the left-hand side of the diagram above can be done with exponential overhead). More involved algorithms than the one we present here can indubitably be made, forcing the upper bounds to be low-degree polynomials.

Certain prior results are known for the representations we consider, but at a much coarser level of granularity; for example, Lehman [17] proved that the Hurwitz characteristic of α is primitive recursive iff the Dedekind cut of α is primitive recursive.

Remark 1. The overhead we consider is polynomial in the *value* of the index of the desired approximation to an irrational; for example, if one wants to compute the nth digit in the base-b expansion of an irrational $\alpha \in (0,1)$ from a Dedekind cut of α, the overhead will be polynomial in n and b (as opposed to polynomial in the binary representations of n and b). Note that accordingly, the conversions we consider are thus computable in exponential time in the binary representations of n and b.

Remark 2. As the representations in the above diagrams are most easily expressed using functions, we believe that the natural formalizations for conversions are Turing machines with oracle access to the representations being converted from. In other work on real number computation, there is a well-developed notion of reducibility between representations that, roughly, requires the representation to be written as an infinite string on one of the input tapes of a type-2 Turing machine [6,12,22,24]. In that setting, for example, a function $f : \mathbb{Q} \cap [0,1] \longrightarrow \{0,\ldots,b-1\}$ is most naturally expressed by imposing a computable ordering on its domain (e.g., rationals appear in non-decreasing order of their denominator), and the function values $f(q)$ appear encoded as bit strings in this order. We strongly conjecture that our results carry over to the type-2 setting *mutatis mutandis*.

1   Preliminaries

We assume basic familiary with computability and computational complexity (standard textbooks are [1,8,20]). We write $f(n) = \mathrm{poly}(n)$ if $f : \mathbb{N} \longrightarrow \mathbb{N}$ is bounded above by a polynomial in n with positive integer coefficients, and $f(n) = \mathrm{polylog}(n)$ if f is bounded above by a polynomial in $\log n$ with positive integer coefficients.

We first define oracle machines in the usual way:

Definition 3. *A (parameterized) function-oracle Turing machine is a (multi-tape) Turing machine $M = (Q, q_0, F, \Sigma, \Gamma, \delta)$ with initial state $q_0 \in Q$, final states $F \subseteq Q$, input and tape alphabets Σ and Γ (with $\Sigma \subseteq \Gamma$ and $\{\_\} \subseteq \Gamma \setminus \Sigma$), and partial transition function δ such that M has a special query tape and two distinct states $q_q, q_a \in Q$ (the query and answer states).*

To be executed, M is provided with a total function $f : \mathbb{N} \longrightarrow \mathbb{N}$ (the oracle) prior to execution on any input. We write M^f for M when f has been fixed–note that M^f is then a usual (non-parameterized) function-oracle machine [8]. The transition relation of M^f is defined as usual for Turing machines, except for the state q_q: If M enters state q_q, let m be (a representation in the tape alphabet of) the value currently on the query tape; M moves to state q_a in a single step, and the contents of the query tape are instantaneously changed to (the representation in the tape alphabet of) $f(m)$. The time- and space complexity of a function-oracle machine is counted as for usual Turing machines, with the transition between q_q and q_a taking $|f(m)|$ time steps, and the space use of the query tape after the transition being $|f(m)|$. The (input) size of a query is the number of symbols on the query tape when M enters state q_q.

Function-oracle machines are in standard use in complexity theory of functions on the set of real numbers (see, e.g., [11]).

Remark 4. All input and output tapes of Turing machines are assumed to have alphabet $\{0,1\}$ in addition to the blank symbol. All work tapes have alphabet $\{0,1,2\}$ in addition to the blank symbol. Unless otherwise stated, all elements of $\mathbb{N}$, $\mathbb{Z}$, and elements of any finite set, are assumed to be written on input, query, and output tapes in their binary representation. Pairs (p,q) of integers are assumed to be written using interleaved notation (i.e., the first bit of the binary representation of p followed by the first bit of the binary representation of q, and so forth). In case the representations have unequal length, the shortest binary representation is padded with '2' in the interleaving. Observe that the length of the representation of a pair (p,q) is then $O(\log \max\{p,q\})$. All elements $p/q \in \mathbb{Q}$ are assumed to be represented by the representation of (p,q).

As expected, using the semantic function of a function-oracle Turing machine M with oracle to f as the oracle of another function-oracle Turing machine N can be made to "cut out the middleman machine M"; that is, we could use a single oracle machine with an oracle to f with bounds on time and oracle use not much higher than the original machines M and N:

Proposition 5 (Compositionality). *Let M and N be parameterized function-oracle machines and let f be a map. Write $g = \phi_{M^f} : A \longrightarrow B$ and $\phi_{N^g} : C \longrightarrow D$ for sets A, B, C, D (all representable by elements of $\{0,1\}^+$). Let $t_M, t_N, q_M, q_N, s : \mathbb{N} \longrightarrow \mathbb{N}$ be maps. Suppose that*

1. *M^f on input $a \in A$ computes $g(a)$ in time $t_M(|a|)$ using $q_M(|a|)$ queries to f, and*
2. *N^g on input $c \in C$ computes ϕ_{N^g} in time $t_N(|c|)$ using $q_N(|c|)$ queries to g, each of input size at most $s(|c|)$.*

Then there is a parameterized function-oracle machine P such that $\phi_P^f = \phi_{N^g}$, and P^f on input $c \in C$ runs in time at most

$$O(q_N(|c|)t_M(s(|c|)) + t_N(|c|))$$

using at most

$$q_N(|c|)q_M(s(|c|))$$

queries to f.

Proof. P is merely N with the original oracle tape replaced by two new work tapes, a new oracle tape added, and each query to g replaced by execution of a copy of M, with the new work tapes functioning as the "input" and "output" tapes of the copy of M, and the new oracle tape as the oracle tape of the copy of M. Every time N would query g, it writes the query a on the new "input" work tape. The copy of M then computes $g(a)$ using time at most $t_M(s(|c|))$ with $q_M(s(|c|))$ queries to f, hence a total of $q_N(|c|)q_M(s(|c|))$ queries to f. The total time spent by P is the time spent by N plus at most $t_M(s(|c|))$ steps per oracle query, for a total of $O(q_N(|c|)t_M(s(|c|)) + t_N(|c|))$ steps. □

In particular, function-oracle machines running in polynomial time and having queries of polynomial input size are composable in the above way and yield new machines running in polynomial time.

1.1   Farey Sequences and the Stern-Brocot Tree

A *Farey sequence* is a strictly increasing sequence of fractions between 0 and 1. The Farey sequence of *order k*, denoted F_k, contains all fractions which when written in their lowest terms, have denominators less than or equal to k.

Let a/b and c/d be fractions in their lowest terms. The fraction $m(a/b, c/d) = (a + c)/(b + d)$ is called the *medium* of a/b and c/d. The ordered pair of two consecutive fractions in a Farey sequence is called a *Farey pair*. It is easy to see that $a/b < m(a/b, c/d) < c/d$ if $a/b \neq c/d$.

We arrange the fractions strictly between 0 and 1 in a binary search tree $\mathcal{T}_F$.

Definition 6. *The* Farey pair tree $\mathcal{T}_F$ *is the complete infinite binary tree where each node has an associated Farey pair $(a/b, c/d)$ defined by recursion on the position $\sigma \in \{0, 1\}^*$ of a node in $\mathcal{T}_F$ as follows: $\mathcal{T}_F(\epsilon) = (0/1, 1/1)$, and if $\mathcal{T}_F(\sigma) = (a/b, c/d)$, then $\mathcal{T}_F(\sigma 0) = (a/b, (a+c)/(b+d))$ and $\mathcal{T}_F(\sigma 1) = ((a+c)/(b+d), c/d)$. The depth of a node in $\mathcal{T}_F$ is the length of its position (with the depth of the root node being 0).*

Abusing notation slightly, we do not distinguish between the pair $\mathcal{T}_F(\sigma) = (a/b, c/d)$ and the open interval $(a/b, c/d)$.

The (left) Stern-Brocot tree[1] $\mathcal{T}_{SB}$ *is the infinite binary tree obtained by the Farey pair tree where each Farey pair $(a/b, c/d)$ has been replaced by its medium $m(a/b, c/d) = (a + c)/(b + d)$.*

Thus, we have, for example:

$$\mathcal{T}_F(0) = \left(\frac{0}{1}, \frac{1}{2}\right), \ \mathcal{T}_F(1) = \left(\frac{1}{2}, \frac{1}{1}\right), \ \mathcal{T}_F(10) = \left(\frac{1}{2}, \frac{2}{3}\right), \ \mathcal{T}_F(0000) = \left(\frac{0}{1}, \frac{1}{5}\right)$$

We shall later need the two following ancillary propositions which we include without proof.

Proposition 7. *Let $p/q \in \mathbb{Q} \cap [0, 1]$ be a fraction in its lowest terms. Then, p/q is a fraction in a Farey pair at depth at most $p + q - 1$ in $\mathcal{T}_F$.*

Efficient computations of the elements of the Stern-Brocot tree (and hence also the Farey pair tree) is possible [2,19]; for our purposes, we simply need the following result:

Proposition 8. *There is a Turing machine M such that for any $\sigma \in \{0, 1\}^*$, $\phi_M(\sigma) = \mathcal{T}_F(\sigma)$ and M runs in time $\mathrm{poly}(1 + |\sigma|)$.*

---

[1] "Left" because the Stern-Brocot tree originally concerns the interval $(0, 2)$ and we are interested only in $(0, 1)$ which corresponds to the left child of the Stern-Brocot tree.

2   Representations

We now introduce a number of well-known representations of real numbers. Representations by Dedekind cuts [4,7], Beatty sequences [3][2], and Hurwitz characteristics [9][3] were known in the 19th century or earlier. The representations by locators and general base expansions are, to our knowledge, new, but natural. In particular, the general base expansion yields the base-b expansions of α in *any* integer base $b \geq 2$ on demand; it turns out that this is the key to subrecursive equivalence with Dedekind cuts (whereas the base-b expansion for any *fixed* b is not subrecursively equivalent to Dedekind cuts, see [14]).

Definition 9. *A function* $D : \mathbb{Q} \longrightarrow \{0,1\}$ *is the (left) Dedekind cut of the irrational number* α *when* $D(q) = 0$ *iff* $q < \alpha$.

Definition 10. *A function* $B : \mathbb{N} \longrightarrow \mathbb{Z}$ *is the (function computing the) Beatty sequence of the irrational number* α *when*

$$\frac{B(q)}{q} \; < \; \alpha \; < \; \frac{B(q)+1}{q}$$

Definition 11. *A function* $L_\alpha : \mathbb{Q} \times \mathbb{Q} \longrightarrow \{0,1\}$ *is the locator of the real number* α *when* $L_\alpha(p,q) = 0$ *iff* α *is in the interval* (p,q).

Definition 12. *Let* $(0.D_1 D_2 \ldots)_b$ *be the base-b expansion of the real number* α. *We define the function* $E_b^\alpha : \mathbb{N} \longrightarrow \{0,\ldots,b-1\}$ *by* $E_b^\alpha(0) = 0$ *and* $E_b^\alpha(i) = D_i$ *(for* $i \geq 1$*).*

A general base expansion *of the real number* α *is the function*

$$E : (\mathbb{N} \setminus \{0,1\}) \times \mathbb{N} \longrightarrow \{0,\ldots,b-1\}$$

where $E(b,q) = E_b^\alpha(q)$.

Definition 13. *The* Hurwitz characteristic *of the irrational number* $\alpha \in (0,1)$ *is the map* $H : \mathbb{N} \longrightarrow \{0,1\}^*$ *such that* $H(0), H(1), H(2), \ldots$ *is a path in the Stern-Brocot three, and:* [4]

$$\alpha = \lim_{q \to \infty} m(\mathcal{T}_{\mathrm{F}}(H(q))) = \lim_{q \to \infty} \mathcal{T}_{\mathrm{SB}}(H(q))$$

---

[2] Apparently, what is now known as Beatty sequences was used earlier by Bernard Bolzano [5], whence this representation of reals could also be called *Bolzano measures*.

[3] Use of the Hurwitz characteristic to represent numbers rather than a stepping stone for other material is a much younger invention [17].

[4] Strictly speaking, the classic Hurwitz characteristic corresponds to a path through the full Stern-Brocot tree (not just the "left" tree as we consider here), and hence the classic Hurwitz characteristic H' of $\alpha \in (0,1)$ is the function defined by $H'(0) = 0$ and $H'(q) = 0 \cdot H(q-1)$ for $q > 0$. This does not change our results in any material way.

3   Representations Subrecursively Equivalent to Dedekind Cuts

The remainder of the paper is devoted to proving the following theorem:

Theorem 14. *For each representation R below there is a parameterized function-oracle machine M such that, for every irrational α between 0 and 1, M when provided with an oracle to the R-representation of α, will compute the Dedekind cut of R, and N when provided with an oracle to the Dedekind cut of α, will compute the R-representation of α. Let n be the largest integer in the input (i.e., n if domain of R is $\mathbb{N}$, $n = \max\{n_1, n_2\}$ if the domain of R is $\mathbb{N} \times \mathbb{N}$, $\max\{p, q\}$ if the domain of R is $\mathbb{Q}$, and $\max\{p_1, q_1, p_2, q_2\}$ if the domain of R is $\mathbb{Q} \times \mathbb{Q}$). Then, M and N will produce their output in time $\mathrm{poly}(n)$ using at most $\mathrm{poly}(n)$ oracle calls of size at most $\mathrm{poly}(n)$.*

- *the locator of α*
- *the Beatty sequence of α*
- *the general base expansion of α*
- *the Hurwitz characteristic of α*

Furthermore, conversion between any two of the above representations (e.g., from the locator of α to the Beatty sequence of α) can be done by function-oracle machines producing their output using exponential (in n) time, exponential (in n) number of oracle calls, and exponential (in n) size of oracle calls.

The proof of conversion from and to Dedekind cuts is contained in the sequence of lemmas below that all relate the various representations to the Dedekind cut. All lemmas assert existence of parameterized function-oracle machines that will convert *from* or *to* the Dedekind cut of α with polynomial overhead (often with smaller overhead, whence the result follows *a fortiori*). The result that we can convert between any of the representations using exponential overhead then follows by an application of Proposition 5.

3.1   Conversion Between General Base Expansions and Dedekind Cuts

Lemma 15. *There is a parameterized function-oracle Turing machine M such that if $D : \mathbb{Q} \longrightarrow \{0, 1\}$ is the Dedekind cut of any irrational number $\alpha \in (0, 1)$, then $\phi_M^D : (\mathbb{N} \setminus \{0, 1\}) \times \mathbb{N} \longrightarrow \{0, \ldots, b - 1\}$ is the general base expansion of α. Moreover, on input $(b, n) \in (\mathbb{N} \setminus \{0, 1\}) \times \mathbb{N}$, M^D runs in time $O(b^2 \mathrm{poly}(n))$, and uses at most $n \log_2 b$ oracle calls, each of input size at most $O(n \log_2 b)$ bits.*

Proof. M constructs the sequence $E_b^\alpha(1), E_b^\alpha(2), \ldots, E_b^\alpha(n)$ inductively by maintaining an open interval $I_i = (v_i, w_i)$ with rational endpoints $v_i, w_i \in \mathbb{Q}$ for each $i \in \{0, \ldots, n-1\}$ such that (i) $\alpha \in I_i$, (ii) v_i is a multiple of b^{-i}, and (iii) $w_i - v_i = b^{-i}$. Initially, $I_0 = (0, 1)$. For each interval $I_i = (v_i, w_i)$, M splits $I_i = (v_i, w_i)$ into b equal-sized intervals

$$(v_i, v_i + b^{-(i+1)}), \ldots, (v_i + (b-1)b^{-(i+1)}, v_i + b^{-i}) = (v_i + (b-1)b^{-(i+1)}, w_i)$$

Observe that, for any interval (r_1, r_2), if $D(r_1) = D(r_2) = 0$, then $\alpha > r_2$, and if $D(r_1) = D(r_2) = 1$, then $\alpha < r_1$ (and the case $D(r_1) = 1 \wedge D(r_2) = 0$ is not possible). Thus, M can use D to perform binary search on (the endpoints of) the above set of intervals to find the (necessarily unique) interval $(v_i + jb^{-(i+1)}, v_i + (j+1)b^{-(i+1)})$ that contains α (observe that, for this interval, $D(v_i + jb^{-(i+1)}) = 0$ and $D(v_i + (j+1)b^{-(i+1)}) = 1$). We then set $(v_{i+1}, w_{i+1}) = (v_i + jb^{-(i+1)}, v_i + (j+1)b^{-(i+1)})$. By construction, we have $E^\alpha(b, i+1) = j$. Clearly, in each step i, there are at most $\log_2 b$ oracle calls to D, and the construction of each of the b intervals and writing on the query tape can be performed in time polynomial in the binary representation of the numbers involved, hence in time $O(\text{polylog}(b^i)) = O(\text{poly}(i)\text{polylog}(b))$. Hence, the total time needed to produce $E(b, n)$ is at most $O(bn\text{poly}(n)\text{polylog}(b)) = O(b^2\text{poly}(n))$ with at most $n\log_2 b$ queries to D. In each oracle call, the rational numbers involved are all endpoints of intervals where the endpoints are sums of negative powers of b and where the exponent of all powers are at most n. Hence, all oracle calls can be represented by rational numbers using at most $O(n \log_2 b)$ bits. $\qquad\square$

Lemma 16. *There is a parameterized function-oracle Turing machine M such that if $E : (\mathbb{N} \setminus \{0, 1\}) \times \mathbb{N} \longrightarrow \{0, \ldots, b-1\}$ is the general base expansion of α, then $\phi_M^E : \mathbb{Q} \longrightarrow \{0, 1\}$ is the Dedekind cut of α. Moreover, on input $p/q \in \mathbb{Q}$, M^E runs in time $O(\log(\max\{p, q\}))$, and uses exactly 1 oracle call of input size at most $O(\log(q))$.*

Proof. On input $p/q \in \mathbb{Q}$, M first checks if $q = 1$, and outputs 0 if $p \le 0$ and 1 if $p \ge 1$. Otherwise, $q > 1$, and M computes $E(q, 1)$; by definition, this is an element of $\{0, \ldots, q-1\}$. Observe that $D(p/q) = 0$ iff $p/q < \alpha$ iff $p \le E(q, 1)$. Hence, M outputs 0 if $p \le E(q, 1)$, and outputs 1 otherwise. M needs to write the (representation of the) pair $(q, 1)$ on the oracle tape and perform a single comparison of numbers of magnitude at most $\max\{p, q\}$, hence M uses time $O(\log \max\{p, q\})$ for the comparison. M uses exactly one oracle call to E with the pair $(q, 1)$, the representation of which uses at most $O(\log q)$ bits. $\qquad\square$

3.2 Conversion Between Locators and Dedekind Cuts

Lemma 17. *There is a parameterized function-oracle Turing machine M such that if $D : \mathbb{Q} \longrightarrow \{0, 1\}$ is the Dedekind cut of any irrational number $\alpha \in (0, 1)$, then $\phi_M^D : \mathbb{Q} \times \mathbb{Q} \longrightarrow \{0, 1\}$ is the locator of α. Moreover, on input $(p_1/q_1, p_2/q_2) \in \mathbb{Q} \times \mathbb{Q}$, M^D runs in time $O(\log(\max\{p_1, q_1, p_2, q_2\}))$, and uses at most 2 oracle calls, each of input size at most $O(\log \max\{p_1, q_1, p_2, q_2\})$.*

Proof. Let L be the locator of α. Observe that for any two rational numbers $p_1/q_1, p_2/q_2 \in \mathbb{Q}$, we have $\alpha \in (p_1/q_1, p_2/q_2)$ iff $L(p_1/q_1, p_2/q_2) = 0$ iff $(D(p_1/q_1) = 0 \wedge D(p_2/q_2) = 1)$. Hence, M simply queries D (using the binary representations of the rationals) twice, outputs 1 if $(D(p_1/q_1) = 0 \wedge D(p_2/q_2) = 1)$, and outputs 1 otherwise. Clearly, the time needed is the time needed to transfer p_1/q_1 and p_2/q_2 to the query tape plus some constant independent of the size of the input, hence M uses time $O(\log(\max\{p_1, q_1, p_2, q_2\}))$. $\qquad\square$

Lemma 18. *There is a parameterized function-oracle Turing machine M such that if $L : \mathbb{Q} \times \mathbb{Q} \longrightarrow \{0,1\}$ is the locator of any irrational number $\alpha \in (0,1)$, then $\phi_M^L : \mathbb{Q} \longrightarrow \{0,1\}$ is the Dedekind cut of α. Moreover, on input $p/q \in \mathbb{Q}$, M^L runs in time $O(\text{polylog}(\max\{p,q\}))$, and uses at most 1 oracle call of input size at most $O(\log \max\{p,q\})$.*

Proof. Observe that for $p/q \in (0,1) \cap \mathbb{Q}$, we have $D(p/q) = L(p/q, 1)$. Hence, M may, on input p/q simply perform a single query to L; this requires copying its input to the oracle tape, i.e. only linear time in the size of the representation of the input. By convention, the input p/q is representable in at most $O(\log \max\{p,q\})$ bits. $\qquad\square$

3.3 Conversion Between Beatty Sequences and Dedekind Cuts

Lemma 19. *There is a parameterized function-oracle Turing machine M such that if $D : \mathbb{Q} \longrightarrow \{0,1\}$ is the Dedekind cut of any irrational number $\alpha \in (0,1)$, then $\phi_M^D : \mathbb{N} \longrightarrow \mathbb{Z}$ is the Beatty sequence of α. Moreover, on input $n \in \mathbb{N}$, M^D runs in time $O(\text{polylog}(n))$, and uses at most $\lceil \log n \rceil$ oracle calls to D, each of input size at most $O(\log n)$.*

Proof. On input n, M finds the least $i \in \{1, \dots, n\}$ such that $D(\frac{i}{n}) = 1$. As $D(\frac{i}{n}) = 1$ and $j > i$ implies $D(\frac{j}{n}) = 1$, the least i can be found by binary search, halving the search range in each step[5]. This can be done by maintaining two integers l and u ranging in $\{1, \dots, n\}$, and requires a maximum of $\log n$ halving steps. In each halving step, M finds the midpoint m between l and u, writes its binary representation on the query tape, queries D, and records the answer. Then, l and u are updated using basic binary arithmetic operations on integers, represented by at most $O(\log n)$ bits—if $D(m/n) = 1$, $u := m$, and if $D(m/n) = 0$, $l := m$. Clearly, in each step, the arithmetic and update operations can be performed in time polynomial in the size of the representation of the integers, hence in time $\text{polylog}(n)$. As $(i-1)/n < \alpha < i/n$, we have $B(n) = i - 1$, and M^D thus returns $i - 1$. $\qquad\square$

Lemma 20. *There is a parameterized function-oracle Turing machine M such that if $B : \mathbb{N} \longrightarrow \mathbb{Z}$ is the Beatty sequence of any irrational number $\alpha \in (0,1)$, then $\phi_M^B : \mathbb{Q} \longrightarrow \{0,1\}$ is the Dedekind cut of α. Moreover, on input $p/q \in \mathbb{Q}$, M^B runs in time $O(\log(\max\{p,q\}))$, and uses exactly one oracle call of input size $O(\log q)$.*

Proof. Observe that the Dedekind cut D of α satisfies $D(p/q) = 0$ if $p \leq B(q)$, and $D(p/q) = 1$ if $p > B(q)$. Thus, M may perform the oracle call $B(q)$ just once, resulting in an integer $B(q)$ (where $B(q) \in \{0, 1 \dots, q-1\}$). The comparison $p \leq B(q)$ can be performed bitwise using the binary representations of p and $B(q)$ which is clearly linear in $\log(\max\{B(q), p\}) \leq \log \max\{p, q\}$. Writing q on the oracle tape clearly also takes time linear in $\log q$. $\qquad\square$

---

[5] Observe that a brute-force search is also possible, yielding at most n oracle calls with input size at most $O(\log n)$ and obviating the need to reason about arithmetic operations.

3.4   Conversion Between Hurwitz Characteristics and Dedekind Cuts

Lemma 21. *There is a parameterized function-oracle Turing machine M such that if $H : \mathbb{N} \longrightarrow \{0,1\}^*$ is the Hurwitz characteristic of any irrational number $\alpha \in (0,1)$, then $\phi_M^H : \mathbb{Q} \longrightarrow \{0,1\}$ is the Dedekind cut of α. Moreover, on input $p/q \in \mathbb{Q}$, M^H runs in time $\operatorname{poly}(\max\{p,q\})$, and uses exactly one oracle call of input size at most $O(\log \max\{p,q\})$.*

Proof. On input $p/q \in \mathbb{Q}$ (where we assume wlog. that p/q is reduced to lowest terms), M computes $H(p+q)$ (using $\operatorname{polylog}(\max\{p,q\})$ operations to compute the binary representation of $p + q$, and then performing a single oracle call; note that the result of the oracle $H(p + q)$ is a bit string of length exactly $p+q = \operatorname{poly}(\max\{p,q\})$. M then computes $\mathcal{T}_F(H(p+q))$ (by Proposition 8 this can be done in time $\operatorname{poly}(1 + |H(p + q)|) = \operatorname{poly}(\max\{p,q\}))$ to obtain a Farey pair $(a/b, c/d)$ such that $a/b < \alpha < c/d$. By Proposition 7, any reduced fraction p/q occurs as one of the fractions in a Farey pair in $\mathcal{T}_F$ at depth at most $p+q-1$, and thus exactly one of (i) $p/q \leq a/b$ and (ii) $c/d \leq p/q$ must hold. Observe that $D(p/q) = 0$ iff $p/q \leq a/b$. Whether (i) or (ii) holds can be tested in time $O(\log \max\{a,b,c,d,p,q\})$. It is an easy induction on the depth d to see that a numerator or denominator in any fraction occurring in a Farey pair at depth d in $\mathcal{T}_F$ is at most 2^d. Hence, $\max\{a,b,c,d,p,q\} \leq 2^{p+q}$, and the test can thus be performed in time $O(p + q) = O(\max\{p,q\})$. Thus, M needs a total time of $\operatorname{poly}(\max\{p,q\})$. □

Lemma 22. *There is a parameterized function-oracle Turing machine M such that if $D : \mathbb{Q} \longrightarrow \{0,1\}$ is the Dedekind cut of any irrational number $\alpha \in (0,1)$, then $\phi_M^D : \mathbb{N} \longrightarrow \{0,1\}^*$ is the Hurwitz characteristic of α. Moreover, on input $n \in \mathbb{N}$, M^D runs in time $\operatorname{poly}(n)$, and uses exactly n oracle calls, each of input size at most $O(n)$.*

Proof. On input n, M constructs a path of length n in the tree $\mathcal{T}_{SB}$ corresponding to the bit string $H(n)$ by building the corresponding path in $\mathcal{T}_F$. M can do this by starting at $i = 0$ and incrementing i, maintaining a *current* Farey pair $(a_i/b_i, c_i/d_i)$ such that $\alpha \in (a_i/b_i; c_i/d_i)$ for $i = 0,\ldots,n$ as the mediant of $(a_i/b_i, c_i/d_i)$ gives rise to the two children $p_L = (a_i/b_i, (a_i + c_i)/(b_i + d_i))$ and $p_R = ((a_i + c_i)/(b_i + d_i), c_i/d_i)$ of $(a_i/b_i, c_i/d_i)$ in $\mathcal{T}_F$. Because α is irrational, it must be in exactly one of the open intervals $(a_i/b_i, (a_i + c_i)/(b_i + d_i))$ and $((a_i + c_i)/(b_i + d_i), c_i/d_i)$, and thus $(a_{i+1}/b_{i+1}, c_{i+1}/d_{i+1})$ must be either p_L or p_R. Clearly, $\alpha \in (a_i/b_i, (a_i+c_i)/(b_i+d_i))$ iff $D(a_i+c_i/b_i+d_i) = 1$ iff the ith bit of $H(n)$ is 0. Hence, M starts with $(a_0/b_0, c_0/d_0) = (0/1, 1/1)$, and constructs the n intervals $(a_i/b_i, c_i/d_i)$ for $i = 1,\ldots,n$ by computing the mediant and querying D in each step. Observe that the query in step i is the (binary representation of the) mediant of a Farey pair at depth $i - 1$, thus its denominator is bounded above by 2^i and its binary representation uses at most $O(\log 2^i) = O(i)$ bits.

As the numerators and denominators at depth i in $\mathcal{T}_F$ are of size at most 2^i (hence representable by i bits), computing the mediant at step i can be done in time at most $O(i) = O(n)$ by two standard schoolbook additions, and the step i contains exactly one query to D. Hence, the total time needed for M to construct $H(n)$ is at most $O(n\text{poly}(n)) = \text{poly}(n)$, with exactly n oracle calls, each of size at most $O(\log 2^n) = O(n)$. □

4   Conclusion and Future Work

We have analyzed conversions between representations equivalent to Dedekind cuts, and we have seen that we can convert efficiently between any two such representations (Theorem 14) . We strongly conjecture that the same efficiency is not possible between representations equivalent to continued fractions. Indeed, we regard the representations equivalent to continued fractions to be the most interesting and challenging ones from a mathematical point of view. Among these representations we find the trace functions and the contractors (see the figure on Page 2). A function $T : \mathbb{Q} \to \mathbb{Q}$ is a *trace function* for the irrational number α when $|\alpha - r| > |\alpha - T(r)|$. A function $F : [0,1] \to (0,1)$ is a *contractor* if we have $|F(r_1) - F(r_2)| < |r_1 - r_2|$ for any rationals r_1, r_2 where $r_1 \neq r_2$. Both trace functions and contractors can be converted to (and from) continued fractions without unbounded search, and converting from a contractor to a trace function is easy as it can be proved that every contractor is a trace function. But conversely, we believe that it is not possible to convert a trace function to a contractor within reasonably small time or space bounds.

Conversions between representations equivalent to rapidly converging Cauchy sequences also deserve a further study. One such representation will be base-2 expansions over the digits 0 (zero), 1 (one) and $\bar{1}$ (minus one). In this representation, the rational number $1/4$ can be written as 0.01, but also as $0.1\bar{1}$. Another interesting representation are the *fuzzy Dedekind cuts*. A fuzzy Dedekind cut for an irrational number α is a function $D : \mathbb{Z} \times \mathbb{N} \to \{0,1\}$ satisfying (i) $D(p,q) = 0$ implies $\alpha < (p+1)/q$ and (ii) $D(p,q) = 1$ implies $(p-1)/q < \alpha$. Thus, each irrational α will have (infinitely) many fuzzy Dedekind cuts. If D is a fuzzy cut for α and we know that $D(3,8) = 0$, then we know that α lies below $4/8$ (but we do not know if α lies below $3/8$). Moreover, if we also know that $D(6,16) = 1$, then we know that α lies in the interval $(3/8 - 1/16, 3/8 + 1/8)$ (but we do not know if α lies below or above $3/8$).

Finally, some well-known representations are not subrecursively equivalent to any of the three representations above, for example the base-b representation for any integer base $b \geq 2$. It is possible to convert a Dedekind cut to a base-b expansion and a base-b expansion into a Cauchy sequence without unbounded search, but not the other way around [14,21]. It is interesting to investigate the set of representations subrecursively equivalent to such expansions.

Acknowledgment. We are grateful for the meticulous comments of one of the referees; these have helped to significantly improve the paper.

References

1. Arora, S., Barak, B.: Computational Complexity: A Modern Approach. Cambridge University Press, Cambridge (2009)
2. Bates, B., Bunder, M., Tognetti, K.: Locating terms in the Stern-Brocot tree. Eur. J. Comb. **31**(3), 1020–1033 (2010)
3. Beatty, S., et al.: Problems for solutions: 3173–3180. Am. Math. Mon. **33**(3), 159–159 (1926)
4. Bertrand, J.: Traité d'arithmétique (1849)
5. Bolzano, B.: Pure Theory of Numbers. Oxford University Press, Oxford (2004). In the Mathematical Works of Bernard Bolzano edited and translated by Steve Russ, pp. 355–428
6. Brattka, V., Hertling, P.: Topological properties of real number representations. Theoret. Comput. Sci. **284**(2), 241–257 (2002)
7. Dedekind, R.: Stetigkeit und irrationale Zahlen. Vieweg, Braunschweig (1872)
8. Du, D.Z., Ko, K.I.: Theory of Computational Complexity. Wiley Interscience (2000)
9. Hurwitz, A.: Ueber die angenäherte Darstellung der Irrationalzahlen durch rationale Brüche. Mathematische Annalen **39**, 279–284 (1891)
10. Ko, K.: On the definitions of some complexity classes of real numbers. Math. Syst. Theory **16**, 95–109 (1983)
11. Ko, K., Friedman, H.: Computational complexity of real functions. Theoret. Comput. Sci. **20**(3), 323–352 (1982)
12. Kreitz, C., Weihrauch, K.: Theory of representations. Theor. Comput. Sci. **38**, 35–53 (1985)
13. Kristiansen, L.: On subrecursive representability of irrational numbers. Computability **6**, 249–276 (2017)
14. Kristiansen, L.: On subrecursive representability of irrational numbers, part ii. Computability **8**, 43–65 (2019)
15. Kristiansen, L.: On subrecursive representability of irrational numbers: continued fractions and contraction maps. In: Proceedings of the 16th International Conference on Computability and Complexity in Analysis (CCA 2019) (2019)
16. Labhalla, S., Lombardi, H.: Real numbers, continued fractions and complexity classes. Ann. Pure Appl. Logic **50**(1), 1–28 (1990)
17. Lehman, R.S.: On primitive recursive real numbers. Fundamenta Mathematica **49**(2), 105–118 (1961)
18. Mostowski, A.: On computable sequences. Fundamenta Mathematica **44**, 37–51 (1957)
19. Niqui, M.: Exact arithmetic on the Stern-Brocot tree. J. Discrete Algorithms **5**(2), 356–379 (2007)
20. Sipser, M.: Introduction to the Theory of Computation. PWS Publishing Company (1997)
21. Specker, E.: Nicht konstruktiv beweisbare Sätze der Analysis. J. Symb. Logic **14**(3), 145–158 (1949)
22. Weihrauch, K.: The degrees of discontinuity of some translators between representations of real numbers. Technical report, Fernuniversität Hagen (1992)
23. Weihrauch, K.: Computable Analysis. Springer, Heidelberg (2000). https://doi.org/10.1007/978-3-642-56999-9
24. Weihrauch, K., Kreitz, C.: Representations of the real numbers and of the open subsets of the set of real numbers. Ann. Pure Appl. Logic **35**, 247–260 (1987)

Deterministic and Nondeterministic Iterated Uniform Finite-State Transducers: Computational and Descriptional Power

Martin Kutrib[1], Andreas Malcher[1], Carlo Mereghetti[2]([☒]) [iD],
and Beatrice Palano[3] [iD]

[1] Institut für Informatik, Universität Giessen, Arndtstr. 2, 35392 Giessen, Germany
{kutrib,andreas.malcher}@informatik.uni-giessen.de
[2] Dipartimento di Fisica "Aldo Pontremoli", Università degli Studi di Milano,
via Celoria 16, 20133 Milan, Italy
carlo.mereghetti@unimi.it
[3] Dipartimento di Informatica "G. degli Antoni", Università degli Studi di Milano,
via Celoria 18, 20133 Milan, Italy
palano@unimi.it

Abstract. An iterated uniform finite-state transducer (IUFST) operates the same length-preserving transduction, starting with a sweep on the input string and then iteratively sweeping on the output of the previous sweep. The IUFST accepts or rejects the input string by halting in an accepting or rejecting state along its sweeps. We consider both the deterministic (IUFST) and nondeterministic (NIUFST) version of this device. We show that constant sweep bounded IUFSTs and NIUFSTs accept all and only regular languages. We study the size cost of removing nondeterminism as well as sweeps on constant sweep bounded NIUFSTs, and the descriptional power of constant sweep bounded IUFSTs and NIUFSTs with respect to classical models of finite-state automata. Finally, we focus on non-constant sweep bounded devices, proving the existence of a proper infinite nonregular language hierarchy depending on the sweep complexity both in the deterministic and nondeterministic case. Also, we show that the nondeterministic devices are always more powerful than their deterministic variant if at least a logarithmic number of sweeps is given.

Keywords: Iterated transducers · State complexity · Sweep complexity · Language hierarchies

1 Introduction

The notion of an iterated uniform finite-state transducer (IUFST) has been introduced in [13] and can be described as a finite transducer that iteratively sweeps from left to right over the input tape while performing the same length-preserving transduction at each sweep. In particular, the output of the previous sweep is

© Springer Nature Switzerland AG 2020
M. Anselmo et al. (Eds.): CiE 2020, LNCS 12098, pp. 87–99, 2020.
https://doi.org/10.1007/978-3-030-51466-2_8

taken as input for every new transduction sweep. (Throughout the paper, the attribute "uniform" indicates that the transduction is identical at each sweep: the transduction always starts from the same initial state on the leftmost tape symbol, and operates the same transduction rules at each computation step.) This model is motivated by typical applications of transducers or cascades of transducers, where the output of one transducer is used as the input for the next transducer. For example, finite-state transducers are used for the lexical analysis of computer programs and the produced output is subsequently processed by pushdown automata for the syntactical analysis. In [7], cascades of finite-state transducers are used in natural language processing. Another example is the Krohn-Rhodes decomposition theorem which shows that every regular language is representable as the cascade of several finite-state transducers, each one having a "simple" algebraic structure [8,10]. Finally, it is shown in [6] that cascades of deterministic pushdown transducers lead to a proper infinite hierarchy in between the deterministic context-free and the deterministic context-sensitive languages with respect to the number of transducers involved.

In contrast to all these examples and other works in the literature (see, e.g., [5,16,18]), where the subsequently applied transducers are in principle different and not necessarily length-preserving, the model of IUFSTs introduced in [13] requires that the same transducer is applied in every sweep and that the transduction is deterministic and length-preserving. More precisely, an IUFST works in several sweeps on a tape which initially contains the input string concatenated with a right endmarker. In every sweep, the finite-state transducer starts in its initial state at the first tape cell, is applied to the tape, and prints its output on the tape. The input is accepted or rejected, if the transducer halts in an accepting or rejecting state. In [13], IUFSTs both with a constant number and a non-constant (in the length of the input) number of sweeps are investigated. In the former case, it is possible to characterize exactly the set of regular languages. Thus, tight upper and lower bounds for converting IUFSTs into deterministic finite automata (DFAs) and vice versa are established. Furthermore, as always done for several models (see, e.g., [1–3]), the state complexity of language operations, that is, the costs in terms of the number of states needed for union, intersection, complementation, and reversal, is investigated in depth. Finally, the usually studied decidability questions such as emptiness, finiteness, equivalence, and inclusion are proved to be NL-complete, showing that these questions have the same computational complexity as for DFAs. For the case of a non-constant number of sweeps, the situation is quite different. It is shown that a logarithmic number of sweeps is sufficient to accept unary non-semilinear languages, while with a sublogarithmic number of sweeps only regular languages can be accepted. Moreover, the existence of a finite hierarchy with respect to the number of sweeps is obtained. Finally, all usually studied decidability questions are shown to be undecidable and not even semidecidable for IUFSTs performing at least a logarithmic number of sweeps.

In this paper, we enhance the model of IUFSTs by *nondeterminism*, thus obtaining their *nondeterministic* version (NIUFSTs). As in [13], we are interest in NIUFSTs exhibiting both a constant and non-constant number of sweeps.

Constant sweep bounded NIUFSTs are proved to accept exactly regular languages. So, their ability of representing regular languages in a very succinct way turns out to be worth investigating, as well as comparing such an ability with that of other more traditional models of finite-state automata. This type of investigation, whose importance is witnessed by a well consolidated trend in the literature, focuses on the *size* of formalisms for representing languages and is usually referred to as descriptional complexity. Being able to have "small" devices representing/accepting certain languages, leads to relevant consequences either from a practical point of view (less hardware needed to construct such devices, less energy absorption, less cooling problems, *etc.*), and from a theoretical point of view (higher manageability of proofs and representations for languages, reductions of difficult problems on general computing devices to the same problems on simpler machines, *etc.*). The reader is referred to, e.g., [11], for a thoughtful survey on descriptional complexity and its consequences.

Non-constant sweep bounded NIUFSTs are then studied for their computational power, i.e., the ability of accepting language families. In particular, such an ability is related to the number of sweeps as a function of the input length.

After defining NIUFSTs in Sect. 2, we discuss in detail an example that demonstrates the size advantages of NIUFSTs with a constant number of sweeps in comparison with its deterministic variant and the classical models of deterministic and nondeterministic finite automata (NFAs). Precisely, we exhibit a language accepted by a NIUFST such that any equivalent IUFST requires exponentially more states and sweeps, while any equivalent NFA (resp., DFA) requires exponentially (resp., double-exponentially) more states.

In Sect. 3, we study size advantages of NIUFSTs with a constant number of sweeps in more generality. By evaluating the state cost of sweep removal, we show that any NIUFST featuring n states and k sweeps can be simulated by an n^k-state NFA, and hence by a 2^{n^k}-state DFA as well. Next, we exhibit a unary (resp., binary) language witnessing the obtained size blow-up for turning a constant sweep NIUFST into an equivalent NFA (resp., DFA) is unavoidable.

In the last two sections, we consider NIUFSTs with a non-constant number of sweeps. First, we establish in Sect. 4 an infinite proper hierarchy with respect to the number of sweeps. Interestingly, this result also extends the known finite hierarchy in the deterministic case to an infinite hierarchy.

Finally, we study in Sect. 5 the question of whether the nondeterministic model is more powerful than the deterministic model. We get that the question can be answered in the affirmative if at least a logarithmic number of sweeps is provided. Moreover, we show that nondeterminism cannot be matched in power by the deterministic paradigm even if a sublinear number of sweeps is given.

2   Definitions and Preliminaries

We denote the set of positive integers and zero by $\mathbb{N}$. Given a set S, we write 2^S for its power set and $|S|$ for its cardinality. Let Σ^* denote the set of all words over the finite alphabet Σ. The *empty word* is denoted by λ, and $\Sigma^+ = \Sigma^* \setminus \{\lambda\}$. The length of a word w is denoted by $|w|$.

Roughly speaking, an iterated uniform finite-state transducer is a finite-state transducer which processes the input in multiple passes (also sweeps). In the first pass it reads the input word followed by an endmarker and emits an output word. In the following passes it reads the output word of the previous pass and emits a new output word. The number of passes taken, the *sweep complexity*, is given as a function of the length of the input. Here, we are interested in weak processing devices: we will consider length-preserving finite-state transducers, also known as Mealy machines [17], to be iterated.

Formally, we define a *nondeterministic iterated uniform finite-state transducer* (NIUFST) as a system $T = \langle Q, \Sigma, \Delta, q_0, \triangleleft, \delta, F_+, F_- \rangle$, where Q is the set of *internal states*, Σ is the set of *input symbols*, Δ is the set of *output symbols*, $q_0 \in Q$ is the initial state, $\triangleleft \in \Delta \setminus \Sigma$ is the *endmarker*, $F_+ \subseteq Q$ is the set of *accepting states*, $F_- \subseteq (Q \setminus F_+)$ is the set of *rejecting states*, and $\delta \colon Q \times (\Sigma \cup \Delta) \to 2^{Q \times \Delta}$ is the *transition function*, which is total on $(Q \setminus (F_+ \cup F_-)) \times (\Sigma \cup \Delta)$ and such that the endmarker is emitted only if it is read (i.e., no transition $(q, \triangleleft) \in \delta(p, x)$ is allowed, with $x \neq \triangleleft$). The NIUFST T *halts* whenever the transition function is undefined (which may happen only for states from $F_+ \cup F_-$) or T enters an accepting or rejecting state at the end of a sweep. Since transduction is applied in multiple passes, that is, in any but the initial pass it operates on an output of the previous pass, the transition function depends on input symbols from $\Sigma \cup \Delta$. We denote by $T(w)$ the set of possible outputs produced by T in a complete sweep on input $w \in (\Sigma \cup \Delta)^*$.

A *computation* of the NIUFST T on input $w \in \Sigma^*$ consists of a sequence of words $w_1, \ldots, w_i, w_{i+1}, \ldots \in (\Sigma \cup \Delta)^*$ satisfying $w_1 \in T(w\triangleleft)$ and $w_{i+1} \in T(w_i)$ for $i \geq 1$. The computation is halting if there exists an $r \geq 1$ such that T halts on w_r, thus performing r sweeps. The input word $w \in \Sigma^*$ is *accepted* by T if *all* computations on w are halting and at least one computation halts in an accepting state. The input word $w \in \Sigma^*$ is *rejected* if *all* computations on w are halting and *none* of the computations halts in an accepting state. Indeed, the output of the last sweep is not used. The language accepted by T is the set $L(T) \subseteq \Sigma^*$ defined as $L(T) = \{ w \in \Sigma^* \mid w$ is accepted by $T \}$.

A NIUFST is said to be *deterministic* (IUFST) if and only if $|\delta(p, x)| \leq 1$, for all $p \in Q$ and $x \in (\Sigma \cup \Delta)$. In this case, we simply write $\delta(p, x) = (q, y)$ instead of $\delta(p, x) = \{(q, y)\}$ assuming that the transition function is a mapping $\delta \colon Q \times (\Sigma \cup \Delta) \to Q \times \Delta$.

We chose to designate our transductors as "uniform" since they perform the same transduction at each sweep: they always start from the same initial state on the leftmost tape symbol, operating the same transduction rules at every computation step. Yet, we quickly observe that a NIUFST is clearly a restricted version of a linear bounded automaton (see, e.g., [12]). So, any language accepted by a NIUFST is context-sensitive. We leave it as an open problem to exhibit context-sensitive languages which cannot be accepted by any NIUFST.

Given a function $s\colon \mathbb{N} \to \mathbb{N}$, a NIUFST is said to be of *sweep complexity* $s(n)$ whenever it accepts or rejects any word of length n in at most $s(n)$ sweeps. In this case, we use the notation $s(n)$-NIUFST. Note that sweep complexity requires that any input is either accepted or rejected, that is, the NIUFST always halts. It is easy to see that a 1-IUFST (resp., 1-NIUFST) is actually a deterministic (resp., nondeterministic) finite automaton (DFA and NFA, respectively).

Throughout the paper, two accepting devices are said to be *equivalent* if and only if they accept the same language.

2.1 Accepting Languages by Iterated Transductions: An Example

In order to clarify the notion of acceptance by iterated transduction, we propose a language and design several accepting devices for such a language. For any integer $k \geq 2$, we define the block language

$$B_k = \{\, u_1 \# u_2 \# \cdots \# u_m \mid u_i \in \{0,1\}^k, m > 1, \exists i < m \colon u_i = u_m \,\}.$$

To accept B_k by a DFA, 2^{2^k+1} states are necessary and sufficient. On the other hand, an exponentially smaller NFA A may accept B_k as follows:

1. In a first phase, on each block in the input string, A stores the block in its finite control and then nondeterministically decides whether to keep the block or to ignore it. Along this phase, A checks the correct block structure of the input so far scanned as well. This phase takes 2^{k+1} states.
2. Once A decides to keep a block, say u, in its finite control, a second phase starts in which A scans the rest of the input checking the correct block structure and guessing another block w to be matched symbol-by-symbol against u. If matching is successful and w is the last block, then A accepts. This phase takes $2^{k+1} \cdot (k+1)$ states.

Globally, the NFA A features $2^{k+1} + 2^{k+1} \cdot (k+1) = 2^{k+1} \cdot (k+2)$ states.

Indeed, A can also be seen as a $2^{k+1} \cdot (k+1)$-state 1-NIUFST which outputs the scanned symbol at each step. However, paying by the number of sweeps (see, e.g., [15]), we can build a k-NIUFST T for B_k with only $O(k)$ states. Informally:

1. In a first sweep, T checks the correct block structure of the input string, nondeterministically chooses two blocks to be matched symbol-by-symbol, and compares the first symbol of the two blocks by storing the first symbol of the first block in its finite control and replacing these two symbols with a blank symbol.
2. At the ith sweep, T checks the ith symbol of the two blocks chosen in the first sweep by storing and blank-replacing symbols as explained at the previous point. To distinguish the first sweep (where both nondeterministic block choices and symbol comparisons take place) from the others (where only symbol comparisons take place), a special symbol can replace the first input symbol at the beginning of the first sweep.

It is not hard to see that $O(k)$ states are needed to check input formatting along the first sweep, and that a constant number of states suffices to blank-replacing and comparing input symbols. Indeed, after k sweeps all nondeterministically chosen blocks symbols are compared so that T may correctly accept or reject. This gives the claimed state and sweep bounds for T.

We remark that: (i) a $2^k(k+4)$-state 2^k-IUFST is designed in [13] for B_k, (ii) 2^{2^k+1} states are necessary and sufficient to accept B_k by a DFA, and that (iii) n^k states are sufficient for a DFA to simulate an n-state k-IUFST [13]. These facts, together with above designed $O(k)$-states k-NIUFST, show that NIUFSTs can be exponentially more succinct than IUFSTs either in the number of states, or the number of sweeps, or possibly both. Indeed, we also have that NIUFSTs can be exponentially more succinct than NFAs and double-exponentially more succinct than DFAs.

In the next section, we approach more generally the analysis of the descriptional power of NIUFSTs with respect to their deterministic counterparts and classical finite-state models.

3   Reducing Sweeps and Removing Nondeterminism

Let us begin by showing how to reduce sweeps from NIUFSTs and evaluate the state cost of reduction. We will then use this construction to reduce to one the sweeps of constant sweep bounded NIUFST, thus obtaining equivalent NFAs whose number of states will be suitably bounded.

Theorem 1. *Let $n, k > 0$ be integers. Every n-state k-NIUFST (resp., k-IUFST) can be converted to an equivalent n^i-state $\lceil \frac{k}{i} \rceil$-NIUFST (resp., $\lceil \frac{k}{i} \rceil$-IUFST).*

Proof. Let $T = \langle Q, \Sigma, \Delta, q_0, \triangleleft, \delta, F_+, F_- \rangle$ be a k-NIUFST with $|Q| = n$. To simplify the proof, we show how to transform a k-IUFST into an equivalent $\lceil \frac{k}{2} \rceil$-NIUFST T' with n^2 states, i.e, we prove the theorem for $i = 2$. We suppose that δ is completely defined. The opposite case is briefly discussed at the end of the proof.

The idea is to simulate two consecutive sweeps of T in one sweep. To this aim, the set of states of T' is defined as Q^2, its initial state is the pair (q_0, q_0), and the set of output symbols is Δ^2. In order to define the transition function δ' of T', we remark that: (i) the simulation of the first two sweeps takes place on the input string in Σ^*, while (ii) the simulations for the other sweeps take place on output strings in $(\Delta^2)^*$ but only the second component of scanned symbols from Δ^2 is to be considered for the computation in T'. Therefore, we use σ to denote either a symbol from Σ for situation (i) and the second component of a symbol from Δ^2 for situation (ii) and, with a slight abuse of notation, we define $\delta' \colon Q^2 \times (\Sigma \cup \Delta) \to 2^{Q^2 \times \Delta^2}$ as

$$\delta'((s_1, s_2), \sigma) \ni ((t_1, t_2), (\tau_1, \tau_2)) \Leftrightarrow \delta(s_1, \sigma) \ni (t_1, \tau_1) \text{ and } \delta(s_2, \tau_1) \ni (t_2, \tau_2).$$

A state $(f_1, f_2) \in Q^2$ is accepting (resp., rejecting) in T' whenever either $f_1 \in F_+$ (resp., $f_1 \in F_-$) or $f_2 \in F_+$ but $f_1 \notin F_-$ (resp, $f_2 \in F_-$ but $f_1 \notin F_+$). The reader may verify that the n^2-state $\lceil \frac{k}{2} \rceil$-NIUFST T' is equivalent to T.

In case δ of T is not completely defined (and this may happen only on states in $F_+ \cup F_-$), the number of states of T' does not increase as well. In fact, suppose T halts in $q \in F_+ \cup F_-$ in the middle of the input string on the jth sweep. We define δ' in such a way that the simulation of the jth sweep of T remains in q at every step and after the endmarker scanning as well.

It is not hard to see that this construction can be suitably adapted to merge $2 < i \leq k$ sweeps into one, thus yielding a NIUFST equivalent to T featuring $\lceil \frac{k}{i} \rceil$ sweeps and n^i states. Yet, it is also easy to see that the construction preserves determinism. □

The sweep reduction presented in Theorem 1 can be directly used to transform constant sweep bounded NIUFSTs into equivalent NFAs:

Theorem 2. *Let $n, k > 0$ be integers. Every n-state k-NIUFST can be converted to an equivalent NFA with at most n^k states.*

Proof. Given an n-state k-NIUFST, by Theorem 1 we can obtain an equivalent n^k-state 1-NIUFST which is actually a NFA. □

We obtain the optimality of the state blow-up in Theorem 2 by establishing an optimality condition for the size cost of sweep reduction proved in Theorem 1. To this aim, for $n, k > 0$, let the unary language

$$L_{n,k} = \{ a^{c \cdot n^k} \mid c \geq 0 \}.$$

In [13], an n-state k-IUFST for $L_{n,k}$ is provided, whereas any equivalent DFA or NFA needs at least n^k states. By using $L_{n,k}$ as witness language, we can show

Theorem 3. *Let $n, k, i > 0$ be integers such that i divides k. There exists an n-state k-NIUFST T such that any equivalent $\frac{k}{i}$-NIUFST T' cannot have less than n^i states.*

Proof. Suppose by contradiction, we can always design T' with $\frac{k}{i}$ sweeps and $s < n^i$ states. By using our construction in Theorem 1, we can obtain from T' an equivalent 1-NIUFST A with $s^{\frac{k}{i}} < n^k$ states. Clearly, having a single sweep, A is a NFA. By using this approach on the n-state k-IUFST above recalled for the language $L_{n,k}$, we could obtain an equivalent NFA featuring less than n^k states, a contradiction. □

By Theorem 3, one may easily obtain

Corollary 4. *For any integers $n, k > 0$, there exists an n-state k-NIUFST which cannot be converted to an equivalent NFA with less than n^k states.*

We conclude this section by discussing the optimal size cost of turning constant sweep bounded NIUFSTs into DFAs, i.e., the cost of removing both nondeterminism and sweeps at once:

Theorem 5. *Let $n, k > 0$ be integers. Every n-state k-NIUFST can be converted to an equivalent DFA with at most 2^{n^k} states.*

Proof. The result follows by first converting, according to Theorem 2, the n-state k-NIUFST into an equivalent n^k-state NFA which, in turn, is converted to an equivalent 2^{n^k}-state DFA by the usual powerset construction (see, e.g., [12]) □

The optimality of the size blow-up in Theorem 5 can be proved by considering the following language for any $n, k > 1$:

$$E_{n,k} = \{\, vbw \mid v, w \in \{a, b\}^*, \ |w| = c \cdot n^k \text{ for } c > 0 \,\}.$$

Theorem 6. *For any integers $m > 1$ and $k > 0$, there is an m-state k-NIUFST which cannot be converted to an equivalent DFA with less than $2^{(m-1)^k}$ states.*

Proof. As above quoted, an n-state k-NIUFST is given in [13], for the unary language $L_{n,k} = \{\, a^{c \cdot n^k} \mid c \geq 0 \,\}$. Such a device can be trivially converted to an n-state k-NIUFST T for the binary language $\{w \in \{a, b\}^* \mid |w| = c \cdot n^k \text{ for } c > 0\}$. We use T as a module for designing an $(n + 1)$-state k-NIUFST T' accepting the language $E_{n,k}$. Informally, T' uses a separate state to scan the input string during the first sweep and, upon reading a symbol b, it nondeterministically decides whether to keep on reading the input string or to call the n-state k-IUFST module T which checks whether the length of the remaining part of the input string is a multiple of n^k.

On the other hand, by suitably using pigeonhole arguments (see, e.g., [4]), we can show that any DFA for $E_{n,k}$ needs at least 2^{n^k} states. In fact, suppose by contradiction a DFA A exists, accepting $E_{n,k}$ with less than 2^{n^k} states. Clearly, by counting arguments, there exist $\alpha, \beta \in \{a, b\}^*$ such that $\alpha \neq \beta$, $|\alpha| = |\beta| = n^k$, and the computation of A on both α and β ends up in the same (non-accepting) state q. Since $\alpha \neq \beta$, without loss of generality, we can assume that $\alpha = xay$ and $\beta = vbw$, for suitable $x, y, v, w \in \{a, b\}^*$ such that $|x| = |v|$ and $|y| = |w|$.

Now, consider any string $z \in \{a, b\}^*$ satisfying $|z| = n^k - |y| = n^k - |w|$, and let the strings $\alpha' = \alpha z$ and $\beta' = \beta z$. Note that $1 \leq |z| \leq n^k$, and so $|\alpha'| = |\beta'| \leq 2 \cdot n^k$. Therefore, the acceptance/rejection by A on α' and β' is only due to the symbol at position $|x| + 1 = |v| + 1$. This clearly means that the string α' does not belong to $E_{n,k}$ while β' does. However, in the computation on both α' and β', the DFA A reaches the same state q before consuming z and, being deterministic, the same state after consuming z. Hence, either A accepts both the two strings or rejects both of them, a contradiction. □

4   An Infinite Sweep Hierarchy

We now consider $s(n)$-NIUFSTs where $s(n)$ is a non-constant function. In what follows, by $\log n$ we denote the logarithm of n to base 2. In [13] it is proved that $o(\log n)$ sweep bounded IUFSTs accept regular languages only, and that such a logarithmic sweep lower bound is tight for nonregular acceptance. Then, a three-level proper language hierarchy is established, where $O(n)$ sweeps are better than $O(\sqrt{n})$ sweeps which, in turn, are better than $O(\log n)$ sweeps for IUFST. Here, we extend the hierarchy to infinitely many levels for both IUFSTs and NIUFSTs.

Let $f : \mathbb{N} \to \mathbb{N}$ be a non-decreasing function. Its inverse is defined as the function $f^{-1}(n) = \min\{ m \in \mathbb{N} \mid f(m) \geq n \}$. To show an infinite hierarchy dependent on some resources, where the limits of the resources are given by some functions in the length of the input, it is often necessary to control the lengths of the input so that they depend on the limiting functions. Usually, this is done by requiring that the functions are *constructible* in a desired sense. The following notion of constructibility expresses the idea that the length of a word relative to the length of a prefix is determined by a function. A non-decreasing computable function $f : \mathbb{N} \to \mathbb{N}$ is said to be *constructible* if there exists an $s(n)$-IUFST T with $s(n) \in O(f^{-1}(n))$ and an input alphabet $\Sigma \cup \{a\}$, such that

$$L(T) \subseteq \{ a^m v \mid m \geq 1, v \in \Sigma^*, |v| = f(m) \}$$

and such that, for all $m \geq 1$, there exists a word of the form $a^m v$ in $L(T)$. The $s(n)$-IUFST T is said to be a *constructor* for f.

Since constructible functions describe the length of the whole word dependent on the length of a prefix, it is obvious that each constructible function must be greater than or equal to the identity. In order to show that the class of functions that are constructible in this sense is sufficiently rich to witness an infinite dense hierarchy, we next show that it is closed under addition and multiplication:

Proposition 7. *Let $f : \mathbb{N} \to \mathbb{N}$ and $g : \mathbb{N} \to \mathbb{N}$ be two constructible functions. Then the functions $f + g$ and $f \cdot g$ are constructible as well.*

In [13] it is shown that the unary language $L_{\text{uexpo}} = \{ a^{2^k} \mid k \geq 0 \}$ is accepted by some $s(n)$-IUFST with $s(n) \in O(\log n)$. The construction can straightforwardly be extended to show that the function $f(n) = 2^n$ is constructible. Moreover, again from [13], we know that $L_{\text{eq}} = \{ u\$v \mid u \in \Sigma_1^*, v \in \Sigma_2^*, \text{ and } |u| = |v| \}$ is a language accepted by some $s(n)$-IUFST with $s(n) \in O(n)$, where Σ_1 is an alphabet not containing the symbol $\$$ and Σ_2 is an arbitrary alphabet. Even in this case, only a tiny modification shows that the identity function is constructible. These facts together with Proposition 7 yield, in particular, that the function $f(n) = n^x$ for all $x \geq 1$ is constructible.

In what follows, we use the fact, proved in [13], that the copy language with center marker $\{ u\$u \mid u \in \{a, b\}^* \}$ is accepted by some $s(n)$-IUFST satisfying $s(n) \in O(n)$. The next theorem provides some language that separates the levels of the hierarchy.

Theorem 8. *Let $f : \mathbb{N} \to \mathbb{N}$ be a constructible function, T_f be a constructor for f with input alphabet $\Sigma \cup \{a\}$, and b be a new symbol not belonging to $\Sigma \cup \{a\}$. Then language $L_f = \{ u\$uv \mid u \in \{a, b\}^*, v \in \Sigma^*, a^{2|u|+1}v \in L(T_f) \}$ is accepted by some $s(n)$-IUFST with $s(n) \in O(f^{-1}(n))$.*

Proof. Since the suffix v of a word $w \in L_f$ must be the suffix of some word $a^{2|u|+1}v$ in $L(T_f)$, we have that $|v| = f(2|u|+1)$ and $|w| = 2|u|+1+f(2|u|+1)$. Since $s(n)$ is claimed to be of order $O(f^{-1}(n))$, an $s(n)$-IUFST accepting L_f may perform at least $O(2|u|+1)$ many sweeps.

An $s(n)$-IUFST T accepting L_f essentially combines in parallel the acceptors for the copy language with center marker and the language $L(T_f)$. To this end, T establishes two tracks in its output. On the first track T simulates an acceptor for the copy language $\{ u\$u \mid u \in \{a,b\}^* \}$, where the first symbol of Σ (i.e., the first symbol of v) acts as endmarker. In this way, the prefix $u\$u$ is verified. The result of the computation is written to the output track. This task takes $O(2|u|+1)$ sweeps. On the second track T simulates the constructor T_f, where all symbols up to the first symbol of Σ (i.e., all symbols of the prefix $u\$u$) are treated as input symbols a. In this way, T verifies that $|v| = f(2|u|+1)$. The result of the computation is written to the output track. This task takes $O(2|u|+1)$ sweeps.

Finally, T rejects whenever one of the above simulations ends rejecting. Instead, T accepts if it detects positive simulation results of the two tasks on the tracks. □

To show that the witness language L_f of Theorem 8 is not accepted by any $s(n)$-NIUFST with $s(n) \in o(f^{-1}(n))$, we use Kolmogorov complexity and incompressibility arguments. General information on this technique can be found, for example, in the textbook [14, Ch. 7]. Let $w \in \{a,b\}^*$ be an arbitrary binary string. Its Kolmogorov complexity $C(w)$ is defined to be the minimal size of a binary program (Turing machine) describing w. The following key fact for using the incompressibility method is well known: there exist binary strings w of *any* length such that $|w| \leq C(w)$.

Theorem 9. *Let $f: \mathbb{N} \to \mathbb{N}$ be a constructible function, T_f be a constructor for f with input alphabet $\Sigma \cup \{a\}$, and b be a new symbol not belonging to $\Sigma \cup \{a\}$. Then language $L_f = \{ u\$uv \mid u \in \{a,b\}^*, v \in \Sigma^*, a^{2|u|+1}v \in L(T_f) \}$ cannot be accepted by any $s(n)$-NIUFST with $s(n) \in o(f^{-1}(n))$.*

Proof. Contrarily, let us assume that L_f is accepted by some $s(n)$-NIUFST $T = \langle Q, \Sigma \cup \{a,b\}, \Delta, q_0, \lhd, \delta, F_+, F_- \rangle$ with $s(n) \in o(f^{-1}(n))$.

We choose a word $u \in \{a,b\}^*$ long enough such that $C(u) \geq |u|$. Then, we consider an accepting computation of T on $u\$uv$, and derive a contradiction by showing that u can be compressed via T. To this end, we describe a program P which reconstructs u from a description of T, the length $|u|$, and the sequence of the $o(f^{-1}(n))$ many states $q_1, q_2, \ldots, q_r$ entered along the accepting computation at that moments in which T reads the first symbol after the $\$$ along its $o(f^{-1}(n))$ sweeps, n being the total length of the input.

Basically, the program P takes the length $|u|$ and enumerates the finitely many words $u'v'$ with $u' \in \{a,b\}^{|u|}$, $v' \in \Sigma^*$, and $|v'| = f(2|u|+1)$. Then, for each word in the list, it simulates by dovetailing all possible computations of T on $u'v'$ and, in particular, it simulates $o(f^{-1}(n))$ successive partial sweeps of T on $u'v'$, where the ith sweep is started in state q_i for $1 \leq i \leq r$. If the simulation ends accepting, we know that $u\$u'v'$ belongs to L_f and, thus, $u' = u$.

Let us consider the size of P. Let $|T|$ denote the constant size of the description of T, and $|P|$ denote the constant size of the program P itself. The binary description of the length $|u|$ takes $O(\log(|u|))$ bits. Each state of T can be encoded by $O(\log(|Q|))$ bits. So, we have

$$C(u) \in |P| + |T| + O(\log(|u|)) + o(f^{-1}(n)) \cdot O(\log(|Q|)) = O(\log(|u|)) + o(f^{-1}(n)).$$

Since $n = 2|u| + 1 + f(2|u| + 1)$ and $f(n) \geq n$, for all $n \geq 1$, we have $n \in \Theta(f(2|u| + 1))$. So, we can conclude that $C(u) \in O(\log(|u|)) + o(|u|) = o(|u|)$. This contradicts our initial assumption $C(u) \geq |u|$, for u long enough. Therefore, T cannot accept L_f with sweep complexity $o(f^{-1}(n))$. $\square$

Finally, we would like to remark that, due to our observation that all functions $f(n) = n^x$ are constructible for $x \geq 1$, it is an easy application of the above theorems to obtain the following infinite hierarchies with regard to the number of sweeps both in the deterministic and the nondeterministic case. Namely: *For every $x \geq 1$ we have that the set of all languages that are accepted by $s(n)$-IUFSTs ($s(n)$-NIUFSTs) with $s(n) \in O(n^{1/(x+1)})$ is properly included in the set of all languages that are accepted by $s(n)$-IUFSTs ($s(n)$-NIUFSTs) with $s(n) \in O(n^{1/x})$.*

5 Nondeterminism Beats Determinism on All Levels

We now turn to compare the computational power of IUFSTs and NIUFSTs. Since for sweep bounds of order $o(\log n)$ both variants accept regular languages only (see [9,13]), it remains to consider sweep bounds beyond $o(\log n)$. Here, we will show that there exist witness languages that are accepted by some nondeterministic $s(n)$-NIUFST with $s(n) \in O(\log n)$, but cannot be accepted by any deterministic $s(n)$-IUFST with $s(n) \in o(n)$, thus separating determinism from nondeterminism for almost all levels of the sweep hierarchy.

For any integer $k \geq 1$, let $\mathrm{bin}_k \colon \{0, 1, 2, \ldots, 2^k - 1\} \to \{0, 1\}^k$ map any integer in the range from 0 to $2^k - 1$ to its binary representation of length k, starting from the left with the least significant bit and possibly completed with zeroes to the right. E.g., $\mathrm{bin}_4(5) = 1010$ and $\mathrm{bin}_4(12) = 0011$. We consider the language

$$D = \{\, a^k b^{2^k} \, \mathrm{bin}_k(0) u_0 \, \mathrm{bin}_k(1) u_1 \cdots \mathrm{bin}_k(2^k - 1) u_{2^k - 1} \, \mathrm{bin}_k(i) u_i \mid$$
$$k \geq 2,\ 1 \leq i \leq 2^k - 1,\ u_j \in \{a, b\}^k \text{ for all } 1 \leq j \leq 2^k - 1 \,\}.$$

Theorem 10. *The language D can be accepted by an $s(n)$-NIUFST satisfying $s(n) \in O(\log n)$.*

Proof. We sketch the construction of an $s(n)$-NIUFST T that accepts D with $s(n) \in O(\log n)$. The basic idea of the construction is to use two output tracks. So, during its first sweep, T splits the input into two tracks, each one getting the original input. In addition, T verifies if the structure of the input is correct, that is, if the input is of the form $a^+ b^+ 0^+ \{a, b\}^+ (\{0, 1\}^+ \{a, b\}^+)^+ 1^+ \{a, b\}^+$ with at least two leading a's. If the form is incorrect, T rejects.

In subsequent sweeps, T behaves as follows. The original input on the first track is kept but the symbols can be marked, while on the second track the input is successively shifted to the right. More precisely, in any sweep the first unmarked symbol a in the leading a-block is marked. In the following b-block, every second unmarked symbol b is marked. In the further course of the sweep, the leftmost unmarked symbol in any $\{0,1\}$-block as well as in any $\{a,b\}$-block is marked. On the second track, the input is shifted to the right by one symbol, whereby the last symbol is deleted and some blank symbol is added at the left.

Let $k \geq 2$ be the length of the leading a-block. When the last of its symbols is marked, T checks in the further course of the sweep whether in the following b-block exactly one symbol remains unmarked, and whether in all remaining blocks the last symbol is being marked. Only in this case the computation continues. In all other cases T halts rejecting.

From the construction so far, we derive that if the computation continues then all but the second block have the same length, namely, length k. Moreover, since in the second block every second unmarked symbol has been marked during a sweep and one symbol is left, the length of the block is 2^k.

Next, T continues to shift right the content of the second track until the $\{0,1\}$-blocks are aligned with their neighboring $\{0,1\}$-blocks (except for the last one). This takes other k sweeps. In the next sweep, T checks if the $\{0,1\}$-block on the second track is an integer that is one less the integer in the aligned block on the first track. This can be done by adding one on the fly and comparing the result with the content on the first track. Only if the check is successful, T continues. Otherwise, it halts rejecting. In the former case, we get that the sequence of $\{0,1\}$-blocks are the numbers from 0 to $2^k - 1$ in ascending order.

In the next sweep, T guesses the $\{0,1\}$-block that has to match the rightmost $\{0,1\}$-block and marks it appropriately. Finally, this block together with its following $\{a,b\}$-block is symbolwise compared with the last $\{0,1\}$-block together with its following $\{a,b\}$-block in another $2k$ sweeps. To this end, note that T can detect that the last block follows when it scans a $\{0,1\}$-block consisting of 1's only. For the comparison, the symbols can further be marked appropriately.

Now, T accepts only if the guessed $\{0,1\}$-block together with its following $\{a,b\}$-block match the last $\{0,1\}$- and $\{a,b\}$-block. Otherwise T rejects. The construction shows that for any word from D there is one accepting computation and that only words from D are accepted. So, T accepts D.

Altogether, T performs at most $1+k+k+1+2k \in O(k)$ sweeps. The length of the input is $k + 2^k + (2^k + 1) \cdot 2k = O(k2^k)$. Since $\log(O(k2^k)) \in O(k)$, the NIUFST T obeys the sweep bound $s(n) \in O(\log n)$. □

To see that the language D is not accepted by any $s(n)$-IUFST with $s(n) \in o(n)$, we use again Kolmogorov complexity and incompressibility

Theorem 11. *The language D cannot be accepted by any $s(n)$-IUFST satisfying $s(n) \in o(n)$.*

Acknowledgements. The authors wish to thank the anonymous referees for useful comments and remarks.

References

1. Bednárová, Z., Geffert, V., Mereghetti, C., Palano, B.: The size-cost of Boolean operations on constant height deterministic pushdown automata. Theoret. Comput. Sci. **449**, 23–36 (2012)
2. Bednárová, Z., Geffert, V., Mereghetti, C., Palano, B.: Boolean language operations on nondeterministic automata with a pushdown of constant height. J. Comput. Syst. Sci. **90**, 99–114 (2017)
3. Bertoni, A., Mereghetti, C., Palano, B.: Trace monoids with idempotent generators and measure-only quantum automata. Nat. Comput. **9**(2), 383–395 (2010)
4. Bianchi, M.P., Mereghetti, C., Palano, B.: Complexity of promise problems on classical and quantum automata. In: Calude, C.S., Freivalds, R., Kazuo, I. (eds.) Computing with New Resources. LNCS, vol. 8808, pp. 161–175. Springer, Cham (2014). https://doi.org/10.1007/978-3-319-13350-8_12
5. Bordihn, H., Fernau, H., Holzer, M., Manca, V., Martín-Vide, C.: Iterated sequential transducers as language generating devices. Theoret. Comput. Sci. **369**(1–3), 67–81 (2006)
6. Citrini, C., Crespi-Reghizzi, S., Mandrioli, D.: On deterministic multi-pass analysis. SIAM J. Comput. **15**(3), 668–693 (1986)
7. Friburger, N., Maurel, D.: Finite-state transducer cascades to extract named entities in texts. Theoret. Comput. Sci. **313**(1), 93–104 (2004)
8. Ginzburg, A.: Algebraic Theory of Automata. Academic Press, Cambridge (1968)
9. Hartmanis, J.: Computational complexity of one-tape turing machine computations. J. ACM **15**(2), 325–339 (1968)
10. Hartmanis, J., Stearns, R.E.: Algebraic Structure Theory of Sequential Machines. Prentice-Hall, Upper Saddle River (1966)
11. Holzer, M., Kutrib, M.: Descriptional complexity - an introductory survey. In: Martín-Vide, C. (ed.) Scientific Applications of Language Methods, pp. 1–58. Imperial College Press (2010)
12. Hopcroft, J.E., Ullman, J.D.: Introduction to Automata Theory, Languages, and Computation. Addison-Wesley, Boston (1979)
13. Kutrib, M., Malcher, A., Mereghetti, C., Palano, B.: Descriptional complexity of iterated uniform finite-state transducers. In: Hospodár, M., Jirásková, G., Konstantinidis, S. (eds.) DCFS 2019. LNCS, vol. 11612, pp. 223–234. Springer, Cham (2019). https://doi.org/10.1007/978-3-030-23247-4_17
14. Li, M., Vitányi, P.M.B.: An Introduction to Kolmogorov Complexity and Its Applications, 3rd edn. Springer, Cham (2019). https://doi.org/10.1007/978-3-030-11298-1
15. Malcher, A., Mereghetti, C., Palano, B.: Descriptional complexity of two-way pushdown automata with restricted head reversals. Theoret. Comput. Sci. **449**, 119–133 (2012)
16. Manca, V.: On the generative power of iterated transductions. In: Words, Semigroups, and Transductions - Festschrift in Honor of G. Thierrin, pp. 315–327. World Scientific (2001)
17. Mealy, G.H.: A method for synthesizing sequential circuits. Bell Syst. Tech. J. **34**, 1045–1079 (1955)
18. Pierce, A.: Decision problems on iterated length-preserving transducers. Bachelor's thesis, SCS Carnegie Mellon University, Pittsburgh (2011)

Computable Analysis and Classification Problems

Rodney G. Downey[1] and Alexander G. Melnikov[2(✉)]

[1] Victoria University of Wellington, Wellington, New Zealand
`rod.downey@vuw.ac.nz`
[2] Massey University, Auckland, New Zealand
`alexander.g.melnikov@gmail.com`

1 Introduction

Suppose we are given a collection of mathematical objects such as the class of connected compact Polish groups or the set of all real numbers which are normal to some base. Is there a *reasonable* classification of these objects (e.g. by invariants)? One nice property expected from a useful classification is that it makes the classified objects easier to handle *algorithmically*. Even if the general classification of a class of objects is impossible, sometimes there is a useful *hierarchy* of objects in this class, e.g., the Cantor-Bendixson rank of a scattered Polish space, the Ulm type of a reduced abelian p-group, etc. We would expect that the algorithmic, algebraic, or topological complexity of objects increase from lower to higher levels of a hierarchy. Can you make this intuition formal? Is it possible to formally *measure* the algorithmic complexity of a given classification or a hierarchy, or perhaps show that there is no reasonable classification at all? Can algorithmic tools help us to define useful hierarchies?

In the present article we discuss several recent works in computable analysis which are related to classification. The main underlying theme here is applying computability theory to a classification problem or a hierarchy; neither the problem nor the hierarchy has to be computability-theoretic. The discussed results can be informally split into three categories:

I. *Local effective classifications.* Given some classical metric or Banach space we ask which elements of the space satisfy a certain property. For example: Which real numbers are normal? Which continuous functions are regular? And so on.

II. *Global effective classifications.* How hard is it to classify, say, compact Polish groups? What about compact Polish spaces? To answer these questions formally we extend methods of computable structure theory to computable separable spaces.

III. *Applications of computability to hierarchies.* Usually elements of higher levels of a given hierarchy are expected to be more complex. Computability theory can be used to make this intuition formal. Such results are clearly related to (I), but there are also explicit connections with Weihrauch reducibility and with theme II.

© Springer Nature Switzerland AG 2020
M. Anselmo et al. (Eds.): CiE 2020, LNCS 12098, pp. 100–111, 2020.
https://doi.org/10.1007/978-3-030-51466-2_9

These three themes above are closely related and no firm line can be drawn between any two of them. We begin our discussion with the second global scheme (II) and a certain related hierarchy. Then we discuss the local theme (I) and finish with more on local hierarchies (III).

2 Global Effective Classifications

There has been a lot of applications of computability theoretic techniques to classification problems in *countable* algebra; see [19,31]. A few years ago Melnikov [26] proposed that methods of computable structure theory can be extended to computable separable spaces. We will see that some of these methods can be used to measure the classification problems for classes of separable spaces. We first look at the simpler case of Polish metric spaces, and then we discuss Banach spaces and Polish groups.

Computable Structures. To apply methods of computable structure theory we need to make the standard notion of a computable Polish space [5,34,37] look more familiar to a computable algebraist. A *structure* on a Polish space $\mathcal{M} = (M, d)$ is any map $\alpha : \omega \to \mathcal{M}$ such that its range is a dense subset of $\mathcal{M}$. The *open diagram* $D^+(\alpha)$ of a structure $\alpha : \omega \to \mathcal{M}$ is the collection of Gödel numbers of all elementary facts of the form $d(\alpha(i), \alpha(j)) < r$ and $d(\alpha(i), \alpha(j)) > q$ which hold on $\mathcal{M}$, where i, j range over ω and q, r over $\mathbb{Q}$. We say that α is *computable* if $D^+(\alpha)$ is a computably enumerable set[1].

We return to classification. Fix some class $\mathcal{K}$ of Polish spaces. First, suppose our task is to classify *computable* members of $\mathcal{K}$. Fix a uniformly computable list of all partially computable structures on Polish spaces: $\alpha_0, \alpha_1, \alpha_2, \dots$, where each computable structure α_i is identified with its c.e. open diagram $D^+(\alpha_i)$.

Definition 1. The complexity of the classification problem for computable members of $\mathcal{K}$ is measured using the following two *index sets*:

1. The characterisation problem for computable members of $\mathcal{K}$:

$$I(\mathcal{K}) = \{e : \alpha_e \in \mathcal{K}\}.$$

2. The isomorphism problem for computable members of $\mathcal{K}$:

$$E(\mathcal{K}) = \{(e, j) : e, j \in I(\mathcal{K}) \text{ and } cl(\alpha_e) \cong cl(\alpha_j)\},$$

where $\cong$ stands for isometric isomorphism. Computable members of a class $\mathcal{K}$ of Polish metric spaces admit an *effective classification* (up to isometric isomorphim) if both $I(\mathcal{K})$ and $E(\mathcal{K})$ are hyperarithmetic.

---

[1] A standard trick can be used to (computably) remove repetitions and replace α with an injective α'; see, e.g., [20]. Thus, without loss of generality the map $\alpha : \omega \to \mathcal{M}$ will be assumed injective.

Remark 2. Note that $\cong$ does *not* stand for *computable* isometric isomorphism. Of course, if we wish to classify our spaces up to homeomorphism or quasi-isometry (etc.) then we should adjust the interpretation of $\cong$ accordingly. Definition 1 is an adaptation of a similar method proposed in [19] for countable discrete computable structures. Cenzer and Remmel [10] used index sets to measure the complexity of various properties of Π_1^0-classes. Although there is not much in common between their results and the theorems discussed in the present article, it is important that index sets had been used in computable analysis long before us.

Relativisation. To extend these methods beyond computable Polish spaces we use relativisation. A structure $\alpha : \omega \to \mathcal{M}$ is computable relative to X if $D^+(\alpha)$ is computably enumerable relative to X. Using structures computable relative to an oracle X, relativise the definitions of $I(\mathcal{K})$ and $E(\mathcal{K})$ to X; the resulting sets will be denoted $I^X(\mathcal{K})$ and $E^X(\mathcal{K})$. Usually, when we establish that, say, $I(\mathcal{K}) \in \Sigma_n^0$, we can apply relativisation to show $I^X(\mathcal{K}) \in \Sigma_n^X$, and similarly for $E^X(K)$. Recall this means that, for some recursive scheme R, we have

$$i \in I^X(\mathcal{K}) \iff (\exists x_1)\ldots(Q_n x_n)R(X; x_1, \ldots, x_n, i),$$

where X is a set-parameter. It follows that the number of alternations of quantifiers in $(\exists x_1)\ldots(Q_n x_n)R(X; x_1, \ldots, x_n, i)$ is an *invariant* of the whole class. This motivates the following:

Definition 3. We say that that $\mathcal{K}$ admits an *effective classification* if, for every oracle X, both $I^X(\mathcal{K})$ and $E^X(\mathcal{K})$ belong to a some (fixed) level of the hyperarithmetical hierarchy relativised to X.

The First Results. Perhaps, the first non-trivial illustration of the proposed approach to classification in the literature is the theorem below.

Theorem 4 (Melnikov and Nies [29]). *The class $\mathcal{K}_{comp}$ of compact Polish metric spaces admits an effective classification.*

The proof of Theorem 4 relies on ideas of of Gromov [21] and $\mathcal{L}_{\omega_1 \omega}$-definability. The most important step of the proof is establishing that every computable compact Polish metric space is Δ_3^0-*categorical*; the categoricity hierarchy will be discussed in the next subsection. Theorem 4 contrasts with:

Theorem 5 (Nies and Solecki [33]). *The characterisation problem for computable locally compact Polish metric spaces is Π_1^1-complete.*

It follows that for a given Polish space $\mathcal{M}$, deciding if it is locally compact is as hard as just checking the Π_1^1 definition of local compactness. Thus, the class of locally compact Polish metric spaces does not admit an effective classification. The result formally confirms our intuition that locally compact spaces are very hard to classify.

The Categoricity Hierarchy. We use the standard notion of a computable map between computable Polish spaces; see, e.g., [26]. The definition below extends the classical notion of computable categoricity [1] to Polish spaces.

Definition 6 ([26]). A Polish space $\mathcal{M}$ is computably categorical if for any two computable structures α and β on $\mathcal{M}$ there is an isometric isomorphism between $cl(\alpha)$ and $cl(\beta)$ computable with respect to α and β[2].

Examples of computably categorical spaces include [26]: every Polish space associated to a separable Hilbert space, Cantor space with the usual ultrametric, and the Urysohn space. If we generalise Definition 6 by allowing the isomorphism to be Δ_α^0, the resulting notion of a Δ_α^0-categorical space is a direct adaptation of the classical notion of Δ_α^0-*categoricity* from computable algebra [1].

How is this technical notion related to classification? If every member of $\mathcal{K}$ is Δ_α^0-categorical and $I(\mathcal{K})$ is hyperarithmetical then $E(\mathcal{K})$ is hyperarithmetical too[3]. Our results tend to be easily relativizable to any oracle X, and thus both $I^X(\mathcal{K})$ and $E^X(\mathcal{K})$ will usually be hyperarithmetical relative to X as well. This approach was used by Melnikov and Nies [29] to prove Theorem 4. Clearly, Δ_α^0-categoricity induces a hierarchy, and using a transformation from graphs to Polish spaces [20] it is not hard to show that the hierarchy is proper in general.

We will return to discussing Δ_α^0-categorcity later. First we discuss another natural relativisation of Definition 6. It reveals a connection between first-order definability and computable categoricity of spaces, in the spirit of [1]. We say that a computable Polish space $\mathcal{M} = cl(\alpha_i)_{i \in \omega}$ is *relatively computably categorical* if any (not necessarily computable) structure $(\beta_i)_{i \in \omega}$ computes an isometry from $cl(\beta_i)_{i \in \omega}$ onto $cl(\alpha_i)_{i \in \omega}$. Greenberg, Knight, Melnikov, and Turetsky [20] showed:

Theorem 7. *A computable Polish metric space $\mathcal{M}$ is relatively computably categorical iff for every $\epsilon \in \mathbb{Q}$ it admits a c.e. ϵ-Scott family consisting of first-order positive $\exists$-formulae[4].*

It is crucial in the proof of Theorem 7 that the positive $\exists$-formulae define open sets.

Problem 8. Extend Theorem 7 to relative Δ_α^0-categoricity. For that, an adequate formal definition of relative Δ_α^0-categoricity must be designed.

Banach spaces. All Banach spaces in this section are separable. What is the most natural general approach to computability on Banach spaces? Pour El and Richards [34] restrict themselves only to computable structures which also compute the standard vector space operations. The well-known result of Mazur

---

[2] Here cl stands for the completion operator. Recall that the inverse of a computable surjective isometry is computable too, thus the notion is symmetric.

[3] Since an isometric image of a complete metric space is closed, saying that a Δ_α^0 map is "onto" is arithmetical.

[4] This means that ϵ-automorphism orbits of tuples in $\mathcal{M}$ can be described by a (uniformly in ϵ) c.e. family of positive first-order existential positive formulae in the language $(d_{<r}, d_{>r})_{r \in \mathbb{Q}}$ with finitely many "stable" parameters; a parameter is stable if there exist a ball such that any point from the ball can be used as a parameter without changing the formula.

and Ulam (can be found in [35]) implies that there is only one way to define the operations consistently with a given norm so that we get a complete normed space. Does the result of Mazur and Ulam hold computably? If the answer was "yes" then the definition from Pour El and Richards [34] would be an overkill. Interestingly, Melnikov and Ng [30] showed that the effective version of Mazur-Ulam fails for the space of continuous functions on the unit interval.

Problem 9. Give an optimal effective analysis of Mazur-Ulam.

In the context of the present article, the above-mentioned result of Melnikov and Ng justifies the use of the definition below which is of course equivalent to the approach in Pour-El and Richards [34].

Definition 10. A separable Banach space $\mathcal{B}$ is computable if the associated metric space admits a computable structure which computes the vector space operations on $\mathcal{B}$.

In the case of Banach spaces (over $\mathbb{R}$) we modify Definitions 1 and 3 using the natural enumeration of all (partially) computable countable normed spaces $B_0, B_1, \ldots$. The relativisation principle still applies to this approach. To save space we omit the definition; see [6]. Now to the results.

Lebesgue Spaces. Suppose we are given a computable Banach space $\mathcal{B}$. How hard is it to determine whether $\mathcal{B}$ is a Lebesgue space? In other words, what is the complexity of the characterisation problem (Definition 1) for Lebesgue spaces? The crude upper bound involves searching for a (separable) measure space Ω and a real p which makes the space look like $L^p(\Omega)$. The well-known Kakutani–Bohnenblus characterisation of Lebesgue spaces in terms of Banach lattices [4,22] does not seem to be of much help either. Brown, McNicholl and Melnikov [6] have proven the following rather surprising result:

Theorem 11 (Brown et al. [6]). *The characterisation problem for Lebesgue spaces is Π_3^0.*

How to reduce the unclassifiable brute-force complexity down to Π_3^0? Using a non-trivial and novel technique, McNicholl [24] proved that for a computable real p, ℓ^p is Δ_2^0-categorical[5]. The proof of Theorem 11 extends these techniques to arbitrary Lebesgue spaces and combines them with new ideas. We see that Δ_α^0-categoricity helps again, though indirectly.

Question 12 ([6]). Is the bound Π_3^0 from Theorem 11 tight?

The main obstacle in simplifying the upper bound is related to:

---

[5] In the case of Banach spaces, we of course must restrict Definition 6 to computable structures which also compute the vector space operations. See [8,12] for further results on categoricity of Lebesgue spaces.

Question 13 (McNicholl). Suppose $\mathcal{B} \cong L^p(\Omega)$ is computable. Does p have to be a computable real? If yes, is it *uniformly* computable[6]?

It also follows from the main results of [6] that the isomorphism problem for computable Lebesgue spaces is arithmetical (see Remark 17). Although we do not know if the upper bounds are tight if p is not held fixed, the results are relativizable to any oracle. Thus, we have:

Corollary 14. *Separable Lebesgue spaces are effectively classifyable.*

Question 15. Estimate the complexity of the effective classification problem for: (1) separable Hardy spaces, and (2) separable Sobolev spaces.

Question 16. For each n, is there Δ^0_{n+1}-categorical but not Δ^0_n-categorical Banach space? Same for Polish groups[7].

A Computable Characterisation of C[0,1]. Suppose we are given a description of a (separable) Banach space $\mathcal{B}$. How hard is it to determine whether it is isomorphic to some fixed Banach space $\mathcal{C}$? Within the proposed framework, we can set $\mathcal{K} = \{\mathcal{C}\}$ and measure the complexity of the effective characterisation problem $\{e : cl(B_e) \cong \mathcal{C}\}$. For example, the separable Hilbert space ℓ_2 admits a low level arithmetical characterisation: use the parallelogram law and compute a basis [34]. Also, various natural Lebesgue spaces such as ℓ_3 admit arithmetical characterisations, with some index sets complete at proper difference levels such as $d\text{-}\Sigma^0_2$ [6].

Remark 17. In effective algebra there are very few natural examples of index sets of structures which are *not* complete at some level of the hyperarithmetical hierarchy; see, e.g., Problem 1 in [19]. Nonetheless, such estimates seem to be more common in computable analysis. For example, suppose $p \geq 1$ is a computable real other than 2. Then the isomorphism problem for the class of L^p spaces is co-3-Σ^0_3-complete [6].

Recall that $C[0,1]$ is universal among all separable Banach spaces. Building on the earlier work in [30] and [7], Franklin et al. [17] have recently announced the following unexpected result:

Theorem 18 (Franklin et al. [17]). *The Banach space $C[0,1]$ admits an effective (arithmetical) characterisation.*

Again, the result can be relativised to any oracle, but here it is not that important since $C[0,1]$ is computable. The main technical lemma in the proof states that $C[0,1]$ is Δ^0_5-categorical.

Question 19 Calculate optimal bounds for the characterisation problem and Δ^0_n-categoricity of $C[0,1]$.

---

[6] Currently, the best known uniform upper bound is Δ^0_2 [6]. McNicholl [25] has recently announced a partial positive solution for the case when $p > 2$, but he also announced that the uniformity of computing p fails. Thus, we conjecture that the upper bound in Theorem 11 is tight.

[7] Such examples exist for Polish spaces, as follows from [20].

Polish Groups. Following Melnikov and Montalbàn [28], we say that a (metrized) Polish group is computable if it admits a computable structure that computes the standard · and $^{-1}$ on the group[8]. Fix a uniform enumeration $(G_e)_{e \in \omega}$ of all partially computable Polish groups. The definitions of the characterisation problem and the isomorphism problem (Definition 1) should be adjusted accordingly. We note however that this approach seems best suited for *compact* groups. This is because compactness and totality of a (potential) group operation are both low-level arithmetical properties [27]. Thus, in the compact case the index sets $I(\mathcal{K})$ and $E(\mathcal{K})$ will reflect the complexity of $\mathcal{K}$ rather than some pathologies of coding.

Recall that compact Polish spaces admit an effective classification (Theorem 4). How hard is it to classify compact Polish groups? Every compact Polish group G contains the largest connected subgroup H which makes G/H profinite. Thus, the classes of connected and profinite groups are central to the general theory of Polish groups.

Theorem 20 (Melnikov [27]).

1. *The characterisation problems for profinite and connected compact Polish groups are both arithmetical.*
2. *The topological isomorphism problems for profinite abelian groups and for connected compact abelian groups are both Σ_1^1-complete.*

As usual, the result can be relativised to any oracle. It follows that recognising whether a given group is profinite or connected compact is not that hard (Π_2^0- and Π_3^0-complete, resp.), but the isomorphism problem is too hard even in the abelian case. In contrast with the previous results, the main tool in the proof of Thm 20 is not Δ_α^0-categoricity[9]. Instead, Melnikov proves a computable version of the celebrated Pontryagin duality which is then used to apply effective algebraic results [14,16] to topological groups. Provably, the duality is effective only when passing from discrete to compact groups [27], but this half-effectivity was sufficient.

Problem 21. (1.) Study Δ_α^0-categoricity of Polish groups. (2.) Measure the effective classification problem for natural subclasses of compact Polish groups. (3.) What is the complexity of the Pontryagin dual of a computable compact connected abelian group? (The profinite case is known.)

---

[8] Computability of the metric is obviously not enough. Unlike the Banach space case, there is nothing like the Mazur-Ulam theorem for topological groups. For example, every separable profinite group is homeomorphic to the computable Cantor space, but obviously not every profinite group is computable. Perhaps, computability of $^{-1}$ can be dropped at least in some cases, but this has not yet been explored.

[9] But there is again a tight connection with categoricity. Melnikov [27] showed that the characterisation problem for computably categorical *recursive* profinite abelian groups Π_4^0-complete; see [27] for the definitions.

A Local Approach to Global Classification. We must emphasise that the subdivision of the discussed results into local and global is somewhat subjective. For instance, Melnikov and Montalbán [28] have suggested an intermediate approach. Recall that a transformation space is a Polish space together with a smooth action of a Polish group on the space. The key observation is that the standard S_∞-transformation space of countable structures in a given finite language is actually *computable* [28]. Thus, the global classification problem for countable structures becomes a local classification problem for points in a transformation space.

The use of topological groups sometimes makes proofs easier. For instance, a technical result of Montalbán [32] characterises uniformly computably categorical structures (categorical relative to all oracles) in terms of infinitary Scott sentences. Using transformation groups and the old result of Effros [15], the rather tricky argument from [32] can be simplified and generalised at the same time:

Theorem 22. *Let (X, G, a) be a (not necessarily computable) transformation group, and $x \in X$. The following are equivalent:*

1. *x is uniformly computably categorical on a cone;*
2. *the orbit of x is $\mathbf{\Pi}_2^0$.*

The theorem above is an explicit link between Polish groups, the Borel hierarchy, the categoricity hierarchy, and classification problems.

Question 23. Is it possible to extend Theorem 22 to a (computable) categoricity-theoretic description of $\mathbf{\Sigma}_3^0$-orbits in a transformation space[10]?

3   Local Classifications and Hierarchies of Functions

Suppose $\mathcal{M}$ is a computable Polish space, such as $C[0, 1]$ or the reals[11]. Fix some property P of points in $\mathcal{M}$, and consider the index set $I(P) = \{e : x_e \text{ satisfies } P\}$ of P, where $(x_e)_{e \in \omega}$ is an effective enumeration of all (partially) computable points in $\mathcal{M}$. If the complexity of $I(P)$ is no simpler than the naive brute-force upper bound derived from the definition of P, then we can conjecture that members of the space having property P do not admit any reasonable classification. Such results are usually relativisable to any oracle.

For example, Becher, Heiber, and Slaman [3] showed that the index set of all computable real numbers normal to base 2 is Π_3^0-complete[12]. Their proof is

---

[10] Melnikov and Montalbán's initial strong conjecture was that such a characterisation in terms of non-uniform categoricity exists; alas, their proof contained an error (spotted by Solecki).

[11] We usually fix a "natural" computable structure on the space, such as the rationals in $\mathbb{R}$. For a computably categorical space (such as $\mathbb{R}$) this assumption is of course not necessary, while for spaces like $C[0, 1]$ it is essential [30].

[12] A number is said to be normal to base b if, for every positive integer n, all possible strings n digits long have density b^{-n}. A number is (absolutely) normal if it is normal to base b for every integer b greater than 1. Equivalently, a sequence is normal if and only if there is no finite-state gambler that succeeds on it [36].

relativizable, and thus implies the earlier result of Ki and Linton [23] who showed that the set of reals normal to base 2 is $\mathbf{\Pi}^0_3$-complete. Becher and Slaman [9] later extended their techniques to show that the index set of numbers computable to *some* base is Σ^0_4-complete, and again the result can be fully relativised to any oracle.

Westrick [38] gives a detailed index set analysis of Kechris and Woodin's differentiability hierarchy for continuous functions in $C[0,1]$. She showed that the index set of rank $\leq \alpha$ computable functions in $C[0,1]$ is $\Pi^0_{2\alpha+1}$-complete, where α is any computable ordinal.

The index set $I(P)$ does not have to use the enumeration of *all* computable members in a class. For example, we could instead start with a uniform enumeration of all polynomial time computable points $(l_e)_{e\in\omega}$ of the space; if we can still prove a completeness result restricted to $(l_e)_{e\in\omega}$ then this means that the property "x is polynomial time computable" does not help to characterise when P holds on x. We illustrate this approach by a non-trivial example. Following [11], we call $f \in C[0,1]$ regular to base 2 if the graph of f coded as a pair of binary strings is recognised by a Büchi automaton. Regular continuous functions are Lipschitz and also map rationals to rationals in linear time. Using quantifier elimination one can express "f is regular" as a Σ^0_2-statement about f. Franklin et al. [17] have recently announced:

Theorem 24. *Given a linear time computable Lipschitz $f : [0,1] \to \mathbb{R}$ with $f(\mathbb{Q} \cap [0,1]) \subseteq \mathbb{Q}$, checking whether f is regular is a Σ^0_2-complete problem.*

Remark 25. Working independently, Gorman et al. [18] have announced that every continuous regular function has to be affine outside a measure 0 nowhere dense set. This property can be added to the list of properties in the theorem above too.

It follows that none of the properties of regular functions known to date helps to reduce the complexity of the definition of regularity.

3.1   Baire Hierarchy and Parallel Weihrauch Reducibility

So far our "local" results were restricted to continuous functions. If we want to extend the ideas beyond continuity, we need more ideas. We might have some hope for reasonable classes of functions such as the *Baire* classes. Recall that f is Baire class 0 iff it is continuous, and f is called Baire $n+1$ iff there is a collection $\{f_i \mid i \in \mathbb{N}\}$ of Baire class n functions such that $f(x) = \lim_n f_n(x)$, and this extends to limit ordinals in the obvious way. Baire 1 functions are precisely the derivatives of differentiable functions. Baire functions are all Lebesgue measureable, and any Lebesgue measureable function is the same as a Baire 2 function except on a set of measure 0. Can we say more? In the below, we will concentrate on *classification* for the Baire class functions.

Recall that *Weihrauch* reducibility [2] is defined as $f \leq_W g$ to mean there are computable h, k such that $f(x) = h(x, g(k(x)))$. A more general version of this is called *parallel* Weihrauch reducibility. The idea is that to compute $f(x)$ to within 2^{-n} we might explore $g(y)$, yet for $f(x)$ to within 2^{-m} perhaps $g(y')$. For this new reduction we will write $f \leq_T g$.

Remark 26. This is motivated as follows. It is not reasonable to expect that for a general reduction we should have a uniform map like $\leq_W$ which takes arguments of f pointwise to arguments of g determining the reduction. Turing reduction on sets, $A \leq_T B$, can use many queries to B to determine $A(n)$. In the case of continuous functions this can be viewed as as the "parallelisation" of g; replace g with ω many copies of g. See [13] for a clarification.

Example 27. Consider the α-th jump function, $j^\alpha(x)(i) = x^{(\alpha)}(i)$, where $x^{(\alpha)}(i)$ is 0 if $\Phi_i^x(i)$ does not halt and 1 otherwise. Then it is possible to show that $j^1(x) \leq_T S_0$ where S_0 is the step function which is 0 below 0 and 1 above.

In [13] Day, Downey and Westrick instigated an analysis of the classification of Baire functions using $\leq_T$. In that paper, they allowed real parameters, the *boldface* version $\leq_{\mathbf{T}}$. It is easy to show that f is Baire α iff $f \leq_{\mathbf{T}} j^\alpha$. Using recursion-theoretic techniques, Day, Downey and Westrick [13] have proved:

Theorem 28. *If α is a constructive ordinal and a Baire function f is not Baire α then $j^{\alpha+1} \leq_{\mathbf{T}} f$, and if f is limit, and f is not Baire β for $\beta < \alpha$, either $f \equiv_{\mathbf{T}} j^\alpha$, or $j^\alpha <_{\mathbf{T}} f$.*

In [13], Day, Downey and Westrick refined $\leq_{\mathbf{T}}$ to look at analogs of m- and tt-reductions; we omit the definitions[13].The idea being that Post's Theorem (for example) puts Σ_m^0 complete sets above the other Σ_m^0 sets via an m-reduction. We state the following satisfying classification result

Theorem 29 ([13]).

1. *For all Baire functions f and g, either $f \leq_{\mathbf{m}} g$ or $g \leq_{\mathbf{m}} - f$. Hence if f is Baire α then $f \leq_{\mathbf{m}} j^{\alpha+1}$.*
2. *If f is Borel and f is not Baire α, then either $j^{\alpha+1} \leq_{\mathbf{m}} f$ or $-j^{\alpha+1} \leq_{\mathbf{m}} f$.*

These new reductions had reflections in results from classical analysis to give computable "explanations" for classical results. In [13] Day et al. showed (classical) Baire and Bourgain hierarchies of functions intertwine with the degree structures above. Kihara (to appear) has recently extended the results above.

As we see the boldface version gives real insight into classical investigations, but it might be argued that a finer classification could be obtained using no parameters. This is still an open challenge. Lots of open problems remain, particularly what would be the correct notion for lightface versions.

References

1. Ash, C., Knight, J.: Computable Structures and the Hyperarithmetical Hierarchy. Studies Logic and the Foundations of Mathematics, vol. 144. North-Holland Publishing Co., Amsterdam (2000)
2. Brattka, V., Gherardi, G.: Borel complexity of topological operations on computable metric spaces. J. Logic Comput. **19**(1), 45–76 (2009)

---

[13] This is not immediate, as the idea of "positive information" in this setting needs clarification. Day et al. use *separation names.*

3. Becher, V., Heiber, P.A., Slaman, T.A.: Normal numbers and the Borel hierarchy. Fund. Math. **226**(1), 63–78 (2014)
4. Bohnenblust, H.F.: An axiomatic characterization of L_p-spaces. Duke Math. J. **6**, 627–640 (1940)
5. Brattka, V., Hertling, P., Weihrauch, K.: A tutorial on computable analysis. In: Cooper, S.B., Löwe, B., Sorbi, A. (eds.) New Computational Paradigms, pp. 425–491. Springer, New York (2008). https://doi.org/10.1007/978-0-387-68546-5_18
6. Brown, T., McNicholl, T., Melnikov, A.: On the complexity of classifying lebesgue spaces. To appear
7. Brown, T.A.: Computable Structure Theory on Banach Spaces. ProQuest LLC, Ann Arbor, MI (2019). Thesis (Ph.D.)-Iowa State University
8. Brown, T., McNicholl, T.: Analytic computable structure theory and L^p spaces-part 2 Archive for Mathematical Logic (to appear)
9. Becher, V., Slaman, T.A.: On the normality of numbers to different bases. J. Lond. Math. Soc. **90**(2), 472–494 (2014)
10. Cenzer, D., Remmel, J.B.: Index sets in computable analysis. Theoret. Comput. Sci. **219**(1–2), 111–150 (1999). Computability and complexity in analysis (Castle Dagstuhl, 1997)
11. Chaudhuri, S., Sankaranarayanan, S., Vardi, M.Y.: Regular real analysis. In: 2013 28th Annual ACM/IEEE Symposium on Logic in Computer Science (LICS 2013), pp. 509–518. IEEE Computer Society, Los Alamitos (2013)
12. Clanin, J., McNicholl, T.H., Stull, D.M.: Analytic computable structure theory and L^p spaces. Fundamenta Mathematicae **244**, 255–285 (2019)
13. Day, A., Downey, R.G., Westrick, L.B.: Three topological reducibilities for discontinuous functions. To appear
14. Downey, R., Montalbán, A.: The isomorphism problem for torsion-free abelian groups is analytic complete. J. Algebra **320**(6), 2291–2300 (2008)
15. Effros, E.G.: Transformation groups and C^*-algebras. Ann. Math. **2**(81), 38–55 (1965)
16. Ershov, Y., Goncharov, S.: Constructive Models. Siberian School of Algebra and Logic. Consultants Bureau, New York (2000)
17. Franklin, J., Adam, R.H., Khoussainov, B., Melnikov, A., Ng, K.M.: Continuous functions and effective classification (2020)
18. Gorman, A., et al.: Continuous regular functions. Preprint (2019)
19. Goncharov, S., Knight, J.: Computable structure and antistructure theorems. Algebra Logika **41**(6), 639–681, 757 (2002)
20. Greenberg, N., Melnikov, A.G., Knight, J.F., Turetsky, D.: Uniform procedures in uncountable structures. J. Symb. Log. **83**(2), 529–550 (2018)
21. Gromov, M.: Metric structures for Riemannian and non-Riemannian spaces. Modern Birkhäuser Classics. Birkhäuser Boston Inc., Boston, MA, english edition (2007). Based on the 1981 French original, With appendices by M. Katz, P. Pansu and S. Semmes, Translated from the French by Sean Michael Bates
22. Kakutani, S.: Concrete representation of abstract L-spaces and the mean ergodic theorem. Ann. Math. **42**, 523–537 (1941)
23. Ki, H., Linton, T.: Normal numbers and subsets of **N** with given densities. Fund. Math. **144**(2), 163–179 (1994)
24. McNicholl, T.H.: Computable copies of ℓ^p. Computability **6**(4), 391–408 (2017)
25. McNicholl, T.H.: Computing the exponent of a Lebesgue space. To appear
26. Melnikov, A.G.: Computably isometric spaces. J. Symbolic Logic **78**(4), 1055–1085 (2013)

27. Melnikov, A.: Computable topological groups and Pontryagin duality. Trans. Amer. Math. Soc. **370**(12), 8709–8737 (2018)
28. Melnikov, A., Montalbán, A.: Computable Polish group actions. J. Symb. Log. **83**(2), 443–460 (2018)
29. Melnikov, A.G., Nies, A.: The classification problem for compact computable metric spaces. In: Bonizzoni, P., Brattka, V., Löwe, B. (eds.) CiE 2013. LNCS, vol. 7921, pp. 320–328. Springer, Heidelberg (2013). https://doi.org/10.1007/978-3-642-39053-1_37
30. Melnikov, A.G., Ng, K.M.: Computable structures and operations on the space of continuous functions. Fund. Math. **233**(2), 101–141 (2016)
31. Montalbán, A.: Computability theoretic classifications for classes of structures. In: Proceedings of the International Congress of Mathematicians-Seoul 2014, vol. II, pp. 79–101. Kyung Moon Sa, Seoul (2014)
32. Montalbán, A.: A robuster Scott rank. Proc. Am. Math. Soc. **143**(12), 5427–5436 (2015)
33. Nies, A., Solecki, S.: Local compactness for computable polish metric spaces is π_1^1-complete. In: Evolving Computability - 11th Conference on Computability in Europe, CiE 2015, Bucharest, Romania, 29 June–3 July 2015, pp. 286–290 (2015)
34. Pour-El, M.B., Richards, J.I.: Computability in Analysis and Physics. Perspectives in Mathematical Logic. Springer, Berlin (1989)
35. Sierpinski, W.: General topology. Mathematical Expositions, No. 7. University of Toronto Press, Toronto (1952). Translated by C. Cecilia Krieger
36. Schnorr, C.-P., Stimm, H.: Endliche Automaten und Zufallsfolgen. Acta Informat. **1**(4), 345–359 (1971)
37. Weihrauch, K.: Computable Analysis. Texts in Theoretical Computer Science. An EATCS Series. Springer, Berlin (2000). https://doi.org/10.1007/978-3-642-56999-9
38. Westrick, L.B.: A lightface analysis of the differentiability rank. J. Symb. Log. **79**(1), 240–265 (2014)

Non-coding Enumeration Operators

Russell Miller[1,2](✉)

[1] Queens College, 65-30 Kissena Blvd, Queens, NY 11367, USA
Russell.Miller@qc.cuny.edu
[2] C.U.N.Y. Graduate Center, 365 Fifth Avenue, New York, NY 10016, USA

Abstract. An *enumeration operator* maps each set A of natural numbers to a set $E(A) \subseteq \mathbb{N}$, in such a way that $E(A)$ can be enumerated uniformly from every enumeration of A. The maximum possible Turing degree of $E(A)$ is therefore the degree of the jump A'. It is impossible to have $E(A) \equiv_T A'$ for all A, but possible to achieve this for all A outside a meager set of Lebesgue measure 0. We consider the properties of two specific enumeration operators: the *HTP operator*, mapping a set W of prime numbers to the set of polynomials realizing Hilbert's Tenth Problem in the ring $\mathbb{Z}[W^{-1}]$; and the *root operator*, mapping the atomic diagram of an algebraic field F of characteristic 0 to the set of polynomials in $\mathbb{Z}[X]$ with roots in F. These lead to new open questions about enumeration operators in general.

Keywords: Computability · Computable structure theory ·
Enumeration operators · Essential lowness · Hilbert's Tenth Problem

1   Introduction

Consider enumeration operators. These are functions E mapping Cantor space 2^ω into itself, in an effective way defined by a computably enumerable set $\mathcal{E}$ of axioms of the form (D_i, j), all with $i, j \in \omega$. The intended meaning of such an axiom is that, if $A \in 2^\omega$ is the input to E and the finite set D_i is a subset of A, then j must lie in $E(A)$. (Here D_i is defined to be the unique finite subset of ω with $i = \sum_{n \in D_i} 2^n$, as in [13, Defn. II.2.4].) Thus $E(A) = \{j : (\exists i) \, [(D_i, j) \in \mathcal{E} \; \& \; D_i \subseteq A]\}$, and so $E(A)$ may be enumerated effectively using any enumeration (computable or not) of A.

There is a natural analogy to Turing operators T, which allow one to decide membership of numbers in $T(A)$ when given a decision procedure (effective or not) for A. Enumeration operators use only positive information about A, and produce only positive information about $E(A)$; whereas Turing operators use and produce both positive and negative information. Turing operators have been the focus of more intense study by computability theorists, but enumeration operators need make no apology for their presence. Indeed, the fundamental

The author was partially supported by Grant #581896 from the Simons Foundation and by the City University of New York PSC-CUNY Research Award Program.

M. Anselmo et al. (Eds.): CiE 2020, LNCS 12098, pp. 112–123, 2020.
https://doi.org/10.1007/978-3-030-51466-2_10

operator used by Kurt Gödel for his Incompleteness Theorems is an enumeration operator: the *deducibility operator* D, which (using a fixed Gödel coding of first-order formulas in a fixed signature) lists the elements of the set $D(A)$ of code numbers of those formulas provable from the formulas with code numbers in A. Thus A is viewed as an axiom set, and $D(A)$ is the set of consequences of A.

Here we focus not on D, but rather on two other enumeration operators. In Sect. 2, we will examine the *Hilbert's-Tenth-Problem operator* HTP, as defined in [3] and [11]. The results there will continue a program connecting the properties of Hilbert's Tenth Problem for the field $\mathbb{Q}$ itself with the corresponding properties for subrings of $\mathbb{Q}$. Then, in Sect. 3, we will consider the *root operator* for algebraic fields, finding it to have some very pleasing properties, but also noting that it is not a true enumeration operator, according to the definition above. Our final section raises and describes the open question of whether an enumeration operator, strictly as defined above, can realize the properties that make the root operator distinctive. We view this question as being of sufficient interest that posing it, rather than proving the results in Sects. 2 and 3, may be the most consequential act of this article.

A set B is *e-reducible* to A if $B = E(A)$ for some enumeration operator E. Therefore it is largely trivial to compare A and $E(A)$, or their jumps, under e-reducibility; the more interesting comparisons involve Turing reducibility. Much of this article concerns the property of *essential lowness*, which we now define.

Definition 1. *An enumeration operator E is* essentially low for Lebesgue measure *if*

$$\mu(\{A \subseteq \omega : (E(A))' \leq_T A'\}) = 1.$$

Likewise E is essentially low for Baire category *if this same set $\{A \subseteq \omega : (E(A))' \leq_T A'\}$ is comeager, in the sense of Baire.*

Thus essential lowness says that, for almost all inputs A, the output $E(A)$ is low relative to A. It is important to specify the context of "almost all," as an operator can be essentially low for Baire category without being so for Lebesgue measure, or vice versa. This definition can apply equally well to Turing operators.

We use standard notation from [13]. For an introduction to computable fields, [5] and [7] are both helpful.

2   Hilbert's Tenth Problem on Subrings of $\mathbb{Q}$

For a subset W of the set $\mathbb{P}$ of all prime numbers, we set $R_W = \mathbb{Z}[W^{-1}]$ to be the subring of the rational numbers $\mathbb{Q}$ generated by the reciprocals of the primes in W. The map $W \mapsto R_W$ is a bijection from the power set of $\mathbb{P}$ onto the space of all subrings of $\mathbb{Q}$. Moreover, if subrings of $\mathbb{Q}$ are considered in a signature with $+$, $\cdot$, and a predicate I for invertibility (with $I(x)$ defined by $\exists y(x \cdot y = 1)$), and $2^{\mathbb{P}}$ has the usual Cantor topology, then this bijection is a computable homeomorphism of topological spaces, in the sense of [10]. Fix a computable list $g_0, g_1, \ldots$ of the polynomials in $\mathbb{Z}[\vec{X}] = \mathbb{Z}[X_1, X_2, \ldots]$ and define the enumeration operator HTP as follows. (We usually write $\mathrm{HTP}(R_W)$ instead of $\mathrm{HTP}(W)$.)

$$W \mapsto \mathrm{HTP}(R_W) = \{n \in \omega : \exists (x_1, \ldots, x_k) \in (R_W)^{<\omega} \;\; g_n(x_1, \ldots, x_k) = 0\}.$$

We recall the definition of the *boundary set* of subrings of $\mathbb{Q}$. For any single polynomial $f \in \mathbb{Z}[\vec{X}]$, the boundary set is

$$\mathcal{B}(f) = \{W \subseteq \mathbb{P} : f \notin \mathrm{HTP}(R_W) \text{ but } (\forall \text{ finite } S_0 \subseteq \overline{W}) \, f \in \mathrm{HTP}(R_{(\mathbb{P}-S_0)})\}.$$

This says that, although $f = 0$ has no solution in R_W, there is no finitary reason for it to lack a solution: no finite set S_0 of primes omitted from W has the property that every solution requires the inversion of some prime from S_0. Examples and explanation appear in [3,8]. Topologically $\mathcal{B}(f)$ is the boundary of the (open) set $\mathcal{A}(f)$ of subrings in which $f = 0$ has a solution. More generally, the *boundary set* of subrings of $\mathbb{Q}$ is the union of these:

$$\mathcal{B} = \bigcup_{f \in \mathbb{Z}[X_0, X_1, \ldots]} \mathcal{B}(f).$$

In Baire category, it is known that $\mathcal{B}$ is meager, i.e., small in the sense given by Baire; see [8]. In Lebesgue measure, on the other hand, it remains open whether this same class $\mathcal{B}$ is small, i.e. of measure 0, or not. This distinction leads us to state Theorems 1 and 2 separately, for these two notions of smallness, as it is essential for the boundary set to be small. Theorem 1 arises very naturally as a kind of extension of [8, Corollary 1], which stated that an arbitrary set C satisfies $C \leq_T \mathrm{HTP}(R_W)$ for a non-meager class of sets W if and only if $C \leq_T \mathrm{HTP}(\mathbb{Q})$. Here lowness (relative to W) replaces the property of computing C.

Theorem 1. *The set* $\mathrm{HTP}(\mathbb{Q})$ *has low Turing degree if and only if the operator* HTP *is essentially low for Baire category.*

If HTP is not essentially low, then by this theorem $\mathrm{HTP}(\mathbb{Q})$ must be undecidable. Essential lowness of HTP would not yield any consequences about the decidability of $\mathrm{HTP}(\mathbb{Q})$ (as the degree $\mathbf{0}$ of the computable sets is considered to be low, along with every other degree $\mathbf{d}$ satisfying $\mathbf{d}' = \mathbf{0}'$), but it would imply a different important result: the existential undefinability of $\mathbb{Z}$ within the field $\mathbb{Q}$.

Proof. One direction is readily seen. Essential lowness of HTP means that all W in a co-meager subclass of $2^{\mathbb{P}}$ satisfy $(\mathrm{HTP}(R_W))' \leq_T W'$. The intersection of this subclass with the co-meager class $\{W : W' \leq_T \emptyset' \oplus W\}$ is also co-meager. By [1, Cor. 5.2], all W have $\mathrm{HTP}(\mathbb{Q}) \leq_T \mathrm{HTP}(R_W)$, so these comeager-many W all also satisfy

$$(\mathrm{HTP}(\mathbb{Q}))' \leq_T (\mathrm{HTP}(R_W))' \leq_T W' \leq_T \emptyset' \oplus W.$$

But this implies $(\mathrm{HTP}(\mathbb{Q}))' \leq_T \emptyset'$, by a standard result (see [8, Lemma 2].)

For the forwards direction, recall that $\mathrm{HTP}(R_W) \leq_T (W \oplus \mathrm{HTP}(\mathbb{Q}))$ uniformly for all HTP-generic W. The procedure is as follows. For each f, HTP-genericity means, by definition, that $W \in \mathcal{A}(f) \cup \mathcal{C}(f)$, where

$$\mathcal{A}(f) = \{W \subseteq \mathbb{P} : f \in \mathrm{HTP}(R_W)\}.$$
$$\mathcal{C}(f) = \{W \subseteq \mathbb{P} : (\exists \text{ finite } S_0)[S_0 \cap W = \emptyset \And f \notin \mathrm{HTP}(R_{(\mathbb{P}-S_0)})]\}.$$

So, to decide whether $f \in \mathrm{HTP}(R_W)$, we simply search for either a solution to $f = 0$ in R_W (using the W-oracle to enumerate R_W), or a finite subset $S_0 \subseteq (\mathbb{P}-W)$ such that $f \notin \mathrm{HTP}(R_{\mathbb{P}-S_0})$. (The latter is decidable from $\mathrm{HTP}(\mathbb{Q})$, uniformly in S_0; see [1, Prop. 5.4] or [8, Prop. 1].) Therefore, for HTP-generic sets W, we have $(\mathrm{HTP}(R_W))' \leq_T (W \oplus \mathrm{HTP}(\mathbb{Q}))'$ (indeed via a 1-reduction, by the Jump Theorem [13, III.2.3]). But with $\mathrm{HTP}(\mathbb{Q})$ assumed low, the class of W for which

$$(W \oplus \mathrm{HTP}(\mathbb{Q}))' \leq_T W' \oplus (\mathrm{HTP}(\mathbb{Q}))' \equiv_T W' \oplus \emptyset' \equiv_T W'$$

is comeager, as is the class of HTP-generic sets W, so we have shown that $\{W : (\mathrm{HTP}(R_W))' \leq_T W'\}$ is comeager.   □

Theorem 2. *If the operator HTP is essentially low in the sense of Lebesgue measure, then the set $\mathrm{HTP}(\mathbb{Q})$ has low Turing degree. As a partial converse, If $\mathrm{HTP}(\mathbb{Q})$ has low Turing degree and the set $\mathcal{B}$ of all boundary rings has Lebesgue measure 0, then HTP is essentially low for Lebesgue measure.*

Proof. The forward direction has much the same proof as in Theorem 1: measure-1-many sets W satisfy

$$(\mathrm{HTP}(\mathbb{Q}))' \leq_T (\mathrm{HTP}(R_W))' \leq_T W' \leq_T \emptyset' \oplus W.$$

Consequently $(\mathrm{HTP}(\mathbb{Q}))' \leq_T \emptyset'$; see Lemmas 2.1 and 2.3 of [9] for details.

The reverse direction requires more care. Notice that with an $\mathrm{HTP}(\mathbb{Q})$-oracle, we can enumerate the finite binary strings σ such that a given f does not lie in $\mathrm{HTP}(R_{(\mathbb{P}-\sigma^{-1}(0))})$. (Again this follows from [1, Prop. 5.4] or [8, Prop. 1].) This in turn allows us to approximate, from below, the measure of the set $\mathcal{C}(f)$. On the other hand, with no oracle at all, we can approximate from below the measure of the set $\mathcal{A}(f)$. By assumption $\mu(\mathcal{A}(f)) + \mu(\mathcal{C}(f)) = 1$, since the rings lying in neither $\mathcal{A}(f)$ nor $\mathcal{C}(f)$ are by definition boundary rings of f. Therefore, we can approximate $\mu(\mathcal{A}(f)) = 1 - \mu(\mathcal{C}(f))$ from above as well, using the $\mathrm{HTP}(\mathbb{Q})$-oracle.

Given any $\varepsilon > 0$, and any f_i in an enumeration $f_0, f_1, \dots$ of $\mathbb{Z}[\vec{X}]$, the $\mathrm{HTP}(\mathbb{Q})$-oracle now allows us to approximate the measure $\mu(\mathcal{A}(f_i))$ to arbitrary precision. Suppose this approximation places $\mu(\mathcal{A}(f_i))$ within an open interval (a_i, b_i) of length $< \frac{\varepsilon}{2^{i+1}}$. We then enumerate solutions to $f_i = 0$ in $\mathbb{Q}$ until we have found finitely many solutions such that the total measure of the set of sub-rings containing any of those solutions is $> a_i$. Among the subrings R that do not contain any of those finitely many solutions, the set of those that do contain some solution to $f_i = 0$ has measure $< \frac{\varepsilon}{2^{i+1}}$. Therefore, for any n and any $\eta \in 2^n$, we can produce a finite set $S_{\eta,\varepsilon}$ of binary strings σ such that, among subsets $W \subseteq \mathbb{P}$, the equivalence

$$\eta \subset \mathrm{HTP}(R_W) \iff (\exists \sigma \in S_{\eta,\varepsilon})\sigma \subset W$$

holds for all W outside a set of measure $< \sum_{i=0}^n \frac{\varepsilon}{2^{i+1}} < \varepsilon$.

Our other hypothesis, for the reverse direction, is that $\mathrm{HTP}(\mathbb{Q})$ is low, meaning that $(\mathrm{HTP}(\mathbb{Q}))' \leq_T \emptyset'$. It follows that

$$(W \oplus \mathrm{HTP}(\mathbb{Q}))' \leq_T W' \oplus (\mathrm{HTP}(\mathbb{Q}))' \leq_T W' \oplus \emptyset' \equiv_T W'$$

uniformly for all W outside a set of measure $\leq \frac{\varepsilon}{2}$. Using this, we can now give a procedure that (uniformly in the given ε), computes $(\mathrm{HTP}(R_W))'$ from W' correctly on a set of measure $\geq 1 - \varepsilon$. Given a number e, we wish to determine whether $e \in \mathrm{HTP}(R_W)$, which is to say, whether $\Phi_e^{\mathrm{HTP}(R_W)}(e)$ halts. The oracle $(W \oplus \mathrm{HTP}(\mathbb{Q}))'$ will decide whether

$$(\exists s)(\exists \eta \in 2^{<\omega})(\exists \sigma \in S_{\eta, \frac{\varepsilon}{2^{e+2}}}) \, [\sigma \subset W \ \& \ \Phi_e^\eta(e) \text{ halts in } \leq s \text{ steps}],$$

since the quantifier-free subformula is decidable from $(W \oplus \mathrm{HTP}(\mathbb{Q}))$, as seen above. Thus, outside a set of measure $< \frac{\varepsilon}{2^{e+2}}$, the $(W \oplus \mathrm{HTP}(\mathbb{Q}))'$-oracle has decided correctly whether

$$(\exists s)(\exists \eta \in 2^{<\omega}) \, [\eta \subset \mathrm{HTP}(R_W) \ \& \ \Phi_e^\eta(e) \text{ halts in } \leq s \text{ steps}],$$

which is to say, whether $\Phi_e^{\mathrm{HTP}(R_W)}(e)$ halts. Outside a set of measure $< \sum_e \frac{\varepsilon}{2^{e+2}} = \frac{\varepsilon}{2}$, this computation will be correct for every e, and since the computation of $(W \oplus \mathrm{HTP}(\mathbb{Q}))'$ from W' was also correct outside a set of measure $< \frac{\varepsilon}{2}$, we have proven the theorem.   □

3   Algebraic Fields and the Root Operator

Let $\Delta(F)$ be the atomic diagram – viewed as a subset of ω, using Gödel coding – of a field F of characteristic 0, in the pure language of rings. Given $\Delta(F)$, the *root operator* Θ enumerates the subset $W_F \subseteq \omega$ called the *index of F*:

$$W_F = \{i \in \omega : (\exists x \in F) \, f_i(x) = 0\},$$

using a fixed computable enumeration $f_0, f_1, \dots$ of $\mathbb{Z}[X]$. (Here all polynomials have just one variable X, as opposed to Sect. 2.) It is clear that W_F can be enumerated uniformly this way if one has an oracle for the atomic diagram, but of course we mean Θ to be given only an (arbitrary) enumeration of that atomic diagram. Fortunately, these are equivalent. To decide whether $a + b = c$ in the field F, just enumerate $\Delta(F)$ until a formula of the form $a + b = d$ (for some d) appears, and check whether this d is the element c or not; similarly for multiplication. So, given any enumeration of the atomic diagram $\Delta(F)$ of any presentation of a field F of characteristic 0, Θ will indeed enumerate W_F. Our specific interest is in *algebraic fields*, i.e., algebraic extensions of $\mathbb{Q}$.

Algebraic fields of characteristic 0 may be viewed in much the same context as subrings of $\mathbb{Q}$. In both cases, the collection of all isomorphism types belonging to the class forms a topological space computably homeomorphic to Cantor space. Also, in both cases, one natural definable predicate must be adjoined to the signature in order for the homeomorphism to exist. In both cases, the original signature is that of rings, with addition and multiplication symbols. For subrings of $\mathbb{Q}$, in the preceding section, we required an invertibility predicate I in the signature, defined by $I(x) \iff (\exists y) \, x \cdot y = 1$: with this predicate it becomes possible to compute the index $W = \{p \in \mathbb{P} : \frac{1}{p} \in R\}$ of an arbitrary subring R of $\mathbb{Q}$ uniformly from the atomic diagram of each presentation of R.

For our algebraic fields – which, since we restrict to characteristic 0, may be viewed as precisely the class of all subfields of the algebraic closure $\overline{\mathbb{Q}}$ – we likewise require some additional information to compute the index of an arbitrary member of the class. The usual strategy is to adjoin *root predicates*: either a single finitary predicate R, or else a d-ary predicate R_d for each $d > 1$. These are defined by

$$R_d(a_0, \ldots, a_{d-1}) \iff (\exists x) \, x^d + a_{d-1}x^{d-1} + \cdots + a_1 x + a_0 = 0.$$

(If a single R is used, it is simply the union of all these R_d.) Given the atomic diagram of a subfield F of $\overline{\mathbb{Q}}$, in the signature with these predicates, one can readily compute the index W_F of F as defined above. Clearly $W_F = W_E$ whenever $E \cong F$, and the converse also holds (cf. [12, Cor. 3.9], with $\mathbb{Q}$ as the ground field). That is, the subfields of $\overline{\mathbb{Q}}$ may be classified up to isomorphism by their indices. On the other hand, it should be noted that not every subset of ω is the index of an algebraic field under this definition: if $(X^4 - 2)$ has a root in F, for example, then $(X^2 - 2)$ cannot fail to have one. Nevertheless, the indices used here do yield a computable homeomorphism from the space of all subfields of $\overline{\mathbb{Q}}$ onto Cantor space. It is decidable uniformly for each $\sigma \in 2^{<\omega}$ whether there exists an algebraic field F for which σ is an initial segment of W_F, and the collection of σ for which such a field does exist is thus a computable subtree T of 2^ω, with no terminal nodes and no isolated paths. The desired homeomorphism then maps the class of isomorphism types of algebraic fields bijectively onto the space of all paths through T, which in turn is computably homeomorphic to 2^ω.

One should note that this homeomorphism is not canonical: it depends heavily on the choice of the computable enumeration of $\mathbb{Z}[X]$ used to define the indices W_F. Using Lebesgue measure on this space is possible but not recommended: different enumerations of $\mathbb{Z}[X]$ will give distinct measures on the space of (isomorphism types of) subfields of $\overline{\mathbb{Q}}$. A better option is to use the *Haar-compatible measure* μ defined in [10], which has the property that, for every normal field extension $K \supseteq \mathbb{Q}$ with finite vector-space dimension $[K : \mathbb{Q}]$,

$$\mu(\{F \subseteq \overline{\mathbb{Q}} : K \subseteq F\}) = \frac{1}{[K : \mathbb{Q}]},$$

independently of the enumeration of $\mathbb{Z}[X]$. (In [10] the index is chosen differently, but this property remains true in both cases.) In fact, though, the sets of Haar-compatible measure 0 are soon seen to be the same sets that have Lebesgue measure 0 for some computable enumeration of $\mathbb{Z}[X]$ – indeed, for all reasonable ones. So it is unnecessary to fuss over the exact choice of the measure to be used.

With this background, we can proceed to consider the root operator Θ, mapping $\Delta(F)$ to W_F. There do exist (isomorphism types of) fields such that every presentation F of the field satisfies $W_F \leq_T \Delta(F)$. The most obvious examples are fields for which W_F itself is computable, but other examples exist. For instance, if $F = \mathbb{Q}[\sqrt{p_n} : n \notin \emptyset']$, then W_F is co-c.e., and so every presentation of F can compute W_F, by enumerating W_F and simultaneously enumerating the c.e. set that is the complement of W_F. (Even more generally, this holds whenever the complement $\overline{W_F}$ is enumeration-reducible to the Σ_1-theory of F.)

However, the isomorphism types F of fields such that all presentations of F compute W_F are rare: they form a meager set of Haar-compatible measure 0 within the class of all subfields of $\overline{\mathbb{Q}}$. To see this, notice first that the ability to enumerate W_F is equivalent to the ability to compute (or enumerate) a presentation of the field F: whenever a polynomial in $\mathbb{Z}[X]$ is found to have a root in F, we first check whether we have already put such a root into our presentation, then consider the possible ways that such a root (if not already present) could sit over the finitely-generated subfield we have built so far, and enumerate W_F further until the minimal polynomial of a primitive generator of one of these possibilities appears in W_F, at which point it is safe to extend our field to be isomorphic to that possibility. All of this is effective, by Kronecker's Theorem (see [6, Thm. 2.2]), allowing us to state our next theorem.

Theorem 3. *Let $\mathcal{L}$ be the set of indices W such that, for some presentation F of the field with $W = W_F$, the set $W = \Theta^{\Delta(F)}$ is not $\Delta(F)$-computable. Then $\mathcal{L}$ has measure 1 as a subset of 2^ω, under the Haar-compatible measure, and is also co-meager in 2^ω. (By analogy to Definition 1, we say that Θ is essentially noncomputable.)*

Proof. By results of Jockusch [2] and Kurtz [4], there is a co-meager, measure-1 class of sets $W \in 2^\omega$ each having the property that there exists some $V \in 2^\omega$ for which W is V-computably enumerable but not V-computable. Whenever W has this property, we can find a presentation of the field F with $W = W_F$ that is computable from such a set V. For this presentation, $\Delta(F)$ fails to compute $W_F = \Theta^{\Delta(F)}$, although it can enumerate it. $\square$

We note that, for each index W in the measure-1 co-meager set described by the theorem, the presentations F that *do* have $W_F \leq_T \Delta(F)$ are (of course) those in the upper cone above W; whereas every set that computes V can compute a presentation of this field. Thus the presentations of F that succeed in computing W_F may be viewed as a small subset of the set of all presentations of F, as every upper cone that is a proper subset of another upper cone has measure 0 within the larger upper cone. Hence Theorem 3 may be viewed as saying that for almost all isomorphism types of algebraic fields, almost all presentations of that field fail to compute the index of the isomorphism type.

Nevertheless, the indices are almost never too far away from the presentation. Of course, since $\Delta(F)$ can enumerate W_F, it is immediate that we always have $W_F \leq_T (\Delta(F))'$. In certain cases the two are Turing-equivalent – e.g., when F is a computable presentation of the field $\mathbb{Q}[\sqrt{p_n} : n \in \emptyset']$, whose W_F has degree $\mathbf{0}'$. However, our next two theorems say that in almost all cases, this fails, and indeed W_F is almost always low relative to $\Delta(F)$ – by which we mean that their jumps satisfy $(W_F)' \leq_T (\Delta(F))'$. (From here on we will mostly write F and F' when we actually mean $\Delta(F)$ and $(\Delta(F))'$.)

Theorem 4. *Let $\mathcal{L}$ be the set of indices W such that, for every presentation F of the field with index W, W is low relative to the presentation F, i.e., $W' \leq_T F'$. Then $\mathcal{L}$ is comeager as a subset of 2^ω, and there is a uniform procedure that, for comeager-many $W \in 2^\omega$ (in particular, for all 1-generic W), computes W' from F' for all F with $W_F = W$.*

Proof. We describe the uniform procedure. Let F be a field with $W_F = W$. Our procedure is given an F'-oracle, from which it can compute both W and $\emptyset'$, uniformly. To decide whether $e \in W'$, it begins running $\Phi_e^W(e)$, and if this computation ever halts, then it outputs 1, knowing this to be correct. Simultaneously, it tests each initial segment $\sigma \subseteq W$, using $\emptyset'$ to decide whether

$$\forall \tau \supseteq \sigma \forall t \ \Phi_{e,t}^\tau(e)\uparrow \ .$$

If the procedure ever finds a $\sigma \subseteq W$ for which this holds, then it outputs 0, since no extension of this σ (including W itself) can ever cause Φ_e to halt on input e.

Now if W is 1-generic, then for every e there must exist some finite initial segment $\sigma \subseteq W$ for which either $\Phi_e^\sigma(e)\downarrow$ or else no $\tau \supseteq \sigma$ ever gives convergence. The procedure eventually discovers such a σ and outputs the correct answer about whether $e \in W'$. (It is clear that it nevers gives an answer except when it has found such a σ.) Thus $W' \leq_T F'$, uniformly in F, for all 1-generic W. The theorem now follows from the co-meagerness of the 1-generic sets in 2^ω. $\square$

The situation for Lebesgue measure is similar but not quite as nice. It remains true that W is almost always low relative to all presentations of the field with index W: in particular, this holds outside a set of measure 0. However, no single uniform procedure can establish this for all W outside a set of measure 0. Instead, we must argue up to sets of arbitrarily small measure. (Notice that in Baire category, there is no natural analogue of "up to arbitrarily small measure": the only divisions are meager-or-not and comeager-or-not. It is fortunate that we had a uniform procedure there!)

Theorem 5. *For each $\varepsilon > 0$, there is a uniform procedure that, for all W in a set of Haar-compatible measure $> 1 - \varepsilon$, computes W' from F' for all fields F with $W_F = W$. Hence, for all W outside a set of measure 0, every presentation F of that field satisfies $W' \leq_T F'$.*

Proof. We fix a rational $\varepsilon > 0$ and give a uniform procedure that computes W' from F' on a set of measure $> 1 - \varepsilon$. For each W in this set, each presentation F of the field with index W, and each e, the procedure will determine whether $\Phi_e^W(e)$ halts.

The procedure goes through each pair $\langle r, n \rangle$ with $r \in [0,1]$ rational and with $n \in \omega$ in turn, asking its F' oracle whether both of the following hold:

$$(\exists k, s)\left(\exists \tau_0, \ldots, \tau_k \in 2^n \text{of total measure} > r - \frac{\varepsilon}{2^{e+1}}\right)(\forall i \leq k)\,\Phi_{e,s}^{\tau_i}(e)\downarrow$$

$$\neg(\exists k, s)(\exists \tau_0, \ldots, \tau_k \in 2^n \text{of total measure} > r)(\forall i \leq k)\Phi_{e,s}^{\tau_i}(e)\downarrow$$

Here "total measure" means the Haar-compatible measure of the set $\{A \in 2^\omega : (\exists i \leq k)\tau_i \subseteq A\}$, which is readily computed (making sure to avoid doublecounting overlaps!). The first sentence is considered to hold vacuously if $r < \frac{\varepsilon}{2^{e+1}}$. Also, with $\tau_i \in 2^n$, we interpret "convergence" of $\Phi_{e,s}^{\tau_i \oplus F}(e)$ to mean that the procedure halts without ever asking whether any number $\geq n$ lies in τ_i. These sentences are existential, hence decidable from the F'-oracle.

Eventually our procedure must find a pair $\langle r, n \rangle$ for which both answers from F' are positive. It then stops asking such questions, and searches until it finds the promised $\tau_0, \ldots, \tau_k \in 2^n$ of total measure $> r - \frac{\varepsilon}{2^{e+1}}$ for which all $\Phi_{e,s}(e)\downarrow$. As soon as it has found such strings, it checks whether any of the (finitely many) τ_i it has found is an initial segment of W. (Here is where the full F'-oracle is needed: it can compute W. For the first step, a $\emptyset'$-oracle would have sufficed.) If so, then it outputs that $\Phi_e^W(e)\downarrow$, knowing this answer to be correct. If not, then it outputs that $\Phi_e^W(e)\uparrow$. The latter answer is not guaranteed to be correct, of course, but it can fail only for those W within a set of measure $< \frac{\varepsilon}{2^{e+1}}$, and so the set of W for which our procedure fails to compute W' correctly has measure $\leq \sum_e \frac{\varepsilon}{2^{e+1}} = \varepsilon$, as required. $\square$

4   Non-coding Enumeration Operators

A careful reading of Sect. 3 will reveal that we avoided ever actually calling the root operator Θ an enumeration operator. Indeed, it was intended specifically to operate on (enumerations of) atomic diagrams of algebraic fields, rather than on (enumerations of) arbitrary subsets of ω. One could run it with an enumeration of an arbitrary $A \subseteq \omega$, but with high probability the result would be that every $f \in \mathbb{Z}[X]$ would be deemed to have a root, as the configuration describing this would appear sooner or later. Even worse, the output would not be independent of the enumeration of A, since for most A the ternary relations $+$ and $\cdot$ described by A would not be functions: $a + b$ would be deemed to equal the first c for which the code number of the statement "$a + b = c$" appeared in A. So this Θ does not satisfy the definition given in Sect. 1: it functions as such an operator only on a specific domain within 2^ω, and that domain is meager with measure 0.

The basic question, therefore, asks whether the pleasing results about Θ can hold of a true enumeration operator. When E satisfies the full definition, can E be essentially low and essentially noncomputable, as Θ was on its domain? The answer to this first question is immediate and positive. To find an essentially noncomputable enumeration operator, one need look no further than the map $A \mapsto (A \oplus \emptyset')$, whose output is Turing-equivalent to A' for a comeager measure-1 collection of sets A. To add essential lowness for Lebesgue measure, we adjust the operator to map A to $(A \oplus L)$, where L is a fixed c.e. set of low (nonzero) Turing degree. It is known that, for such an L,

$$\mu(\{A \subseteq \omega : (A \oplus L)' \equiv_T A \oplus L'\}) = 1,$$

and since $L' \equiv_T \emptyset'$, this implies that $(A \oplus L)$ is low relative to A for almost all sets A. Moreover, every A that is 1-generic relative to L lies in this set, so the set is comeager. On the other hand, A computes $(A \oplus L)$ only if $A \geq_T L$, and since L is noncomputable, the upper cone of sets $\geq_T L$ has measure 0 and is meager. (All measure-theoretic results used here were proven by Stillwell in [14].)

In brief, one can accomplish the same goals achieved by Θ, simply by coding an appropriate c.e. set into the output of the enumeration operator. However, Θ accomplished the goal by different means: it produced an output $\Theta^{\Delta(F)} >_T \Delta(F)$ (in almost all cases) simply because the problem of determining existence of roots of polynomials in a field seems to be inherently noncomputable. The following lemma makes this more specific.

Lemma 1. *Fix a noncomputable subset $C \subseteq \omega$, and let $\mathcal{L}$ be the set of indices W such that, for every F with $W_F = W$, $C \not\leq_T \Theta^{\Delta(F)}$. Then $\mathcal{L}$ is comeager of measure 1.*

Proof. $\Theta^{\Delta(F)}$ is just W_F itself, and since C is noncomputable, only a comeager measure-0 collection of sets W satisfy $C \leq_T W$. $\qquad\square$

The simplicity of the proof exposes the overstatement of this lemma, which really just says that upper cones above noncomputable sets are meager of measure 0, as has been long known. Part of the difficulty of working with Θ is that, since its domain is only a small subset of 2^ω, we used different means to measure what it did. In particular, for a subset $\mathcal{S}$ of the domain, we measured the image of $\mathcal{S}$ under the operator, rather than measuring $\mathcal{S}$ itself. Since the image of Θ (on its intended domain: atomic diagrams of algebraic fields) is all of 2^ω, this was reasonable, but it makes Lemma 1 trivial.

Nevertheless, Lemma 1 was stated this way for a reason: it introduces the notion of a non-coding operator.

Definition 2. *An enumeration operator E is* non-coding *for Lebesgue measure if, for every noncomputable $C \subseteq \omega$,*

$$\mu(\{A \subseteq \omega : C \leq_T E(A)\}) = 0.$$

Likewise, E is non-coding *for Baire category if this set is meager whenever $C >_T \emptyset$.*

Thus this is the same idea we noted above for Θ, but now defined using measure and category on the domain 2^ω of E, rather than on the image of a specific smaller domain.

All operators of the form $A \mapsto A \oplus C$ (with C noncomputable) clearly fail this definition. Indeed, any enumeration operator E for which $A' \leq_T E(A)$ holds on a set of positive measure must fail the definition, as $\emptyset' \leq_T A'$ always holds. So a noncoding operator must avoid producing the jump of its input, at least in almost all cases. This brings us to our main open questions, which can be seen as a sort of uniform version of Post's Problem for enumeration operators.

Question 1. Does there exist a non-coding enumeration operator E for which

$$\mu(\{A \subseteq \omega : E(A) \leq_T A\}) = 0?$$

Indeed, does there exist such an operator for which this set has measure < 1? Likewise, can this set be meager? Or at least, can it fail to be co-meager?

We point out the following operator. Let R, S be two noncomputable c.e. sets that form a minimal pair (as in [13, §IX.1], e.g.), and define the enumeration operator E by the union of the following c.e. sets of enumeration axioms:

$$\{(\emptyset, 2x) : x \in R\} \quad \cup \quad \{(\{5\}, 2x) : x \in \omega\} \quad \cup \quad \{(\{5\}, 2x + 1) : x \in S\}.$$

Thus, if $5 \in A$, then $E(A) \equiv_T \omega \oplus S$, while otherwise $E(A) \equiv_T R \oplus \emptyset$. Therefore, in almost all cases we have $E(A) \not\leq_T A$. However, if $C \leq_T E(A)$ on a co-meager or measure-1 collection of sets A, then C must be computable, as it would be computed by both R and S. So it is readily possible to answer Question 1 if one relaxes the definition of non-coding enumeration operators to require only that $\{A \subseteq \omega : C \leq_T E(A)\}$ have measure < 1, or that it not be co-meager. In fact, Definition 2 requires that these sets be small (as opposed to "not-large").

A significant part of the difficulty in answering Question 1 is that enumeration operators E must enumerate the same set $E(A)$ for all enumerations of the set A. This precludes the use of many of the techniques employed in studying the theory of the c.e. degrees, such as the Friedberg-Muchnik method, or the strategy for the Sacks Density Theorem. The latter, for example, starts with two c.e. sets $C <_T D$, and immediately fixes computable enumerations of each. Its strategy succeeds in enumerating a set strictly between C and D, but if different computable enumerations were used, it would generally enumerate a different set in that interval. Therefore that method would require significant refinement – a kind of uniformization – to succeed in answering Question 1.

Our last proposition illustrates the importance of Question 1.

Proposition 1. *Suppose that the answer to the strong form of Question 1 is negative for Baire category. (That is, assume that for all non-coding enumeration operators E, $\{A \subseteq \omega : E(A) \leq_T A\}$ is not meager.) Then the existence of a co-meager collection of subsets $W \subseteq \mathbb{P}$ satisfying $\mathrm{HTP}(R_W) \not\leq_T W$ is equivalent to the undecidability of $\mathrm{HTP}(\mathbb{Q})$.*

Likewise, if the answer to the weak form of Question 1 for Baire category is negative, then the existence of a non-meager collection of subsets W satisfying $\mathrm{HTP}(R_W) \not\leq_T W$, is equivalent to the undecidability of $\mathrm{HTP}(\mathbb{Q})$.

Proof. First of all, every $W \subseteq \mathbb{P}$ satisfies $\mathrm{HTP}(\mathbb{Q}) \leq_T \mathrm{HTP}(R_W)$. Therefore, every W outside the upper cone above $\mathrm{HTP}(\mathbb{Q})$ satisfies $\mathrm{HTP}(R_W) \not\leq W$. If $\mathrm{HTP}(\mathbb{Q})$ were undecidable, then the upper cone above it would be meager, so the backwards direction in each paragraph of the proposition holds even without knowing any answers to Question 1.

The forwards direction is where Question 1 comes into play. In each paragraph, the hypotheses imply that the HTP-operator cannot be non-coding for Baire category. Therefore there is some non-meager subset $\mathcal{S} \subseteq 2^{\mathbb{P}}$ on which it codes some noncomputable information: some $C >_T \emptyset$ satisfies

$$(\forall W \in \mathcal{S}) \; C \leq_T \mathrm{HTP}(R_W).$$

But then it follows from Corollary 1 in [8] that $C \leq_T \mathrm{HTP}(\mathbb{Q})$.   □

A negative answer to one of the two alternative forms of Question 1 essentially says that if $HTP(R_W) \not\leq_T W$ holds for a non-meager (alternatively, co-meager) class of sets W, then there must be some specific non-computable information that HTP is coding into those sets. The result [8, Corollary 1] then shows that this specific information can be derived from just $HTP(\mathbb{Q})$. Similar statements would hold for Lebesgue measure if the boundary set $\mathcal{B}$ defined earlier has measure 0, but this remains an open question.

References

1. Eisenträger, K., Miller, R., Park, J., Shlapentokh, A.: As easy as $\mathbb{Q}$: Hilbert's Tenth Problem for subrings of the rationals. Trans. Am. Math. Soc. **369**(11), 8291–8315 (2017)
2. Jockusch, C.G.: Degrees of generic sets. Recursion theory: its generalisation and applications (Proc. Logic Colloq., Univ. Leeds, Leeds, 1979). London Mathematical Society Lecture Note Series, vol. 45, pp. 110–139 (1981)
3. Kramer, K., Miller, R.: The Hilbert's-Tenth-Problem operator. Isr. J. Math. **230**(2), 693–713 (2019). https://doi.org/10.1007/s11856-019-1833-2
4. Kurtz, S.: Randomness and genericity in the degrees of unsolvability. Ph.D. thesis, University of Illinois at Urbana-Champaign (1981)
5. Miller, R.: Computable fields and Galois theory. Not. Am. Math. Soc. **55**(7), 798–807 (2008)
6. Miller, R.: $\mathbb{Q}$-computable categoricity for algebraic fields. J. Symb. Log. **74**(4), 1325–1351 (2009)
7. Miller, R.: An introduction to computable model theory on groups and fields. Groups Complex. Cryptol. **3**(1), 25–46 (2011)
8. Miller, R.: Baire category theory and Hilbert's Tenth Problem inside $\mathbb{Q}$. In: Beckmann, A., Bienvenu, L., Jonoska, N. (eds.) CiE 2016. LNCS, vol. 9709, pp. 343–352. Springer, Cham (2016). https://doi.org/10.1007/978-3-319-40189-8_35
9. Miller, R.: Measure theory and Hilbert's Tenth Problem inside $\mathbb{Q}$. In: Friedman, S.D., Raghavan, D., Yang, Y. (eds.) Sets and Computations. Lecture Note Series, vol. 33, pp. 253–269. Institute for Mathematical Sciences, National University of Singapore (2017)
10. Miller, R.: Isomorphism and classification for countable structures. Computability **8**(2), 99–117 (2019)
11. Miller, R.: HTP-complete rings of rational numbers. Submitted for publication
12. Miller, R., Shlapentokh, A.: Computable categoricity for algebraic fields with splitting algorithms. Trans. Am. Math. Soc. **367**(6), 3981–4017 (2015)
13. Soare, R.I.: Recursively Enumerable Sets and Degrees. Springer, New York (1987)
14. Stillwell, J.: Decidability of the "almost all" theory of degrees. J. Symb. Log. **37**(3), 501–506 (1972)

On the Interplay Between Inductive Inference of Recursive Functions, Complexity Theory and Recursive Numberings

Thomas Zeugmann[(✉)]

Division of Computer Science, Hokkaido University, Sapporo 060-0814, Japan
thomas@ist.hokudai.ac.jp

Abstract. The present paper surveys some results from the inductive inference of recursive functions, which are related to the characterization of inferrible function classes in terms of complexity theory, and in terms of recursive numberings. Some new results and open problems are also included.

1 Introduction

Inductive inference of recursive functions goes back to Gold [11], who considered *learning in the limit*, and has attracted a large amount of interest ever since. In learning in the limit an inference strategy S is successively fed the graph $f(0), f(1), \ldots$ of a recursive function in natural order and on every initial segment of it, the strategy has to output a hypothesis, which is a natural number. These numbers are interpreted as programs in a fixed Gödel numbering φ of all partial recursive functions over the natural numbers. The sequence of all hypotheses output on f has then to stabilize on a number i such that $\varphi_i = f$. A strategy infers a class $\mathcal{U}$ of recursive functions, if it infers every function from $\mathcal{U}$.

One of the most influential papers has been Blum and Blum [6], who introduced two types of characterizations of learnable function classes in terms of computational complexity. The gist behind such characterizations is that classes $\mathcal{U}$ of recursive functions are learnable with respect to a given learning criterion if and only if all functions in $\mathcal{U}$ possess a particular complexity theoretic property.

The learning criterion considered is reliable inference on the set $\mathcal{R}$, where $\mathcal{R}$ denotes the set of all recursive functions. We denote the family of all classes $\mathcal{U}$ which are reliably inferable by $\mathcal{R}$-REL. Here *reliability* means the learner converges on any function f from $\mathcal{R}$ iff it learns f in the limit. In the first version, operator honesty classes are used. If $\mathfrak{O}$ is a total effective operator then a function f is said to be $\mathfrak{O}$-honest if $\mathfrak{O}(f)$ is an upper bound for the complexity Φ_i for all but finitely many arguments of φ_i, where $\varphi_i = f$. Then Blum and Blum [6] showed that a class $\mathcal{U}$ is in $\mathcal{R}$-REL iff there is a total effective operator $\mathfrak{O}$ such that every function $f \in \mathcal{U}$ is $\mathfrak{O}$-honest. Operator honest characterizations are also called *a priori* characterizations.

© Springer Nature Switzerland AG 2020
M. Anselmo et al. (Eds.): CiE 2020, LNCS 12098, pp. 124–136, 2020.
https://doi.org/10.1007/978-3-030-51466-2_11

In the second version, one considers functions which possess a fastest program modulo a general recursive operator $\mathfrak{O}$ (called $\mathfrak{O}$-compression index). Now, the *a posteriori* characterization is as follows: A class $\mathcal{U}$ is in $\mathcal{R}$-REL iff there is a general recursive operator $\mathfrak{O}$ such that every function from $\mathcal{U}$ has an $\mathfrak{O}$-compression index.

Combining these two characterizations yields that the family of operator honesty classes coincides with the family of the operator compressed classes.

While operator honesty characterizations have been obtained for many learning criteria (cf. [29] and the references therein), the situation concerning *a posteriori* characterizations is much less satisfactory. Some results were shown in [27], but many problems remain open. In particular, it would be quite interesting to have an *a posteriori* characterization for $\mathfrak{T}$-REL. The learning criterion $\mathfrak{T}$-REL is defined as $\mathcal{R}$-REL, but reliability is required on the set $\mathfrak{T}$ of all total functions. Note that $\mathfrak{T}$-REL $\subset$ $\mathcal{R}$-REL. Also, we shall present an *a posteriori* characterization for the function classes which are $\mathcal{T}$-consistently learnable with δ-delay (cf. [1] for a formal definition). Intuitively speaking, a $\mathcal{T}$-consistent δ-delayed learning strategy correctly reflects all inputs seen so far except the last δ ones, where δ is a natural number.

Note that there are also prominent examples of learning criterions for which even *a priori* characterizations are missing. These include the *behaviorally correct* learnable functions classes (cf. [8,9] for more information). So in both cases we also point to the open problem whether or not one can show the non-existence of such desired characterizations.

Moreover, in Blum and Blum [6] the *a posteriori* characterization of $\mathcal{R}$-REL has been used to show that some interesting function classes are in $\mathcal{R}$-REL, e.g., the class of approximations of the halting problem. Stephan and Zeugmann [22] extended these results to several classes based on approximations to non-recursive functions. Besides these results, our knowledge concerning the learnability of interesting function classes is severely limited, except the recursively enumerable functions classes (or subsets thereof), and with respect to function classes used to achieve separations.

Finally, the problem of suitable hypothesis spaces is considered. That is, instead of Gödel numberings one is interested in numberings having learner-friendly properties. Again, we survey some illustrative results, present some new ones, and outline open problems. Note that one can also combine the results obtained in this setting with the results mentioned above, i.e., one can derive some complexity theoretic properties of such numberings.

2   Preliminaries

Unspecified notations follow Rogers [21]. By $\mathbb{N} = \{0, 1, 2, \ldots\}$ we denote the set of all natural numbers. The set of all finite sequences of natural numbers is denoted by $\mathbb{N}^*$. For a, $b \in \mathbb{N}$ we define $a \dotminus b$ to be $a - b$ if $a \geq b$ and 0, otherwise.

The cardinality of a set S is denoted by $|S|$. We write $\wp(S)$ for the power set of set S. Let $\emptyset$, $\in$, $\subset$, $\subseteq$, $\supset$, $\supseteq$, and $\#$ denote the empty set, element of, proper subset, subset, proper superset, superset, and incomparability of sets, respectively.

By $\mathfrak{P}$ and $\mathfrak{T}$ we denote the set of all partial and total functions of one variable over $\mathbb{N}$, respectively. The classes of all partial recursive and recursive functions of one, and two arguments over $\mathbb{N}$ are denoted by $\mathcal{P}$, $\mathcal{P}^2$, $\mathcal{R}$, and $\mathcal{R}^2$, respectively. Furthermore, for any $f \in \mathfrak{P}$ we use $\mathrm{dom}(f)$ to denote the *domain* of the function f, i.e., $\mathrm{dom}(f) =_{df} \{x \mid x \in \mathbb{N},\ f(x)$ is defined$\}$. Additionally, by $\mathrm{range}(f)$ we denote the *range* of f, i.e., $\mathrm{range}(f) =_{df} \{f(x) \mid x \in \mathrm{dom}(f)\}$. Let $f, g \in \mathfrak{P}$ be any partial functions. We write $f \subseteq g$ if for all $x \in \mathrm{dom}(f)$ the condition $f(x) = g(x)$ is satisfied. By $\mathcal{R}_{0,1}$ and $\mathcal{R}_{mon}$ we denote the set of all $\{0,1\}$-valued recursive functions (recursive predicates) and of all monotone recursive functions, respectively.

Every function $\psi \in \mathcal{P}^2$ is said to be a *numbering*. Let $\psi \in \mathcal{P}^2$, then we write ψ_i instead of $\lambda x.\psi(i, x)$, set $\mathcal{P}_\psi = \{\psi_i \mid i \in \mathbb{N}\}$ and $\mathcal{R}_\psi = \mathcal{P}_\psi \cap \mathcal{R}$. Consequently, if $f \in \mathcal{P}_\psi$, then there is a number i such that $f = \psi_i$. If $f \in \mathcal{P}$ and $i \in \mathbb{N}$ are such that $\psi_i = f$, then i is called a ψ-*program for* f. Let ψ be any numbering, and let $i \in \mathbb{N}$; if $\psi_i(x)$ is defined (abbr. $\psi_i(x)\downarrow$) then we also say that $\psi_i(x)$ *converges*. Otherwise, $\psi_i(x)$ is said to *diverge* (abbr. $\psi_i(x)\uparrow$).

A numbering $\varphi \in \mathcal{P}^2$ is called a Gödel numbering (cf. Rogers [21]) if $\mathcal{P}_\varphi = \mathcal{P}$, and for every numbering $\psi \in \mathcal{P}^2$, there is a $c \in \mathcal{R}$ such that $\psi_i = \varphi_{c(i)}$ for all $i \in \mathbb{N}$. *Göd* denotes the set of all Gödel numberings. Furthermore, we write (φ, Φ) to denote any *complexity measure* as defined in Blum [7]. That is, $\varphi \in Göd$, $\Phi \in \mathcal{P}^2$ and (1) $\mathrm{dom}(\varphi_i) = \mathrm{dom}(\Phi_i)$ for all $i \in \mathbb{N}$ and (2) the predicate "$\Phi_i(x) = y$" is uniformly recursive for all $i, x, y \in \mathbb{N}$.

Moreover, let $\mathrm{NUM} = \{\mathcal{U} \mid \exists \psi[\psi \in \mathcal{R}^2 \wedge \mathcal{U} \subseteq \mathcal{P}_\psi]\}$ denote the family of all subsets of all recursively enumerable classes of recursive functions.

Furthermore, using a fixed encoding $\langle \ldots \rangle$ of $\mathbb{N}^*$ onto $\mathbb{N}$ we write f^n instead of $\langle(f(0), \ldots, f(n))\rangle$, for any $n \in \mathbb{N}$, $f \in \mathcal{R}$.

The quantifier $\forall^\infty$ stands for "almost everywhere" and means "all but finitely many." Finally, a sequence $(j_n)_{j \in \mathbb{N}}$ of natural numbers is said to *converge* to the number j if all but finitely many numbers of it are equal to j. Next we define some concepts of learning.

Definition 1 (Gold [11,12]). *Let $\mathcal{U} \subseteq \mathcal{R}$ and let $\psi \in \mathcal{P}^2$. The class $\mathcal{U}$ is said to be* learnable in the limit *with respect to ψ if there is a strategy $S \in \mathcal{P}$ such that for each function $f \in \mathcal{U}$,*

(1) *for all $n \in \mathbb{N}$, $S(f^n)$ is defined,*
(2) *there is a $j \in \mathbb{N}$ such that $\psi_j = f$ and the sequence $(S(f^n))_{n \in \mathbb{N}}$ converges to j.*

If a class $\mathcal{U}$ is learnable in the limit with respect to ψ by a strategy S, then we write $\mathcal{U} \in \mathrm{LIM}_\psi(S)$. Let $\mathrm{LIM}_\psi = \{\mathcal{U} \mid \mathcal{U} \text{ is learnable in the limit w.r.t. } \psi\}$, and define $\mathrm{LIM} = \bigcup_{\psi \in \mathcal{P}^2} \mathrm{LIM}_\psi$.

As far as the semantics of the hypotheses output by a strategy S is concerned, whenever S is defined on input f^n, then we always interpret the number $S(f^n)$ as a ψ–number. This convention is adopted to all the definitions below.

Furthermore, note that $\mathrm{LIM}_\varphi = \mathrm{LIM}$ for any $\varphi \in G\ddot{o}d$. In the above definition LIM stands for "limit."

Looking at Definition 1 one may be tempted to think that it is too general. Maybe we should add some requirements that seem very natural. Since it may be hard for a strategy to know which inputs may occur, it could be very convenient to require $S \in \mathcal{R}$. Furthermore, if the strategy outputs a program i such that $\varphi_i \notin \mathcal{R}$, then this output cannot be correct. Hence, it seems natural to require the strategy to output exclusively hypotheses describing recursive functions. These demands directly yield the following definition:

Definition 2 (Wiehagen [23]). *Let $\mathcal{U} \subseteq \mathcal{R}$ and let $\psi \in \mathcal{P}^2$. The class $\mathcal{U}$ is said to be $\mathcal{R}$–totally learnable with respect to ψ if there is a strategy $S \in \mathcal{R}$ such that*

(1) *$\psi_{S(n)} \in \mathcal{R}$ for all $n \in \mathbb{N}$,*
(2) *for each $f \in \mathcal{U}$ there is a $j \in \mathbb{N}$ such that $\psi_j = f$, and $(S(f^n))_{n \in \mathbb{N}}$ converges to j.*

$\mathcal{R}$-TOTAL$_\psi(S)$, $\mathcal{R}$-TOTAL$_\psi$, and $\mathcal{R}$-TOTAL are defined in analogy to the above.

However, now it is not difficult to show that $\mathcal{R}$-TOTAL = NUM (cf. Zeugmann and Zilles [29, Theorem 2]). This is the first characterization of a learning type in terms of recursive numberings. This characterization shows how $\mathcal{R}$-total learning can be achieved, i.e., by using the well-known *identification by enumeration* technique.

Next, we recall the definition of reliable learning introduced by Blum and Blum [6] and Minicozzi [19]. Intuitively, a learner M is reliable provided it converges if and only if it learns.

Definition 3 (Blum and Blum [6], Minicozzi [19]). *Let $\mathcal{U} \subseteq \mathcal{R}$, $\mathcal{M} \subseteq \mathfrak{T}$ and let $\varphi \in G\ddot{o}d$. The class $\mathcal{U}$ is said to be reliably learnable on $\mathcal{M}$ if there is a strategy $S \in \mathcal{R}$ such that*

(1) *$\mathcal{U} \in \mathrm{LIM}_\varphi(S)$, and*
(2) *for all functions $f \in \mathcal{M}$, if the sequence $(S(f^n))_{n \in \mathbb{N}}$ converges, say to j, then $\varphi_j = f$.*

Let $\mathcal{M}$-REL denote the family of all classes $\mathcal{U}$ that are reliably learnable on $\mathcal{M}$.

Note that neither in Definition 1 nor in Definition 3 a requirement is made concerning the intermediate hypotheses output by the strategy S. The following definition is obtained from Definition 1 by adding the requirement that S correctly reflects all but the last δ data seen so far.

Definition 4 (Akama and Zeugmann [1]). *Let $\mathcal{U} \subseteq \mathcal{R}$, let $\psi \in \mathcal{P}^2$ and let $\delta \in \mathbb{N}$. The class $\mathcal{U}$ is called consistently learnable in the limit with δ-delay with respect to ψ if there is a strategy $S \in \mathcal{P}$ such that*

(1) $\mathcal{U} \in \mathrm{LIM}_\psi(\mathrm{S})$,
(2) $\psi_{S(f^n)}(x) = f(x)$ for all $f \in \mathcal{U}$, $n \in \mathbb{N}$ and all x such that $x + \delta \le n$.

We define $\mathrm{CONS}_\psi^\delta(\mathrm{S})$, $\mathrm{CONS}_\psi^\delta$, and CONS^δ analogously to the above.

We note that for $\delta = 0$ Barzdin's [4] original definition of CONS is obtained. We therefore usually omit the upper index δ if $\delta = 0$. This is also done for the other version of consistent learning defined below. We use the term δ-*delay*, since a consistent strategy with δ-delay correctly reflects all but at most the last δ data seen so far. If a strategy S learns a function class $\mathcal{U}$ in the sense of Definition 4, then we refer to S as a δ-*delayed consistent* strategy.

In Definition 4 consistency with δ-delay is only demanded for inputs that correspond to some function f from the target class $\mathcal{U}$. Note that for $\delta = 0$ the following definition incorporates Wiehagen and Liepe's [24] requirement on a strategy to work consistently on all inputs.

Definition 5 (Akama and Zeugmann [1]). *Let* $\mathcal{U} \subseteq \mathcal{R}$, *let* $\psi \in \mathcal{P}^2$ *and let* $\delta \in \mathbb{N}$. *The class* $\mathcal{U}$ *is called* $\mathcal{T}$-*consistently learnable in the limit with* δ-*delay with respect to* ψ *if there is a strategy* $S \in \mathcal{R}$ *such that*

(1) $\mathcal{U} \in \mathrm{CONS}_\psi^\delta(\mathrm{S})$,
(2) $\psi_{S(f^n)}(x) = f(x)$ for all $f \in \mathcal{R}$, $n \in \mathbb{N}$ and all x such that $x + \delta \le n$.

We define $\mathcal{T}$-$\mathrm{CONS}_\psi^\delta(\mathrm{S})$, $\mathcal{T}$-$\mathrm{CONS}_\psi^\delta$, and $\mathcal{T}$-CONS^δ analogously to the above.

We note that for all $\delta \in \mathbb{N}$ and all learning types $\mathrm{LT} \in \{\mathrm{CONS}^\delta, \mathcal{T}\text{-}\mathrm{CONS}^\delta\}$ we have $\mathrm{LT}_\varphi = \mathrm{LT}$ for every $\varphi \in G\ddot{o}d$.

Finally, we look at another mode of convergence which goes back to Feldman [9], who called it *matching in the limit* and considered it in the setting of learning languages. The difference to the mode of convergence used in Definition 1, which is actually *syntactic convergence*, is to relax the requirement that the sequence of hypotheses has to converge to a correct program, by *semantic convergence*. Here by semantic convergence we mean that after some point *all hypotheses are correct* but not necessarily identical. Nowadays, the resulting learning model is usually referred to as *behaviorally correct* learning. This term was coined by Case and Smith [8]. As far as learning of recursive functions is concerned, behaviorally correct learning was formalized by Barzdin [2,3].

Definition 6 (Barzdin [2,3]). *Let* $\mathcal{U} \subseteq \mathcal{R}$ *and let* $\psi \in \mathcal{P}^2$. *The class* $\mathcal{U}$ *is said to be* behaviorally correctly learnable *with respect to* ψ *if there is a strategy* $S \in \mathcal{P}$ *such that for each function* $f \in \mathcal{U}$,

(1) *for all* $n \in \mathbb{N}$, $S(f^n)$ *is defined,*
(2) $\psi_{S(f^n)} = f$ *for all but finitely many* $n \in \mathbb{N}$.

If $\mathcal{U}$ *is behaviorally correctly learnable with respect to* ψ *by a strategy* S, *we write* $\mathcal{U} \in \mathrm{BC}_\psi(\mathrm{S})$. BC_ψ *and* BC *are defined analogously to the above above.*

3   Characterizations in Terms of Complexity

We continue with characterizations in terms of computational complexity. Characterizations are a useful tool to get a better understanding of what different learning types have in common and where the differences are. They may also help to overcome difficulties that arise in the design of powerful learning algorithms.

Let us recall the needed definitions of several types of computable operators. Let $(F_x)_{x \in \mathbb{N}}$ be the canonical enumeration of all finite functions.

Definition 7 (Rogers [21]). *A mapping $\mathfrak{O} \colon \mathfrak{P} \mapsto \mathfrak{P}$ from partial functions to partial functions is called a* partial recursive operator *if there is a recursively enumerable set $W \subset \mathbb{N}^3$ such that for any $y, z \in \mathbb{N}$ it holds that $\mathfrak{O}(f)(y) = z$ if there is $x \in \mathbb{N}$ such that $(x, y, z) \in W$ and f extends the finite function F_x.*

Furthermore, $\mathfrak{O}$ is said to be a general recursive operator *if $\mathfrak{T} \subseteq dom(\mathfrak{O})$, and $f \in \mathfrak{T}$ implies $\mathfrak{O}(f) \in \mathfrak{T}$.*

A mapping $\mathfrak{O} \colon \mathcal{P} \mapsto \mathcal{P}$ is called an effective operator *if there is a function $g \in \mathcal{R}$ such that $\mathfrak{O}(\varphi_i) = \varphi_{g(i)}$ for all $i \in \mathbb{N}$. An effective operator $\mathfrak{O}$ is said to be* total effective *provided that $\mathcal{R} \subseteq dom(\mathfrak{O})$, and $\varphi_i \in \mathcal{R}$ implies $\mathfrak{O}(\varphi_i) \in \mathcal{R}$.*

For more information about general recursive operators and effective operators we refer the reader to [14, 20, 28]. If $\mathfrak{O}$ is an operator which maps functions to functions, we write $\mathfrak{O}(f, x)$ to denote the value of the function $\mathfrak{O}(f)$ at the argument x.

Definition 8. *A partial recursive operator $\mathfrak{O} \colon \mathfrak{P} \mapsto \mathfrak{P}$ is said to be* monotone *if for all functions $f, g \in dom(\mathfrak{O})$ the following condition is satisfied:*
  If $\forall^\infty x[f(x) \le g(x)]$ then $\forall^\infty x[\mathfrak{O}(f, x) \le \mathfrak{O}(g, x)]$.

Let $\mathfrak{O}$ be any arbitrarily fixed operator and let $M \subseteq \mathfrak{P}$. Then the abbreviation "$\mathfrak{O}(M) \subseteq M$" stands for "$M \subseteq dom(\mathfrak{O})$ and $f \in \mathcal{M}$ implies that $\mathfrak{O}(f) \in M$."

Any computable operator can be realized by a 3-tape Turing machine T which works as follows: If for an arbitrary function $f \in dom(\mathfrak{O})$, all pairs $(x, f(x))$, $x \in dom(f)$ are written down on the input tape of T (repetitions are allowed), then T will write exactly all pairs $(x, \mathfrak{O}(f, x))$ on the output tape of T (under unlimited working time).

Let $\mathfrak{O}$ be a partial recursive operator, a general recursive operator or a total effective operator. Then, for $f \in dom(\mathfrak{O})$, $m \in \mathbb{N}$ we set: $\Delta\mathfrak{O}(f, m) =$ "the least n such that, for all $x \le n$, $f(x)$ is defined and, for the computation of $\mathfrak{O}(f, m)$, the Turing machine T only uses the pairs $(x, f(x))$ with $x \le n$; if such an n does not exist, we set $\Delta\mathfrak{O}(f, m) = \infty$."

For any function $u \in \mathcal{R}$ we define Ω_u to be the set of all partial recursive operators $\mathfrak{O}$ satisfying $\Delta\mathfrak{O}(f, m) \le u(m)$ for all $f \in dom(\mathfrak{O})$. For the sake of notation, below we shall use $id + \delta$, $\delta \in \mathbb{N}$, to denote the function $u(x) = x + \delta$ for all $x \in \mathbb{N}$.

Blum and Blum [6] initiated the characterization of learning types in terms of computational complexity. Here they distinguished between *a priori characterizations* and *a posteriori characterizations*. In order to obtain an *a priori characterization* one starts from classes of *operator honesty complexity classes*, which are defined as follows: Let $\mathfrak{O}$ be a computable operator. Then we define

$$\mathcal{C}_{\mathfrak{O}} =_{df} \{f \mid \exists i[\varphi_i = f \wedge \forall^\infty x[\Phi_i(x) \leq \mathfrak{O}(f,x)]]\} \cap \mathcal{R}. \tag{1}$$

That is, every function in $\mathcal{C}_{\mathfrak{O}}$ possesses a program i such that the complexity of program i is in a computable way bounded by its function values, namely by $\mathfrak{O}(f,x)$ almost everywhere. So let LT be any learning type, e.g., learning in the limit. Then the general form of an *a priori characterization* of LT looks as follows:

Theorem 1. *Let $\mathcal{U} \subseteq \mathcal{R}$ be any class. Then we have $\mathcal{U} \in$ LT if and only if there is a computable operator $\mathfrak{O}$ such that $\mathcal{U} \subseteq \mathcal{C}_{\mathfrak{O}}$, where the operator $\mathfrak{O}$ has to fulfill some additional properties.*

Example 1. Consider the set of all operators which can be defined as follows: For any $t \in \mathcal{R}$ we define $\mathfrak{O}(f,x) =_{df} t(x)$ for every $f \in \mathcal{R}$ and $x \in \mathbb{N}$. Then the complexity classes defined in (1) have the form

$$\mathcal{C}_t = \{f \mid \exists i[\varphi_i = f \wedge \forall^\infty x[\Phi_i(x) \leq t(x)]]\} \cap \mathcal{R}. \tag{2}$$

and Theorem 1 yields the following *a priori* characterization of $\mathcal{R}$-TOTAL:

Let $\mathcal{U} \subseteq \mathcal{R}$ be any class. Then we have $\mathcal{U} \in \mathcal{R}$-TOTAL if and only if there is a recursive function $t \in \mathcal{R}$ such that $\mathcal{U} \subseteq \mathcal{C}_t$.

Since this theorem holds obviously also in case that $\mathcal{U} = \mathcal{C}_t$, we can directly use the fact that $\mathcal{R}$-TOTAL $=$ NUM and conclude that $\mathcal{C}_t \in$ NUM for every $t \in \mathcal{R}$. Thus, using the *a priori* characterization of $\mathcal{R}$-TOTAL we could easily reprove $\mathcal{C}_t \in$ NUM, which was originally shown by McCreight and Meyer [17].

Example 2. Note that for every general recursive operator $\mathfrak{O}$ there is a monotone general recursive operator $\mathfrak{M}$ such that $\mathfrak{O}(f,x) \leq \mathfrak{M}(f,x)$ for every function $f \in \mathfrak{T}$ and almost all $x \in \mathbb{N}$ (cf. Meyer and Fischer [18]). Furthermore, Grabowski [13] proved the following *a priori* characterization of $\mathfrak{T}$-REL:

Let $\mathcal{U} \subseteq \mathcal{R}$ be any class. Then we have $\mathcal{U} \in \mathfrak{T}$-REL if and only if there exists a general recursive operator $\mathfrak{O}$ such that $\mathcal{U} \subseteq \mathcal{C}_{\mathfrak{O}}$.

Using that every function $f \in \mathcal{R}_{0,1}$ satisfies $f(x) \leq 1$ for all $x \in \mathbb{N}$ we directly see by an easy application of Meyer and Fischer's [18] result that

$$\mathfrak{T}\text{-REL} \cap \wp(\mathcal{R}_{0,1}) = \mathcal{R}\text{-TOTAL} \cap \wp(\mathcal{R}_{0,1}) = \text{NUM} \cap \wp(\mathcal{R}_{0,1}); \tag{3}$$

i.e., reliable learning on the total functions restricted to classes of recursive predicates is exactly as powerful as $\mathcal{R}$-total learning restricted to classes of recursive predicates. On the other hand, $\mathcal{R}$-TOTAL $\subset \mathfrak{T}$-REL (cf. Grabowski [13]).

Moreover, Stephan and Zeugmann [22] showed that

$$\text{NUM} \cap \wp(\mathcal{R}_{0,1}) \subset \mathcal{R}\text{-REL} \cap \wp(\mathcal{R}_{0,1}). \tag{4}$$

The latter result was already published in Grabowski [13], but the new proof is much easier. It uses the class of approximations to the halting problem that has been considered in [6]. This class is defined as follows: Let (φ, Φ) be any complexity measure, and let $\tau \in \mathcal{R}$ be such that for all $i \in \mathbb{N}$

$$\varphi_{\tau(i)}(x) =_{df} \begin{cases} 1, & \text{if } \Phi_i(x)\downarrow \text{ and } \Phi_x(x) \leq \Phi_i(x); \\ 0, & \text{if } \Phi_i(x)\downarrow \text{ and } \neg[\Phi_x(x) \leq \Phi_i(x)]; \\ \uparrow, & \text{otherwise.} \end{cases}$$

Now, we set $\mathcal{B} = \{\varphi_{\tau(i)} \mid i \in \mathbb{N} \text{ and } \Phi_i \in \mathcal{R}_{mon}\}$. Then in [22, Theorems 2 and 3] both results were proved $\mathcal{B} \notin$ NUM and $\mathcal{B} \in \mathcal{R}$-REL yielding (4). We shall come back to this class.

Now, one can combine this with the *a priori* characterization of $\mathcal{R}$-REL obtained by Blum and Blum [6], which is as follows:

Let $\mathcal{U} \subseteq \mathcal{R}$ be any class. Then we have $\mathcal{U} \in \mathcal{R}$-REL if and only if there exists a total effective operator $\mathfrak{O}$ such that $\mathcal{U} \subseteq \mathcal{C}_{\mathfrak{O}}$.

Note that the difference between the *a priori* characterization of $\mathfrak{T}$-REL and $\mathcal{R}$-REL is that the operator $\mathfrak{O}$ is general recursive and total effective, respectively.

Putting this all together, we directly see that Meyer and Fischer's [18] boundability theorem cannot be strengthened by replacing "general recursive operator" by "total effective operator." And it also allows to show that there is an operator honesty complexity class $\mathcal{C}_{\mathfrak{O}}$ generated by a total effective operator $\mathfrak{O}$ such that $\mathcal{C}_{\mathfrak{O}} \not\subseteq \mathcal{C}_{\widetilde{\mathfrak{O}}}$ for every general recursive operator $\widetilde{\mathfrak{O}}$. For an explicit construction of such an operator $\mathfrak{O}$ we refer the reader to [14, 28].

Furthermore, Theorem 1 can be precisely stated for LIM, CONS$^\delta$, and $\mathcal{T}$-CONS$^\delta$ by using techniques from Blum and Blum [6], Wiehagen [23] and Akama and Zeugmann [1]. The proofs can be found in [29, Theorems 37, 35, 34].

Let $\mathcal{U} \subseteq \mathcal{R}$, then we have Let $\mathcal{U} \subseteq \mathcal{R}$, then we have $\mathcal{U} \in$ LIM if and only if there exists an effective operator $\mathfrak{O}$ such that $\mathfrak{O}(\mathcal{U}) \subseteq \mathcal{R}$ and $\mathcal{U} \subseteq \mathcal{C}_{\mathfrak{O}}$.

Let $\mathcal{U} \subseteq \mathcal{R}$ and let $\delta \in \mathbb{N}$; then we have

(1) $\mathcal{U} \in$ CONS$^\delta$ *if and only if there exists an effective operator $\mathfrak{O} \in \Omega_{id+\delta}$ such that $\mathfrak{O}(\mathcal{U}) \subseteq \mathcal{R}$ and $\mathcal{U} \subseteq \mathcal{C}_{\mathfrak{O}}$.*

(2) $\mathcal{U} \in \mathcal{T}$-CONS$^\delta$ *if and only if there is a general recursive operator $\mathfrak{O} \in \Omega_{id+\delta}$ such that $\mathcal{U} \subseteq \mathcal{C}_{\mathfrak{O}}$.*

These *a priori* characterizations shed also additional light to the fact that the learning types $\mathfrak{T}$-REL, $\mathcal{R}$-REL, and $\mathcal{T}$-CONS$^\delta$ are closed under union, while LIM and CONS$^\delta$ are not. In the former the operator $\mathfrak{O}$ maps $\mathcal{R}$ to $\mathcal{R}$, and in the latter we only have $\mathfrak{O}(\mathcal{U}) \subseteq \mathcal{R}$.

As we have seen, operator honesty characterizations have been found for many learning types, but some important ones are missing. These include BC, TOTAL, and conform learning. The learning criterion TOTAL is obtained from Definition 2 by replacing $S \in \mathcal{R}$ by $S \in \mathcal{P}$ and adding $S(f^n) \in \mathcal{R}$ for all $f \in \mathcal{U}$ and all $n \in \mathbb{N}$. Conform learning is a modification of consistent learning, where the requirement to correctly reflect all the functions values seen so far is replaced by the demand that the hypothesis output does never convergently contradict inputs already seen (cf. [29, Definition 22]).

Blum and Blum [6] also initiated the study of *a posteriori* characterizations of learning types. In particular they showed that any class $\mathcal{U} \in \mathcal{R}$-REL can be characterized in a way such that there exists a general recursive operator $\mathfrak{O}$ for which every function from $\mathcal{U}$ is *everywhere $\mathfrak{O}$-compressed*.

For the sake of completeness we include here the definition of everywhere $\mathfrak{O}$-compressed.

Definition 9 (Blum and Blum [6]). *Let (φ, Φ) be a complexity measure, let $f \in \mathcal{R}$, and let $\mathfrak{O}$ be a general recursive operator. Then a program $i \in \mathbb{N}$ is said to be an $\mathfrak{O}$-compression index of f (relative to (φ, Φ)) if*

(1) $\varphi_i = f$,
(2) $\forall j[\varphi_j = f \rightarrow \forall x\, \Phi_i(x) \leq \mathfrak{O}(\Phi_j, \max\{i, j, x\})]$.

In this case we also say that the function f is everywhere $\mathfrak{O}$-compressed.

We note that Definition 9 formalizes the concept of a fastest program (modulo an operator $\mathfrak{O}$) in a useful way. The $\mathfrak{O}$-compression index i satisfies the condition $\Phi_i(x) \leq \mathfrak{O}(\Phi_j, x)$ for all but finitely many $x \in \mathbb{N}$ and all programs j computing the same function as program i does. Additionally, it also provides an upper bound for the least argument n such that $\Phi_i \leq \mathfrak{O}(\Phi_j, x)$ for all $x > n$, i.e., $\max\{j, i\}$, and a computable majorante for those values $m \leq n$ for which possibly $\Phi_i(m) > \mathfrak{O}(\Phi_j, m)$; i.e., the value $\mathfrak{O}(\Phi_j, \max\{i, j\}$.

Of course, one can also consider the notion of everywhere $\mathfrak{O}$-compressed functions for total effective operators or any other type of computable operator $\mathfrak{O}$ provided that all considered complexity functions Φ_j are in $\text{dom}(\mathfrak{O})$.

Theorem 2 (Blum and Blum [6]). *For every class $\mathcal{U} \subseteq \mathcal{R}$ we have the following: $\mathcal{U} \in \mathcal{R}$-REL if and only if there is a general recursive operator $\mathfrak{O}$ such that every function from $\mathcal{U}$ is everywhere $\mathfrak{O}$-compressed.*

However, in [6] it remained open whether or not one can also reliably learn on $\mathcal{R}$ an $\mathfrak{O}$-compression index for every function f in the target class $\mathcal{U}$. We were able to show (cf. [26]) that this is not always the case, when using the algorithm described in [6]. Furthermore, in [27] we provided a suitable modification of Definition 9 resulting in a *reliable $\mathfrak{O}$-compression index*, and then showed that such reliable $\mathfrak{O}$-compression indices are reliable learnable on $\mathcal{R}$.

On the other hand, Blum and Blum [6, Section 8] used Theorem 2 to show that several interesting functions classes are contained in $\mathcal{R}$-REL including the class $\mathcal{B}$ of approximations to the halting problem. Using different techniques, this result was extend in [22]. Conversely, one can also consider any particular general recursive operator op and ask for the resulting function class of everywhere op-compressed functions, which are, via Theorem 2, known to be in $\mathcal{R}$-REL. Unfortunately, almost nothing is known in this area. Therefore, we would like to encourage research along these two lines, i.e., considering interesting function classes and figuring out to which learning type they belong, or to study special general recursive operators with respect to the learning power they generate.

In order to characterize the learning type $\mathcal{T}$-CONS$^\delta$, the following modification of Definition 9 turned out to be suitable:

Definition 10. *Let* (φ, Φ) *be a complexity measure, let* $f \in \mathcal{R}$, *and let* $\mathfrak{O}$ *be a general recursive operator. Then a program* $i \in \mathbb{N}$ *is said to be an* absolute $\mathfrak{O}$-*compression index of* f *(relative to* (φ, Φ)*) if*

(1) $\varphi_i = f$,
(2) $\forall j \forall x [\varphi_j(y) = f(y)$ *for all* $y \leq \Delta \mathfrak{O}(\Phi_j, \max\{i, x\})$
$\quad \rightarrow \Phi_i(x) \leq \mathfrak{O}(\Phi_j, \max\{i, x\})]$.

In this case we also say that the function f *is* absolutely $\mathfrak{O}$-*compressed.*

To show the following lemma we have to restrict the class of complexity measures a bit. We shall say that a complexity measure (φ, Φ) satisfies Property (+) if for all $i, x \in \mathbb{N}$ such that $\Phi_i(x)$ is defined the condition $\Phi_i(x) \geq \varphi_i(x)$ is satisfied.

Note that Property (+) is not very restrictive, since various "natural" complexity measures satisfy it.

Lemma 1. *Let* (φ, Φ) *be a complexity measure satisfying Property* (+), *and let* $\delta \in \mathbb{N}$ *be arbitrarily fixed. Furthermore, let* $\mathcal{U} \in \mathcal{T}\text{-CONS}^\delta$. *Then there is a general recursive operator* $\mathfrak{O} \in \Omega_{\mathrm{id}+\delta}$ *such that every function from* $\mathcal{U}$ *is absolutely* $\mathfrak{O}$-*compressed.*

The following lemma shows that the condition presented in Lemma 1 is also sufficient. Furthermore, this lemma holds for all complexity measures.

Lemma 2. *Let* (φ, Φ) *be any complexity measure, let* $\delta \in \mathbb{N}$ *be arbitrarily fixed, let* $\mathfrak{O} \in \Omega_{\mathrm{id}+\delta}$, *and let* $\mathcal{U} \subseteq \mathcal{R}$ *such that every function from* $\mathcal{U}$ *is absolutely* $\mathfrak{O}$-*compressed. Then there is a strategy* $S \in \mathcal{R}$ *such that*

(1) $\mathcal{U} \in \mathcal{T}\text{-CONS}^\delta_\varphi(S)$,
(2) *for every* $f \in \mathcal{U}$ *the sequence* $(S(f^n))_{n \in \mathbb{N}}$ *converges to an absolute* $\mathfrak{O}$-*compression index of* f.

Furthermore, Lemmata 1 and 2 directly allow for the following theorem:

Theorem 3. *Let* (φ, Φ) *be a complexity measure satisfying Property* (+), *let* $\delta \in \mathbb{N}$ *be arbitrarily fixed, and let* $\mathcal{U} \subseteq \mathcal{R}$. *Then we have*
$\mathcal{U} \in \mathcal{T}\text{-CONS}^\delta_\varphi(S)$ *if and only if there is an operator* $\mathfrak{O} \in \Omega_{\mathrm{id}+\delta}$ *such that every function* f *from* $\mathcal{U}$ *is absolutely* $\mathfrak{O}$-*compressed. Furthermore, for every* $f \in \mathcal{U}$ *the sequence* $(S(f^n))_{n \in \mathbb{N}}$ *converges to an absolute* $\mathfrak{O}$-*compression index of* f.

Though we succeeded to show an *a posteriori* characterization for the learning type $\mathcal{T}\text{-CONS}^\delta$, it is not completely satisfactory, since it restricts the class of admissible complexity measures. Can this restriction be removed?

Nevertheless, combining the *a priori* characterization of $\mathcal{T}\text{-CONS}^\delta$ with the *a posteriori* characterization provided in Theorem 3 shows that the family of operator honesty classes coincides for every $\delta \in \mathbb{N}$ with the family of absolutely operator compressed classes.

In this regard, it would be very nice to have also an *a posteriori* characterization of $\mathfrak{T}\text{-REL}$.

4   Characterizations in Terms of Computable Numberings

The reader may be curious why our definitions of learning types include a numbering ψ with respect to which we aim to learn. After all, if one can learn a class $\mathcal{U}$ with respect to some numbering ψ, then one can also infer $\mathcal{U}$ with respect to any Gödel numbering φ. However, ψ may possess properties which facilitate learning. For example, since $\mathcal{R}$-TOTAL = NUM, for every class $\mathcal{U} \in \mathcal{R}$-TOTAL there is numbering $\psi \in \mathcal{R}^2$ such that $\mathcal{U} \subseteq \mathcal{R}_\psi$, and so the identification by enumeration technique over ψ always succeeds.

Next, one may consider *measurable* numberings, which are defined as follows: A numbering $\psi \in \mathcal{P}^2$ is said to be *measurable* if the predicate "$\psi_i(x) = y$" is uniformly recursive in i, x, y (cf. Blum [7]). So, if $\psi \in \mathcal{P}^2$ is a measurable numbering and $\mathcal{U} \subseteq \mathcal{R}$ is such that $\mathcal{U} \subseteq \mathcal{P}_\psi$, then the identification by enumeration technique is still applicable. A prominent example is the class $\mathcal{U} = \{\Phi_i \mid i \in \mathbb{N}\} \cap \mathcal{R}$, where (φ, Φ) is a complexity measure. Note that the halting problem for the numbering $\Phi \in \mathcal{P}^2$ is undecidable (cf. [25, Lemma 3]).

Blum and Blum [6] also considered $\mathcal{P}$-REL and $\mathfrak{P}$-REL (cf. Definition 3 for $\mathcal{M} = \mathcal{P}$ and $\mathcal{M} = \mathfrak{P}$, respectively) and showed that $\mathcal{P}$-REL = $\mathfrak{P}$-REL. Furthermore, they proved that a class $\mathcal{U} \subseteq \mathcal{R}$ is in $\mathcal{P}$-REL if and only if there is measurable numbering ψ such that $\mathcal{U} \subseteq \mathcal{P}_\psi$. Furthermore, they reliably on $\mathcal{P}$ learnable function classes are characterized as the h-honesty function classes, i.e., $\mathcal{U} \subseteq \mathcal{C}_h$, where the operator $\mathfrak{O}$ is defined as $\mathfrak{O}(f, n) = h(n, f(n))$ (for a more detailed proof see [29, Theorems 12, 27]).

Note that these results also allow for a first answer of how inductive inference strategies discover their errors. This problem was studied in detail in Freivalds, Kinber, and Wiehagen [10]. The results obtained clearly show the importance of characterizations in terms of computable numberings and related techniques.

One such technique is the *amalgamation* technique, which is given implicitly in Barzdin and Podnieks [5] and then formalized in Wiehagen [23]. It was also independently discovered by Case and Smith [8], who gave it its name. Let amal be a recursive function mapping any finite set I of ψ-programs to a φ-program such that for any $x \in \mathbb{N}$, $\varphi_{\mathrm{amal}(I)}(x)$ is defined by running $\varphi_i(x)$ for every $i \in I$ in parallel and taking the first value obtained, if any.

In order to have a further example, let us take a closer look at $\mathcal{R}$-REL. Here we have the additional problem that the strategy S has to diverge on input initially growing finite segments of any function f it cannot learn. We are interested in learning of how this can be achieved. We need the following notation: For every $f \in \mathcal{R}$ and $n \in \mathbb{N}$ we write $f[n]$ to denote the tuple $(f(0), \cdots, f(n))$. Moreover, for any $f \in \mathcal{R}$, $d \in \mathcal{R}$, and $\psi \in \mathcal{P}^2$ we define $H_f = \{i \mid i \in \mathbb{N}, f[d(i)] \subseteq \psi_i\}$. In [15, Theorem 44] the following was shown:

Let $\mathcal{U} \subseteq \mathcal{R}$ be any function class. Then $\mathcal{U} \in \mathcal{R}$-REL if and only if there is a numbering $\psi \in \mathcal{P}^2$ and a function $d \in \mathcal{R}$ such that

(1) *for every $f \in \mathcal{R}$, if H_f is finite, then H_f contains a ψ-program of the function f, and*

(2) *for every $f \in \mathcal{U}$, the set H_f is finite.*

The proof of this theorem instructively answers where the ability to infer a class $\mathcal{U}$ reliably on $\mathcal{R}$ may be come from. On the one hand, it comes from a well chosen hypothesis space ψ. For any function $f \in \mathcal{U}$ there are only finitely many "candidates" in the set H_f including a ψ-program of f. So, in this case, the amalgamation technique succeeds. On the other hand, the infinity of this set H_f for every function which is not learned, then ensures that the strategy provided in the proof has to diverge. This is also guaranteed by the amalgamation technique, since the sets of "candidates" forms a proper chain of finite sets and so arbitrary large hypotheses are output on every function $f \in \mathcal{R}$ with H_f being infinite.

There are many more characterization theorems in terms of computable numberings including some for LIM and BC (cf., e.g., [29, Section 8] and the references therein), and consistent learning with δ-delay (cf. [1, Section 3.2]).

However, there are also many open problems. For example, Kinber and Zeugmann [16] generalized reliable learning in the limit as defined in this paper to reliable behaviorally correct learning and reliable frequency inference. All these learning types share the useful properties of reliable learning such as closure under recursively enumerable unions and finite invariance (cf. Minicozzi [19]). But we are not aware of any characterization of reliable behaviorally correct learning and reliable frequency inference in terms of computable numberings or in terms of computational complexity.

References

1. Akama, Y., Zeugmann, T.: Consistent and coherent learning with δ-delay. Inf. Comput. **206**(11), 1362–1374 (2008)
2. Barzdin, J.M.: Prognostication of automata and functions. In: Freiman, C.V., Griffith, J.E., Rosenfeld, J.L. (eds.) Information Processing 71, Proceedings of IFIP Congress 71, Volume 1 - Foundations and Systems, Ljubljana, Yugoslavia, 23–28 August 1971, pp. 81–84. North-Holland (1972)
3. Barzdin, J.M.: Две теоремы о предельном синтезе функций. In: Barzdin, J.M. (ed.) Теория Алгоритмов и Программ, vol. I, pp. 82–88. Latvian State University (1974)
4. Barzdin, J.M.: Inductive inference of automata, functions and programs. In: Proceedings of the 20-th International Congress of Mathematicians, Vancouver, Canada, pp. 455–460 (1974). (Republished in Amer. Math. Soc. Transl. **109**(2), 107–112 (1977))
5. Барздинь, Я.М., Подниекс, К.М.: К теорию индуктивного вывода. In: Mathematical Foundations of Computer Science: Proceedings of Symposium and Summer School, Strbské Pleso, High Tatras, Czechoslovakia, 3–8 September 1973, pp. 9–15. Mathematical Institute of the Slovak Academy of Sciences (1973)
6. Blum, L., Blum, M.: Toward a mathematical theory of inductive inference. Inf. Control **28**(2), 125–155 (1975)
7. Blum, M.: A machine-independent theory of the complexity of recursive functions. J. ACM **14**(2), 322–336 (1967)
8. Case, J., Smith, C.: Comparison of identification criteria for machine inductive inference. Theor. Comput. Sci. **25**(2), 193–220 (1983)
9. Feldman, J.: Some decidability results on grammatical inference and complexity. Inf. Control **20**(3), 244–262 (1972)

10. Freivalds, R., Kinber, E.B., Wiehagen, R.: How inductive inference strategies discover their errors. Inf. Comput. **118**(2), 208–226 (1995)
11. Gold, E.M.: Limiting recursion. J. Symb. Log. **30**, 28–48 (1965)
12. Gold, E.M.: Language identification in the limit. Inf. Control **10**(5), 447–474 (1967)
13. Grabowski, J.: Starke Erkennung. In: Linder, R., Thiele, H. (eds.) Strukturerkennung diskreter kybernetischer Systeme, vol. 82, pp. 168–184. Seminarberichte der Sektion Mathematik der Humboldt-Universität zu Berlin (1986)
14. Helm, J.: On effectively computable operators. Zeitschrift für mathematische Logik und Grundlagen der Mathematik (ZML) **17**, 231–244 (1971)
15. Jain, S., Kinber, E., Wiehagen, R., Zeugmann, T.: On learning of functions refutably. Theor. Comput. Sci. **298**(1), 111–143 (2003)
16. Kinber, E., Zeugmann, T.: One-sided error probabilistic inductive inference and reliable frequency identification. Inf. Comput. **92**(2), 253–284 (1991)
17. McCreight, E.M., Meyer, A.R.: Classes of computable functions defined by bounds on computation: preliminary report. In: Proceedings of the 1st Annual ACM Symposium on Theory of Computing, Marina del Rey, California, United States, pp. 79–88. ACM Press (1969)
18. Meyer, A., Fischer, P.C.: On computational speed-up. In: Conference Record of 1968 Ninth Annual Symposium on Switching and Automata Theory, Papers Presented at the Ninth Annual Symposium Schenectady, New York, 15–18 October 1968, pp. 351–355. IEEE Computer Society (1968)
19. Minicozzi, E.: Some natural properties of strong identification in inductive inference. Theor. Comput. Sci. **2**, 345–360 (1976)
20. Odifreddi, P.: Classical Recursion Theory. North Holland, Amsterdam (1989)
21. Rogers, H.: Theory of Recursive Functions and Effective Computability. MIT Press, Cambridge (1987). Reprinted McGraw-Hill 1967
22. Stephan, F., Zeugmann, T.: Learning classes of approximations to non-recursive functions. Theor. Comput. Sci. **288**(2), 309–341 (2002)
23. Wiehagen, R.: Zur Theorie der Algorithmischen Erkennung. Dissertation B, Humboldt-Universität zu Berlin (1978)
24. Wiehagen, R., Liepe, W.: Charakteristische Eigenschaften von erkennbaren Klassen rekursiver Funktionen. Elektronische Informationsverarbeitung und Kybernetik **12**(8/9), 421–438 (1976)
25. Wiehagen, R., Zeugmann, T.: Ignoring data may be the only way to learn efficiently. J. Exp. Theor. Artif. Intell. **6**(1), 131–144 (1994)
26. Zeugmann, T.: Optimale Erkennung. Diplomarbeit, Sektion Mathematik, Humboldt-Universität zu Berlin (1981)
27. Zeugmann, T.: A-posteriori characterizations in inductive inference of recursive functions. Elektronische Informationsverarbeitung und Kybernetik **19**(10/11), 559–594 (1983)
28. Zeugmann, T.: On the nonboundability of total effective operators. Zeitschrift für mathematische Logik und Grundlagen der Mathematik (ZML) **30**, 169–172 (1984)
29. Zeugmann, T., Zilles, S.: Learning recursive functions: a survey. Theor. Comput. Sci. **397**(1–3), 4–56 (2008)

PRAWF: An Interactive Proof System for Program Extraction

Ulrich Berger[1]([✉]), Olga Petrovska[1], and Hideki Tsuiki[2]

[1] Swansea University, Swansea, UK
{u.berger,olga.petrovska}@swansea.ac.uk
[2] Kyoto University, Kyoto, Japan
tsuiki.hideki.8e@kyoto-u.ac.jp

Abstract. We present an interactive proof system dedicated to program extraction from proofs. In a previous paper [5] the underlying theory IFP (Intuitionistic Fixed Point Logic) was presented and its soundness proven. The present contribution describes a prototype implementation and explains its use through several case studies. The system benefits from an improvement of the theory which makes it possible to extract programs from proofs using unrestricted strictly positive inductive and coinductive definitions, thus removing the previous admissibility restrictions.

1 Introduction

One of the salient features of constructive proofs is the fact that they carry computational content which can be extracted by a simple automatic procedure. Examples of formal systems providing constructive proofs are intuitionistic (Heyting) arithmetic or (varieties of) constructive type theory. There exist several computer implementations of these systems, which support program extraction based on Curry-Howard correspondence (e.g. Minlog [4], Nuprl [8,10], Coq [7,9], Isabelle [1], Agda [2]). However, none of them has program extraction as their main raison d'être.

In [5] the system IFP (Intuitionistic Fixed Point Logic) was introduced whose primary goal is program extraction. IFP is first-order logic extended with least and greatest fixed points of strictly positive predicate transformers. Program extraction in IFP is based on a refined realizability interpretation that permits arbitrary classically true disjunction-free formulas as axioms and ignores the (trivial) computational content of proofs of Harrop formulas thus leading to programs without formal garbage. The main purpose of [5] was to show soundness of this realizability interpretation, that is, the correctness of extracted programs.

In the present paper we present PRAWF[1], the first prototype of an implementation of IFP as an interactive proof system with program extraction feature. PRAWF is based on a (compared with [5]) simplified notion of program and

---

[1] 'Prawf' (pronounced /prauv/) is Welsh for 'Proof'.

© Springer Nature Switzerland AG 2020
M. Anselmo et al. (Eds.): CiE 2020, LNCS 12098, pp. 137–148, 2020.
https://doi.org/10.1007/978-3-030-51466-2_12

an improved Soundness Theorem that admits least and greatest fixed points of arbitrary strictly positive predicate transformers, removing the admissibility restriction in [5].

The paper is structured as follows: In Sect. 2 we briefly recap IFP and program extraction in IFP, explaining in some detail the above mentioned changes and improvements. In Sect. 3 we describe PRAWF and its basic use through some simple examples involving real and natural numbers. Section 4 contains an advanced case study about exact real number representations: The well-known signed digit representation and infinite Gray-code [12] are represented by coinductive predicates and inclusion between the predicates is proven in PRAWF thus enabling the extraction of a program transforming the signed digit representation into infinite Gray-code. In the conclusion we reflect on what we achieved and compare our work with related approaches.

2   Program Extraction in IFP

We briefly summarize the system IFP and its associated program extraction procedure. For full details we refer to [5].

$$\frac{}{a : A \vdash a : A} \text{ assumption}$$

$$\frac{p : A \qquad q : B}{\mathbf{Pair}(p, q) : A \wedge B} \wedge^+ \qquad \frac{p : A \wedge B}{\mathbf{proj}_L(p) : A} \wedge_L^- \qquad \frac{p : A \wedge B}{\mathbf{proj}_R(p) : B} \wedge_R^-$$

$$\frac{p : A}{\mathbf{Lt}(p) : A \vee B} \vee_L^+ \qquad \frac{p : B}{\mathbf{Rt}(p) : A \vee B} \vee_R^+$$

$$\frac{p : A \vee B \qquad q : A \to C \qquad r : B \to C}{\mathbf{case}\, p\, \mathbf{of}\, \{\mathbf{Lt}(a) \to q\, a; \mathbf{Rt}(b) \to r\, b\} : C} \vee^-$$

$$\frac{a : A \vdash p : B}{\lambda a\, p : A \to B} \to^+ \qquad \frac{p : A \to B \qquad q : A}{p\, q : B} \to^-$$

$$\frac{p : A}{p : \forall x\, A} \forall^+ \qquad \frac{p : \forall x\, A}{p : A[t/x]} \forall^-$$

$$\frac{p : A[t/x]}{p : \exists x\, A} \exists^+ \qquad \frac{p : \exists x\, A \qquad q : \forall x\, (A \to B)}{p\, q : B} \exists^-$$

$$\frac{}{\lambda a\, a : \Phi(\square(\Phi)) \subseteq \square(\Phi)} cl \qquad \frac{}{\lambda a\, a : \square(\Phi) \subseteq \Phi(\square(\Phi))} cocl \qquad (\square \in \{\mu, \nu\})$$

$$\frac{p : \Phi(\mathcal{P}) \subseteq \mathcal{P}}{\mathbf{rec}(\lambda f\, (p \circ m_\Phi\, f)) : \mu(\Phi) \subseteq \mathcal{P}} ind$$

$$\frac{p : \mathcal{P} \subseteq \Phi(\mathcal{P})}{\mathbf{rec}(\lambda f\, (m_\Phi\, f \circ p)) : \mathcal{P} \subseteq \nu(\Phi)} coind$$

Fig. 1. Proofs and their extracted programs

IFP *Syntax and Proofs.* The syntax of IFP has *terms, formulas, predicates* and *operators*, the latter describing strictly positive (s.p.) and hence monotone predicate transformers. For every s.p. operator Φ there are predicates $\mu\Phi$ and $\nu\Phi$ denoting the predicates defined inductively resp. coinductively from Φ:

$Terms \ni r, s, t \quad ::= x$ (variables) $\mid f(t_1, \ldots, t_n)$ (f function constant)

$Formulas \ni A, B \quad ::= \mathcal{P}(\vec{t})$ ($\mathcal{P}$ not an abstraction, $\vec{t}$ arity($\mathcal{P}$)-many terms)
$\qquad\qquad\qquad \mid A \wedge B \mid A \vee B \mid A \rightarrow B \mid \forall x\, A \mid \exists x\, A$

$Predicates \ni \mathcal{P}, Q ::= X$ (predicate variables) $\mid P$ (predicate constants)
$\qquad\qquad\qquad \mid \lambda\vec{x}\, A \mid \mu\Phi \mid \nu\Phi$
$\qquad\qquad\qquad$ (arity($\lambda\vec{x}\, A$) $= |\vec{x}|$, arity($\mu\Phi$) $=$ arity($\nu\Phi$) $=$ arity(Φ))

$Operators \ni \Phi, \Psi \quad ::= \lambda X\, \mathcal{P} \quad$ (P s.p. in X, arity($\lambda X\, \mathcal{P}$) $=$ arity(X) $=$ arity($\mathcal{P}$))

where P *is s.p. in* X if every free occurrence of X in P is at a *s.p. position*, i.e. not in the premise of an implication.

The proof rules of IFP are the usual rules of intuitionistic first-order logic with equality (regarding equality as binary predicate constant) augmented by rules stating that $\mu\Phi$ and $\nu\Phi$ are the least and greatest fixed points of Φ (see Fig. 1, ignoring for the moment the expressions to the left of the colon).

Programs. The programs extracted from proofs are terms in an untyped λ-calculus enriched by constructors for pattern matching and recursion.

$$Programs \ni p, q ::= a, b \text{ (program variables)}$$
$$\mid \mathbf{Nil} \mid \mathbf{Lt}(p) \mid \mathbf{Rt}(p) \mid \mathbf{Pair}(p, q)$$
$$\mid \mathbf{case}\, p \,\mathbf{of}\, \{Cl_1; \ldots; Cl_n\}$$
$$\mid \lambda a\, p$$
$$\mid p\, q$$
$$\mid \mathbf{rec}\, p$$

where in the case-construct each Cl_i is a *clause* of the form $C(a_1, \ldots, a_k) \rightarrow q$ in which C is a *constructor*, i.e. one of $\mathbf{Nil}, \mathbf{Lt}, \mathbf{Rt}, \mathbf{Pair}$, and the a_i are pairwise different program variables binding the free occurrences of the a_i in q. $\mathbf{rec}\, p$ computes the (least) fixed point of p, hence $p\, \mathbf{rec}(p) = \mathbf{rec}(p)$. It is well-known that an essentially equivalent calculus can be defined within the pure untyped λ-calculus, however, the enriched version is more convenient to work with. For the sake of readability we slightly simplify our notion of program compared to the one in [5] by no longer distinguishing between programs and functions.

Program Extraction. In its raw form the extracted program of a proof is simply obtained by replacing the proof rules by corresponding program constructs following the Curry-Howard correspondence. This is summarized in Fig. 1 where $p : A$ means that p is the program extracted from a proof of A. In the assumption rule and the rule $\rightarrow^+$, '$a : A \vdash$'indicates that the assumption A in the proof has been assigned the program variable a. In the rule $\wedge_L^-$, $\mathbf{proj}_L(p)$ stands for the program $\mathbf{case}\, p\, \mathbf{of}\, \{\mathbf{Pair}(a, b) \rightarrow a\}$. Similarly for $\wedge_R^-$. The rules $\forall^+$ and $\exists^-$

are subject to the usual provisos. In the rules ind and coind the program m_Φ realizes (in a sense explained below) the monotonicity of Φ, that is the formula $X \subseteq Y \to \Phi(X) \subseteq \Phi(Y)$ (with fresh predicate variables X, Y).

The correctness of programs is expressed through a realizability relation $p \, \mathbf{r} \, A$ between programs p and formulas A which is defined by recursion on formulas (see [5]). Formally, realizability is defined as a family of unary predicates $\mathbf{R}(A)$ on a Scott domain D of 'potential realizers'. $p \, \mathbf{r} \, A$ means that the denotation of p in D satisfies the predicate $\mathbf{R}(A)$. The Soundness Theorem [5] shows that if p is extracted from a proof of A, then p realizes A. The Soundness Theorem is formalised in a theory RIFP that extends IFP by a sort of realizers and axioms that describe the behaviour of programs. The denotational semantics of programs is linked to the operational one through the Adequacy Theorem, stating that programs with non-$\bot$ value terminate and reduce to that value [6].

Refinements. Program extraction in its raw form (as sketched above) produces correct programs which, however, contain a lot of garbage and are therefore practically useless. This is due to programs extracted from sub proofs of *Harrop formulas*, that is, formulas which do not contain a disjunction at a strictly positive position. These programs contain no useful information and should therefore be contracted to a trivial program, say **Nil**. In a refined realizability interpretation, which was presented in [5] and which is implemented in PRAWF, this contraction is carried out. It is based on a refined notion of realizability and a refined program extraction procedure. The proof of the soundness theorem becomes considerably more complicated and could only be accomplished in [5] by subjecting induction and coinduction to a certain admissibility condition. In [6] a soundness proof without this restriction is given. It uses an intermediate system IFP' whose induction and coinduction rules require as additional premise a proof of the monotonicity of Φ, e.g.,

$$\frac{p : \Phi(\mathcal{P}) \subseteq \mathcal{P} \qquad m : X \subseteq Y \to \Phi(X) \subseteq \Phi(Y)}{\mathbf{rec}(\lambda f \, (p \circ m \, f)) : \mu(\Phi) \subseteq \mathcal{P}} \; ind'$$

Soundness is then proven for IFP' and transferred to IFP via an embedding of IFP into IFP'. Minlog [4] has a similar refined realizability interpretation but treats disjunction-free formulas and Harrop formulas in the same way. This simplifies program extraction but seems to restrict the validity of the Soundness Theorem to a constructive framework (see also the remarks in Sect. 5).

Axioms. For a proof with assumptions the soundness theorem states that the extracted program computes a realizer of the proven formula from realizers of the assumptions. If the assumptions contain no disjunctions at all - we call such assumptions *non-computational (nc)* - then they are Harrop formulas and hence their realizers are trivial but, even more, they are equivalent to their realizability interpretations. This fact is extremely useful since it implies that a program extracted from a proof that uses nc-assumptions (regarded as axioms specifying a class of structures) will not depend on realizers of these axioms and will be correct in any model of the axioms. For example, in a proof about real numbers (see Sect. 3) the arithmetic operations may be given abstractly and specified by nc-axioms (e.g. $\forall x \, (x + 0 = x)$ and $\forall x \, (x \neq 0 \rightarrow \exists y \, (x * y = 1)))$.

Computation vs. Equational Reasoning. In the systems Nuprl, Coq, Agda, and Minlog computation is built into the notion of proof by considering terms, formulas or types up to normal form with respect to certain rewrite rules. As a consequence, each of these systems has various (decidable or undecidable) notions of equality, which may make proof checking (deciding the correctness of a proof) algorithmically hard if not undecidable. The motivations for interweaving logic and computation are partly philosophical and partly practical since in this way a fair amount of (otherwise laborious) equational reasoning can be automatized. In contrast, the system IFP strictly separates computation from reasoning. Its proof calculus is free of computation and there is only one notion of equality obeying the usual rules of equational logic. This makes proof checking a nearly trivial task. Equational reasoning can be to a large extent (or even completely) externalised by stating the required equations (which are nc-formulas) as axioms which can be proven elsewhere (or believed). Computation is confined to programs and is given through rewrite rules which enjoy an Adequacy Theorem stating that the operational and the denotational semantics of programs match [3,6].

3   Prawf

Prawf [11] is a prototype implementation in Haskell, which allows users to write IFP proofs and extract executable programs from them. It follows pretty closely the theory of IFP sketched in the previous section but extends it in several respects:

- the logical language of Prawf is many-sorted;
- names for predicates and operators can be introduced through declarations;
- induction and coinduction come in three variations, the original ones presented in Sect. 2, and two strengthenings (half-strong and strong (co)induction) which are explained and motivated below.

The software has two modes: a *prover mode* and an *execution mode*. The prover mode enables users to create a proof environment, consisting of a language, a context, declarations and axioms.

The proof rules in Prawf correspond to those of IFP and include the usual natural deduction rules for predicate logic, rules for (co)induction, half-strong (co)induction and strong (co)induction, as well as the equality rules such as symmetry, reflexivity and congruence.

A theorem can be proven by applying these rules step by step or by using a tactic. A tactic consists of a sequence of proof commands that allows users to re-run a proof either partially or fully. Once proven, a theorem can be saved in a theory and used as a part of another proof.

The execution mode allows running extracted programs. In this mode a user can take advantage of the standard Prelude commands as well as special functions for running and showing programs.

An Introductory Example: Natural Numbers and Addition. We explain the working of PRAWF by means of a simple example based on the language of real numbers with the constants 0 and 1 and the operations + and − for addition and subtraction. We first give the idea in ordinary mathematical language and then show how to do it in PRAWF. We define the natural numbers as the least subset (predicate) of the reals that contains 0 and that contains x whenever it contains $x - 1$:

$$\mathbf{N} \stackrel{\mathrm{Def}}{=} \mu(\varPhi), \text{ where } \varPhi \stackrel{\mathrm{Def}}{=} \lambda X \lambda x(x = 0 \vee X(x - 1))$$

We prove that $\mathbf{N}$ is closed under addition:

$$\forall x(\mathbf{N}(x) \rightarrow \forall y(\mathbf{N}(y) \rightarrow \mathbf{N}(x + y)))$$

Hence assume $\mathbf{N}(x)$. We have to show $\forall y(\mathbf{N}(y) \rightarrow \mathbf{N}(x + y))$, that is $\mathbf{N} \subseteq \mathcal{P}$ where $\mathcal{P} \stackrel{\mathrm{Def}}{=} \lambda y\, \mathbf{N}(x + y)$. By the induction rule it suffices to show $\varPhi(\mathcal{P}) \subseteq \mathcal{P}$, that is,

$$\forall y\, ((y = 0 \vee \mathbf{N}(x + (y - 1)) \rightarrow \mathbf{N}(x + y))$$

If $y = 0$ then $\mathbf{N}(x + y)$ holds since $x + 0 = x$ (using an axiom) and $\mathbf{N}(x)$ holds by assumption. If $\mathbf{N}(x + (y - 1))$, then $\mathbf{N}((x + y) - 1)$ since $x + (y - 1) = x + (y - 1)$ (using an axiom) and hence $\mathbf{N}(x + y)$ by the closure rule.

In order to carry out this example in PRAWF one first needs to define the language. This can be done by creating in the directory batches a subdirectory real (the name can be freely chosen), and in that directory the text file lang.txt (the name is prescribed) with the contents

```
<sorts>
R
<end sorts>

<constants>
0,1:R
<end constants>

<functions>
+ : (R,R) -> R;
- : (R,R) -> R;
<end functions>

<predicates>
= : (R,R);
<end predicates>
```

Note that we do not need to give definitions of + and −. For the proofs it is sufficient to know their properties that are expressed through axioms.

In the same subdirectory one creates the file decls.txt (name prescribed) containing the definition of $\mathbf{N}$:

```
Phi:(R) = lambda Y:(R) lambda (z:R) (z=0 v Y(z-1))
N:(R) = Mu(Phi)
```

Finally, one creates (again in the directory **real**) the file **axi.txt** (name prescribed) containing the axioms one wishes to use. The axioms must be nc-formulas that is, not contain any disjunctions (for example the predicate **N** must not occur since its definition contains ∨ as part of the definition of Φ).

```
ax1 . all x:R x+0 = x
ax2 . all x:R all y:R (x+y)-1 = x+(y-1)
```

Now we are set to start our proof. We load the Haskell file **Mode.hs**, execute **main**, load our batch by typing **real** (after which the contents of the files we created will be displayed) and type at the prompt our goal formula

```
Enter goal formula> all x:R (N(x) -> all y:R (N(y) -> N(x+y)))
```

Proving in PRAWF proceeds in the usual goal-directed backwards reasoning style. In our example the first two steps are easy: **alli** (for ∀-introduction backwards), then **impi v1** (for →-introduction backwards creating the assumption **v1 : N(x)**). After these two steps one arrives at

```
Assumptions:
 v1 : N(x)
Context of the goal:
 variables: x: R
Current goal:
 |-  ?2 : all (y:R) (N(y) -> N(x+y))
```

at which point we use induction by typing **ind**. This brings us to the premise of induction

```
Assumptions:
 v1 : N(x)
Context of the goal:
 variables: x: R
Current goal:
 |-  ?3 : all (y:R) (Phi(lambda (y:R) N(x+y))(y) -> N(x+y))
```

The command **unfold Phi** yields

```
Assumptions:
 v1 : N(x)
Context of the goal:
 variables: x: R
Current goal:
 |-  ?3 : all (y:R) ((y=0 v N(x+(y-1))) -> N(x+y))
```

which easily follows from our axioms and some equality reasoning. The necessary steps in PRAWF (and much more) can be found in a tutorial on the PRAWF website.

Program Extraction. After completing the proof above one can extract a program by typing `extract addition` (`addition` is a name we chose). This will write the extracted program into the file `progs.txt`:

```
addition . ProgAbst "v1" (ProgRec (ProgAbst "f_mu"
(ProgAbst "a_comp" (ProgCase (ProgVar "a_comp")
[(Lt,["a_ore"],ProgVar "v1"),(Rt,["b_ore"],ProgCon
Rt [ProgApp (ProgVar "f_mu") (ProgVar "b_ore")])])))))
```

The program transforms realizers of $N(x)$ and $N(y)$ into a realizer of $N(x + y)$. By the inductive definition of the predicate N its elements are realized in unary notation where `Lt(_)` plays the role of 0 and `Rt` plays the role of the successor function. The extracted program can be rewritten in more readable form as follows (using Haskell notation):

```
addition v1 a_comp = case a_comp of
                     {
                          Lt _ -> v1 ;
                          Rt b_ore -> Rt (addition v1 b_ore)
                     }
```

which clearly is the usual algorithm for addition of unary natural numbers. How to run this program is described in detail in the tutorial.

Half-Strong and Strong Induction. The premise of the induction rule for the predicate N is logically equivalent to the conjunction of the *induction base*, $\mathcal{P}(0)$, and the *induction step*, $\forall x\,(\mathcal{P}(x-1) \to \mathcal{P}(x))$. The induction step slightly differs from the usual induction step since it lacks the additional assumption $N(x)$. This discrepancy disappears in the following rule of *half-strong induction* and its associated extracted program

$$\frac{p : \mu(\Phi) \cap \Phi(\mathcal{P}) \subseteq \mathcal{P}}{\mathbf{rec}(\lambda f\,(p \circ \langle \mathbf{id}, m_\Phi\, f\rangle)) : \mu(\Phi) \subseteq \mathcal{P}} \ hsind$$

where $\langle f, g\rangle \overset{\text{Def}}{=} \lambda a\, \mathbf{Pair}(f\,a, g\,a)$ and $\mathbf{id} \overset{\text{Def}}{=} \lambda a\,a$. In the following example, half-strong induction appears to be needed to extract a good program. We aim to prove that the distance of two natural numbers is a natural number

$$\forall x(N(x) \to \forall y(N(y) \to N(x - y) \vee N(y - x)))$$

It is possible to prove this by (ordinary) induction on $N(x)$, however, the proof is complicated and the extracted program contrived and inefficient. On the other hand, with half-strong induction (command `hsind`) the goal reduces to proving $\forall y(N(y) \to N(x-y) \vee N(y-x))$ from the assumptions $N(x)$ and $x = 0 \vee \forall y(N(y) \to N((x - 1) - y) \vee N(y - (x - 1)))$ which, because of the extra assumption $N(x)$, is relatively straightforward. Moreover, the extracted program is the expected one which removes successors from (realizers of) x and y until one of the two becomes 0 after which what remains of the other one is returned as result. The interested reader is invited to try this example on their own.

Strong induction is similar to half-strong induction, however, the intersection with $\mu(\Phi)$ is taken 'inside' Φ:

$$\frac{p : \Phi(\mu(\Phi) \cap \mathcal{P}) \subseteq \mathcal{P}}{\mathbf{rec}(\lambda f\,(p \circ m_\Phi\,\langle \mathbf{id}, f \rangle)) : \mu(\Phi) \subseteq \mathcal{P}}\ \textit{sind}$$

For the case of the natural numbers its effect is that the step becomes logically equivalent to $\forall x\,(\mathbf{N}(x) \wedge \mathcal{P}(x) \rightarrow \mathcal{P}(x+1))$, that is, precisely the step in Peano induction. The extracted program corresponds exactly to primitive recursion.

Half-strong and strong coinduction will be discussed in Sect. 4.

4   Case Study: Exact Real Number Representations

As a rather large example, we formalize the existence of various exact representations of real numbers and prove the existence of conversions between them in PRAWF.

We continue to work in the theory of real numbers implemented in PRAWF through the batch `real` introduced in the previous section, but extend language, declarations and axioms as needed.

The structure of real numbers represented by the sort R and various constants and functions does not support any kind of computation on real numbers. For computation, we need representations. These can be provided in IFP through suitable predicates and their realizability interpretation, in a similar style as we represented unary natural numbers through the predicate **N**. In general, a representation is provided by defining a predicate $\mathcal{P}$ such that a realizer of $\mathcal{P}(x)$ is a representation of x.

Exact representations of real numbers are typically infinite sequences or streams which are naturally expressed through coinductively defined predicates. For example, the predicate S(x) meaning the existence of the signed digit representation of x, which is one of the standard representations of real numbers for computation, is expressed as follows. To give an impression how this looks in PRAWF we use in the following machine notation where v stands for $\vee$, ex for $\exists$, m is a constant for -1, * is multiplication, and <= means 'less or equal'.

```
SD:(R) = lambda (x:R) ((x = m v x = 1) v x = 0)
PhiS:(R) = lambda X:(R) lambda (x:R) ex (d:R)
                (SD(d) and  (abs(2*x-d)<= 1) and X(2*x-d))
S:(R) = Nu(PhiS)
```

This defines S as the largest predicate on the reals satisfying S = PhiS(S). A realizer of S(x) is an infinite stream of signed digits where a digit is a realizer of a formula of the form SD(y) that is either Lt(Lt(Nil)) or Lt(Rt(Nil)) or R(Nil) (representing $-1, 0, 1$). Streams are given as infinitely nested pairs, e.g. (writing a:b for Pair(a,b)) Lt(Lt(Nil)) : Rt(Nil) : Rt(Nil) : ... (that is, $-1 : 0 : 0 : ...$) which represents the real number -0.5.

Another representation, called infinite Gray-code [12], is defined through a coinductive predicate G(x) defined in PRAWF by

```
B:(R) = lambda (x:R) (x <= 0 v 0 <= x)
D:(R) = lambda (x:R) (not (x = 0)) -> B(x)
PhiG:(R) = lambda X:(R) lambda (x:R)
                (m <= x and x <= 1) and (D(x) and X(t(x)))
G:(R) = Nu(PhiG)
```

Here, `t:(R)->R` is the tent function defined as `t(x)=1-2*abs(x)`. An interesting and challenging aspects of the infinite Gray-code is the fact that it is partial, more precisely, a realizer of `G(x)` is a stream that may have one undefined element. This is due to the premise `not (x = 0)` in the definition of D which, if false (that is, `x=0`), will admit as realizer a program whose value is undefined (e.g. a program that loops infinitely).

Following [6], we proved in PRAWF

```
Theorem . all (x:R) (S(x) -> G(x))
```

The proof is rather involved in that it consists of two coinductions, half-strong coinduction, and Archimedean induction, which is a special form of induction suitable for proving predicates with a premise $x \neq 0$ like D (see below). Due to space restrictions we can only highlight the most interesting aspects of the proof. The main parts of the proof are

```
Claim1 . all x:R (S(x) -> D(x))
Claim2 . all x:R (S(x) -> S(t(x)))
```

which immediately implies the **Theorem** by coinduction.

The proof of **Claim1** uses the inductive predicate **Accp** defined by

```
PhiAccp:(R) = lambda X:(R) lambda (x:R) (all y:R y << x -> X(y))
Accp:(R) = Mu(PhiAccp)
```

where `x << y` is defined as `2*abs(x) <= 1 and y = 2*x`. Accp is the accessible or wellfounded part of the relation `<<`. Using Brouwer's Thesis, which states that induction on a well-founded relation is valid, and the Archimedean property of the reals (see [6]) one can show that `Accp(x)` holds for all nonzero x. Therefore, induction on `Accp(x)` turns into an induction principle for nonzero real numbers which in [6] is dubbed *Archimedean induction*. It is logically equivalent to the rule

$$\frac{\forall x \neq 0 \, ((|x| \leq 1/2 \rightarrow \mathcal{P}(2x)) \rightarrow \mathcal{P}(x))}{\forall x \neq 0 \, \mathcal{P}(x)} \; AI$$

Archimedean induction is used to prove `all x:R (S(x) -> B(x))` which is the essential step in the proof of **Claim1**.

The proof of **Claim2** uses *half-strong coinduction* which is the rule

$$\frac{p : \mathcal{P} \subseteq \nu(\Phi) \cup \Phi(\mathcal{P})}{\mathbf{rec}(\lambda f \, ([\mathbf{id} + m_\Phi \, f] \circ p)) : \mathcal{P} \subseteq \nu(\Phi)} \; hscoind$$

where $[f + g] \stackrel{\text{Def}}{=} \lambda a \, \mathbf{case} \, a \, \mathbf{of} \, \{\mathbf{Lt}(b) \rightarrow f \, b; \mathbf{Rt}(c) \rightarrow g \, c\}$. Similarly, *strong coinduction* is the rule

$$\frac{p : \mathcal{P} \subseteq \Phi(\nu(\Phi) \cup \mathcal{P})}{\mathbf{rec}(\lambda f\,(m_\Phi\,[\mathbf{id} + f] \circ p)) : \mathcal{P} \subseteq \nu(\Phi)} \; scoind$$

This rule can be used to give a short proof of `G(-x) -> G(x)` and extract a simple program which negates the head of the input stream and leaves its tail untouched (instead of recursively reproducing the tail, which would happen with ordinary coinduction).

5   Conclusion

We presented PRAWF, a first prototype implementation of the logical system IFP and its associated program extraction procedure. The successful formalization in PRAWF of exact real number representations and formal proofs of their relationships guarantee the correctness of the proofs in [6]. This advanced case study also gives us evidence that our approach scales to substantial nontrivial problems.

The examples also demonstrate the enormous advantage gained from the possibility of describing different data representation in an abstract setting using only first-order logic, and postulating arbitrary true nc-axioms. In the formalization of infinite Gray-code it was also essential that our method is able to produce partial extracted programs.

We would like to point out that the Soundness Theorem, that is, the correctness proof for extracted programs, though constructive, is valid with respect to a classical semantics. This is in line with the attitude in constructive mathematics to produce only results that are constructively *and* classically valid, which is not necessary the case in other approaches to program extraction.

Despite its successful maiden voyage PRAWF has some loose ends that need to be tied up. The most urgent one is an implementation of the soundness proof, that is, the enhancement of program extraction so that not only extracted programs but also their correctness proofs are created automatically. Currently, correctness relies on soundness as a meta theorem that has not been formalized yet. Other necessary improvements concern support for schematic theorems (Π_1^1-theorems, essentially), advanced proof tactics and interpretations between different languages.

We also plan to extend PRAWF by sequent calculus rules and rules that permit the extraction of concurrent programs. The latter will be needed to prove, conversely, that `G` (infinite Gray-code) is included in `S` (signed digit representation). We know from [12] that the extracted translation program has to be concurrent and nondeterministic.

Acknowledgements. This work was supported by the International Research Staff Exchange Scheme (IRSES) No. 612638 CORCON and No. 294962 COMPUTAL of the European Commission, the JSPS Core-to-Core Program, A. Advanced research Networks and JSPS KAKENHI Grant Number 15K00015 as well as the European Union's Horizon 2020 research and innovation programme under the Marie Skłodowska-Curie grant agreement No. 731143.

References

1. Isabelle. https://isabelle.in.tum.de/
2. Agda official website. http://wiki.portal.chalmers.se/agda/
3. Berger, U.: Realisability for induction and coinduction with applications to constructive analysis. J. Univ. Comput. Sci. **16**(18), 2535–2555 (2010)
4. Berger, U., Miyamoto, K., Schwichtenberg, H., Seisenberger, M.: Minlog - a tool for program extraction supporting algebras and coalgebras. In: Corradini, A., Klin, B., Cîrstea, C. (eds.) CALCO 2011. LNCS, vol. 6859, pp. 393–399. Springer, Heidelberg (2011). https://doi.org/10.1007/978-3-642-22944-2_29
5. Berger, U., Petrovska, O.: Optimized program extraction for induction and coinduction. In: Manea, F., Miller, R.G., Nowotka, D. (eds.) CiE 2018. LNCS, vol. 10936, pp. 70–80. Springer, Cham (2018). https://doi.org/10.1007/978-3-319-94418-0_7
6. Berger, U., Tsuiki, H.: Intuitionistic fixed point logic (2019). Unpublished manuscript available on ArXiv
7. Bertot, Y., Castéran, P.: Interactive theorem proving and program development (2004)
8. Constable, R.: Implementing Mathematics with the Nuprl Proof Development System. Prentice-Hall, Upper Saddle River (1986)
9. The Coq Proof Assistant. https://coq.inria.fr
10. Lockwood, J.: Nuprl: an open logical programming environment: a practical framework for sharing formal models and tools. Program extraction (1998). http://www.nuprl.org
11. Prawf official website. https://prawftree.wordpress.com/
12. Tsuiki, H.: Real number computation through gray code embedding. Theor. Comput. Sci. **284**(2), 467–485 (2002)

ASNP: A Tame Fragment of Existential Second-Order Logic

Manuel Bodirsky$^{(\boxtimes)}$, Simon Knäuer, and Florian Starke

Institute of Algebra, TU Dresden, Dresden, Germany
manuel.bodirsky@tu-dresden.de

Abstract. *Amalgamation SNP (ASNP)* is a fragment of existential second-order logic that strictly contains binary connected MMSNP of Feder and Vardi and binary connected guarded monotone SNP of Bienvenu, ten Cate, Lutz, and Wolter; it is a promising candidate for an expressive subclass of NP that exhibits a complexity dichotomy. We show that ASNP has a complexity dichotomy if and only if the infinite-domain dichotomy conjecture holds for constraint satisfaction problems for first-order reducts of binary finitely bounded homogeneous structures. For such CSPs, powerful universal-algebraic hardness conditions are known that are conjectured to describe the border between NP-hard and polynomial-time tractable CSPs. The connection to CSPs also implies that every ASNP sentence can be evaluated in polynomial time on classes of finite structures of bounded treewidth. We show that the syntax of ASNP is decidable. The proof relies on the fact that for classes of finite binary structures given by finitely many forbidden substructures, the amalgamation property is decidable.

1 Introduction

Feder and Vardi in their groundbreaking work [15] formulated the famous *dichotomy conjecture* for finite-domain constraint satisfaction problems, which has recently been resolved [11,26]. Their motivation to study finite-domain CSPs was the question which fragments of existential second-order logic might exhibit a complexity dichotomy in the sense that every problem that can be expressed in the fragment is either in P or NP-complete. Existential second-order logic without any restriction is known to capture NP [14] and hence does not have a complexity dichotomy by an old result of Ladner [24]. Feder and Vardi proved that even the fragments of *monadic SNP* and *monotone SNP* do not have a complexity dichotomy since every problem in NP is polynomial-time equivalent to a problem that can be expressed in these fragments. However, the dichotomy for finite-domain CSPs implies that *monotone monadic SNP (MMSNP)* has a dichotomy, too [15,23].

This work has been supported by the European Research Council (ERC) under the European Union's Horizon 2020 research and innovation programme (Grant Agreement No 681988, CSP-Infinity) and by DFG Graduiertenkolleg 1763 (QuantLA).

M. Anselmo et al. (Eds.): CiE 2020, LNCS 12098, pp. 149–162, 2020.
https://doi.org/10.1007/978-3-030-51466-2_13

MMSNP is also known to have a tight connection to a certain class of infinite-domain CSPs [7]: an MMSNP sentence is equivalent to a *connected* MMSNP sentence if and only if it describes an infinite-domain CSP. Moreover, every problem in MMSNP is equivalent to a finite disjunction of connected MMSNP sentences. The infinite structures that appear in this connection are tame from a model-theoretic perspective: they are reducts of finitely bounded homogeneous structures (see Sect. 4.1). CSPs for such structures are believed to have a complexity dichotomy, too; there is even a known hardness condition such that all other CSPs in the class are conjectured to be in P [8]. The hardness condition can be expressed in several equivalent forms [1,2].

In this paper we investigate another candidate for an expressive logic that has a complexity dichotomy. Our minimum requirement for what constitutes a *logic* is relatively liberal: we require that the syntax of the logic should be decidable. The same requirement has been made for the question whether there exists a logic that captures the class of polynomial-time solvable decision problems (see, e.g., [19,20]). The idea of our logic is to modify monotone SNP so that only CSPs for model-theoretically tame structures can be expressed in the logic; the challenge is to come up with a definition of such a logic which has a decidable syntax. We would like to require that the (universal) first-order part of a monotone SNP sentence describes an *amalgamation class*. We mention that the *Joint Embedding Property (JEP)*, which follows from the *Amalgamation Property (AP)*, has recently been shown to be undecidable [10]. In contrast, we use the fact that the AP for binary signatures is decidable (Sect. 5). We call our new logic *Amalgamation SNP (ASNP)*. This logic contains binary connected MMSNP; it also contains the more expressive logic of *binary connected guarded monotone SNP*. Guarded monotone SNP (GMSNP) has been introduced in the context of knowledge representation [3] (see Sect. 6). We show that ASNP has a complexity dichotomy if and only if the infinite-domain dichotomy conjecture holds for constraint satisfaction problems for first-order reducts of binary finitely bounded homogeneous structures. In particular, every problem that can be expressed in ASNP is a CSP for some countably infinite ω-categorical structure $\mathfrak{B}$. In Sect. 7 we present an example application of this fact: every problem that can be expressed in one of these logics can be solved in polynomial time on instances of bounded treewidth.

2   Constraint Satisfaction Problems

Let $\mathfrak{A}, \mathfrak{B}$ be structures with a finite relational signature τ; each symbol $R \in \tau$ is equipped with an *arity* $\mathrm{ar}(R) \in \mathbb{N}$. A function $h\colon A \to B$ is called a *homomorphism from $\mathfrak{A}$ to $\mathfrak{B}$* if for every $R \in \tau$ and $(a_1, \dots, a_{\mathrm{ar}(R)}) \in R^{\mathfrak{A}}$ we have $(h(a_1), \dots, h(a_{\mathrm{ar}(R)})) \in R^{\mathfrak{B}}$; in this case we write $\mathfrak{A} \to \mathfrak{B}$. We write $\mathrm{CSP}(\mathfrak{B})$ for the class of all finite τ-structures $\mathfrak{A}$ such that $\mathfrak{A} \to \mathfrak{B}$.

Example 1. If $\mathfrak{B} = K_3$ is the 3-clique, i.e., the complete undirected graph with three vertices, then $\mathrm{CSP}(\mathfrak{B})$ is the graph 3-colouring problem, which is NP-complete [18].

Example 2. If $\mathfrak{B} = (\mathbb{Q}; <)$ then $\text{CSP}(\mathfrak{B})$ is the digraph acyclicity problem, which is in P.

Example 3. If $\mathfrak{B} = (\mathbb{Q}; \text{Betw})$ for $\text{Betw} := \{(x, y, z) \mid x < y < z \vee z < y < x\}$ then $\text{CSP}(\mathfrak{B})$ is the Betweenness problem, which is NP-complete [18].

A homomorphism h from $\mathfrak{A}$ to $\mathfrak{B}$ is called an *embedding of $\mathfrak{A}$ into $\mathfrak{B}$* if h is injective and for every $R \in \tau$ and $a_1, \dots, a_{\text{ar}(R)} \in A$ we have $(a_1, \dots, a_{\text{ar}(R)}) \in R^{\mathfrak{A}}$ if and only if $(h(a_1), \dots, h(a_{\text{ar}(R)})) \in R^{\mathfrak{B}}$; in this case we write $\mathfrak{A} \hookrightarrow \mathfrak{B}$. The *union* of two τ-structures $\mathfrak{A}, \mathfrak{B}$ is the τ-structure $\mathfrak{A} \cup \mathfrak{B}$ with domain $A \cup B$ and the relation $R^{\mathfrak{A} \cup \mathfrak{B}} := R^{\mathfrak{A}} \cup R^{\mathfrak{B}}$ for every $R \in \tau$. The *intersection* $\mathfrak{A} \cap \mathfrak{B}$ is defined analogously. A *disjoint union* of $\mathfrak{A}$ and $\mathfrak{B}$ is the union of isomorphic copies of $\mathfrak{A}$ and $\mathfrak{B}$ with disjoint domains. As disjoint unions are unique up to isomorphism, we usually speak of *the* disjoint union of $\mathfrak{A}$ and $\mathfrak{B}$, and denote it by $\mathfrak{A} \uplus \mathfrak{B}$. A structure is *connected* if it cannot be written as a disjoint union of at least two structures with non-empty domain. A class of structures $\mathcal{C}$ is *closed under inverse homomorphisms* if whenever $\mathfrak{B} \in \mathcal{C}$ and $\mathfrak{A}$ homomorphically maps to $\mathfrak{B}$ we have $\mathfrak{A} \in \mathcal{C}$. If τ is a finite relational signature, then it is well-known and easy to see [5] that $\mathcal{C} = \text{CSP}(\mathfrak{B})$ for a countably infinite τ-structure $\mathfrak{B}$ if and only if $\mathcal{C}$ is closed under inverse homomorphisms and disjoint unions.

3   Monotone SNP

Let τ be a finite relational signature, i.e., τ is a set of relation symbols R, each equipped with an *arity* $\text{ar}(R) \in \mathbb{N}$. An *SNP ($\tau$-) sentence* is an existential second-order (τ-) sentence with a universal first-order part, i.e., a sentence of the form

$$\exists R_1, \dots, R_k \, \forall x_1, \dots, x_n \colon \phi$$

where ϕ is a quantifier-free formula over the signature $\tau \cup \{R_1, \dots, R_k\}$. We make the additional convention that the equality symbol, which is usually allowed in first-order logic, is not allowed in ϕ (see [15]). We write $[\![\Phi]\!]$ for the class of all finite models of Φ.

Example 4. $\text{CSP}(\mathbb{Q}; <) = [\![\Phi]\!]$ for the SNP $\{<\}$-sentence Φ given below.

$$\exists T \, \forall x, y, z \big((\neg(x < y) \vee T(x, y))$$
$$\wedge \big(\neg T(x, y) \vee \neg T(y, z) \vee T(x, z)\big) \wedge \neg T(x, x)\big)$$

A class $\mathcal{C}$ of finite τ-structures is said to be *in SNP* if there exists an SNP τ-sentence Φ such that $[\![\Phi]\!] = \mathcal{C}$; we use analogous definitions for all logics considered in this paper. We may assume that the quantifier-free part of SNP sentences is written in conjunctive normal form, and then use the usual terminology (*clauses*, *literals*, etc).

Definition 1. *An SNP τ-sentence Φ with quantifier-free part ϕ and existentially quantified relation symbols σ is called*

- monotone *if each literal of ϕ with a symbol from τ is* negative, *i.e., of the form $\neg R(\bar{x})$ for $R \in \tau$.*
- monadic *if all the existentially quantified relations are unary.*
- connected *if each clause of ϕ is connected, i.e., the following $\tau \cup \sigma$-structure $\mathfrak{C}$ is connected: the domain of $\mathfrak{C}$ is the set of variables of the clause, and $t \in R^{\mathfrak{C}}$ if and only if $\neg R(t)$ is a disjunct of the clause.*

The SNP sentence from Example 4 is monotone, but not monadic, and it can be shown that there does not exist an equivalent MMSNP sentence [4].

Theorem 1 ([5])**.** *Every sentence in connected monotone SNP describes a problem of the form* $\mathrm{CSP}(\mathfrak{B})$ *for some relational structure $\mathfrak{B}$. Conversely, for every structure $\mathfrak{B}$, if $\mathrm{CSP}(\mathfrak{B})$ is in SNP then it is also in connected monotone SNP.*

4   Amalgamation SNP

In this section we define the new logic *Amalgamation SNP (ASNP)*. We first revisit some basic concepts from model theory.

4.1   The Amalgamation Property

Let τ be a finite relational signature and let $\mathcal{C}$ be a class of τ-structures. We say that $\mathcal{C}$ is *finitely bounded* if there exists a finite set of finite τ-structures $\mathcal{F}$ such that $\mathfrak{A} \in \mathcal{C}$ if and only if no structure in $\mathcal{F}$ embeds into $\mathfrak{A}$; in this case we also write $\mathcal{C} = \mathrm{Forb}(\mathfrak{A})$. Note that $\mathcal{C}$ is finitely bounded if and only if there exists a universal τ-sentence ϕ (which might involve the equality symbol) such that for every finite τ-structure $\mathfrak{A}$ we have $\mathfrak{A} \models \phi$ if and only if $\mathfrak{A} \in \mathcal{C}$. We say that $\mathcal{C}$ has

- the *Joint Embedding Property (JEP)* if for all structures $\mathfrak{B}_1, \mathfrak{B}_2 \in \mathcal{C}$ there exists a structure $\mathfrak{C} \in \mathcal{C}$ that embeds both $\mathfrak{B}_1$ and $\mathfrak{B}_2$.
- the *Amalgamation Property (AP)* if for any two structures $\mathfrak{B}_1, \mathfrak{B}_2 \in \mathcal{C}$ such that $B_1 \cap B_2$ induce the same substructure in $\mathfrak{B}_1$ and in $\mathfrak{B}_2$ (a so-called *amalgamation diagram*) there exists a structure $\mathfrak{C} \in \mathcal{C}$ and embeddings $e_1 \colon \mathfrak{B}_1 \hookrightarrow \mathfrak{C}$ and $e_2 \colon \mathfrak{B}_2 \hookrightarrow \mathfrak{C}$ such that $e_1(a) = e_2(a)$ for all $a \in B_1 \cap B_2$.

Note that since τ is relational, the AP implies the JEP. A class of finite τ-structures which has the AP and is closed under induced substructures and isomorphisms is called an *amalgamation class*.

The *age* of $\mathfrak{B}$ is the class of all finite τ-structures that embed into $\mathfrak{B}$. We say that $\mathfrak{B}$ is *finitely bounded* if $\mathrm{Age}(\mathfrak{B})$ is finitely bounded. A relational τ-structure $\mathfrak{B}$ is called *homogeneous* if every isomorphism between finite substructures of $\mathfrak{B}$ can be extended to an automorphism of $\mathfrak{B}$. Fraïssé's theorem implies that for every amalgamation class $\mathcal{C}$ there exists a countable homogeneous τ-structure $\mathfrak{B}$ with $\mathrm{Age}(\mathfrak{B}) = \mathcal{C}$; the structure $\mathfrak{B}$ is unique up to isomorphism, also called the *Fraïssé-limit* of $\mathcal{C}$. Conversely, it is easy to see that the age of a homogeneous τ-structure is an amalgamation class. A structure $\mathfrak{A}$ is called a *reduct* of a structure

$\mathfrak{B}$ if $\mathfrak{A}$ is obtained from $\mathfrak{B}$ by restricting the signature. It is called a *first-order reduct* of $\mathfrak{B}$ if $\mathfrak{A}$ is obtained from $\mathfrak{B}$ by first expanding by all first-order definable relations, and then restricting the signature. An example of a first-order reduct of $(\mathbb{Q}; <)$ is the structure $(\mathbb{Q}; \mathrm{Betw})$ from Example 3.

4.2   Defining Amalgamation SNP

As we have mentioned in the introduction, the idea of our logic is to require that a certain class of finite structures associated to the first-order part of an SNP sentence is an amalgamation class. We then use the fact that for binary signatures, the amalgamation property is decidable (Sect. 5).

Definition 2. *Let τ be a finite relational signature. An* Amalgamation SNP *τ-sentence is an SNP sentence Φ of the form $\exists R_1, \ldots, R_k \, \forall x_1, \ldots, x_n \colon \phi$ where*

- $R_1, \ldots, R_k$ *are binary;*
- *ϕ is a conjunction of $\{R_1, \ldots, R_k\}$-formulas and of conjuncts of the form $S(x_1, \ldots, x_k) \Rightarrow \psi(x_1, \ldots, x_k)$ where $S \in \tau$ and ψ is a $\{R_1, \ldots, R_k\}$-formula;*
- *the class of $\{R_1, \ldots, R_k\}$-reducts of the finite models of ϕ is an amalgamation class.*

Note that ASNP inherits from SNP the restriction that equality symbols are not allowed. Also note that Amalgamation SNP sentences are necessarily monotone. This implies in particular that the class of $\{R_1, \ldots, R_k\}$-reducts of the finite models of ϕ is precisely the class of finite $\{R_1, \ldots, R_k\}$-structures that satisfy the conjuncts of ϕ that are $\{R_1, \ldots, R_k\}$-formulas (i.e., that do not contain any symbol from τ).

Example 5. The monotone SNP sentence from Example 4 describing $\mathrm{CSP}(\mathbb{Q}; <)$ is in ASNP. The problem $\mathrm{CSP}(\mathbb{Q}; \mathrm{Betw})$ from Example 3 can be expressed by the ASNP sentence

$$\exists T \, \forall x, y, z \big((\mathrm{Betw}(x, y, z) \Rightarrow ((T(x, y) \wedge T(y, z)) \vee (T(z, y) \wedge T(y, x)))$$
$$\wedge ((T(x, y) \wedge T(y, z)) \Rightarrow T(x, z)) \wedge \neg T(x, x) \big).$$

Note that every finite-domain CSP can be expressed in ASNP; this can be seen similarly as in the argument of Feder and Vardi that finite-domain CSPs can be expressed in MMSNP [15].

Then the class of finite models of the first-order part of Φ has the JEP, and since equality is not allowed in SNP the class is even closed under disjoint unions; it follows that also Φ is closed under disjoint unions. It can be shown as in the proof of Theorem 1 that every Amalgamation SNP sentence can be rewritten into an equivalent connected Amalgamation SNP sentence.

4.3 ASNP and CSPs

We present the link between ASNP and infinite-domain CSPs.

Theorem 2. *For every ASNP τ-sentence Φ there exists a first-order reduct $\mathfrak{C}$ of a binary finitely bounded homogeneous structure such that $\mathrm{CSP}(\mathfrak{C}) = [\![\Phi]\!]$.*

Proof. Let ρ be the set of existentially quantified relation symbols of Φ. Let $\phi = \forall x_1, \ldots, x_n \colon \psi$, for a quantifier-free formula ψ in conjunctive normal form, be the first-order part of Φ. Let $\mathcal{C}$ be the class of ρ-reducts of the finite models of ϕ; by assumption, $\mathcal{C}$ is an amalgamation class. Moreover, $\mathcal{C}$ is finitely bounded because it is the class of models of a universal ρ-sentence. Let $\mathfrak{B}$ be the Fraïssé-limit of $\mathcal{C}$; then $\mathfrak{B}$ is a finitely bounded homogeneous structure. Let $\mathfrak{C}$ be the τ-structure which is the first-order reduct of the structure $\mathfrak{B}$ where the relation $S^{\mathfrak{C}}$ for $S \in \tau$ is defined as follows: if $\phi_1, \ldots, \phi_s$ are all the ρ-formulas such that ψ contains the conjunct $S(x_1, \ldots, x_k) \Rightarrow \phi_i(x_1, \ldots, x_k)$ for all $i \in \{1, \ldots, s\}$, then the first-order definition of S is given by $S(x_1, \ldots, x_k) \Leftrightarrow (\phi_1 \wedge \cdots \wedge \phi_s)$.

Claim 1. If $\mathfrak{A}$ is a finite τ-structure such that $\mathfrak{A} \to \mathfrak{C}$, then $\mathfrak{A} \models \Phi$.

Let $h \colon \mathfrak{A} \to \mathfrak{C}$ be a homomorphism. Let $\mathfrak{A}'$ be the $(\tau \cup \rho)$-expansion of $\mathfrak{A}$ where $R \in \rho$ of arity l denotes $\{(a_1, \ldots, a_l) \mid (h(a_1), \ldots, h(a_l)) \in R^{\mathfrak{B}}\}$. Then $\mathfrak{A}'$ satisfies ϕ: to see this, let $a_1, \ldots, a_n \in A$ and let ψ' be a conjunct of ψ. Since $\mathfrak{C} \models \forall x_1, \ldots, x_n \colon \psi$ we have in particular that $\mathfrak{C} \models \psi'(h(a_1), \ldots, h(a_n))$ and so there must be a disjunct ψ'' of ψ' such that $\mathfrak{C} \models \psi''(h(a_1), \ldots, h(a_n))$. Then one of the following cases applies.

- ψ'' is a τ-literal and hence must be negative since Φ is a monotone SNP sentence. In this case $\mathfrak{C} \models \psi''(h(a_1), \ldots, h(a_n))$ implies $\mathfrak{A}' \models \psi''(a_1, \ldots, a_n)$ since h is a homomorphism.
- ψ'' is a ρ-literal. Then by the definition of $\mathfrak{A}'$ we have that $\mathfrak{A}' \models \psi''(a_1, \ldots, a_n)$ if and only if $\mathfrak{C} \models \psi''(h(a_1), \ldots, h(a_n))$.

Hence, $\mathfrak{A}' \models \psi'(a_1, \ldots, a_n)$. Since the conjunct ψ' of ψ and $a_1, \ldots, a_n \in A$ were arbitrarily chosen, we have that $\mathfrak{A}' \models \forall x_1, \ldots, x_n \colon \psi$. Hence, $\mathfrak{A}$ satisfies Φ.

Claim 2. If $\mathfrak{A}$ is a finite τ-structure such that $\mathfrak{A} \models \Phi$, then $\mathfrak{A} \to \mathfrak{C}$.

If $\mathfrak{A}$ has a $(\tau \cup \rho)$-expansion $\mathfrak{A}'$ that satisfies ϕ, then there exists an embedding from the ρ-reduct $\mathfrak{A}''$ of $\mathfrak{A}'$ into $\mathfrak{B}$ by the definition of $\mathfrak{B}$. This embedding is in particular a homomorphism from $\mathfrak{A}$ to $\mathfrak{C}$. □

Theorem 3. *Let $\mathfrak{C}$ be a first-order reduct of a binary finitely bounded homogeneous structure $\mathfrak{B}$. Then $\mathrm{CSP}(\mathfrak{C})$ can be expressed in ASNP.*

Proof. Let σ be the signature of $\mathfrak{B}$ and τ the signature of $\mathfrak{C}$. We may assume without loss of generality that $\mathfrak{B}$ contains a binary relation E that denotes the equality relation; it is easy to see that an expansion by the equality relation preserves finite boundedness. Consider the structure $\mathfrak{B}^*$ with the domain $B \times \mathbb{N}$ where

$$R^{\mathfrak{B}^*} := \{((b_1, n_1), \ldots, (b_k, n_k)) \mid n_1, \ldots, n_k \in \mathbb{N}, (b_1, \ldots, b_k) \in R^{\mathfrak{B}}\}.$$

To show that $\mathfrak{B}^*$ is homogeneous, let h be an isomorphism between finite substructures of $\mathfrak{B}^*$. Let $T \subseteq B$ be the set of all first entries of elements of the first structure. Define $g : T \to B$ by picking for $b \in T$ an element of the form $(b, n) \in S$ and defining by $g(b) := h(b, n)_1$. This is well-defined: if h is defined on (b, n_1) and on (b, n_2), then $((b, n_1), (b, n_2)) \in E^{\mathfrak{B}^*}$, and hence $h(b, n_1)_1 = h(b, n_2)_1$. The same consideration for h^{-1} shows that g is a bijection, and in fact an isomorphism between finite substructures of $\mathfrak{B}$. By the homogeneity of $\mathfrak{B}$ there exists an extension $g^* \in \mathrm{Aut}(\mathfrak{B})$ of g. For each $b \in B$ pick a permutation f_b of $\mathbb{N}$ that extends the bijection given by $n \mapsto h(b, n)_2$. Then the map $h^* : B^* \to B^*$ given by $h(b, n) := (g^*(b), f_b(n))$ is an automorphism of $\mathfrak{B}^*$ that extends h. Since $\mathfrak{B}$ is finitely bounded, there exists a universal σ-formula ϕ such that $\mathrm{Age}(\mathfrak{B}) = [\![\phi]\!]$. Note that ϕ might contain the equality symbol (which we do not allow in SNP sentences). Let ϕ^* be the formula obtained from ϕ by

- replacing each occurrence of the equality symbol by the symbol $E \in \sigma$;
- joining conjuncts that imply that E denotes an equivalence relation;
- joining for every $R \in \sigma$ of arity n the conjunct

$$\forall x_1, \ldots, x_n, y_1, \ldots, y_n \big(R(x_1, \ldots, x_n) \vee \neg R(y_1, \ldots, y_n) \vee \bigvee_{i \leq n} \neg E(x_i, y_i) \big)$$

(implementing indiscernibility of identicals for the relation E).

We claim that $\mathrm{Age}(\mathfrak{B}^*) = [\![\phi^*]\!]$. To see this, let $\mathfrak{A}^*$ be a finite σ-structure. If $\mathfrak{A}^*$ satisfies ϕ^*, then every induced substructure $\mathfrak{A}$ of $\mathfrak{A}^*$ with the property that $(x, y) \in E^{\mathfrak{A}}$ implies that at most one of x and y is an element of A, satisfies ϕ, and hence is a substructure of $\mathfrak{B}$. This in turn means that $\mathfrak{A}^*$ is in $\mathrm{Age}(\mathfrak{B}^*)$. The implications in this statement can be reversed which shows the claim.

Let ϕ' be the formula obtained from ϕ^* as follows. For each $S \in \tau$ let χ_S be the first-order definition of $S^{\mathfrak{C}}$ in $\mathfrak{B}$; since $\mathfrak{B}$ is homogeneous we may assume that χ_S is quantifier-free [21]. Furthermore, we may assume that χ_S is given in conjunctive normal form. Let k be the arity of S. We then add for each conjunct χ'_S of χ_S the conjunct

$$\forall x_1, \ldots, x_k \big(S(x_1, \ldots, x_k) \Rightarrow \chi'_S(x_1, \ldots, x_k) \big)$$

By construction, the sentence Φ obtained from ϕ' by quantifying all relation symbols of σ is an ASNP τ-sentence. $\qquad\square$

Corollary 1. *ASNP has a complexity dichotomy if and only if the infinite-domain dichotomy conjecture is true for first-order reducts of binary finitely bounded homogeneous structures.*

5   Deciding Amalgamation

In this section we show how to algorithmically decide whether a given existential second-order sentence is in ASNP. The following is a known fact in the model theory of homogeneous structures (the first author has learned the fact from Gregory Cherlin), but we are not aware of any published proof in the literature.

Theorem 4. *Let $\mathcal{F}$ be a finite set of finite binary relational τ-structures. There is an algorithm that decides whether* $\mathrm{Forb}(\mathcal{F})$ *has the amalgamation property.*

Proof. Let m be the maximal size of a structure in $\mathcal{F}$, and let ℓ be the number of isomorphism types of two-element structures in $\mathcal{C} := \mathrm{Forb}(\mathcal{F})$. It is well-known and easy to prove that $\mathcal{C}$ has the amalgamation property if and only if it has the so-called *1-point amalgamation property*, i.e., the amalgamation property restricted to diagrams $(\mathfrak{B}_1, \mathfrak{B}_2)$ where $|B_1| = |B_2| = |B_1 \cap B_2| + 1$. Suppose that $(\mathfrak{B}_1, \mathfrak{B}_2)$ is such an amalgamation diagram without amalgam. Let $B_0 := B_1 \cap B_2$. Let $B_1 \setminus B_0 = \{p\}$ and $B_2 \setminus B_0 = \{q\}$. Let $\mathfrak{D}$ be a τ-structure $\mathfrak{D}$ with domain $B_1 \cup B_2$ such that $\mathfrak{B}_1$ and $\mathfrak{B}_2$ are substructures of $\mathfrak{D}$. Since $\mathfrak{D}$ by assumption is not an amalgam for $(\mathfrak{B}_1, \mathfrak{B}_2)$, there must exist $A = \{a_1, \ldots, a_{m-2}\} \in B_0$ such that the substructure of $\mathfrak{D}$ induced by $\{a_1, \ldots, a_{m-2}, p, q\}$ embeds a structure from $\mathcal{F}$.

Note that the number of such τ-structures $\mathfrak{D}$ is bounded by ℓ since they only differ by the substructure induced by p and q. So let $A_1, \ldots, A_\ell \subseteq B_0$ be a list of sets witnessing that all of these structures $\mathfrak{D}$ embed a structure from $\mathcal{F}$. Let $\mathfrak{C}_1$ be the substructure of $\mathfrak{B}_1$ induced by $\{p\} \cup A_1 \cup \cdots \cup A_\ell$ and $\mathfrak{C}_2$ be the substructure of $\mathfrak{B}_2$ induced by $\{q\} \cup A_1 \cup \cdots \cup A_\ell$. Suppose for contradiction that $(\mathfrak{C}_1, \mathfrak{C}_2)$ has an amalgam $\mathfrak{C}$; we may assume that this amalgam is of size at most $(m-2) \cdot \ell$. Depending on the two-element structure induced by $\{p, q\}$ in $\mathfrak{C}$, there exists an $i \le \ell$ such that the structure induced by $\{p, q\} \cup A_i$ in $\mathfrak{C}$ embeds a structure from $\mathcal{F}$, a contradiction.   □

Corollary 2. *There is an algorithm that decides for a given existential second-order sentence Φ whether it is in ASNP.*

Proof. Let k be the maximal number of variables per clause in the first-order part ϕ of Φ, and let $\mathcal{F}$ be the set of all structures at most the elements $\{1, \ldots, k\}$ that do not satisfy ϕ. Then $\mathrm{Forb}(\mathcal{F}) = [\![\phi]\!]$ and the result follows from Theorem 2.   □

6   Guarded Monotone SNP

In this section we revisit an expressive generalisation of MMSNP introduced by Bienvenu, ten Cate, Lutz, and Wolter [3] in the context of ontology-based data access, called *guarded monotone SNP (GMSNP)*. It is equally expressive as the logic MMSNP$_2$ introduced by Madelaine [25][1]. We will see that every GMSNP sentence is equivalent to a finite disjunction of *connected* GMSNP sentences (Proposition 1), each of which lies in ASNP if the signature is binary (Theorem 5).

---

[1] MMSNP$_2$ relates to MMSNP as Courcelle's MSO$_2$ relates to MSO [13].

Definition 3. *A monotone SNP τ-sentence Φ with existentially quantified relations ρ is called* guarded *if each conjunct of Φ can be written in the form*

$$\alpha_1 \wedge \cdots \wedge \alpha_n \Rightarrow \beta_1 \vee \cdots \vee \beta_m, \quad where$$

- $\alpha_1, \ldots, \alpha_n$ *are atomic $(\tau \cup \rho)$-formulas, called* body atoms*,*
- $\beta_1, \ldots, \beta_m$ *are atomic ρ-formulas, called* head atoms*,*
- *for every head atom β_i there is a body atom α_j such that α_j contains all variables from β_i (such clauses are called* guarded*).*

We do allow the case that $m = 0$, i.e., the case where the head consists of the empty disjunction, which is equivalent to $\bot$ (false).

The next proposition extends a well-known fact for MMSNP to guarded SNP.

Proposition 1. *Every GMSNP sentence Φ is equivalent to a finite disjunction $\Phi_1 \vee \cdots \vee \Phi_k$ of connected GMSNP sentences.*

Proof. We prove Proposition 1. Let Φ be a guarded SNP sentence. Suppose that the quantifier-free part of Φ has a disconnected clause ψ (Definition 1). By definition the variable set can be partitioned into non-empty variable sets X_1 and X_2 such that for every negative literal $\neg R(x_1, \ldots, x_r)$ of the clause either $\{x_1, \ldots, x_r\} \subseteq X_1$ or $\{x_1, \ldots, x_r\} \subseteq X_2$. The same is true for every positive literal, since otherwise the definition of guarded clauses would imply a negative literal on a set that contains $\{x_1, \ldots, x_r\}$, contradicting the property above. Hence, ψ can be written as $\psi_1(\bar{x}) \vee \psi_2(\bar{y})$ for non-empty disjoint tuples of variables $\bar{x}$ and $\bar{y}$. Let ϕ_1 be the formula obtained from ϕ by replacing ψ by ψ_1, and let ϕ_2 be the formula obtained from ϕ by replacing ψ by ψ_2.

Let $P_1, \ldots, P_k$ be the existential predicates in Φ, and let τ be the input signature of Φ. It suffices to show that for every $(\tau \cup \{P_1, \ldots, P_k\})$-expansion $\mathfrak{A}'$ of $\mathfrak{A}$ we have that $\mathfrak{A}'$ satisfies ϕ if and only if $\mathfrak{A}'$ satisfies ϕ_1 or ϕ_2. If $\mathfrak{A}'$ falsifies a clause of ϕ, there is nothing to show since then $\mathfrak{A}'$ satisfies neither ϕ_1 nor ϕ_2. If $\mathfrak{A}'$ satisfies all clauses of ϕ, it in particular satisfies a literal from ψ; depending on whether this literal lies in ψ_1 or in ψ_2, we obtain that $\mathfrak{A}'$ satisfies ψ_1 or ψ_2, and hence ϕ_1 or ϕ_2. Iterating this process for each disconnected clause of ϕ, we eventually arrive at a finite disjunction of connected guarded SNP sentences. $\square$

It is well-known and easy to see [17] that each of $\Phi_1, \ldots, \Phi_k$ can be reduced to Φ in polynomial time. Conversely, if each of $\Phi_1, \ldots, \Phi_k$ is in P, then Φ is in P, too. It follows in particular that if connected GMSNP has a complexity dichotomy into P and NP-complete, then so has GMSNP.

Theorem 5. *For every sentence Φ in connected GMSNP there exists a reduct $\mathfrak{C}$ of a finitely bounded homogeneous structures such that $[\![\Phi]\!] = \mathrm{CSP}(\mathfrak{C})$. If all existentially quantified relation symbols in Φ are binary then it is equivalent to an ASNP sentence.*

In the proof of Theorem 5 we use a result of Cherlin, Shelah, and Shi [12] in a strengthened form due to Hubička and Nešetřil [22], namely that for every finite set $\mathcal{F}$ of finite σ-structures, for some finite relational signature σ, there exists a finitely bounded homogeneous $(\sigma \cup \rho)$-structure $\mathfrak{B}$ such that a finite σ-structure $\mathfrak{A}$ homomorphically maps to $\mathfrak{B}$ if none of the structures in $\mathcal{F}$ homomorphically maps to $\mathfrak{B}$. We now prove Theorem 5.

Proof. Let Φ be a τ-sentence in connected guarded monotone SNP with existentially quantified relation symbols $\{E_1, \ldots, E_k\}$. Let σ be the signature which contains for every relation symbol $R \in \{E_1, \ldots, E_k\}$ two new relation symbols R^+ and R^- of the same arity and for every relation symbol $R \in \tau$ a new relation symbol R'. Let ϕ be the first-order part of Φ, written in conjunctive normal form, and let n be the number of variables in the largest clause of ϕ. Let ϕ' be the sentence obtained from ϕ by replacing each occurrence of $R \in \{E_1, \ldots, E_k\}$ by R^+ and each occurrence of $\neg R$ by R^-, and finally each occurrence of $R \in \tau$ by R'. Let $\mathcal{F}$ be the (finite) class of all finite σ-structures with at most n elements that do not satisfy ϕ'. We apply the mentioned theorem of Hubička and Nešetřil to $\mathcal{F}$, and obtain a finitely bounded homogeneous $\sigma \cup \rho$-structure $\mathfrak{B}$ such that the age of the σ-reduct $\mathfrak{C}$ of $\mathfrak{B}$ equals $\text{Forb}(\mathcal{N})$. We say that $S \subseteq B$ is *correctly labelled* if for every $R \in \{E_1, \ldots, E_k\}$ of arity m and $s_1, \ldots, s_m \in S$ we have $R^-(s_1, \ldots, s_m)$ if and only if $\neg R(s_1, \ldots, s_m)$. Let $\mathfrak{B}'$ the $\tau \cup \sigma \cup \rho$-expansion of $\mathfrak{B}$ where $R \in \tau$ of arity m denotes

$$\{(t_1, \ldots, t_m) \in (R')^{\mathfrak{B}} \mid \{t_1, \ldots, t_m\} \text{ is correctly labelled}\}.$$

Since $\mathfrak{B}$ is finitely bounded homogeneous, $\mathfrak{B}'$ is finitely bounded homogeneous, too. Let $\mathfrak{C}$ be the τ-reduct of $\mathfrak{B}'$. We claim that $\llbracket \Phi \rrbracket = \text{CSP}(\mathfrak{C})$. First suppose that $\mathfrak{A}$ is a finite τ-structure that satisfies Φ. Then it has an $\{E_1, \ldots, E_k\}$-expansion $\mathfrak{A}'$ that satisfies ϕ. Let $\mathfrak{A}''$ be the σ-structure with the same domain as $\mathfrak{A}'$ where

- R' denotes $R^{\mathfrak{A}'}$ for each $R \in \tau$;
- R^+ denotes $R^{\mathfrak{A}'}$ for each $R \in \{E_1, \ldots, E_k\}$;
- R^- denotes $\neg R^{\mathfrak{A}'}$ for each $R \in \{E_1, \ldots, E_k\}$.

Then $\mathfrak{A}''$ satisfies ϕ', and hence embeds into $\mathfrak{B}$. This embedding is a homomorphism from $\mathfrak{A}$ to $\mathfrak{C}$ since the image of the embedding is correctly labelled by the construction of $\mathfrak{A}''$.

Conversely, suppose that $\mathfrak{A}$ has a homomorphism h to $\mathfrak{C}$. Let $\mathfrak{A}'$ be the $\tau \cup \{E_1, \ldots, E_k\}$-expansion of $\mathfrak{A}$ by defining $(a_1, \ldots, a_n) \in R^{\mathfrak{A}}$ if and only if $(h(a_1), \ldots, h(a_n)) \in R^{\mathfrak{B}'}$, for every n-ary $R \in \{E_1, \ldots, E_k\}$. Then each clause of ϕ is satisfied, because each clause of ϕ is guarded: let $x_1, \ldots, x_m$ be the variables of some clause of ϕ. If $a_1, \ldots, a_m \in A$ satisfy the body of this clause, and $\psi(a_{i_1}, \ldots, a_{i_l})$ is a head atom of such a clause, then the set $\{h(a_{i_1}), \ldots, h(a_{i_l})\}$ is correctly labelled. This implies that some of the head atoms of the clause must be true in $\mathfrak{A}'$ because $\mathfrak{B}'$ satisfies ϕ'. The second statement follows from Theorem 3. □

The following example shows that GMSNP does not contain ASNP.

Example 6. CSP$(\mathbb{Q}; <)$ is in ASNP (see Example 5) but not in GMSNP. Indeed, suppose that Φ is a GMSNP sentence which is true on all finite directed paths. We assume that the quantifier-free part ϕ of Φ is in conjunctive normal form. Let ρ be the existentially quantified relation symbols of Φ, let $k := |\rho|$, and let l be the number of variables in Φ. Every directed path, viewed as a $\{<\}$-structure, satisfies Φ, and therefore has an $\{<\} \cup \rho$-expansion $\mathfrak{A}$ that satisfies ϕ. Note that there are finitely many different $\{<\} \cup \rho$-expansions of a path of length $l \in \mathbb{N}$; let $p \in \mathbb{N}$ be this number. Hence, for a path of length $L := (p + 1)l$, there must be $i, j \in \{0, \ldots, p\}$ with $i < j$ such that the substructures of $\mathfrak{A}$ induced by $il+1, il+2, \ldots, il+l$ and by $jl+1, jl+2, \ldots, jl+l$ are isomorphic. We then claim that the directed cycle $(i + 1)l + 1, (i + 1)l + 2, \ldots, jl + 1, \ldots, jl + l, (i + 1)l + 1$ satisfies Φ: this is witnessed by the $\{<\} \cup \rho$-expansion inherited from $\mathfrak{A}$ which satisfies ϕ since each clause in ϕ is guarded. Hence, Φ does not express digraph acyclicity.

7 Application: Instances of Bounded Treewidth

If a computational problem can be formulated in ASNP or in GMSNP, then this has remarkable consequences besides a potential complexity dichotomy. In this section we show that every problem that can be formulated in ASNP or in GMSNP is in P when restricted to instances of bounded treewidth. The corresponding result for Monadic Second-Order Logic (MSO) instead of ASNP is a famous theorem of Courcelle [13]. We strongly believe that ASNP is not contained in MSO (consider for instance the Betweenness Problem from Example 3), so our result appears to be incomparable to Courcelle's.

In the proof of our result, we need the following concepts from model theory. A first-order theory T is called *ω-categorical* if all countable models of T are isomorphic [21]. A structure $\mathfrak{B}$ is called ω-categorical if its first-order theory (i.e., the set of first-order sentences that hold in $\mathfrak{B}$) is ω-categorical. Note that with this definition, finite structures are ω-categorical. Another classic example is the structure $(\mathbb{Q}; <)$. The definition of treewidth can be treated as a black box in our proof, and we refer the reader to [6].

Theorem 6. *Let Φ be an ASNP or a connected GMSNP τ-sentence and let $k \in \mathbb{N}$. Then the problem to decide whether a given finite τ-structure $\mathfrak{A}$ of treewidth at most k satisfies Φ can be decided in polynomial time with a Datalog program (of width k).*

Proof. Since structures that are homogeneous in a finite relational language are ω-categorical [21] and first-order reducts of ω-categorical structures are ω-categorical [21], Theorem 2 and Theorem 5 imply that the problem to decide whether a finite τ-structure satisfies ϕ can be formulated as CSP$(\mathfrak{B})$ for an ω-categorical structure $\mathfrak{B}$. Then the statement follows from Corollary 1 in [6]. □

Remark 1. In Theorem 6 it actually suffices to assume that the *core* of $\mathfrak{A}$ has treewidth at most k.

Corollary 3. *Let Φ be a GMSNP τ-sentence and let $k \in \mathbb{N}$. Then there is a polynomial-time algorithm that decides whether a given τ-structure of treewidth at most k satisfies Φ.*

Proof. Immediate from Theorem 1 and Theorem 6.   □

8   Conclusion and Open Problems

ASNP is a candidate for an expressive logic with a complexity dichotomy: every problem in ASNP is NP-complete or in P if and only if the infinite-domain dichotomy conjecture for first-order reducts of binary finitely bounded homogeneous structures holds. See Fig. 1 for the relation to other candidate logics that are known to have a dichotomy, might have a complexity, or provably do not have a dichotomy.

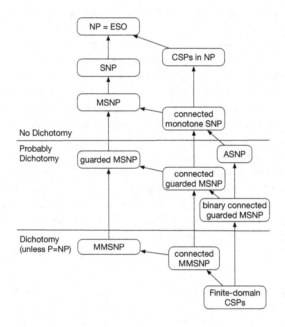

Fig. 1. Fragments of existential second-order logic and complexity dichotomies.

We presented an application of ASNP concerning the evaluation of computational problems on classes of structures of bounded treewidth. We also proved that the syntax of ASNP is algorithmically decidable. The following problems concerning ASNP are open.

1. Is the Amalgamation Property decidable for (not necessarily binary) classes given by finitely many forbidden substructures?
2. Is every binary CSP in Monadic Second-Order Logic (MSO) also in ASNP?

3. Is every problem in NP polynomial-time equivalent to a problem in Amalgamation SNP if we drop the monotonicity assumption?
4. Is there a natural logic (which in particular has an effective syntax) that contains both ASNP and connected GMSNP and which describes CSPs for reducts of finitely bounded homogeneous structures?

References

1. Barto, L., Kompatscher, M., Olšák, M., Pham, T.V., Pinsker, M.: Equations in oligomorphic clones and the constraint satisfaction problem for ω-categorical structures. J. Math. Logic, **19**(2), #1950010 (2019). An extended abstract appeared at the Proceedings of the 32nd Annual ACM/IEEE Symposium on Logic in Computer Science - LICS'17
2. Barto, L., Pinsker, M.: The algebraic dichotomy conjecture for infinite domain constraint satisfaction problems. In: Proceedings of the 31th Annual IEEE Symposium on Logic in Computer Science, LICS 2016, pp. 615–622 (2016). Preprint arXiv:1602.04353
3. Bienvenu, M., ten Cate, B., Lutz, C., Wolter, F.: Ontology-based data access: a study through disjunctive datalog, CSP, and MMSNP. ACM Trans. Database Syst. **39**(4), 33 (2014)
4. Bodirsky, M.: Constraint satisfaction with infinite domains. Dissertation, Humboldt-Universität zu Berlin (2004)
5. Bodirsky, M.: Complexity classification in infinite-domain constraint satisfaction. Mémoire d'habilitation à diriger des recherches, Université Diderot - Paris 7. arXiv:1201.0856 (2012)
6. Bodirsky, M., Dalmau, V.: Datalog and constraint satisfaction with infinite templates. J. Comput. Syst. Sci. **79**, 79–100 (2013). A preliminary version appeared in the proceedings of the Symposium on Theoretical Aspects of Computer Science (STACS'05)
7. Bodirsky, M., Madelaine, F., Mottet, A.: A universal-algebraic proof of the complexity dichotomy for Monotone Monadic SNP. In: Proceedings of the Symposium on Logic in Computer Science (LICS), pp. 105–114 (2018). Preprint available under arXiv:1802.03255
8. Bodirsky, M., Pinsker, M., Pongrácz, A.: Projective clone homomorphisms. J. Symb. Logic (2014). Preprint arXiv:1409.4601
9. Bodirsky, M., Rydval, J.: Temporal constraint satisfaction problems in fixed point logic (2019). Preprint
10. Braunfeld, S.: Towards the undecidability of atomicity for permutation classes via the undecidability of joint embedding for hereditary graph classes (2019). Preprint available at arXiv:1903.11932
11. Bulatov, A.A.: A dichotomy theorem for nonuniform CSPs. In: 58th IEEE Annual Symposium on Foundations of Computer Science, FOCS 2017, Berkeley, CA, USA, 15–17 October 2017, pp. 319–330 (2017)
12. Cherlin, G., Shelah, S., Shi, N.: Universal graphs with forbidden subgraphs and algebraic closure. Adv. Appl. Math. **22**, 454–491 (1999)
13. Courcelle, B., Engelfriet, J.: Graph Structure and Monadic Second-Order Logic: A Language-Theoretic Approach. Cambridge University Press, New York (2012)
14. Fagin, R.: Generalized first-order spectra and polynomial-time recognizable sets. Complex. Comput. **7**, 43–73 (1974)

15. Feder, T., Vardi, M.Y.: The computational structure of monotone monadic SNP and constraint satisfaction: a study through Datalog and group theory. SIAM J. Comput. **28**, 57–104 (1999)
16. Feder, T., Vardi, M.Y.: Homomorphism closed vs. existential positive. In: Proceedings of the Symposium on Logic in Computer Science (LICS), pp. 311–320 (2003)
17. Foniok, J., Nešetřil, J.: Generalised dualities and maximal finite antichains in the homomorphism order of relational structures. Eur. J. Combin. **29**(4), 881–899 (2008)
18. Garey, M., Johnson, D.: A Guide to NP-Completeness. CSLI Press, Stanford (1978)
19. Grohe, M.: The quest for a logic capturing PTIME. In: Proceedings of ACM/IEEE Symposium on Logic in Computer Science (LICS), pp. 267–271 (2008)
20. Gurevich, Y.: Toward logic tailored for computational complexity. In: Börger, E., Oberschelp, W., Richter, M.M., Schinzel, B., Thomas, W. (eds.) Computation and Proof Theory. LNM, vol. 1104, pp. 175–216. Springer, Heidelberg (1984). https://doi.org/10.1007/BFb0099486
21. Hodges, W.: Model Theory. Cambridge University Press, Cambridge (1993)
22. Hubička, J., Nešetřil, J.: Homomorphism and embedding universal structures for restricted classes. Multiple-Valued Logic Soft Comput. **27**(2–3), 229–253 (2016). arXiv:0909.4939
23. Kun, G.: Constraints, MMSNP, and expander relational structures. Combinatorica **33**(3), 335–347 (2013). https://doi.org/10.1007/s00493-013-2405-4
24. Ladner, R.E.: On the structure of polynomial time reducibility. J. ACM **22**(1), 155–171 (1975)
25. Madelaine, F.: Universal structures and the logic of forbidden patterns. In: Ésik, Z. (ed.) CSL 2006. LNCS, vol. 4207, pp. 471–485. Springer, Heidelberg (2006). https://doi.org/10.1007/11874683_31
26. Zhuk, D.: A proof of CSP dichotomy conjecture. In: 58th IEEE Annual Symposium on Foundations of Computer Science, FOCS 2017, Berkeley, CA, USA, 15–17 October 2017, pp. 331–342 (2017). https://arxiv.org/abs/1704.01914

Ackermannian Goodstein Sequences of Intermediate Growth

David Fernández-Duque[(✉)] and Andreas Weiermann

Ghent University, Ghent, Belgium
{David.FernandezDuque,Andreas.Weiermann}@UGent.be

Abstract. The original Goodstein process proceeds by writing natural numbers in nested exponential k-normal form, then successively raising the base to $k+1$ and subtracting one from the end result. Such sequences always reach zero, but this fact is unprovable in Peano arithmetic. In this paper we instead consider notations for natural numbers based on the Ackermann function. We define two new Goodstein processes, obtaining new independence results for ACA_0' and ACA_0^+, theories of second order arithmetic related to the existence of Turing jumps.

Keywords: Goodstein sequences · Independence proofs · Ordinal notation systems

1 Introduction

Goodstein's principle [6] is arguably the oldest example of a purely number-theoretic statement known to be independent of PA, as it does not require the coding of metamathematical notions such as Gödel's provability predicate [4]. The proof proceeds by transfinite induction up to the ordinal ε_0 [5]. PA does not prove such transfinite induction, and indeed Kirby and Paris later showed that Goodstein's principle is unprovable in PA [8].

Goodstein's original principle involves the termination of certain sequences of numbers. Say that m is in *nested (exponential) base-k normal form* if it is written in standard exponential base k, with each exponent written in turn in base k. Thus for example, 20 would become $2^{2^2} + 2^2$ in nested base-2 normal form. Then, define a sequence $(g_k(0))_{m \in \mathbb{N}}$ by setting $g_0(m) = m$ and defining $g_{k+1}(m)$ recursively by writing $g_k(m)$ in nested base-$(k+2)$ normal form, replacing every occurrence of $k + 2$ by $k + 3$, then subtracting one (unless $g_k(m) = 0$, in which case $g_{k+1}(m) = 0$).

In the case that $m = 20$, we obtain

$$g_0(20) = 20 = 2^{2^2} + 2^2$$

$$g_1(20) = 3^{3^3} + 3^3 - 1 = 3^{3^3} + 3^2 \cdot 2 + 3 \cdot 2 + 2$$

$$g_2(20) = 4^{4^4} + 4^2 \cdot 2 + 4 \cdot 2 + 2 - 1 = 4^{4^4} + 4^2 \cdot 2 + 4 \cdot 2 + 1,$$

© Springer Nature Switzerland AG 2020
M. Anselmo et al. (Eds.): CiE 2020, LNCS 12098, pp. 163–174, 2020.
https://doi.org/10.1007/978-3-030-51466-2_14

and so forth. At first glance, these numbers seem to grow superexponentially. It should thus be a surprise that, as Goodstein showed, for every m there is k^* for which $g_{k^*}(m) = 0$.

By coding finite Goodstein sequences as natural numbers in a standard way, Goodstein's principle can be formalized in the language of arithmetic, but this formalized statement is unprovable in PA. Independence can be shown by proving that the Goodstein process takes at least as long as stepping down the *fundamental sequences* below ε_0; these are canonical sequences $(\xi[n])_{n<\omega}$ such that $\xi[n] < \xi$ for all ξ and for limit ξ, $\xi[n] \to \xi$ as $n \to \infty$. For standard fundamental sequences below ε_0, PA does not prove that the sequence $\xi > \xi[1] > \xi[1][2] > \xi[1][2][3] \cdots$ is finite.

Exponential notation is not suitable for writing very big numbers (e.g. Graham's number [7]), in which case it may be convenient to use systems of notation which employ faster-growing functions. In [2], T. Arai, S. Wainer and the authors have shown that the Ackermann function may be used to write natural numbers, giving rise to a new Goodstein process which is independent of the theory ATR$_0$ of *arithmetical transfinite recursion;* this is a theory in the language of second order arithmetic which is much more powerful than PA. The main axiom of ATR$_0$ states that for any set X and ordinal α, the α-Turing jump of X exists; we refer the reader to [13] for details.

The idea is, for each $k \geq 2$, to define a notion of Ackermannian normal form for each $m \in \mathbb{N}$. Having done this, we can define Ackermannian Goodstein sequences analogously to Goodstein's original version. The normal forms used in [2] are defined using an elaborate 'sandwiching' procedure first introduced in [14], approximating a number m by successive branches of the Ackermann function. In this paper, we consider simpler, and arguably more intuitive, normal forms, also based on the Ackermann function. We show that these give rise to two different Goodstein-like processes, independent of ACA$_0'$ and ACA$_0^+$, respectively. As was the case for ATR$_0$, these are theories of second order arithmetic which state that certain Turing jumps exist. ACA$_0'$ asserts that, for all $n \in \mathbb{N}$ and $X \subseteq \mathbb{N}$, the n-Turing jump of X exists, while ACA$_0^+$ asserts that its ω-jump exists; see [13] for details. The proof-theoretic ordinal of ACA$_0'$ is ε_ω [1], and that of ACA$_0^+$ is $\varphi_2(0)$ [9]; we will briefly review these ordinals later in the text, but refer the reader to standard texts such as [10, 12] for a more detailed treatment of proof-theoretic ordinals.

2   Basic Definitions

Let us fix $k \geq 2$ and agree on the following version of the Ackermann function.

Definition 1. *For $a, b \in \mathbb{N}$ we define $A_a(k, b)$ by the following recursion.*

1. $A_0(k, b) := k^b$,
2. $A_{a+1}(k, 0) := A_a^k(k, \cdot)(0)$,
3. $A_{a+1}(k, b+1) := A_a^k(k, \cdot)(A_{a+1}(k, b))$.

Here, the notation $A_a^k(k, \cdot)$ refers to the k-fold composition of the function $x \mapsto A_a(k, x)$. It is well known that for every fixed a, the function $b \mapsto A_a(k, b)$ is primitive recursive and the function $a \mapsto A_a(k, 0)$ is not primitive recursive. We use the Ackermann function to define k normal forms for natural numbers. These normal forms emerged from discussions with Toshiyasu Arai and Stan Wainer, which finally led to the definition of a more powerful normal form defined in [14] and used to prove termination in [2].

Lemma 1. *Let $k \geq 2$. For all $c > 0$, there exist unique $a, b, m, n \in \mathbb{N}$ such that*

1. $c = A_a(k, b) \cdot m + n$,
2. $A_a(k, 0) \leq c < A_{a+1}(k, 0)$,
3. $A_a(k, b) \leq c < A_a(k, b + 1)$, *and*
4. $n < A_a(k, b)$.

We write $c =_{\mathrm{NF}} A_a(k, b) \cdot m + n$ in this case. This means that we have in mind an underlying context fixed by k and that for the number c we have uniquely associated the numbers a, b, m, n. Note that it could be possible that $A_{a+1}(k, 0) = A_a(k, b)$, so that we have to choose the right representation for the context; in this case, item 2 guarantees that a is chosen to take the maximal possible value.

By rewriting iteratively b and n in such a normal form, we arrive at the Ackermann k-normal form of c. If we also rewrite a iteratively, we arrive at the nested Ackermann k-normal form of c. The following properties of normal forms are not hard to prove from the definitions.

Lemma 2. *1. $A_a^\ell(k, 0)$ is in k-normal form for every ℓ such that $0 < \ell < k$.*
2. if $A_a(k, b)$ is in k-normal form, then for every $\ell < b$, the number $A_a(k, \ell)$ is also in k-normal form.

In the sequel we work with standard notations for ordinals. We use the function $\xi \mapsto \varepsilon_\xi$ to enumerate the fixed points of $\xi \mapsto \omega^\xi$. With $\alpha, \beta \mapsto \varphi_\alpha(\beta)$ we denote the binary Veblen function, where $\beta \mapsto \varphi_\alpha(\beta)$ enumerates the common fixed points of all $\varphi_{\alpha'}$ with $\alpha' < \alpha$. We often omit parentheses and simply write $\varphi_\alpha \beta$. Then $\varphi_0 \xi = \omega^\xi$, $\varphi_1 \xi = \varepsilon_\xi$, $\varphi_2 0$ is the first fixed point of the function $\xi \mapsto \varphi_1 \xi$, $\varphi_\omega 0$ is the first common fixed point of the function $\xi \mapsto \varphi_n \xi$, and Γ_0 is the first ordinal closed under $\alpha, \beta \mapsto \varphi_\alpha \beta$. In fact, not much ordinal theory is presumed in this article; we almost exclusively work with ordinals less than $\varphi_2 0$, which can be written in terms of addition and the functions $\xi \mapsto \omega^\xi$, $\xi \mapsto \varepsilon_\xi$. For more details, we refer the reader to standard texts such as [10,12].

3   Goodstein Sequences for ACA$'_0$

In this section we define a Goodstein process that is independent of ACA$'_0$. We do so by working with unnested Ackermannian normal forms. Such normal forms give rise to the following notion of base change.

Definition 2. *Given $k \geq 2$ and $c \in \mathbb{N}$, define $c[k \leftarrow k + 1]$ by:*

1. $0[k\leftarrow k+1] := 0$.
2. $c[k\leftarrow k+1] := A_a(k+1, b[k\leftarrow k+1]) \cdot m + n[k\leftarrow k+1]$ *if* $c =_{\mathrm{NF}} A_a(k, b) \cdot m + n$.

With this, we may define a new Goodstein process, based on unnested Ackermannian normal forms.

Definition 3. *Let $\ell < \omega$. Put $b_0(\ell) := \ell$. Assume recursively that $b_k(\ell)$ is defined and $b_k(\ell) > 0$. Then $b_{k+1}(\ell) = b_k(\ell)[k + 2\leftarrow k + 3] - 1$. If $b_k(\ell) = 0$, then $b_{k+1}(\ell) := 0$.*

We will show that for every ℓ there is i with $b_i(\ell) = 0$. In order to prove this, we first establish some natural properties of the base-change operation.

Lemma 3. *Fix $k \geq 2$ and let $c, d \in \mathbb{N}$. Then:*

1. $c \leq c[k\leftarrow k+1]$.
2. *If $c < d$, then $c[k\leftarrow k+1] < d[k\leftarrow k+1]$.*

Proof. The first assertion is proved by induction on c. It clearly holds for $c = 0$. If $c =_{\mathrm{NF}} A_a(k, b) \cdot m + n$ then the induction hypothesis yields $c = A_a(k, b) \cdot m + n \leq A_a(k, b[k\leftarrow k+1]) \cdot m + n[k\leftarrow k+1] = c[k\leftarrow k+1]$.

The second assertion is harder to prove. The proof is by induction on d with a subsidiary induction on c. The assertion is clear if $c = 0$. Let $c =_{\mathrm{NF}} A_a(k, b) \cdot m + n$ and $d =_{\mathrm{NF}} A_{a'}(k, b') \cdot m' + n'$. We distinguish cases according to the position of a relative to a', the position of b relative to b', etc.

CASE 1 $(a < a')$. We sub-divide into two cases.
CASE 1.1 $(A_{a+1}(k, 0) < d)$. Then, the induction hypothesis applied to $c < A_{a+1}(k, 0)$ yields $c[k\leftarrow k+1] < A_{a+1}(k+1, 0) < A_{a'}(k+1, b'[k\leftarrow k+1]) \cdot m' + n'[k\leftarrow k+1] = d[k\leftarrow k+1]$.
CASE 1.2 $(A_{a+1}(k, 0) = d)$. In this case, $a + 1 = a'$, $b' = 0$, $m' = 1$, and $n' = 0$. We have $A_a(k, b) \leq c < A_{a+1}(k, 0) = A_a(k, A_a^{k-1}(k, \cdot)(0))$. For $\ell < k$ we have that $A_a^\ell(k, 0)$ is in k-normal form by Lemma 2. Thus the induction hypothesis yields $b[k\leftarrow k+1] < A_a^{k-1}(k+1, \cdot)(0)$. The number $A_a(k, b)$ is in k-normal form and so the induction hypothesis applied to $n < A_a(k, b)$ yields $n[k\leftarrow k+1] < A_a(k+1, b[k\leftarrow k+1])$. Moreover we have that $m < A_{a+1}(k, 0)$. This yields

$$c[k\leftarrow k+1] = A_a(k+1, b[k\leftarrow k+1]) \cdot m + n[k\leftarrow k+1]$$
$$\leq A_a(k+1, A_a^{k-1}(k+1, \cdot)(0)) \cdot A_{a+1}(k, 0) + A_a(k+1, A_a^{k-1}(k+1, \cdot)(0))$$
$$\leq (A_a^k(k+1, \cdot)(0))^2 + A_a^k(k+1, \cdot)(0)$$
$$\leq A_a(k+1, A_a^k(k+1, \cdot)(0)) = A_{a+1}(k+1, 0),$$

where the second inequality follows from

$$A_{a+1}(k, 0) = A_a^k(k, \cdot)(0) \leq A_a^k(k+1, \cdot)(0)$$

and the last from

$$A_a(k+1, x) \geq A_0(k+1, x) \geq 3^x \geq x^2 + x. \tag{1}$$

CASE 2 ($a' < a$). This case does not occur since then $d < A_{a'+1}(k,0) \leq A_a(k,0) \leq c$.

CASE 3 ($a = a'$ and $b < b'$). The induction hypothesis yields $b[k \leftarrow k+1] < b'[k \leftarrow k+1]$ and $n[k \leftarrow k+1] < A_a(k+1, b[k \leftarrow k+1])$. Now, consider two sub-cases.

CASE 3.1 ($A_a(k, b+1) < d$). Since d is in k-normal form and $b+1 \leq b'$ we see that $A_a(k, b+1)$ is in k-normal form by Lemma 2. Then, the induction hypothesis yields $c[k \leftarrow k+1] < A_a(k+1, (b+1)[k \leftarrow k+1]) \leq A_a(k+1, b'[k \leftarrow k+1]) \leq d[k \leftarrow k+1]$.

CASE 3.2 ($A_a(k, b+1) = d$). We know that $c = A_a(k,b) \cdot m + n < A_a(k, b+1) = d$. Consider two further sub-cases.

CASE 3.2.1 ($a = 0$). This means that $c = k^b \cdot m + n < k^{b+1} = d$, $m < k$, and $n < k^b$, where d has k-normal form k^{b+1}. The induction hypothesis yields $b[k \leftarrow k+1] < (b+1)[k \leftarrow k+1]$ and $n[k \leftarrow k+1] < (k+1)^{b[k \leftarrow k+1]}$. We then have that $c[k \leftarrow k+1] = (k+1)^{b[k \leftarrow k+1]} \cdot m + n[k \leftarrow k+1] < (k+1)^{b[k \leftarrow k+1]+1} \leq (k+1)^{(b+1)[k \leftarrow k+1]} = d$.

CASE 3.2.2 ($a > 0$). Then,

$$
\begin{aligned}
c[k \leftarrow k+1] &= A_a(k+1, b[k \leftarrow k+1]) \cdot m + n[k \leftarrow k+1] \\
&\leq A_a(k+1, b[k \leftarrow k+1])) \cdot A_a(k, b+1) + A_a(k+1, b[k \leftarrow k+1]) \\
&\leq (A_{a-1}^k(k+1, \cdot)(A_a(k+1, b[k \leftarrow k+1])))^2 \\
&\quad + A_{a-1}^k(k+1, \cdot)(A_a(k+1, b[k \leftarrow k+1])) \\
&< A_a(k+1, b'[k \leftarrow k+1]) \text{ by (1)},
\end{aligned}
$$

where the second inequality uses

$$
A_a(k, b+1) = A_{a-1}^k(k, \cdot)(A_a(k,b)) \leq A_{a-1}^k(k+1, \cdot)(A_a(k+1, b[k \leftarrow k+1])).
$$

CASE 4 ($a = a'$ and $b' < b$). This case does not appear since otherwise $d \leq A_a(k, b'+1) \leq c$.

CASE 5 ($a = a'$ and $b' = b$ and $m < m'$). Then the induction hypothesis yields

$$
\begin{aligned}
c[k \leftarrow k+1] &= A_a(k+1, b[k \leftarrow k+1]) \cdot m + n[k \leftarrow k+1] \\
&< A_a(k+1, b[k \leftarrow k+1])) \cdot m + A_a(k+1, b[k \leftarrow k+1]) \\
&\leq A_a(k+1, b[k \leftarrow k+1])) \cdot m' \leq d[k \leftarrow k+1].
\end{aligned}
$$

CASE 6 ($a = a'$ and $b' = b$ and $m' < m$). This case is not possible given the assumptions.

CASE 7 ($a = a'$ and $b' = b$ and $m' = m$). Then $n < n'$ and the induction hypothesis yields

$$
\begin{aligned}
c[k \leftarrow k+1] &= A_a(k+1, b[k \leftarrow k+1]) \cdot m + n[k \leftarrow k+1] \\
&< A_a(k+1, b[k \leftarrow k+1]) \cdot m + n'[k \leftarrow k+1] = d[k \leftarrow k+1].
\end{aligned}
$$

Thus, the base-change operation is monotone. Next we see that it also preserves normal forms.

Lemma 4. *If $c = A_a(k, b) \cdot m + n$ is in k-normal form, then $c[k \leftarrow k + 1] = A_a(k + 1, b[k \leftarrow k + 1]) \cdot m + n[k \leftarrow k + 1]$ is in $k + 1$ normal form.*

Proof. Assume that $c =_{\text{NF}} A_a(k, b) \cdot m + n$. Then, $c < A_{a+1}(k, 0)$, $c < A_a(k, b+1)$, and $n < A_a(k, b)$. Clearly, $A_a(k + 1, 0) \le c[k \leftarrow k + 1]$. By Lemma 2, $A_{a+1}(k, 0)$ is in k-normal form, so that by Lemma 3, $c < A_{a+1}(k, 0)$ yields $c[k \leftarrow k + 1] < A_{a+1}(k+1, 0)$. Since $A_a(k, b)$ is in k-normal form, Lemma 3 yields $n[k \leftarrow k + 1] < A_a(k + 1, b[k \leftarrow k + 1])$. It remains to check that we also have $c[k \leftarrow k + 1] < A_a(k + 1, b[k \leftarrow k + 1] + 1)$.

If $a = 0$, then $c =_{\text{NF}} A_a(k, b) \cdot m + n$ means that $c = k^b \cdot m + n$ with $m < k$ and $n < k^b$. Then, $m < k + 1$ and $n[k \leftarrow k + 1] < (k + 1)^{b[k \leftarrow k+1]}$. Thus $c[k \leftarrow k + 1] = (k + 1)^{b[k \leftarrow k+1]} \cdot m + n[k \leftarrow k + 1] < (k + 1)^{b[k \leftarrow k+1]+1}$ and thus $c[k \leftarrow k + 1] =_{\text{NF}} (k + 1)^{b[k \leftarrow k+1]} \cdot m + n[k \leftarrow k + 1]$. In the remaining case, we have for $a > 0$ that

$$
\begin{aligned}
c[k \leftarrow k + 1] &= A_a(k + 1, b[k \leftarrow k + 1]) \cdot m + n[k \leftarrow k + 1] \\
&< A_a(k + 1, b[k \leftarrow k + 1]) \cdot A_a(k, b + 1) + A_a(k + 1, b[k \leftarrow k + 1]) \\
&\le A_a(k + 1, b[k \leftarrow k + 1]) \cdot A_a(k, b[k \leftarrow k + 1] + 1) + A_a(k + 1, b[k \leftarrow k + 1]) \\
&\le (A_{a-1}^k(k, \cdot) A_a(k + 1, b[k \leftarrow k + 1]))^2 + A_{a-1}^k(k, \cdot) A_a(k + 1, b[k \leftarrow k + 1]) \\
&< A_{a-1}^{k+1}(k + 1, \cdot) A_a(k + 1, b[k \leftarrow k + 1]) \text{ by (1)} \\
&= A_a(k + 1, b[k \leftarrow k + 1] + 1).
\end{aligned}
$$

So $A_a(k + 1, b[k \leftarrow k + 1]) \cdot m + n[k \leftarrow k + 1]$ is in $k + 1$-normal form.

These Ackermannian normal forms give rise to a new Goodstein process. In order to prove that this process is terminating, we must assign ordinals to natural numbers, in such a way that the process gives rise to a decreasing (hence finite) sequence. For each k, we define a function $\psi_k \colon \mathbb{N} \to \Lambda$, where Λ is a suitable ordinal, in such a way that $\psi_k m$ is computed from the k-normal form of m. Unnested Ackermannian normal forms correspond to ordinals below $\Lambda = \varepsilon_\omega$, as the following map shows.

Definition 4. *For $k \ge 2$, define $\psi_k \colon \mathbb{N} \to \varepsilon_\omega$ as follows:*

1. $\psi_k 0 := 0$.
2. $\psi_k c := \omega^{\varepsilon_a + \psi_k b} \cdot m + \psi_k n$ if $c =_{\text{NF}} A_a(k, b) \cdot m + n$.

Lemma 5. *If $c < d < \omega$ then $\psi_k c < \psi_k d$.*

Proof. Proof by induction on d with subsidiary induction on c. The assertion is clear if $c = 0$. Let $c =_{\text{NF}} A_a(k, b) \cdot m + n$ and $d =_{\text{NF}} A_{a'}(k, b') \cdot m' + n'$. We distinguish cases according to the position of a relative to a', the position of b relative to b', etc.

CASE 1 $(a < a')$. We have $n < c < A_{a+1}(k, 0) \le A_{a'}(k, 0)$ and, since $A_{a'}(k, 0) \le d$, the induction hypothesis yields $\psi_k n < \omega^{\varepsilon_{a'} + \psi_k 0} = \varepsilon_{a'}$. We have $b < c < A_{a+1}(k, 0) \le A_{a'}(k, 0)$ and the induction hypothesis yields $\psi_k b < \omega^{\varepsilon_{a'} + \psi_k 0} = \varepsilon_{a'}$. It follows that $\varepsilon_a + \psi_k b < \varepsilon_{a'}$, hence $\psi_k c = \omega^{\varepsilon_a + \psi_k b} \cdot m + \psi_k n < \varepsilon_{a'} \le \psi_k d$.

CASE 2 ($a > a'$). This case is not possible since this would imply that $d < A_{a'+1}(k,0) \leq A_a(k,0) \leq c < d$.

CASE 3 ($a = a'$). We consider several sub-cases.

CASE 3.1 ($b < b'$). The induction hypothesis yields $\psi_k b < \psi_k b'$. Hence $\omega^{\varepsilon_a + \psi_k b} < \omega^{\varepsilon_a + \psi_k b'}$. We have $n < A_a(k,b)$, and the subsidiary induction hypothesis yields $\psi_k n < \omega^{\varepsilon_a + \psi_k b} < \omega^{\varepsilon_a + \psi_k b'}$. Putting things together we see $\psi_k c = \omega^{\varepsilon_a + \psi_k b} \cdot m + \psi_k n < \omega^{\varepsilon_a + \psi_k b'} \leq \psi_k d$.

CASE 3.2 ($b > b'$). This case is not possible since this would imply $d < A_a(k, b' + 1) \leq A_a(k,b) \leq c < d$.

CASE 3.3 ($b = b'$). This case is divided into further sub-cases.

CASE 3.3.1 ($m < m'$). We have $n < A_a(k,b)$ and the subsidiary induction hypothesis yields $\psi_k n < \omega^{\varepsilon_a + \psi_k b}$. Hence $\psi_k c = \omega^{\varepsilon_a + \psi_k b} \cdot m + \psi_k n < \omega^{\varepsilon_a + \psi_k b} \cdot m' \leq \psi_k d$.

CASE 3.3.2 ($m > m'$). This case is not possible since this would imply $d = A_a(k,b) \cdot m' + n' \leq A_a(k,b) \cdot m \leq c < d$.

CASE 3.3.3 ($m = m'$). The inequality $c < d$ yields $n < n'$ and the induction hypothesis yields $\psi_k n < \psi_k n'$. Hence $\psi_k c = \omega^{\varepsilon_a + \psi_k b} \cdot m + \psi_k n < \omega^{\varepsilon_a + \psi_k b} \cdot m + \psi_k n' = \psi_k d$.

Our ordinal assignment is invariant under base change, in the following sense.

Lemma 6. $\psi_{k+1}(c[k{\leftarrow}k+1]) = \psi_k c$.

Proof. Proof by induction on c. The assertion is clear for $c = 0$. Let $c =_{\mathrm{NF}} A_a(k,b) \cdot m + n$. Then, $c[k{\leftarrow}k+1] =_{\mathrm{NF}} A_a(k+1, b[k{\leftarrow}k+1]) \cdot m + n[k{\leftarrow}k+1]$, and the induction hypothesis yields

$$\psi_{k+1}(c[k{\leftarrow}k+1]) = \psi_{k+1}(A_a(k+1, b[k{\leftarrow}k+1]) \cdot m + n[k{\leftarrow}k+1])$$
$$= \omega^{\varepsilon_a + \psi_{k+1}(b[k{\leftarrow}k+1])} \cdot m + \psi_{k+1}(n[k{\leftarrow}k+1])$$
$$= \omega^{\varepsilon_a + \psi_k b} \cdot m + \psi_k n = \psi_k c.$$

It is well-known that the so-called slow-growing hierarchy at level $\varphi_\omega 0$ matches up with the Ackermann function, so one might expect that the corresponding Goodstein process can be proved terminating in $\mathsf{PA} + \mathrm{TI}(\varphi_\omega 0)$. This is true but, somewhat surprisingly, much less is needed here. We can lower $\varphi_\omega 0$ to $\varepsilon_\omega = \varphi_1 \omega$.

Theorem 1. *For all $\ell < \omega$, there exists a $k < \omega$ such that $b_k(\ell) = 0$. This is provable in $\mathsf{PA} + \mathrm{TI}(\varepsilon_\omega)$.*

Proof. Define $o(\ell, k) := \psi_{k+2} b_k(\ell)$. If $b_k(\ell) > 0$, then, by the previous lemmata,

$$o(\ell, k+1) = \psi_{k+3} b_{k+1}(\ell) = \psi_{k+3}(b_k(\ell)[k{\leftarrow}k+1] - 1)$$
$$< \psi_{k+3}(b_k(\ell)[k{\leftarrow}k+1]) = \psi_{k+2}(b_k(\ell)) = o(\ell, k).$$

Since $(o(\ell, k))_{k<\omega}$ cannot be an infinite decreasing sequence of ordinals, there must be some k with $o(\ell, k) = 0$, yielding $b_k(\ell) = 0$.

Now we are going to show that for every $\alpha < \varepsilon_\omega$, $\mathsf{PA}+\mathsf{TI}(\alpha) \nvdash \forall \ell \exists k \, b_k(\ell) = 0$. This will require some work with fundamental sequences.

Definition 1. *Let Λ be an ordinal. A* system of fundamental sequences *on Λ is a function $\cdot[\cdot]\colon \Lambda \times \mathbb{N} \to \Lambda$ such that $\alpha[n] \leq \alpha$ with equality holding if and only if $\alpha = 0$, and $\alpha[n] \leq \alpha[m]$ whenever $n \leq m$. The system of fundamental sequences is* convergent *if $\lambda = \lim_{n \to \infty} \lambda[n]$ whenever λ is a limit, and has the* Bachmann property *if whenever $\alpha[n] < \beta < \alpha$, it follows that $\alpha[n] \leq \beta[1]$.*

It is clear that if Λ is an ordinal then for every $\alpha < \Lambda$ there is n such that $\alpha[1][2]\ldots[n] = 0$, but this fact is not always provable in weak theories. The Bachmann property that will be useful due to the following.

Proposition 1. *Let Λ be an ordinal with a system of fundamental sequences satisfying the Bachmann property, and let $(\xi_n)_{n \in \mathbb{N}}$ be a sequence of elements of Λ such that, for all n, $\xi_n[n+1] \leq \xi_{n+1} \leq \xi_n$. Then, for all n, $\xi_n \geq \xi_0[1][2]\ldots[n]$.*

Proof. Let $\preceq_k$ be the reflexive transitive closure of $\{(\alpha[k], \alpha) : \alpha < \varphi_2(0)\}$. We need a few properties of these orderings. Clearly, if $\alpha \preceq_k \beta$, then $\alpha \leq \beta$. It can be checked by a simple induction and the Bachmann property that, if $\alpha[n] \leq \beta < \alpha$, then $\alpha[n] \preceq_1 \beta$. Moreover, $\preceq_k$ is monotone in the sense that if $\alpha \preceq_k \beta$, then $\alpha \preceq_{k+1} \beta$, and if $\alpha \preceq_k \beta$, then $\alpha[k] \preceq_k \beta[k]$ (see, e.g., [11] for details).

We claim that for all n, $\xi_n \succeq_n \xi_0[1]\ldots[n]$, from which the desired inequality immediately follows. For the base case, we use the fact that $\succeq_0$ is transitive by definition. For the successor, note that the induction hypothesis yields $\xi_0[1]\ldots[n] \preceq_n \xi_n$, hence $\xi_0[1]\ldots[n+1] \preceq_{n+1} \xi_n[n+1]$. Then, consider three cases.

CASE 1 ($\xi_{n+1} = \xi_n$). By transitivity and monotonicity, $\xi_0[1]\ldots[n+1] \preceq_{n+1} \xi_0[1]\ldots[n] \preceq_n \xi_n = \xi_{n+1}$ yields $\xi_0[1]\ldots[n+1] \preceq_{n+1} \xi_{n+1}$.
CASE 2 ($\xi_{n+1} = \xi_n[n+1]$). Then, $\xi_0[1]\ldots[n+1] \preceq_{n+1} \xi_n[n+1] = \xi_{n+1}$.
CASE 3 ($\xi_n[n+1] < \xi_{n+1} < \xi_n$). The Bachmann property yields $\xi_n[n+1] \preceq_1 \xi_{n+1}$, and since $\xi_0[1]\ldots[n+1] \preceq_{n+1} \xi_n[n+1]$, monotinicity and transitivity yield $\xi_0[1]\ldots[n+1] \preceq_{n+1} \xi_{n+1}$.

Let $\omega_0(\alpha) := \alpha$ and $\omega_{k+1}(\alpha) = \omega^{\omega_k(\alpha)}$. Let us define the standard fundamental sequences for ordinals less than $\varphi_2 0$ as follows.

1. If $\alpha = \omega^\beta + \gamma$ with $0 < \gamma < \alpha$, then $\alpha[k] := \omega^\beta + \gamma[k]$.
2. If $\alpha = \omega^\beta > \beta$, then we set $\alpha[k] := 0$ if $\beta = 0$, $\alpha[k] := \omega^\gamma \cdot k$ if $\beta = \gamma + 1$, and $\alpha[k] := \omega^{\beta[k]}$ if $\beta \in \mathrm{Lim}$.
3. If $\alpha = \varepsilon_\beta > \beta$, then $\alpha[k] := \omega_k(1)$ if $\beta = 0$, $\alpha[k] := \omega_k(\varepsilon_\gamma + 1)$ if $\beta = \gamma + 1$, and $\alpha[k] := \varepsilon_{\beta[k]}$ if $\beta \in \mathrm{Lim}$.

This system of fundamental sequences enjoys the Bachmann property [11].

In view of Proposition 1, the following technical lemma will be crucial for proving our main independence result for ACA_0'.

Lemma 7. *Given $k, c < \omega$ with $k \geq 2$, $\psi_{k+1}(c[k\leftarrow k+1] - 1) \geq (\psi_k c)[k]$.*

Proof. We prove the claim by induction on c. Let $c =_{\mathrm{NF}} A_a(k,b) \cdot m + n$.

CASE 1 $(n > 0)$. Then the induction hypothesis and Lemma 5 yield

$$\psi_{k+1}(c[k{\leftarrow}k+1] - 1) = \omega^{\varepsilon_a + \psi_{k+1}(b[k{\leftarrow}k+1])} \cdot m + \psi_{k+1}(n[k{\leftarrow}k+1] - 1)$$
$$\geq \omega^{\varepsilon_a + \psi_k(b)} \cdot m + (\psi_k(n))[k] = (\omega^{\varepsilon_a + \psi_k(b)} \cdot m + \psi_k(n))[k]$$
$$= (\psi_k(A_a(k,b) \cdot m + n))[k] = (\psi_k c)[k].$$

CASE 2 $(n = 0$ and $m > 1)$. Then the induction hypothesis and Lemma 5 yield

$$\psi_{k+1}(c[k{\leftarrow}k+1] - 1)$$
$$= \psi_{k+1}(A_a(k+1, b[k{\leftarrow}k+1])) \cdot (m - 1) + \psi_{k+1}(A_a(k+1, b[k{\leftarrow}k+1])) - 1)$$
$$\geq \psi_k(A_a(k,b) \cdot (m - 1)) + (\psi_k(A_a(k,b)))[k] = (\psi_k(A_a(k,b) \cdot m))[k] = (\psi_k c)[k].$$

CASE 3 $(n = 0$ and $m = 1)$. We consider several sub-cases.
CASE 3.1 $(a > 0$ and $b > 0)$. The induction hypothesis yields

$$\psi_{k+1}(c[k{\leftarrow}k+1] - 1) = \psi_{k+1}(A_a(k+1, b[k{\leftarrow}k+1]) - 1)$$
$$\geq \psi_{k+1}(A_a(k+1, (b[k{\leftarrow}k+1]) - 1) \cdot k) = \omega^{\varepsilon_a + \psi_{k+1}(b[k{\leftarrow}k+1]-1)} \cdot k$$
$$\geq \omega^{\varepsilon_a + (\psi_k(b))[k]} \cdot k \geq (\omega^{\varepsilon_a + \psi_k(b)})[k] = (\psi_k c)[k],$$

since $A_a(k+1, (b[k{\leftarrow}k+1]) - 1) \cdot k$ is in $k + 1$ normal form by Lemma 2 and Lemma 4.
CASE 3.2 $(a > 0$ and $b = 0)$. Then, the induction hypothesis yields

$$\psi_{k+1}(c[k{\leftarrow}k+1] - 1) = \psi_{k+1}(A_a(k+1, 0) - 1) = \psi_{k+1}(A_{a-1}^{k+1}(k, \cdot)(0) - 1)$$
$$= \psi_{k+1}(A_{a-1}(k+1, A_{a-1}^k(k+1, \cdot)(0) - 1))$$
$$\geq \psi_{k+1}(A_{a-1}^k(k+1, \cdot)(0)) = \omega^{\varepsilon_{a-1} + \psi_{k+1}((A_{a-1}^{k-1}(k+1, \cdot)(0)))}$$
$$\geq \omega^{\psi_{k+1}((A_{a-1}^{k-1}(k+1, \cdot)(0)))} \geq \omega^{\omega_{k-1}(\varepsilon_{a-1}+1)}$$
$$= (\varepsilon_a)[k] = (\psi_k(A_a(k,0)))[k] = (\psi_k c)[k],$$

since $A_{a-1}^\ell(k+1, \cdot)(0)$ is in $k+1$ normal form for $\ell \leq k$ by Lemma 2 and Lemma 4.
CASE 3.3 $(a = 0$ and $b > 0)$. Then the induction hypothesis yields similarly as in CASE 3.1:

$$\psi_{k+1}(c[k{\leftarrow}k+1] - 1) = \psi_{k+1}(A_0(k+1, b) - 1)$$
$$= \psi_{k+1}((k+1)^{(b[k{\leftarrow}k+1]-1)} \cdot k + \cdots + (k+1)^0 \cdot k)$$
$$\geq \psi_{k+1}((k+1)^{(b[k{\leftarrow}k+1]-1)} \cdot k)$$
$$\geq \omega^{\psi_{k+1}(b[k{\leftarrow}k+1]-1)} \cdot k \geq \omega^{(\psi_k b)[k]} \cdot k \geq (\psi_k c)[k],$$

since $(k+1)^{(b[k{\leftarrow}k+1]-1)} \cdot k$ is in $k + 1$ normal form.
CASE 3.4 $(a = 0$ and $b = 0)$. The assertion follows trivially since then $c = 1$.

Theorem 2. *Let* $\alpha < \varepsilon_\omega$. *Then* $\mathsf{PA} + \mathsf{TI}(\alpha) \nvdash \forall \ell \exists k\ b_k(\ell) = 0$. *Hence* $\mathsf{ACA}_0' \nvdash \forall \ell \exists k b_k(\ell) = 0$.

Proof. Assume for a contradiction that $\mathsf{PA} + \mathsf{TI}(\alpha) \vdash \forall \ell \exists k\ b_k(\ell) = 0$. Then $\mathsf{PA} + \mathsf{TI}(\alpha) \vdash \forall \ell \exists k\ b_k(A_\ell(2,0)) = 0$. Recall that $o(A_\ell(2,0),k) = \psi_{k+2}(b_k(A_\ell(2,0)))$. We have $o(A_\ell(2,0),0) = \varepsilon_n$. Lemma 7 and Lemma 5 yield $o(A_\ell(2,0),k)[k+1] \leq o(A_\ell(2,0),k+1) < o(A_\ell(2,0),k)$, hence Proposition 1 yields $o(A_\ell(2,0),k) \geq o(A_\ell(2,0))[1] \ldots [k]$. So the least k such that $b_k(A_\ell(2,0)) = 0$ is at least as big as the least k such that $\varepsilon_\ell[1] \ldots [k] = 0$. But by standard results in proof theory [3], $\mathsf{PA} + \mathsf{TI}(\alpha)$ does not prove that this k is always defined as a function of ℓ. This contradicts $\mathsf{PA} + \mathsf{TI}(\alpha) \vdash \forall \ell \exists k\ b_k(A_\ell(2,0))) = 0$.

4   Goodstein Sequences for ACA_0^+

In this section, we indicate how to extend our approach to a situation where the base change operation can also be applied to the first argument of the Ackermann function. The resulting Goodstein principle will then be independent of ACA_0^+. The key difference is that the base-change operation is now performed recursively on the first argument, as well as the second.

Definition 5. *For* $k \geq 2$ *and* $c \in \mathbb{N}$, *define* $c[k \leftarrow k+1]$ *by:*

1. $0[k \leftarrow k+1] := 0$
2. $c[k \leftarrow k+1] := A_{a[k \leftarrow k+1]}(k+1, b[k \leftarrow k+1]) \cdot m + n[k \leftarrow k+1]$ *if* $c =_{\mathrm{NF}} A_a(k,b) \cdot m + n$.

Note that in this section, $c[k \leftarrow k+1]$ will always indicate the operation of Definition 5. We can then define a Goodstein process based on this new base change operator.

Definition 6. *Let* $\ell < \omega$. *Put* $c_0(\ell) := \ell$. *Assume recursively that* $c_k(\ell)$ *is defined and* $c_k(\ell) > 0$. *Then,* $c_{k+1}(\ell) = c_k(\ell)[k + 2 \leftarrow k + 3] - 1$. *If* $c_k(\ell) = 0$, *then* $c_{k+1}(\ell) := 0$.

Termination and independence results can then be obtained following the same general strategy as before. We begin with the following lemmas, whose proofs are similar to those for their analogues in Sect. 3.

Lemma 8. *If* $c < d$ *and* $k \geq 2$, *then* $c[k \leftarrow k+1] < d[k \leftarrow k+1]$.

Lemma 9. *If* $c = A_a(k,b) \cdot m + n$ *is in k-normal form, then* $c[k \leftarrow k+1] = A_{a[k \leftarrow k+1]}(k+1, b[k \leftarrow k+1]) \cdot m + n[k \leftarrow k+1]$ *is in $k+1$ normal form.*

It is well-known that the so-called slow-growing hierarchy at level Γ_0 matches up with the functions which are elementary in the Ackermann function, so one might expect that the corresponding Goodstein process can be proved terminating in $\mathsf{PA} + \mathsf{TI}(\Gamma_0)$. This is true but, somewhat surprisingly, much less is needed here. Indeed, nested Ackermannian normal forms are related to the much smaller ordinal $\varphi_2(0)$ by the following mapping.

Definition 7. *Given $k \geq 2$, define a function $\chi_k \colon \mathbb{N} \to \varphi_2(0)$ given by:*

1. $\chi_k 0 := 0$.
2. $\chi_k c := \omega^{\varepsilon_{\chi_k a} + \chi_k b} \cdot m + \psi_k n$ *if* $c =_{\mathrm{NF}} A_a(k, b) \cdot m + n$.

As was the case for the mappings ψ_k, the maps χ_k are strictly increasing and invariant under base change, as can be checked using analogous proofs to those in Sect. 3.

Lemma 10. *Let $c, d, k < \omega$ with $k \geq 2$.*

1. *If $c < d$, then $\chi_k c < \chi_k d$.*
2. $\chi_{k+1}(c[k \leftarrow k + 1]) = \chi_k c$.

Theorem 3. *For all $\ell < \omega$, there exists a $k < \omega$ such that $c_k(\ell) = 0$. This is provable in $\mathsf{PA} + \mathrm{TI}(\varphi_2 0)$.*

Next, we show that for every $\alpha < \varphi_2 0$, $\mathsf{PA} + \mathrm{TI}(\alpha) \not\vdash \forall \ell \exists k\ c_k(\ell) = 0$. For this, we need the following analogue of Lemma 7.

Lemma 11. $\chi_{k+1}(c[k \leftarrow k + 1] - 1) \geq (\chi_k c)[k]$.

Proof. We proceed by induction on c. Let $c =_{\mathrm{NF}} A_a(k, b) \cdot m + n$. Let us concentrate on the critical case $m = 1$ and $n = 0$, where $a > 0$ and $b = 0$.

The induction hypothesis yields

$$\chi_{k+1}(c[k \leftarrow k + 1] - 1) = \chi_{k+1}(A_a(k + 1, 0) - 1)$$
$$= \chi_{k+1}(A_{a[k \leftarrow k+1]-1}^{k+1}(k + 1, \cdot)(0) - 1) \geq \chi_{k+1}(A_{a[k \leftarrow k+1]-1}^{k}(k + 1, \cdot)(0))$$
$$= \omega^{\varepsilon_{\chi_{k+1}(a[k \leftarrow k+1]-1)} + \omega^{\chi_{k+1}(A_{a[k \leftarrow k+1]-1}^{k-1}(k+1, \cdot)(0))}} \geq \omega_k(\varepsilon_{\chi_{k+1}(a[k \leftarrow k+1]-1)} + 1)$$
$$\geq \omega_k(\varepsilon_{(\chi_k a)[k]} + 1) \geq (\varepsilon_{\chi_k a})[k] = (\chi_k(A_a(k, 0)))[k],$$

since $A_{a[k \leftarrow k+1]-1}^{k}(k + 1, \cdot)(0)$ is in $k + 1$ normal form.

The remaining details of the proof of the theorem can be carried out similarly as before.

Theorem 4. *For every $\alpha < \varphi_2 0$, $\mathsf{PA} + \mathrm{TI}(\alpha) \not\vdash \forall \ell \exists k\ c_k(\ell) = 0$. Hence $\mathsf{ACA}_0^+ \not\vdash \forall \ell \exists k c_k(\ell) = 0$.*

References

1. Afshari, B., Rathjen, M.: Ordinal analysis and the infinite Ramsey theorem. In: Cooper, S.B., Dawar, A., Löwe, B. (eds.) CiE 2012. LNCS, vol. 7318, pp. 1–10. Springer, Heidelberg (2012). https://doi.org/10.1007/978-3-642-30870-3_1
2. Arai, T., Fernández-Duque, D., Wainer, S., Weiermann, A.: Predicatively unprovable termination of the Ackermannian Goodstein process. In: Proceedings of the AMS (2019). https://doi.org/10.1090/proc/14813

3. Cichon, E., Buchholz, W., Weiermann, A.: A uniform approach to fundamental sequences and hierarchies. Math. Logic Q. **40**, 273–286 (1994)
4. Gödel, K.: Über Formal Unentscheidbare Sätze der Principia Mathematica und Verwandter Systeme. I. Monatshefte für Mathematik und Physik **38**, 173–198 (1931). https://doi.org/10.1007/BF01700692
5. Goodstein, R.: On the restricted ordinal theorem. J. Symb. Logic **9**(2), 33–41 (1944)
6. Goodstein, R.: Transfinite ordinals in recursive number theory. J. Symb. Logic **12**(4), 123–129 (1947)
7. Graham, R.L., Rothschild, B.L.: Ramsey's theorem for n-dimensional arrays. Bull. Am. Math. Soc. **75**(2), 418–422 (1969)
8. Kirby, L., Paris, J.: Accessible independence results for Peano arithmetic. Bull. Lond. Math. Soc. **14**(4), 285–293 (1982)
9. Leigh, G., Rathjen, M.: An ordinal analysis for theories of self-referential truth. Arch. Math. Log. **49**(2), 213–247 (2010). https://doi.org/10.1007/s00153-009-0170-2
10. Pohlers, W.: Proof Theory, The First Step into Impredicativity. Springer, Heidelberg (2009). https://doi.org/10.1007/978-3-540-69319-2
11. Schmidt, D.: Built-up systems of fundamental sequences and hierarchies of number-theoretic functions. Arch. Math. Log. **18**(1), 47–53 (1977)
12. Schütte, K.: Proof Theory. Springer, Heidelberg (1977). https://doi.org/10.1007/978-3-642-66473-1
13. Simpson, S.: Subsystems of Second Order Arithmetic. Cambridge University Press, New York (2009)
14. Weiermann, A.: Ackermannian Goodstein principles for first order Peano arithmetic. In: Sets and computations. Lecture Notes Series, Institute for Mathematical Sciences, National University of Singapore, vol. 33, pp. 157–181. World Scientific, Hackensack (2018). English summary

On the Complexity of Validity Degrees
in Łukasiewicz Logic

Zuzana Haniková[(✉)] [ID]

Institute of Computer Science of the Czech Academy of Sciences,
Prague, Czechia
hanikova@cs.cas.cz

Abstract. Łukasiewicz logic is an established formal system of many-valued logic. Decision problems in both propositional and first-order case have been classified as to their computational complexity or degrees of undecidability; for the propositional fragment, theoremhood and provability from finite theories are coNP complete. This paper extends the range of results by looking at validity degree in propositional Łukasiewicz logic, a natural optimization problem to find the minimal value of a term under a finite theory in a fixed complete semantics interpreting the logic. A classification for this problem is provided using the oracle class $\mathrm{FP}^{\mathrm{NP}}$, where it is shown complete under metric reductions.

1 Introduction

Łukasiewicz logic originated in the 1920s as a semantically motivated formal system for many-valued logic. This paper works with the *infinite-valued* Łukasiewicz logic Ł, introduced by Łukasiewicz and Tarski [20]. As with some other non-classical systems, such as intuitionistic logic, the syntax is similar to classical logic, while the valid inferences form a strict subset of those of classical logic.

Validity/provability degrees as a concept in Łukasiewicz logic stem from a research line proposed by Goguen [11]. The paper set the challenge to develop a formal approach allowing to derive *partly true* conclusions from *partly true* assumptions. In [26] the task was taken up by Pavelka, who offered a comprehensive formalism based on complete residuated lattices, using essentially diagrams of arbitrary but fixed residuated lattices to capture provability degrees in the syntax. Pavelka used *graded* terms[1] and his formal system incorporated rules that explicitly use the algebra on degrees/grades alongside syntactic derivations. For example, a *graded modus ponens* reads $\{\langle r, \varphi \rangle, \langle s, \varphi \to \psi \rangle\}/\langle r \odot s, \psi \rangle$ with r and s truth constants, φ and ψ terms, and $\odot$ the monoidal operation of the residuated lattice. Pavelka's approach was later simplified by Hájek [12], who proposed an expansion of Łukasiewicz infinite-valued logic with constants for *rational elements* of $[0, 1]$, and rendered each graded term $\langle r, \varphi \rangle$ as the implication $r \to \varphi$. This was an elegant example of embedding the graded syntax

---

[1] We use *term* and *(propositional) formula* interchangeably in this paper.

M. Anselmo et al. (Eds.): CiE 2020, LNCS 12098, pp. 175–188, 2020.
https://doi.org/10.1007/978-3-030-51466-2_15

approach in what turns out to be a conservative expansion of Łukasiewicz logic. The resulting logic was named Rational Pavelka logic (RPL); see [12, 14, 4, 7].

Assume truth values range in a complete lattice. The *validity degree* of a term φ under a theory T is the infimum of values φ can get under assignments that make T true. No constants are needed to define this notion. Still, the constants provide a canonical way of introducing provability degrees, the syntactic counterpart; thus we look at the language of RPL next to that of Ł.

Both Ł and RPL have an equivalent algebraic semantics (in the sense of [5]). In particular, Ł corresponds to the variety of MV-algebras; [6, 9, 24] and the references therein provide resources for their well-developed theory. MV-algebras are strongly linked to Abelian ℓ-groups ([22]); this is manifest in the choice of algebraic language, and we follow MV-algebraists and use the language $\oplus$ and $\neg$ as a reference language for our complexity results. This is also a matter of convenience since some previous results are framed in this language.

We shall use the real-valued (*standard*) MV-semantics, with the unit interval as the domain and piecewise linear functions as interpretations of the function symbols; one can prove strong completeness for finite theories over Ł w.r.t. this algebra. The algebra has been useful for obtaining complexity results for Ł, since Mundici's pioneering NP completeness result on its SAT problem [23], which also gives coNP completeness for theoremhood in Ł. Other complexity results for propositional logic Ł include [1, 2] reducing the decision problems in Ł to the setting of finite MV-chains, [17, 18] dealing with admissible rules, [25], [3], or [8]. All these works target decision problems.

The validity degree task (to determine the validity degree of a term φ under a finite theory T) is a natural optimization problem induced by the many-valued setting and the purpose of this paper is to see where it sits among other optimization problems. Using tools of complexity theory, we classify the validity degree task in propositional Łukasiewicz logic Ł and its extension RPL, for instances that pair a finite theory T with a term φ. Our emphasis is on Ł rather than RPL: it is far better known, and the existing algebraic methods for Ł provide us with tools. In fact, the few complexity results available for RPL rely on reductions to Ł. In [12] Hájek proved that for finite theories in RPL, validity degrees are rational; his method inspires ours in eliminating the constants, relying on their implicit definability in Ł. Hájek also provided complexity classification for the decision version of the problem in [13], showing that provability from finite theories in propositional RPL is coNP complete, using mixed integer programming. In [15], the same result is obtained from analogous results for Ł, using the implicit definability of constants directly.

We fill the gap of a basic classification for the optimization problem. Our upper bounds are based on improving Hájek's rationality proof for validity degrees with establishing an explicit polynomial bound on denominator size, relying on Aguzzoli and Ciabattoni's paper [2]. Their paper uses the language of Ł; however, the methods of [12, 15] allow us to tackle the rational constants and to derive analogous upper bounds for RPL, and we do that in Sect. 4; such upper bounds then apply also to any fragments of language, i.a., the MV-language. For lower bounds (Sect. 5), we work with the language of Ł, whereby the hardness result applies also to RPL.

The decision version of the validity degree is coNP complete, and the SAT problem for $[0,1]_Ł$ is NP complete. Looking at these and similar results on NP completeness of decision versions for other common optimization problems, one might ask what would the appropriate (many-one, poly-time) reduction notion be between the optimization versions, and indeed if such reductions always exist. Krentel [19] defines *metric reductions* in response to the former question and shows that as far as these reductions are concerned, the answer to the latter is *negative* unless P = NP (an outline of relevant results is in Sect. 3). Thus there is a sense in which a mere fact that the decision version of a problem is NP complete does not provide enough information about the optimization version.

Under standard complexity assumptions, one cannot even approximate the validity degree efficiently: [16, Theorem 7.4] says that no efficient algorithm can compute the validity degree for an empty theory within a distance of $\delta < 1/2$ unless P = NP.

The combined results of Sects. 4 and 5 yield the following statement.

Theorem 1. *The validity degree task, considered in either Ł or RPL, is complete for the class* $\mathrm{FP}^{\mathrm{NP}}$ *under metric reductions.*

This appears to be the first work to shift the focus from decision to optimization problems as regards complexity of fuzzy logics, identifying a relevant complexity class. We find it compelling to investigate complexity problems for non-classical logics that have no counterpart in classical logic, and the validity degree problem, discussed here for Ł, presents one such research direction. (While, e.g., admissible rules present another, now well established one.)

This work is about the *propositional* fragments of Ł and RPL, so notions such as language, term/formula, or assignment need to be read appropriately.

2 Łukasiewicz Logic and Rational Pavelka Logic

The basic language of propositional Łukasiewicz logic Ł has two function symbols: unary $\neg$ (negation) and binary $\oplus$ (strong disjunction or sum). Other function symbols are definable: 1 as $x \oplus \neg x$ and 0 as $\neg 1$; further, $x \odot y$ is $\neg(\neg x \oplus \neg y)$ (strong conjunction or product); $x \to y$ is $\neg x \oplus y$; $x \equiv y$ is $(x \to y) \odot (y \to x)$; $x \vee y$ is $(x \to y) \to y$ or $(y \to x) \to x$; and $x \wedge y$ is $\neg(\neg x \vee \neg y)$.

The interpretations of $\oplus$, $\odot$, $\wedge$ and $\vee$ are commutative and associative, so one can write, e.g., $x_1 \oplus \cdots \oplus x_n$ without worrying about order and parentheses. We write x^n for $\underbrace{x \odot \cdots \odot x}_{n \text{ times}}$ and nx for $\underbrace{x \oplus \cdots \oplus x}_{n \text{ times}}$. Also, $\vee$ and $\wedge$ distribute over each other and $\odot$ distributes over $\vee$.

Well-formed Ł-terms are defined as usual. The basic language is a point of reference for complexity considerations in this paper, however we may at times use the expanded language for clarity (as in classical logic).

Definition 1. ([2]) *For any term* $\varphi(x_1, \ldots, x_n)$, $\sharp(x)\varphi$ *denotes the number of occurrences of the variable* x *in* φ, *and* $\sharp\varphi = \Sigma_{i=1}^n \sharp(x_i)\varphi$.

The $\sharp$ function is a good notion of *length* for terms without iterated $\neg$ symbols ($\neg\neg\varphi \equiv \varphi$ is a theorem of Ł). Our complexity results apply also to the language of the Full Lambek calculus with exchange and weakening (FL$_{ew}$), i.e., $\{\odot, \rightarrow, \vee, \wedge, 0, 1\}$ (and the MV-symbol $\oplus$). Indeed one observes that rendering $\odot$ and $\rightarrow$ in the basic language does not affect length; for $\vee$ and $\wedge$, any occurrence of these defined symbols can be expanded to the basic language in two different ways (due to commutativity), and this can be used to rewrite any term with these symbols with only polynomial increase in length.

2.1 MV-algebras

The general MV-algebraic semantics will not be needed in this paper, anymore than a formal calculus for Ł. We will work with the *standard* MV-algebra $[0,1]_Ł$: the domain is the real interval $[0,1]$ and with each MV-term $\varphi(x_1, \ldots, x_n)$ we associate a function $f_\varphi : [0,1]^n \longrightarrow [0,1]$, defined by induction on term structure with $f_{\neg\varphi}$ defined as $1 - f_\varphi$, $f_{\varphi\oplus\psi}$ as $\min(1, f_\varphi + f_\psi)$. 1 is the only designated element, accounting for the notions of *truth/validity*. For any assignment v in $[0,1]_Ł$, $v(\varphi \rightarrow \psi) = 1$ iff $v(\varphi) \leq v(\psi)$, and thus $v(\varphi \equiv \psi) = 1$ iff $v(\varphi) = v(\psi)$.

The class of MV-algebras is generated by $[0,1]_Ł$ as a quasivariety; it is also generated by the class of *finite MV-chains*, the $(k+1)$-element MV-chain being the subalgebra of $[0,1]_Ł$ on the domain $\{0, 1/k, \ldots, (k+1)/k, 1\}$.

Provability from finite theories in Ł coincides with the finite consequence relation of $[0,1]_Ł$. We have bypassed introducing the formal calculus; to provide a meaning to the references to Ł within this paper, let us adopt this as a definition. We lose little since the algorithmic approach only tackles finite theories anyway.

A function $f: [0,1]^n \rightarrow [0,1]$ is a *McNaughton function* if it is continuous and *piecewise linear with integer coefficients*: there are finitely many linear polynomials $\{p_i\}_{i \in I}$, with $p_i(\bar{x}) = \Sigma_{j=1}^n a_{ij} x_j + b_i$ and $\bar{a}_i, b_i$ integers for each i, such that for any $\bar{u} \in [0,1]^n$ there is an $i \in I$ with $f(\bar{u}) = p_i(\bar{u})$. McNaughton theorem ([21]) says that term-definable functions of $[0,1]_Ł$ coincide with McNaughton functions. The theorem highlights the fact that one can provide a countably infinite array of pairwise non-equivalent MV-terms for any fixed number of variables starting with one, as opposed to the case of Boolean functions.

A *polyhedral complex* C is a set of polyhedra (*cells*) such that if A is in C, so are all faces of A, and for A, B in C, $A \cap B$ is a common face of A and B. Given an MV-term $\varphi(x_1, \ldots, x_n)$ one can build canonically a polyhedral complex $C(\varphi)$ such that $[0,1]^n = \bigcup C(\varphi)$ and f_φ is linear over each n-dimensional cell of $C(\varphi)$. The minimum of f_φ is attained at a vertex of an n-dimensional cell of $C(\varphi)$. [2] derives the upper bound $(\frac{\sharp\varphi}{n})^n$ for the least common denominator of any vertex of any n-dimensional cell of $C(\varphi)$ (see also [23]). By [1] this is a *tight* bound on cardinality of MV-chains witnessing non-validity of MV-terms.

For any MV-term φ, the *1-region* of f_φ is the union of cells of $C(\varphi)$ such that f_φ attains the value 1 on all points in the cell. (The highest dimension of the cells in the 1-region of φ can range anywhere between 0 and n.) The 1-region of f_φ is compact for any φ. One can investigate the minimum of f_ψ relative to the 1-region of an f_φ; details in [2].

2.2 RMV-algebras

The language of RPL expands the language $\{\oplus, \neg\}$ of Ł with a set $\mathcal{Q} = Q \cap [0,1]$ of constants. The constants are represented as ordered pairs of coprime integers in binary. The size of the binary representation of an integer n is denoted $|n|$.

The standard RMV-algebra $[0,1]_{\mathrm{L}}^{\mathcal{Q}}$ has $[0,1]_{\mathrm{L}}$ as its MV-reduct and interprets rational constants as themselves. As for Ł above, we identify RPL with the finite consequence relation of $[0,1]_{\mathrm{L}}^{\mathcal{Q}}$. If φ is an RMV-term, f_φ is the function defined by φ in $[0,1]_{\mathrm{L}}^{\mathcal{Q}}$.

Let us extend the $\sharp$ function to obtain a good length notion for RMV-terms. Rational constants can be viewed as atoms but the number of atom occurrences is not a suitable length notion since it ignores the space needed to represent each constant, which can be arbitrary with respect to the term structure.

Definition 2. *Let an RMV-term φ have constants $p_1/q_1, \ldots, p_m/q_m$ and variables $x_1, \ldots, x_n$. For a rational $p/q \in [0,1]$, let $\sharp(p/q)\varphi$ denote the number of occurrences of p/q in φ. Define $\sharp\varphi = \Sigma_{i=1}^n \sharp(x_i)\varphi + \Sigma_{j=1}^m \sharp(p_j/q_j)\varphi(|p_j| + |q_j|)$.*

Each rational r in $[0,1]$ is implicitly definable by an MV-term in $[0,1]_{\mathrm{L}}$[2]: i.e., there is an MV-term $\varphi(x_1, \ldots, x_k)$ and an $i \in \{1, \ldots, k\}$ such that, for each assignment v in $[0,1]_{\mathrm{L}}$, we have $v(x_i) = r$ whenever $v(\varphi) = 1$ (cf. [12,15]). To implicitly define a rational p/q, with $1 \leq p \leq q$, in $[0,1]_{\mathrm{L}}$, first define $1/q$, using the one-variable term $z_{1/q} \equiv (\neg z_{1/q})^{q-1}$, whereupon p/q becomes term-definable under a theory containing this definition of $1/q$, namely we have $z_{p/q} \equiv p z_{1/q}$ (cf. the technical results in [28,10,16]). With p and q in binary, these implicit definitions are exponential-size in $|p|$ and $|q|$. One can make them polynomial-size on pain of introducing (a linear number of) new variables.

Lemma 1. *([15, Lemma 4.1]) For $q \in \mathbb{N}$, $q \geq 2$, take the binary representation of $q - 1$, i.e., let $q - 1 = \Sigma_{i=0}^m p_i 2^i$ with $p_i \in \{0,1\}$ and $p_m = 1$. Let $I = \{i \mid p_i = 1\}$. In $[0,1]_{\mathrm{L}}$, the set*

$$\{y_0 \equiv \neg z_{1/q}, y_1 \equiv y_0^2, y_2 \equiv y_1^2, \ldots, y_m \equiv y_{m-1}^2, z_{1/q} \equiv \Pi_{i \in I} y_i\}$$

has a unique satisfying assignment, sending $z_{1/q}$ to $1/q$.

To define $1/q$, we need $|q - 1| + 1$ variables, and the length of the product in the last equivalence is linear in $|q|$. Similarly one can achieve a polynomial-size variant of an implicit definition for p/q.

It is shown in [12] how to obtain finite strong completeness of RPL w.r.t. $[0,1]_{\mathrm{L}}^{\mathcal{Q}}$ from finite strong completeness of Ł w.r.t. $[0,1]_{\mathrm{L}}$, based on the

---

[2] On the other hand, no rationals beyond 0 and 1 are term-definable in $[0,1]_{\mathrm{L}}$, as a consequence of McNaughton theorem.

following statement ([12, Lemma 3.3.13]). Let $\delta_{p/q}$ be an MV-term that implicitly defines the value p/q in a variable $z_{p/q}$ in $[0,1]_L$. First, given an RMV-term φ in variables $x_1, \ldots, x_n$ and constants $p_1/q_1, \ldots, p_m/q_m$, let δ_φ stand for $\delta_{p_1/q_1} \odot \cdots \odot \delta_{p_m/q_m}{}^3$, and let $\varphi^\star$ result from φ by replacing each constant p_i/q_i with the variable z_{p_i/q_i}. Now let $\{\psi_1, \ldots, \psi_k\} \cup \{\varphi\}$ be a finite set of RMV-terms (in some variables $x_1, \ldots, x_n$, particularly, with no occurrences of y-variables or z-variables) and let τ denote $\{\psi_1 \odot \cdots \odot \psi_k\}$. The statement says that $\tau \vdash_{RPL} \varphi$ iff $\tau^\star \odot \delta_{\tau \odot \varphi} \vdash_L \varphi^\star$. The reason is that under $\delta_{\tau \odot \varphi}$, the variables that correspond to the implicitly defined constants behave exactly as the constants would. Moreover, $\delta_{\tau \odot \varphi}$ is an MV-term.

Lemma 2. *Let τ and φ be RMV-terms with rational constants $(p_1/q_1, \ldots, p_m/q_m)$. Using the δ notation as above, we have:*

1. *$\delta_{\tau \odot \varphi}$ has $\Sigma_{j=1}^m(|p_j| + |q_j - 1|) + 2m$ variables.*
2. *the length of $\delta_{\tau \odot \varphi}$, written as an MV-term featuring $\oplus$ and $\neg$, is at most $\Sigma_{j=1}^m(8|p_j| + 8|q_j - 1| + 4)$.*

Finally we are ready to define the *validity degree* of a term φ in a theory T:

$$\|\varphi\|_T = \inf\{v(\varphi) \mid v \text{ model of } T\},$$

where a valuation v is a model of T if it assigns the value 1 to all terms in T. We only consider finite theories; for $T = \{\psi_1, \ldots, \psi_k\}$ write $\tau = \psi_1 \odot \cdots \odot \psi_k$; then $\|\varphi\|_\tau = \min\{v(\varphi) \mid v \text{ model of } \tau\}$. For τ inconsistent, $\|\varphi\|_\tau = 1$. In the rest of this paper, T is finite and represented by the term τ as above. We define the optimization problem.

VALIDITY DEGREE
Instance: RMV-terms τ and φ (possibly without constants).
Output: $\|\varphi\|_\tau$.

Lemma 3. *$\|\varphi\|_\tau = \|\varphi^\star\|_{(\tau^\star \odot \delta_{\tau \odot \varphi})}$.*

3   Optimization Problems and Metric Reductions

This section briefly sketches our computational paradigm, reproducing some notions and results on the structure of the oracle class FP^{NP} as given in Krentel [19], with a wider framework as provided in [27]. We also introduce an optimization problem from [19] that will be used in Sect. 5.

In this paper we use the term *optimization problem* for what is sometimes called an *evaluation* or *cost* version of a function problem (cf. [27]). In our setting, the output is the validity degree $\|\varphi\|_\tau$ (as an extremal value of f_φ on the 1-region of f_τ), rather than an *assignment* at which the extremal value is attained.

---

[3] It is assumed that the collections of auxiliary variables for the implicit definitions of p_i, q_i with $1 \leq i \leq n$ are pairwise disjoint and also disjoint from the variables $x_1, \ldots, x_n$.

Let $z : \mathbb{N} \longrightarrow \mathbb{N}$ be smooth.[4] $\mathrm{FP}^{\mathrm{NP}}[z(n)]$ is the class of functions computable in polynomial time with an NP oracle with at most $z(|x|)$ oracle calls for instance x. In particular, $\mathrm{FP}^{\mathrm{NP}}$ stands for $\mathrm{FP}^{\mathrm{NP}}[n^{O(1)}]$.

Definition 3. ([19]) *Let Σ be a finite alphabet and $f, g : \Sigma^* \longrightarrow \mathbb{N}$. A* metric reduction *from f to g is a pair (h_1, h_2) of polynomial-time computable functions where $h_1 : \Sigma^* \longrightarrow \Sigma^*$ and $h_2 : \Sigma^* \times \mathbb{N} \longrightarrow \mathbb{N}$, such that $f(x) = h_2(x, g(h_1(x)))$ for all $x \in \Sigma^*$.*

The concept of a *metric reduction* is a natural generalization of polynomial-time many-one reduction to optimization problems. It follows from the definition that for each function z as above, $\mathrm{FP}^{\mathrm{NP}}[z(n)]$ is closed under metric reductions. The paper [19] provides examples of problems that are complete for $\mathrm{FP}^{\mathrm{NP}}$ under metric reductions. We define one such problem (see [19]).

WEIGHTED MAX-SAT
Instance: Boolean CNF term $(C_1 \wedge \cdots \wedge C_n)(x_1, \ldots, x_k)$ with weights on clauses $w_1, \ldots, w_n$, each w_i positive integer in binary.
Output: the maximal sum of weights of true clauses over all (Boolean) assignments to the variables $x_1, \ldots, x_k$.

Theorem 2. ([19]) WEIGHTED MAX-SAT *is $\mathrm{FP}^{\mathrm{NP}}$ complete.*

The paper [19] provides a separation result for problems in $\mathrm{FP}^{\mathrm{NP}}$, a simple form of which is given below. In particular, under standard complexity assumptions there are no metric reductions from $\mathrm{FP}^{\mathrm{NP}}$ complete problems (such as WEIGHTED MAX-SAT) to some problems in $\mathrm{FP}^{\mathrm{NP}}[O(\log n)]$, such as the VERTEX COVER problem.

Theorem 3. ([19]) *Assume $\mathrm{P} \neq \mathrm{NP}$.*
Then $\mathrm{FP}^{\mathrm{NP}}[O(\log \log n)] \subsetneq \mathrm{FP}^{\mathrm{NP}}[O(\log n)] \subsetneq \mathrm{FP}^{\mathrm{NP}}[n^{O(1)}]$.

4 Upper Bound: Validity Degree is in $\mathrm{FP}^{\mathrm{NP}}$

We present a polynomial-time oracle computation for VALIDITY DEGREE, using a coNP complete decision version of the problem as an oracle; this yields membership of VALIDITY DEGREE in $\mathrm{FP}^{\mathrm{NP}}$. The instances of the problem are pairs (τ, φ) of RMV-terms, i.e., terms with the MV-symbols $\oplus$ and $\neg$ where atoms are variables and rational constants. The following oracle will be used.

D-RPL-GRADED-PROVABILITY
Instance: (τ, φ, k) with τ, φ RMV-terms and k a rational number in $[0, 1]$.
Output: $\tau \vdash_{\mathrm{RPL}} k \to \varphi$?

---

[4] I.e., z is nondecreasing and the function $1^n \mapsto 1^{z(n)}$ is polynomial-time computable.

Note that $\tau \vdash_{\text{RPL}} k \to \varphi$ iff $k \le \|\varphi\|_\tau$. By [13], RPL-provability from finite theories (given RMV terms τ and φ, it is the case that $\tau \vdash_{\text{RPL}} \varphi$?) is coNP complete. Hence, so is D-RPL-GRADED-PROVABILITY.

The oracle computation can employ a binary search, given an explicit upper bound on denominators. To obtain a polynomial-time (oracle) computation, the result of [12] that $\|\varphi\|_\tau$ is rational is not enough: we need an upper bound $N(\tau, \varphi)$ on the denominator that is in itself of polynomial size (in binary).

To expose the algebraic methods employed in this section, let us start with a simpler related problem, interesting in its own right: the natural optimization version of the term satisfiability problem in the standard MV-algebra $[0, 1]_{\text{L}}$.

MAX VALUE
Instance: MV-term $\varphi(x_1, \ldots, x_n)$.
Output: $\max f_\varphi$ on $[0, 1]^n$.

This problem reduces to VALIDITY DEGREE: one maximizes f_φ by minimizing $f_{\neg\varphi}$ under an empty theory. As mentioned in Sect. 1, even this simpler problem cannot be efficiently approximated (see [16, Theorem 7.4]).

Lemma 4. *Let p_1/q_1 and p_2/q_2 be two distinct rational numbers and N a positive integer, let $q_1, q_2 \le N$. Then $\left| \frac{p_1}{q_1} - \frac{p_2}{q_2} \right| \ge \frac{1}{N^2}$.*

Lemma 5. *Let $a < b$ be rationals and N a positive integer. Assume the interval $[a, b)$ contains exactly one rational c with denominator at most N, and other rationals with denominator at most N are at a distance greater than $b - a$ from c. There is a poly-time algorithm that finds c on input a,b, and N in binary.*

Theorem 4. MAX VALUE *is in* FP^{NP}.

Proof. Let $\varphi(x_1, \ldots, x_n)$ be an MV-term. Then f_φ is maximal on a rational vector $\langle p_1/q_1, \ldots, p_n/q_n \rangle$; the least common denominator of the vector is at most $(\frac{\#\varphi}{n})^n$ ([2, Theorem 14]). It follows that the denominator of $f_\varphi(p_1/q_1, \ldots, p_n/q_n)$ is at most $N(\varphi) = (\frac{\#\varphi}{n})^n$.

We sketch a polynomial-time algorithm computing $\max(f_\varphi)$ using binary search on rationals in $[0, 1]$ with denominators at most $N(\varphi)$, using the *generalized satisfiability* (GENSAT), known to be NP complete ([25]), as oracle: given MV-term φ and a rational $r \in [0, 1]$, is $\max(f_\varphi) \ge r$?

Test GENSAT$(\varphi, 1)$. If so, output 1 and terminate.
Otherwise, let $a = 0$, $b = 1$, and $k = 0$.
Repeat
$k = k + 1$; if GENSAT$(\varphi, a + b/2)$, let $a = a + b/2$, otherwise let $b = a + b/2$
until $2^k > (N(\varphi))^2$.
Finally, find $\|\varphi\|_\tau$ in $[a, b)$ relying on Lemma 5.

Assume the algorithm runs through the loop at least once. After the search terminates, k is the least integer s.t. $2^k > (N(\varphi))^2$, i.e., $k > 2\log(N(\varphi)) \ge k - 1$. hence the number k of passes through the loop is polynomial. Also, the semi-closed interval $[a, b]$ of length $1/2^k < 1/(N(\varphi))^2$ contains $\max f_\varphi$, and by Lemma 4, $\max f_\varphi$ is the only value in $[a, b]$ with denominator at most $N(\varphi)$. The values of a and b are $l/2^k$ and $(l+1)/2^k$ respectively, so $|a|$ and $|b|$ are polynomial in k.

Let us address the VALIDITY DEGREE problem. The binary search will be analogous, we need to establish an upper bound for the denominators. The following lemma can be obtained from the proof of [2, Theorem 17], a result on finite consequence relation in Ł.

Lemma 6. *Let τ and φ be MV-terms and let n be the number of variables in these terms. Assume $M, N \in \mathbb{N}$ are coprime non-negative integers such that $\|\varphi\|_\tau = M/N$. Then*

$$N \leq \left(\frac{\sharp\tau + \sharp\varphi}{n} \right)^n$$

Proof. Following [2] and the references therein, one can build, in a canonical way, (n-dimensional[5]) polyhedral complexes $C(\tau)$ and $C(\varphi)$ such that $\bigcup C(\tau) = [0, 1]^n = \bigcup C(\varphi)$, with f_τ linear over each n-dimensional cell of $C(\tau)$ and f_φ linear over each n-dimensional cell of $C(\varphi)$.

It follows from the analysis of [2] that the minimum of f_φ on the 1-region of τ is attained at a vertex (of an n-dimensional cell) of the common refinement of $C(\tau)$ and $C(\varphi)$. It can further be derived from that paper that the least common denominator of any vertex in this common refinement is bounded by $(\frac{\sharp\tau + \sharp\varphi}{n})^n$; the proof is analogous to the case when τ is void.

Hence, there is a rational vector $\langle p_1/q_1, \ldots, p_n/q_n \rangle$ on which f_τ is 1, f_φ attains the value $\|\varphi\|_\tau$, and the least common denominator of $\langle p_1/q_1, \ldots, p_n/q_n \rangle$ is $(\frac{\sharp\tau + \sharp\varphi}{n})^n$. It follows that $N \leq \left(\frac{\sharp\tau + \sharp\varphi}{n} \right)^n$. $\square$

Denote by $N(\tau, \varphi)$ the obtained upper bound on the denominator of $\|\varphi\|_\tau$ for MV-terms τ and φ. To provide an upper bound $N^\star(\tau, \varphi)$ on the denominator of $\|\varphi\|_\tau$ in case τ and φ are RMV-terms, we rely on Lemma 3 in order to apply the existing results for MV-terms: namely, we use the upper bounds on $\|\varphi^\star\|_{(\tau^\star \odot \delta_{\tau \odot \varphi})}$.

Lemma 7. *Let τ and φ be RMV-terms. $N^\star(\tau, \varphi) = N(\tau^\star \odot \delta_{\tau \odot \varphi}, \varphi^\star) = $* $= (\frac{\sharp\tau^\star + \sharp\delta_{\tau \odot \varphi} + \sharp\varphi^\star}{n})^n$, *where n denotes the number of variables in the terms $\tau^\star$, $\delta_{\tau \odot \varphi}$, and $\varphi^\star$.*

Lemma 8. *For τ and φ RMV-terms, $N^\star(\tau, \varphi)$ is polynomial size in $\sharp\tau$ and $\sharp\varphi$.*

Theorem 5. VALIDITY DEGREE *is in* $\mathrm{FP}^{\mathrm{NP}}$.

---

[5] The dimension of f_τ and f_φ can be extended to n in a number of ways, e.g., supplying dummy variables. This will modify the length by a linear function of n.

Proof. We provide a polynomial-time Turing reduction of VALIDITY DEGREE to D-RPL-GRADED-PROVABILITY; i.e., for RMV-terms τ and φ the algorithm computes $\|\varphi\|_\tau$ in time polynomial in $\sharp\tau + \sharp\varphi$, relying on the oracle. The algorithm is based on a binary search analogous to the algorithm for MAX VALUE from Theorem 4.

The initial test is D-RPL-GRADED-PROVABILITY$(1, \tau, \varphi)$, where a positive answer yields $\|\varphi\|_\tau = 1$.

If this is not the case, the binary search is initiated. The upper bound $N = N^\star(\tau, \varphi)$ on denominator of $\|\varphi\|_\tau$ is as in Lemma 7 and 8. This provides discrete structure to search in and the terminating condition $2^k > N^2$.

The final application of Lemma 5 is analogous to the proof of Theorem 4.

5   Lower Bound: Validity Degree Is FP^{NP} Hard

We give a metric reduction of WEIGHTED MAX-SAT to VALIDITY DEGREE. In this section the VALIDITY DEGREE problem is considered for MV-terms τ and φ, i.e., we work in the MV-fragment of the RMV language. The lower bound obtained for the MV-language then applies also to RMV-language.

Theorem 6. VALIDITY DEGREE *is* FP^{NP} *hard under metric reductions.*

Proof. For clarity, the proof is divided in two parts. First, we reduce WEIGHTED MAX-SAT to VALIDITY DEGREE in an MV-language with the definable symbols. Subsequently we show how to polynomially translate general MV-terms that occur in the range of the metric reduction to MV-terms in the basic language.

We define the function h_1 from Definition 3, which takes inputs to WEIGHTED MAX-SAT and transforms them to inputs to VALIDITY DEGREE. Consider a classical CNF-term (with language $\wedge$, $\vee$, and $\neg$) φ with variables $x_1, \ldots, x_k$ and weights $w_1, \ldots, w_n$ for the clauses $C_1, \ldots, C_n$ of φ. One obtains the solution to WEIGHTED MAX-SAT by maximizing $\Sigma_{i=1}^n v(C_i) w_i$ over all Boolean assignments v to $x_1, \ldots, x_k$. To utilize VALIDITY DEGREE, we need to render this expression in the MV-language and to isolate the Boolean semantics among the broader semantics of $[0, 1]_{\mathrm{L}}$.

We define a finite theory T and a term Φ in stages by making several observations. At any stage, T is assumed to include terms specified in the earlier stages.

(a) On any input $\langle \tau, \varphi \rangle$, VALIDITY DEGREE gives the minimum of f_φ in $[0, 1]_{\mathrm{L}}$ over the 1-region of f_τ. The routine can also compute the *maximum* of f_φ on the same domain if the input is $\langle \tau, \neg\varphi \rangle$ and the output is subtracted from 1.

(b) To force Boolean assignments, for each $1 \leq j \leq k$ put $x_j \vee \neg x_j$ in T. Since $\vee$ evaluates as max in $[0, 1]_{\mathrm{L}}$, this condition is true only under (standard MV-) assignments where either x_j is 1, or $\neg x_j$ is 1, i.e., x_j is 0.

(c) The algebra $[0,1]_L$ can only correctly add up to the sum 1.[6] Thus the weights $w_1, \ldots, w_n$ need to be scaled. The computations with weights are bounded by $w = \Sigma_{i=1}^{n} w_i$, which is the output of WEIGHTED MAX-SAT in case φ is satisfiable, so an appropriate factor to scale by is $1/w$. The new weights are $w_i' = w_i/w$ for each $i \in \{1, \ldots, n\}$ This is an order-preserving transformation of the weights and the new weights are of poly-size in the input size.

(d) Multiplication is not available, so $e(C_i)w_i'$ cannot be expressed with an MV-term. One can however *implicitly* define some rational expressions as follows.

   – Introduce a new variable b. To implicitly define $1/w$ in variable b, include in T the system from Lemma 1 that polynomially renders the condition $b \equiv (\neg b)^{w-1}$; now any model v of T will have $v(b) = 1/w$.

   – For $1 \leq i \leq n$, introduce a new variable y_i. Include $y_i \rightarrow b$ in T; any model v of T will have $v(y_i) \leq 1/w$. Further, include in T a polynomial rendering of $\underbrace{y_i \oplus y_i \oplus \cdots \oplus y_i}_{w \text{ times}} \equiv C_i$, using Lemma 1; then for any model v of T we have that $v(C_i) = 0$ implies $v(y_i) = 0$, whereas $v(C_i) = 1$ implies $v(y_i) \geq 1/w$, which in combination with the other condition in this item gives $v(y_i) = v(C_i)/w$.

   – For $1 \leq i \leq n$, introduce a new variable z_i. Include in T a polynomial rendering of $\underbrace{y_i \oplus y_i \oplus \cdots \oplus y_i}_{w_i \text{ times}} \equiv z_i$, again relying on Lemma 1. Any model v of T will have $v(z_i) = v(C_i)w_i'$.

To recap, we define T as the following set of MV-terms:

   – $x_j \vee \neg x_j$ for each $j \in \{1, \ldots, k\}$;
   – a polynomial-sized rendering of $b \equiv (\neg b)^{w-1}$ (cf. Lemma 1);
   – for $1 \leq i \leq n$, $y_i \rightarrow b$ and a poly-sized rendering of $wy_i \equiv C_i$ (Lemma 1);
   – for $1 \leq i \leq n$, a poly-sized rendering of $w_i y_i \equiv z_i$ (Lemma 1).

Let a term τ represent T, let Φ stand for $\neg(z_1 \oplus z_2 \oplus \cdots \oplus z_n)$. Let $m = \|\Phi\|_\tau$, i.e., m is the rational number that VALIDITY DEGREE returns on input τ and Φ. We claim that $(1 - m)w$ (the function h_2 from Definition 3) is the solution to the instance $C_1, \ldots, C_n$ and $w_1, \ldots, w_n$ of WEIGHTED MAX-SAT on input.

To see this, observe that the models of τ feature precisely all Boolean assignments to variables $\{x_1, \ldots, x_k\}$. Each such model v extends to the new variables b, y_i and z_i ($1 \leq i \leq n$), namely $v(b) = 1/w$, $v(y_i) = (1/w)v(C_i)$, and $v(z_i) = (w_i/w)v(C_i)$. In particular, if v models T, then the values of b, y_i and z_i under v are determined by the values that v assigns to the x-variables (i.e., the "Boolean" variables). Except for b, the sets of variables introduced for each i are pairwise disjoint.

It follows from the construction of τ and Φ that any Boolean assignment that yields an extremal value of WEIGHTED MAX-SAT also produces an extremal value of VALIDITY DEGREE and vice versa. It is easy to check that the order-reversing operations (taking $1 - y$ back and forth) and the scaling and descaling

---

[6] Addition, represented by the strong disjunction $\oplus$, is truncated at 1.

work as expected (both are order-preserving). Hence, the reduction correctly computes an input to VALIDITY DEGREE and correctly renders the result of this routine as an output of WEIGHTED MAX-SAT.

Finally, both functions involved are clearly polynomial-time functions.

For the second part of the proof, we notice that Φ is a term in the basic language. As for τ, recall that one can render $\varphi \odot \psi$ and $\varphi \to \psi$ in the basic language, using the definitions, without changing the number of variable occurrences; this includes the nested occurrences of $\odot$ in (a rendering of) $(\neg b)^{w-1}$ (recall that the product in the p-size variant is of cardinality $|w|$). To rewrite each disjunction C_i in the basic language, we apply to the following claim.[7]

Claim: let $\alpha = (\alpha_1 \vee \cdots \vee \alpha_n)$, where α_i are terms in the basic language. There is a term β in the basic language L-equivalent to α and such that $\sharp\beta = 2\sharp\alpha$.

To justify the claim, let $\alpha^l = \alpha' \vee \alpha_n$, where $\alpha' = (\alpha_1 \vee \cdots \vee \alpha_{n-1})$. Then α^l is equivalent to $(\alpha' \to \alpha_n) \to \alpha_n$. Repeat this process for α' unless it coincides with α_1. This produces a term equivalent to α, with $\to$ as the only symbol; then rewrite $\to$ in the basic language.

6   Closing Remarks

This result attests a key role of algebraic methods for computational complexity upper bounds in propositional Łukasiewicz logic. Syntactic derivations are not even discussed; indeed at present we have no idea how to employ them.

A proof-theoretic counterpart of a validity degree is the *provability degree*: $|\varphi|_T = \sup\{r \mid T \vdash_{\text{RPL}} r \to \varphi\}$, with the provability relation defined by extending Łukasiewicz logic with suitable axioms. Hájek proved *Pavelka completeness* for RPL in [12]: for any choice of T and φ, $|\varphi|_T$ coincides with $\|\varphi\|_T$. Our results thereby apply also to provability degrees (for finite theories).

To our knowledge there are no works explicitly dealing with the more pragmatical tasks of providing algorithms computing the validity degree (or maximal value), identifying fragments where they might be efficient, or similar.

We have obtained hardness for FP^{NP} under metric reductions for VALIDITY DEGREE but not MAX VALUE. A somewhat similar reduction of WEIGHTED MAX-SAT to a 0-1 integer programming problem was presented in [19], where roughly speaking, some conditions in the matrix correspond to some of our conditions in the theory. We do not know how to avoid employing the theory, and cannot supply a FP^{NP} hardness proof for MAX VALUE at present.

Acknowledgements. This work was supported partly by the grant GA18-00113S of the Czech Science Foundation and partly by the long-term strategic development financing of the Institute of Computer Science (RVO:67985807).

The author is indebted to Stefano Aguzzoli, Tommaso Flaminio, Lluís Godo, Jirka Hanika, and Petr Savický for references, discussion and advice on the material of Sect. 4. Moreover, the feedback provided by anonymous referees simplified and improved the presentation of the material. Any shortcomings in the text remain with the author.

---

[7] Slightly more general claim was made, without proof, at the beginning of Sect. 2.

References

1. Aguzzoli, S.: An asymptotically tight bound on countermodels for Łukasiewicz logic. Int. J. Approx. Reason. **43**(1), 76–89 (2006)
2. Aguzzoli, S., Ciabattoni, A.: Finiteness in infinite-valued Łukasiewicz logic. J. Logic Lang. Inform. **9**(1), 5–29 (2000). https://doi.org/10.1023/A:1008311022292
3. Aguzzoli, S., Mundici, D.: Weirstrass approximation theorem and Łukasiewicz formulas. In: Fitting, M.C., Orlowska, E. (eds.) Beyond Two: Theory and Applications of Multiple-Valued Logic. Studies in Fuzziness and Soft Computing, vol. 114, pp. 251–272. Physica-Verlag, Heidelberg (2003)
4. Bělohlávek, R.: Pavelka-style fuzzy logic in retrospect and prospect. Fuzzy Sets Syst. **281**, 61–72 (2015)
5. Blok, W.J., Pigozzi, D.L.: Algebraizable Logics, Memoirs of the American Mathematical Society, vol. 396. American Mathematical Society, Providence (1989)
6. Cignoli, R., D'Ottaviano, I.M., Mundici, D.: Algebraic Foundations of Many-Valued Reasoning, Trends in Logic, vol. 7. Kluwer, Dordrecht (1999)
7. Cintula, P.: A note on axiomatizations of Pavelka-style complete fuzzy logics. Fuzzy Sets Syst. **292**, 160–174 (2016)
8. Cintula, P., Hájek, P.: Complexity issues in axiomatic extensions of Łukasiewicz logic. J. Logic Comput. **19**(2), 245–260 (2009)
9. Di Nola, A., Leuştean, I.: Łukasiewicz logic and MV-algebras. In: Cintula, P., Hájek, P., Noguera, C. (eds.) Handbook of Mathematical Fuzzy Logic, vol. 2, pp. 469–583. College Publications (2011)
10. Gispert, J.: Universal classes of MV-chains with applications to many-valued logic. Math. Logic Q. **42**, 581–601 (2002)
11. Goguen, J.A.: The logic of inexact concepts. Synthese **19**(3–4), 325–373 (1969)
12. Hájek, P.: Metamathematics of Fuzzy Logic, Trends in Logic, vol. 4. Kluwer, Dordrecht (1998)
13. Hájek, P.: Computational complexity of t-norm based propositional fuzzy logics with rational truth constants. Fuzzy Sets Syst. **157**(5), 677–682 (2006)
14. Hájek, P., Paris, J., Shepherdson, J.C.: Rational Pavelka logic is a conservative extension of Łukasiewicz logic. J. Symb. Logic **65**(2), 669–682 (2000)
15. Haniková, Z.: Implicit definability of truth constants in Łukasiewicz logic. Soft. Comput. **23**(7), 2279–2287 (2019). https://doi.org/10.1007/s00500-018-3461-x
16. Haniková, Z., Savický, P.: Term satisfiability in FL_{ew}-algebras. Theoret. Comput. Sci. **631**, 1–15 (2016)
17. Jeřábek, E.: Admissible rules of Łukasiewicz logic. J. Logic Comput. **20**(2), 425–447 (2010)
18. Jeřábek, E.: The complexity of admissible rules of Łukasiewicz logic. J. Logic Comput. **23**(3), 693–705 (2013)
19. Krentel, M.W.: The complexity of optimization problems. J. Comput. Syst. Sci. **36**, 490–509 (1988)
20. Łukasiewicz, J., Tarski, A.: Untersuchungen über den Aussagenkalkül. Comptes Rendus des Séances de la Société des Sciences et des Lettres de Varsovie, cl. III **23**(iii), 30–50 (1930)
21. McNaughton, R.: A theorem about infinite-valued sentential logic. J. Symb. Logic **16**(1), 1–13 (1951)
22. Mundici, D.: Mapping Abelian ℓ-groups with strong unit one-one into MV-algebras. J. Algebra **98**(1), 76–81 (1986)

23. Mundici, D.: Satisfiability in many-valued sentential logic is NP-complete. Theoret. Comput. Sci. **52**(1–2), 145–153 (1987)
24. Mundici, D.: Advanced Łukasiewicz Calculus and MV-Algebras, Trends in Logics, vol. 35. Springer, Heidelberg (2011). https://doi.org/10.1007/978-94-007-0840-2
25. Mundici, D., Olivetti, N.: Resolution and model building in the infinite-valued calculus of Łukasiewicz. Theoret. Comput. Sci. **200**, 335–366 (1998)
26. Pavelka, J.: On fuzzy logic I, II, III. Zeitschrift für Mathematische Logik und Grundlagen der Mathematik **25**, 45–52, 119–134, 447–464 (1979)
27. Stockmeyer, L.J.: Computational complexity. In: Barnhart, C., Laporte, G., et al. (ed.) Handbooks in OR & MS, vol. 3, pp. 455–517. Elsevier Science Publishers (1992)
28. Torrens, A.: Cyclic elements in MV-algebras and post algebras. Math. Logic Q. **40**(4), 431–444 (1994)

Degrees of Non-computability of Homeomorphism Types of Polish Spaces

Mathieu Hoyrup[1], Takayuki Kihara[2(✉)], and Victor Selivanov[3,4]

[1] Université de Lorraine, CNRS, Inria, LORIA, 54000 Nancy, France
`mathieu.hoyrup@inria.fr`
[2] Nagoya University, Nagoya 464-8601, Japan
`kihara@i.nagoya-u.ac.jp`
[3] A.P. Ershov Institute of Informatics Systems SB RAS, Novosibirsk, Russia
`vseliv@iis.nsk.su`
[4] Kazan Federal University, Kazan, Russia

Abstract. There are continuum many homeomorphism types of Polish spaces. In particular, there is a Polish space which is not homeomorphic to any computably presented Polish space. We examine the details of degrees of non-computability of presenting homeomorphic copies of Polish spaces.

Keywords: Computable topology · Computable presentation · Computable Polish space · Degree spectrum

How difficult is it to describe an explicit presentation of an abstract mathematical structure? In computable structure theory, there are a large number of works on degrees of non-computability of presenting isomorphism types (known as degree spectra) of algebraic structures such as groups, rings, fields, linear orders, lattices, Boolean algebras, and so on, cf. [1,5,6,8]. In this article, we deal with its topological analogue, i.e., degrees of non-computability of presenting homeomorphism types of certain topological spaces, and mainly focus on presentations of Polish spaces (i.e., completely metrizable separable spaces). We present some of the results obtained in Hoyrup-Kihara-Selivanov [9] on this topic.

The notion of a presentation plays a central role, not only in computable structure theory, but also in computable analysis [2,3,16]. In this area, one of the most crucial problems was how to present large mathematical objects (which possibly have the cardinality of the continuum) such as metric spaces, topological spaces and so on, and then researchers have obtained a number of reasonable answers to this question. In particular, the notion of a computable presentation of

The second-named author was partially supported by JSPS KAKENHI Grant 19K03602, 15H03634, and the JSPS Core-to-Core Program (A. Advanced Research Networks). The third-named author was supported by the Russian Science Foundation project No 18-11-00028.

M. Anselmo et al. (Eds.): CiE 2020, LNCS 12098, pp. 189–192, 2020.
https://doi.org/10.1007/978-3-030-51466-2_16

a Polish space has been introduced around 1950-60s, cf. [12], and since then this notion has been widely studied in several areas including computable analysis [2,14,16] and descriptive set theory [13].

In recent years, several researchers succeeded to obtain various results on Turing degrees of presentations of *isometric isomorphism types* of Polish metric spaces, separable Banach spaces, and so on, cf. [4,10,11]. However, most of works are devoted to metric structures, and there seem almost no works on presentations on homeomorphism types of Polish spaces. The investigation of Turing degrees of *homeomorphism types* of topological spaces (not necessarily Polish) was initiated in [15]. Some results were obtained for domains but the case of Polish spaces was apparently not investigated seriously so far.

Every Polish space is homeomorphic to the Cauchy completion of a metric on (an initial segment of) the natural numbers ω, so one may consider any distance function $d \colon \omega^2 \to \mathbb{Q}$ as a presentation of a Polish space. Then, observe that there are continuum many homeomorphism types of Polish spaces. In particular, by cardinality argument, there is a Polish space which is not homeomorphic to any computably presented Polish space. Surprisingly however, it was unanswered until very recently even whether the following holds:

Question 1. *Does there exist a $0'$-computably presented Polish space which is not homeomorphic to a computably presented one?*

The solution to Question 1 was very recently obtained by the authors of this article, and independently by Harrison-Trainor, Melnikov, and Ng [7]. One possible approach to solve this problem is using Stone duality between countable Boolean algebras and zero-dimensional compact metrizable spaces, but they also noticed that every *low_4-presented* zero-dimensional compact metrizable space is homeomorphic to a computable one. This reveals certain limitations of Stone duality techniques. Our next step is to develop new techniques other than Stone duality. More explicitly, the next question is the following:

Question 2. *Does there exist a low_4-presented Polish space which is not homeomorphic to a computably presented one?*

One of the main results in Hoyrup-Kihara-Selivanov [9] is that there exists a $0'$-computable low_3 infinite dimensional compact metrizable space which is not homeomorphic to a computable one. This solves Question 2. Indeed, Hoyrup-Kihara-Selivanov [9] showed more general results by considering two types of presentations:

A *Polish presentation* (or simply a *presentation*) of a Polish space $\mathcal{X}$ is a distance function d on ω whose Cauchy completion is homeomorphic to $\mathcal{X}$. If $\mathcal{X}$ is moreover compact, then by a *compact presentation* of $\mathcal{X}$ we mean a presentation of $\mathcal{X}$ equipped with an enumeration of (codes of) all finite rational open covers of $\mathcal{X}$, where a rational open set is a finite union of rational open balls. For readers who are familiar with the abstract theory of modern computable analysis (cf. [2]), we note that a Polish (compact, resp.) presentation of $\mathcal{X}$ is just a name of a homeomorphic copy of $\mathcal{X}$ in the space of overt (compact overt, resp.) subsets of Hilbert cube $[0,1]^\omega$.

Theorem 1 ([9]). *For any Turing degree* $\mathbf{d}$ *and natural number* $n > 0$, *there exists a compact metrizable space* $\mathcal{Z}_{\mathbf{d},n}$ *such that for any Turing degree* $\mathbf{x}$,

$$\mathbf{d} \leq \mathbf{x}^{(2n-1)} \iff \mathcal{Z}_{\mathbf{d},n} \text{ has an } \mathbf{x}\text{-computable compact presentation.}$$

Theorem 2 ([9]). *For any Turing degree* $\mathbf{d}$ *and natural number* $n > 0$, *there exists a compact metrizable space* $\mathcal{P}_{\mathbf{d},n}$ *such that for any Turing degree* $\mathbf{x}$,

$$\mathbf{d} \leq \mathbf{x}^{(2n)} \iff \mathcal{P}_{\mathbf{d},n} \text{ has an } \mathbf{x}\text{-computable compact presentation.}$$
$$\mathbf{d} \leq \mathbf{x}^{(2n+1)} \iff \mathcal{P}_{\mathbf{d},n} \text{ has an } \mathbf{x}\text{-computable Polish presentation.}$$

Corollary 1 ([9]). *There exists a Polish space which is* $\mathbf{0}'$-*computably presentable, but not computably presentable.*

Indeed, for any $n > 0$, *there exists a compact metrizable space* $\mathcal{X}$ *such that*

$$\mathbf{x} \text{ is high}_{2n} \iff \mathcal{X} \text{ has an } \mathbf{x}\text{-computable compact presentation.}$$
$$\mathbf{x} \text{ is high}_{2n+1} \iff \mathcal{X} \text{ has an } \mathbf{x}\text{-computable Polish presentation.}$$

Corollary 2 ([9]). *There is a compact Polish space* $\mathcal{X}$ *which has a computable presentation, but has no presentation which makes* $\mathcal{X}$ *computably compact.*

Another important question is whether the homeomorphism type of a Polish space can have an easiest presentation.

Question 3. *Does there exist a homeomorphism type of a Polish space which is not computably presentable, but have an easiest presentation with respect to Turing reducibility?*

Hoyrup-Kihara-Selivanov [9] partially answered Question 3 in negative. More precisely, we show the cone-avoidance theorem for *compact* Polish spaces, which states that, for any non-c.e. set $A \subseteq \omega$, every compact Polish space has a presentation that does not enumerate A.

Some other side results in [9] are:

Theorem 3 ([9]). *There exists a compact metrizable space* $\mathcal{X}$ *which has a computably compact presentation, but its* nth *Cantor-Bendixson derivative* $\mathcal{X}^n$ *has no* $\mathbf{0}^{(2n-1)}$-*computably compact presentation.*

Indeed, there exists a compact metrizable space $\mathcal{X}$ *such that the following are equivalent for any Turing degree* $\mathbf{x}$:

1. $\mathcal{X}$ *has an* $\mathbf{x}$-*computable compact presentation*
2. *The* nth *derivative* $\mathcal{X}^n$ *has an* $\mathbf{x}^{(2n)}$-*computable compact presentation*

Theorem 4 ([9]). *There exists a computably presentable compact metrizable space whose perfect kernel is not computably presentable.*

Similarly, there exists a computably compact, computably presented, compact metrizable space whose perfect kernel is compact, but has no presentation which makes it computably compact.

Our key idea is using *dimension* (more explicitly, *high-dimensional holes*, i.e., a cycle which is not a boundary) to code a given Turing degree. As a result, all of our examples in the above results are *infinite dimensional*. We do not know if there are finite dimensional examples satisfying our main results. We also note that all of our examples are disconnected, and it is not known if there are connected examples.

References

1. Ash, C.J., Knight, J.: Computable Structures and the Hyperarithmetical Hierarchy, Studies in Logic and the Foundations of Mathematics, vol. 144. North-Holland Publishing Co., Amsterdam (2000)
2. Brattka, V., Hertling, P.: Handbook of Computability and Complexity in Analysis (202x)
3. Brattka, V., Hertling, P., Weihrauch, K.: A tutorial on computable analysis. In: Cooper, S.B., Löwe, B., Sorbi, A. (eds.) New Computational Paradigms, pp. 425–491. Springer, New York (2008). https://doi.org/10.1007/978-0-387-68546-5_18
4. Clanin, J., McNicholl, T.H., Stull, D.M.: Analytic computable structure theory and L^p spaces. Fund. Math. **244**(3), 255–285 (2019)
5. Ershov, Y.L., Goncharov, S.S., Nerode, A., Remmel, J.B., Marek, V.W. (eds.): Handbook of recursive mathematics. In: vol. 2, Studies in Logic and the Foundations of Mathematics, vol. 139. North-Holland, Amsterdam (1998). Recursive algebra, analysis and combinatorics
6. Fokina, E.B., Harizanov, V., Melnikov, A.: Computable model theory. In: Turing's Legacy: Developments from Turing's Ideas in Logic. Lecture Notes in Logic, vol. 42, pp. 124–194. Association for Symbolic Logic, La Jolla (2014)
7. Harrison-Trainor, M., Melnikov, A., Ng, K.M.: Computability up to homeomorphism (2020)
8. Hirschfeldt, D.R., Khoussainov, B., Shore, R.A., Slinko, A.M.: Degree spectra and computable dimensions in algebraic structures. Ann. Pure Appl. Logic **115**(1–3), 71–113 (2002)
9. Hoyrup, M., Kihara, T., Selivanov, V.: Degree spectra of Polish spaces (2020, in preparation)
10. McNicholl, T.H., Stull, D.M.: The isometry degree of a computable copy of ℓ^{p1}. Computability **8**(2), 179–189 (2019)
11. Melnikov, A.G.: Computably isometric spaces. J. Symb. Logic **78**(4), 1055–1085 (2013)
12. Moschovakis, Y.N.: Recursive metric spaces. Fund. Math. **55**, 215–238 (1964)
13. Moschovakis, Y.N.: Descriptive Set Theory, Mathematical Surveys and Monographs, vol. 155, 2nd edn. American Mathematical Society, Providence (2009)
14. Pour-El, M.B., Richards, J.I.: Computability in Analysis and Physics. Perspectives in Mathematical Logic (1989)
15. Selivanov, V.: On degree spectra of topological spaces (2019)
16. Weihrauch, K.: Computable Analysis, An Introduction. Texts in Theoretical Computer Science. An EATCS Series. Springer, Berlin (2000). https://doi.org/10.1007/978-3-642-56999-9

Time-Aware Uniformization of Winning Strategies

Stéphane Le Roux[(✉)]

LSV, Université Paris-Saclay, ENS Paris-Saclay, CNRS, Gif sur Yvette, France
leroux@lsv.fr

Abstract. Two-player win/lose games of infinite duration are involved in several disciplines including computer science and logic. If such a game has deterministic winning strategies, one may ask how simple such strategies can get. The answer may help with actual implementation, or to win despite imperfect information, or to conceal sensitive information especially if the game is repeated.

Given a concurrent two-player win/lose game of infinite duration, this article considers equivalence relations over histories of played actions. A classical restriction used here is that equivalent histories have equal length, hence *time awareness*. A sufficient condition is given such that if a player has winning strategies, she has one that prescribes the same action at equivalent histories, hence *uniformization*. The proof is fairly constructive and preserves finiteness of strategy memory, and counterexamples show relative tightness of the result. Several corollaries follow for games with states and colors.

Keywords: Two-player win/lose games · Imperfect information · Criterion for existence of uniform winning strategies · Finite memory

1 Introduction

In this article, two-player win/lose games of infinite duration are games where two players *concurrently and deterministically* choose one action each at every of infinitely many rounds, and "in the end" exactly one player wins. Such games (especially their simpler, turn-based variant) have been used in various fields ranging from social sciences to computer science and logic, e.g. in automata theory [5,15] and in descriptive set theory [12].

Given such a game and a player, a fundamental question is whether she has a winning strategy, i.e. a way to win regardless of her opponent's actions. If the answer is positive, a second fundamental question is whether she has a simple winning strategy. More specifically, this article investigates the following *strategy uniformization* problem: consider an equivalence relation $\sim$ over histories, i.e. over sequences of played actions; if a player has a winning strategy, has she a winning $\sim$-strategy, i.e. a strategy prescribing the same action after equivalent histories? This problem is relevant to imperfect information games and beyond.

© Springer Nature Switzerland AG 2020
M. Anselmo et al. (Eds.): CiE 2020, LNCS 12098, pp. 193–204, 2020.
https://doi.org/10.1007/978-3-030-51466-2_17

This article provides a sufficient condition on $\sim$ and on the winning condition of a player such that, if she has a winning strategy, she has a winning $\sim$-strategy. The sufficient condition involves time awareness of the player, but *perfect recall* (rephrased in Sect. 2) is not needed. On the one hand, examples show the tightness of the sufficient condition in several directions; on the other hand, further examples show that the sufficient condition is not strictly necessary.

The proof of the sufficient condition has several features. First, from any winning strategy s, it derives a winning $\sim$-strategy $s \circ f$. The map f takes as input the true history of actions, and outputs a well-chosen *virtual history* of equal length. Second, the derivation $s \mapsto s \circ f$ is 1-Lipschitz continuous, i.e., *reactive*, as in reactive systems. (Not only the way of playing is reactive, but also the synthesis of the $\sim$-strategy.) Third, computability of $\sim$ and finiteness of the opponent action set make the derivation *computable*. As a consequence in this restricted context, if the input strategy is computable, so is the uniformized output strategy. Fourth, *finite-memory* implementability of the strategies is preserved, if the opponent action set is finite. Fifth, strengthening the sufficient condition by assuming *perfect recall* makes the virtual-history map f definable incrementally (i.e. by mere extension) as the history grows. This simplifies the proofs and improves the memory bounds.

The weaker sufficient condition, i.e. when not assuming perfect recall, has an important corollary about concurrent games with states and colors: if *any* winning condition (e.g. not necessarily Borel) is defined via the colors, a player who can win can do so by only taking the history of colors and the current state into account, instead of the full history of actions. Finiteness of the memory is also preserved, if the opponent action set is finite. Two additional corollaries involve the energy winning condition or a class of winning conditions laying between Büchi and Muller.

Both the weaker and the stronger sufficient conditions behave rather well algebraically. In particular, they are closed under arbitrary intersections. This yields a corollary involving the conjunction of the two aforementioned winning conditions, i.e., energy and (sub)Muller.

Finding sufficient conditions for strategy uniformization may help reduce the winning strategy search space; or help simplify the notion of strategy: instead of expecting a precise history as an input, it may just expect an equivalence class, e.g. expressed as a simpler trace.

The strategy uniformization problem is also relevant to *protagonist-imperfect-information* games, where the protagonist cannot distinguish between equivalent histories; and also to *antagonist-imperfect-information* games, where the protagonist wants to behave the same after as many histories as possible to conceal information from her opponent or anyone (partially) observing her actions: indeed the opponent, though losing, could try to lose in as many ways as possible over repeated plays of the game, to learn the full strategy of Player 1, i.e. her capabilities. In connection with the latter, the longer version [11] of this article studies the strategy *maximal* uniformization problem: if there is a winning strategy, is there a maximal $\sim$ such that there is a winning $\sim$-strategy? A basic result is proved (there but not here) and examples show its relative tightness.

Related Works. The distinction between perfect and imperfect information was already studied in [16] for finite games. Related concepts were clarified in [9] by using terms such as information partition and perfect recall: this article is meant for a slightly more general setting and thus may use different terminologies.

I am not aware of results similar to my *sufficient condition for universal existence*, but there is an extensive literature, starting around [18], that studies related *decision problems of existence*: in some class of games, is the existence of a uniform winning strategy decidable and how quickly? Some classes of games come from strategy logic, introduced in [4] and connected to information imperfectness, e.g., in [2]. Some other classes come from dynamic epistemic logic, introduced in [8] and connected to games, e.g., in [20] and to decision procedures, e.g., in [14]. Among these works, some [13] have expressed the need for general frameworks and results about uniform strategies; others [3] have studied subtle differences between types of information imperfectness.

Imperfect information games have been also widely used in the field of security, see e.g. the survey [19, Section 4.2]. The aforementioned strategy maximal uniformization problem could be especially relevant in this context.

Structure of the Article. Section 2 presents the main results on the strategy uniformization problem; Sect. 3 presents various corollaries about games with states and colors; and Sect. 4 presents the tightness of the sufficient condition in several directions. Proofs and additional sections can be found in [11].

2   Main Definitions and Results

The end of this section discusses many aspects of the forthcoming definitions and results.

Definitions on Game Theory. In this article, a **two-player win/lose game** is a tuple $\langle A, B, W \rangle$ where A and B are non-empty sets and W is a subset of infinite sequences over $A \times B$, i.e. $W \subseteq (A \times B)^{\omega}$. Informally, Player 1 and Player 2 concurrently choose one action in A and B, respectively, and repeat this ω times. If the produced sequence is in W, Player 1 wins and Player 2 loses, otherwise Player 2 wins and Player 1 loses. So W is called the winning condition (of Player 1).

The **histories** are the finite sequences over $A \times B$, denoted by $(A \times B)^*$. The **opponent-histories** are B^*. The **runs** and **opponent-runs** are their infinite versions.

A Player 1 **strategy** is a function from B^* to A. Informally, it tells Player 1 which action to choose depending on the opponent-histories, i.e. on how Player 2 has played so far.

The **induced history** function $h : ((B^* \to A) \times B^*) \to (A \times B)^*$ expects a Player 1 strategy and an opponent-history as inputs, and outputs a history.

It is defined inductively: $h(s, \epsilon) := \epsilon$ and $h(s, \beta \cdot b) := h(s, \beta) \cdot (s(\beta), b)$ for all $(\beta, b) \in B^* \times B$. Informally, h outputs the very sequence of pairs of actions that are chosen if Player 1 follows the given strategy while Player 2 plays the given opponent-history. Note that $\beta \mapsto h(s, \beta)$ preserves the length and the prefix relation, i.e. $\forall \beta, \beta' \in B^*, |h(s, \beta)| = |\beta| \wedge (\beta \sqsubseteq \beta' \Rightarrow h(s, \beta) \sqsubseteq h(s, \beta'))$.

The function h is extended to accept opponent-runs (in B^ω) and then to output runs: $h(s, \boldsymbol{\beta})$ is the only run whose prefixes are the $h(s, \boldsymbol{\beta}_{\leq n})$ for $n \in \mathbb{N}$, where $\boldsymbol{\beta}_{\leq n}$ is the prefix of $\boldsymbol{\beta}$ of length n. A Player 1 strategy s is a **winning strategy** if $h(s, \boldsymbol{\beta}) \in W$ for all $\boldsymbol{\beta} \in B^\omega$.

Definitions on Equivalence Relations over Histories. Given a game $\langle A, B, W \rangle$, a **strategy constraint** (constraint for short) is an equivalence relation over histories. Given a constraint $\sim$, a strategy s is said to be a $\sim$-**strategy** if $h(s, \beta) \sim h(s, \beta') \Rightarrow s(\beta) = s(\beta')$ for all opponent-histories $\beta, \beta' \in B^*$. Informally, a $\sim$-strategy behaves the same after equivalent histories that are compatible with s.

Useful predicates on constraints, denoted $\sim$, are defined below.

1. Time awareness: $\rho \sim \rho' \Rightarrow |\rho| = |\rho'|$, where $|\rho|$ is the length of the sequence/word ρ.
2. : $\rho \sim \rho' \Rightarrow \rho\rho'' \sim \rho'\rho''$.
3. Perfect recall: $(\rho \sim \rho' \wedge |\rho| = |\rho'|) \Rightarrow \forall n \leq |\rho|, \rho_{\leq n} \sim \rho'_{\leq n}$
4. Weak W-closedness: $\forall \boldsymbol{\rho}, \boldsymbol{\rho}' \in (A \times B)^\omega, (\forall n \in \mathbb{N}, \boldsymbol{\rho}_{\leq n} \sim \boldsymbol{\rho}'_{\leq n}) \Rightarrow (\boldsymbol{\rho} \in W \Leftrightarrow \boldsymbol{\rho}' \in W)$
5. Strong W-closedness: $\forall \boldsymbol{\rho}, \boldsymbol{\rho}' \in (A \times B)^\omega,$
   $(\forall n \in \mathbb{N}, \exists \gamma \in (A \times B)^*, \boldsymbol{\rho}_{\leq n}\gamma \sim \boldsymbol{\rho}'_{\leq n + |\gamma|}) \Rightarrow (\boldsymbol{\rho} \in W \Rightarrow \boldsymbol{\rho}' \in W)$

Note that the first three predicates above constrain only (the information available to) the strategies, while the last two constrain also the winning condition.

Definitions on Automata Theory. The **automata** in this article have the classical form (Σ, Q, q_0, δ) where $q_0 \in Q$ and $\delta : Q \times \Sigma \to Q$, possibly with additional accepting states $F \subseteq Q$ in the definition. The state space Q may be infinite, though. The transition function is lifted in two ways by induction. First, to compute the current state after reading a word: $\delta^+(\epsilon) := q_0$ and $\delta^+(ua) := \delta(\delta^+(u), a)$ for all $(u, a) \in \Sigma^* \times \Sigma$. Second, to compute the sequence of visited states while reading a word: $\delta^{++}(\epsilon) := q_0$ and $\delta^{++}(ua) = \delta^{++}(u)\delta^+(ua)$. Note that $|\delta^{++}(u)| = |u|$ for all $u \in \Sigma^*$.

Given a game $\langle A, B, W \rangle$, a **memory-aware implementation** of a strategy s is a tuple $(M, m_0, \Sigma, \mu,)$ where M is a (in)finite set (the memory), $m_0 \in M$ (the initial memory state), $\Sigma : M \to A$ (the choice of action depending on the memory state), and $\mu : M \times B \to M$ (the memory update), such that $s = \Sigma \circ \mu^+$, where $\mu^+ : B^* \to M$ (the "cumulative" memory update) is defined inductively: $\mu^+(\epsilon) := m_0$ and $\mu^+(\beta b) = \mu(\mu^+(\beta), b)$ for all $(\beta, b) \in B^* \times B$. If M is finite, s is said to be a **finite-memory strategy**.

Word Pairing: for all $n \in \mathbb{N}$, for all $u, v \in \Sigma^n$, let $\mathbf{u}\|\mathbf{v} := (u_1, v_1)\dots(u_n, v_n) \in (\Sigma^2)^n$.

A time-aware constraint is 2-**tape-recognizable** using memory states Q, if there is an automaton $((A \times B)^2, Q, q_0, F, \delta)$ such that $u \sim v$ iff $\delta^+(u\|v) \in F$. (It implies $q_0 = \delta^+(\epsilon\|\epsilon) \in F$.) If moreover Q is finite, the constraint is said to be 2-**tape-regular**. Recognition of relations by several tapes was studied in, e.g., [17]. Note that 2-tape regularity of $\sim$ was called indistinguishability-based in [3].

Main Results. Let us recall additional notions first. Two functions of domain Σ^* that coincide on inputs of length less than n but differ for some input of length n are said to be at distance $\frac{1}{2^n}$. In this context, a map from strategies to strategies (or to $B^* \to B^*$) is said to be 1-Lipschitz continuous if from any input strategy that is partially defined for opponent-histories of length up to n, one can partially infer the output strategy for opponent-histories of length up to n.

Theorem 1. *Consider a game $\langle A, B, W \rangle$ and a constraint $\sim$ that is time-aware and closed by adding a suffix. The two results below are independent.*

1. **Stronger assumptions and conclusions**: *If $\sim$ is also perfectly recalling and weakly W-closed, there exists a map $f : (B^* \to A) \to (B^* \to B^*)$ satisfying the following.*
   (a) *For all $s : B^* \to A$ let f_s denote $f(s)$; we have $s \circ f_s$ is a $\sim$-strategy, and $s \circ f_s$ is winning if s is winning.*
   (b) *The map f is 1-Lipschitz continuous.*
   (c) *For all s the map f_s preserves the length and the prefix relation.*
   (d) *The map $s \mapsto s \circ f_s$ is 1-Lipschitz continuous.*
   (e) *If B is finite and $\sim$ is computable, f is also computable; and as a consequence, so is $s \circ f_s$ for all computable s.*
   (f) *If s has a memory-aware implementation using memory states M, and if $\sim$ is 2-tape recognizable by an automaton with accepting states F, then $s \circ f_s$ has a memory-aware implementation using memory states $M \times F$.*
2. **Weaker assumptions (no perfect recall) and conclusions**: *If $\sim$ is also strongly W-closed, there exists a self-map of the* Player 1 *strategies that satisfies the following.*
   (a) *It maps strategies to $\sim$-strategies, and winning strategies to winning strategies.*
   (b) *It is 1-Lipschitz continuous.*
   (c) *If B is finite and $\sim$ is computable, the self-map is also computable.*
   Moreover, if $\sim$ is 2-tape recognizable using memory $M_\sim$, and if there is a winning strategy with memory M_s, there is also a winning $\sim$-strategy using memory $\mathcal{P}(M_s \times M_\sim)$.

Lemma 1 below shows that the five constraint predicates behave rather well algebraically. It will be especially useful when handling Boolean combinations of winning conditions.

Lemma 1. *Let A, B, and I be non-empty sets. For all $i \in I$ let $W_i \subseteq (A \times B)^\omega$ and $\sim_i$ be a constraint over $(A \times B)^*$. Let $\sim$ be another constraint.*

1. If $\sim_i$ is time-aware (resp. closed by adding a suffix, resp. perfectly recalling) for all $i \in I$, so is $\cap_{i \in I} \sim_i$.
2. If $\sim_i$ is weakly (strongly) W_i-closed for all $i \in I$, then $\cap_{i \in I} \sim_i$ is weakly (strongly) $\cap_{i \in I} W_i$-closed.
3. If $\sim$ is weakly (strongly) W_i-closed for all $i \in I$, then $\sim$ is weakly (strongly) $\cup_{i \in I} W_i$-closed.

Comments on the Definitions and Results. In the literature, Player 1 **strategies** sometimes have type $(A \times B)^* \to A$. In this article, they have type $B^* \to A$ instead. Both options would work here, but the latter is simpler.

In the literature, Player 1 **winning strategies** are often defined as strategies winning against all Player 2 strategies. In this article, they win against all opponent runs instead. Both options would work here, but the latter is simpler.

Consider a game $\langle A, B, W \rangle$ and its **sequentialized version** where Player 1 plays first at each round. It is well-known and easy to show that a Player 1 strategy wins the **concurrent version** iff she wins the sequential version. I have two reasons to use concurrent games here, though. First, the notation is nicer for the purpose at hand. Second, concurrency does not rule out (semi-)deterministic determinacy of interesting classes of games[1] as in [1] and [10], and using a sequentialized version of the main result to handle these concurrent games would require cumbersome back-and-forth game sequentialization that would depend on the winner. That being said, many examples in this article are, morally, sequential/turn-based games.

Strong W-closedness is indeed stronger than **weak W-closedness**, as proved in [11]. Besides these two properties, which relate $\sim$ and W, the other predicates on $\sim$ alone are classical when dealing with information imperfectness, possibly known under various names. For example, **Closedness by adding a suffix** is sometimes called the no-learning property.

However strong the **strong W-closedness** may seem, it is strictly weaker than the conjunction of **perfect recall** and **weak W-closedness**, as proved in [11]. This justifies the attributes stronger/weaker assumptions in Theorem 1. Note that the definition of strong W-closedness involves only the implication $\rho \in W \Rightarrow \rho' \in W$, as opposed to an equivalence.

The update functions of memory-aware implementations have type $M \times B \to M$, so, informally, they observe only the memory internal state and the opponent's action. In particular they do not observe for free any additional state of some system.

The notion of **2-tape recognizability** of equivalence relations is natural indeed, but so is the following. An equivalence relation $\sim$ over Σ^* is said to be **1-tape recognizable** using memory Q if there exists an automaton (Σ, Q, q_0, δ) such that $u \sim v$ iff $\delta^+(u) = \delta^+(v)$. In this case there are at most $|Q|$ equivalence classes. If Q is finite, $\sim$ is said to be **1-tape regular**. When considering

---

[1] A class of games enjoys determinacy if all games therein are determined. A game is (deterministically) determined if one player has a (deterministic) winning strategy.

time-aware constraints, 2-tape recognizability is strictly more general, as proved in [11], and it yields more general results. A detailed account can be read in [3].

Here, the two notions of **recognizability** require nothing about the cardinality of the state space: what matters is the (least) cardinality that suffices. The intention is primarily to invoke the results with finite automata, but allowing for infinite ones is done at no extra cost.

In the **memory** part of Theorem 1.1, the Cartesian product involves only the accepting states F, but it only spares us one state: indeed, in a automaton that is 2-tape recognizing a perfectly recalling $\sim$, the non-final states can be safely merged into a trash state. In Theorem 1, however, the full $M_\sim$ is used and followed by a powerset construction. So by Cantor's theorem, the memory bound in Theorem 1.2 increases strictly, despite finiteness assumption for B.

Generally speaking, 1-**Lipschitz continuous** functions from infinite words to infinite words correspond to (real-time) reactive systems; **continuous** functions correspond to reactive systems with unbounded delay; and **computable functions** to reactive systems with unbounded delay that can be implemented via Turing machines. Therefore both 1-Lipschitz continuity and computabilty are desirable over continuity (and both imply continuity).

In Theorem 1.1, the derived $\sim$-strategy is of the form $s \circ f_s$, i.e. it is essentially the original s fed with modified inputs, which are called virtual opponent-histories. Theorem 1.1b means that it suffices to know s for opponent-history inputs up to some length to infer the corresponding virtual history map f_s for inputs up to the same length. Theorem 1.1c means that for each fixed s, the virtual opponent-history is extended incrementally as the opponent-history grows. The assertations 1b and 1c do not imply one another a priori, but that they both hold implies Theorem 1.1d indeed; and Theorem 1.1d means that one can start synthesizing a $\sim$-strategy and playing accordingly on inputs up to length n already when knowing s on inputs up to length n. This process is even computable in the setting of Theorem 1.1e.

In Theorem 1.2, the derived $\sim$-strategy has a very similar form, but the f_s no longer preserves the prefix relation since the perfect recall assumption is dropped. As a consequence, the virtual opponent-history can no longer be extended incrementally: backtracking is necessary. Thus there is no results that correspond to Theorems 1.1b and 1.1c, yet one retains both 1-Lipschitz continuity of the self-map and its computability under suitable assumptions: Theorems 1.2b and 1.2c correspond to Theorems 1.1d and 1.1e, respectively.

In Lemma 1, **constraints intersection** makes sense since the intersection of equivalence relations is again an equivalence relation. This is false for unions; furthermore, taking the equivalence relation generated by a union of equivalence relations would not preserve weak or strong W-closedness.

3   Application to Concurrent Games with States and Colors

It is sometimes convenient, for intuition and succinctness, to define a winning condition not as a subset of the runs, but in several steps via states and colors.

Given the current state, a pair of actions chosen by the players produces a color and determines the next state, and so on. The winning condition is then defined in terms of infinite sequences of colors.

Definition 1. *An initialized arena is a tuple $\langle A, B, Q, q_0, \delta, C, \mathrm{col}\rangle$ such that*

- *A and B are non-empty sets (of actions of Player 1 and Player 2),*
- *Q is a non-empty set (of states),*
- *$q_0 \in Q$ (is the initial state),*
- *$\delta : Q \times A \times B \to Q$ (is the state update function).*
- *C is a non-empty set (of colors),*
- *$\mathrm{col} : Q \times A \times B \to C$ (is a coloring function).*

Providing an arena with some $W \subseteq C^\omega$ (a winning condition for Player 1) defines a game.

In such a game, a triple in $Q \times A \times B$ is informally called an edge because it leads to a(nother) state via the udpate function δ. Between two states there are $|A \times B|$ edges. Note that the colors are on the edges rather than on the states. This is generally more succinct and it is strictly more expressive in the following sense: in an arena with finite Q, infinite A or B, and colors on the edges, infinite runs may involve infinitely many colors. However, it would never be the case if colors were on the states.

The coloring function col is naturally extended to finite and infinite sequences over $A \times B$. By induction, $\mathrm{col}^{++}(\epsilon) := \epsilon$, and $\mathrm{col}^{++}(a, b) := \mathrm{col}(q_0, a, b)$, and $\mathrm{col}^{++}(\rho(a, b)) := \mathrm{col}^{++}(\rho)\mathrm{col}(\delta^+(\rho), a, b)$. Then $\mathrm{col}^\infty(\rho)$ is the unique sequence in C^ω such that $\mathrm{col}^{++}(\rho_{\leq n})$ is a prefix of $\mathrm{col}^\infty(\rho)$ for all $n \in \mathbb{N}$. Note that $|\mathrm{col}^{++}(\rho)| = |\rho|$ for all $\rho \in (A \times B)^*$.

The histories, strategies, and winning strategies of the game with states and colors are then defined as these of $\langle A, B, (\mathrm{col}^\infty)^{-1}[W]\rangle$, which is a game as defined in Sect. 2. Conversely, a game $\langle A, B, W\rangle$ may be seen as a game with states and colors $\langle A, B, Q, q_0, \delta, C, \mathrm{col}, W\rangle$ where $C = A \times B$, and $Q = \{q_0\}$, and $\mathrm{col}(q_0, a, b) = (a, b)$ for all $(a, b) \in A \times B$.

Recall that the update functions of memory-aware implementations have type $M \times B \to M$, so they do not observe the states in Q for free. This difference with what is customary in some communities is harmless in terms of finiteness of the strategy memory, though.

A Universal Result for Concurrent Games. Corollary 1 below considers games with states and colors. Corollary 1.1 (resp. 1.2) is a corollary of Theorem 1.1 (resp. Theorem 1.2). It says that if there is a winning strategy, there is also one that behaves the same after histories of pairs of actions that yield the same sequence of states (resp. the same current state) and the same sequence of colors. Note that no assumption is made on the winning condition in Corollary 1: *it need not be even Borel.*

Corollary 1. *Consider a game with states and colors $G = \langle A, B, Q, q_0, \delta, C, \mathrm{col}, W\rangle$ where Player 1 has a winning strategy s. The two results below are independent.*

1. *Then* Player 1 *has a winning strategy* s' *(obtained in a Lipschitz manner from* s*) that satisfies the following for all* $\beta, \beta' \in B^*$.

$$\delta^{++} \circ h(s', \beta) = \delta^{++} \circ h(s', \beta') \wedge \mathrm{col}^{++} \circ h(s', \beta) = \mathrm{col}^{++} \circ h(s', \beta') \Rightarrow s'(\beta) = s'(\beta')$$

Furthermore, if s *can be implemented via memory space* M*, so can* s'*; and if* B *is finite and* $\sim$ *is computable,* s' *is obtained in a computable manner from* s.

2. *Then* Player 1 *has a winning strategy* s' *(obtained in a Lipschitz manner from* s*) that satisfies the following for all* $\beta, \beta' \in B^*$.

$$\delta^{+} \circ h(s', \beta) = \delta^{+} \circ h(s', \beta') \wedge \mathrm{col}^{++} \circ h(s', \beta) = \mathrm{col}^{++} \circ h(s', \beta') \Rightarrow s'(\beta) = s'(\beta')$$

Furthermore, if s *can be implemented via memory* M*, then* s' *can be implemented via memory size* $2^{|M|(|Q|^2+1)}$*; and if* B *is finite and* $\sim$ *is computable,* s' *is obtained in a computable manner from* s.

On the one hand, Corollary 1 exemplifies the benefit of dropping the perfect recall assumption to obtain winning strategies that are significantly more uniform. On the other hand, it exemplifies the memory cost of doing so, which corresponds to the proof-theoretic complexification from Theorem 1.1 to Theorem 1.2, as is discussed in [11].

To prove Corollary 1.2 directly, a natural idea is to "copy-paste", i.e., rewrite the strategy at equivalent histories. If done finitely many times, it is easy to prove that the derived strategy is still winning, but things become tricky if done infinitely many times, as it should.

Note that in Corollary 1, assumptions and conclusions apply to both players: indeed, since no assumption is made on W, its complement satisfies all assumptions, too.

A consequence of Corollary 1 is that one could define *state-color strategies* as functions in $(Q^* \times C^*) \to A$ or even $(Q \times C^*) \to A$, while preserving existence of winning strategies. How much one would benefit from doing so depends on the context.

In the remainder of this section, only the weaker sufficient condition, i.e., Theorem 1.2 is invoked instead of Theorem 1.1.

Between Büchi and Muller. In Corollary 1 the exact sequence of colors mattered, but in some cases from formal methods, the winning condition is invariant under shuffling of the color sequence. Corollary 2 below provides an example where Theorem 1 applies (but only to Player 1).

Corollary 2. *Consider a game with states and colors* $G = \langle A, B, Q, q_0, \delta, C, \mathrm{col}, W \rangle$ *with finite* Q *and* C*, and where* W *is defined as follows: let* $(C_i)_{i \in I}$ *be subsets of* C*, and let* $\gamma \in W$ *if there exists* $i \in I$ *such that all colors in* C_i *occur infinitely often in* γ.

If Player 1 *has a winning strategy, she has a finite-memory one that behaves the same if the current state and the multiset of seen colors are the same.*

Note that the games defined in Corollary 2 constitute a subclass of the concurrent Muller games, where finite-memory strategies suffice [7], and a superclass of the concurrent Büchi games, where positional (aka memoryless) strategies suffice. In this intermediate class from Corollary 2, however, positional strategies are not sufficient: indeed, consider the three-state one-player game in Fig. 1 where q_1 and q_2 must be visited infinitely often. As far as I know, Corollary 2 is not a corollary of well-known results, although the complement of the winning condition therein can be expressed by a generalised Büchi automaton.

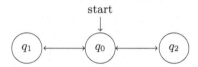

Fig. 1. A game where positional strategies do not suffice: q_1 and q_2 must be visited.

Energy Games. The energy winning condition relates to real-valued colors. It requires that at every finite prefix of a run, the sum of the colors seen so far is non-negative. More formally, $\forall \rho \in (A \times B)^\omega$, $\rho \in W \Leftrightarrow \forall n \in \mathbb{N}, 0 \leq \sum \mathrm{col}^{++}(\rho_{\leq n})$.

Corollary 3 is weaker than the well-known positional determinacy of turn-based energy games, but its proof will be reused in that of Corollary 4.

Corollary 3. *In an energy game $G = \langle A, B, Q, q_0, \delta, \mathbb{R}, \mathrm{col}, W \rangle$, if* Player 1 *has a winning strategy, she has one that behaves the same if the current time, state, and energy level are the same.*

Conjunction of Winning Conditions. Corollary 4 below strengthens Corollary 2 (finite-memory aside) by considering the conjunction of the original Muller condition and the energy condition, which works out by Lemma 1.2.

Corollary 4. *Consider a game with states and colors $G = \langle A, B, Q, q_0, \delta, C \times \mathbb{R}, \mathrm{col}, W \rangle$ with finite Q and C and where $W \subseteq (C \times \mathbb{R})^\omega$ is defined as follows: let $(C_i)_{i \in I}$ be subsets of C, and let $\gamma \in W$ if there exists $i \in I$ such that all colors in C_i occur infinitely often in $\pi_1(\gamma)$ (the sequence of the first components) and if the energy level (on the second component) remains non-negative throughout the run, i.e. $\sum \pi_2 \circ \mathrm{col}^{++}(\gamma_{\leq n})$ for all $n \in \mathbb{N}$.*

If Player 1 *has a winning strategy, she has one that behaves the same if the current state, the multiset of seen colors, and the energy level are the same.*

The diversity of the above corollaries which follow rather easily from Theorem 1, especially Theorem 1.2, should suggest its potential range of application.

4 Tightness Results

This section shows tightness results for Theorem 1. Dropping perfect recall falsifies Theorem 1.1. Dropping either time awareness or closedness by adding a suffix falsifies Theorem 1.2. Despite this relative tightness, the end of the section shows well-known examples that are not captured by Theorem 1.2.

Given a game with a Player 1 winning strategy, given a constraint $\sim$, if there are no Player 1 winning $\sim$-strategies, then $\sim$ is said to be harmful. Otherwise it is said to be harmless.

Below, Proposition 1.1 shows that perfect recall cannot be simply dropped in Theorem 1.1. Proposition 1.2 shows that time awareness cannot be simply dropped in Theorem 1.2. Proposition 1.3 shows that the assumption of closedness by adding a suffix cannot be simply dropped in Theorem 1.2.

Proposition 1. *1. There exist a game $\langle\{0,1\},\{0,1\},W\rangle$ and a constraint that is time-aware, closed by adding a suffix, weakly W-closed, and yet harmful.*
2. There exist a game $\langle\{0,1\},\{0\},W\rangle$ and a constraint that is closed by adding a suffix, strongly W-closed, and yet harmful.
3. There exist a game $\langle\{0,1\},\{0,1\},W\rangle$ and a constraint that is time-aware, strongly W-closed, and yet harmful.

Limitations and Opportunity for Meaningful Generalizations. Despite its relative tightness, Theorem 1 does not imply all known results that can be seen as instances of strategy uniformization problems, so there is room for meaningful generalizations. E.g., due to time awareness requirement, Theorem 1 does not imply positional determinacy of parity games [5,15], where two histories are equivalent if they lead to the same state. Nor does it imply countable compactness of first-order logic [6], which is also an instance of a uniformization problem: Let $(\varphi_n)_{n\in\mathbb{N}}$ be first-order formulas, and define a turn-based game: *Spoiler* plays only at the first round by choosing $m\in\mathbb{N}$. Then *Verifier* gradually builds a countable structure over the signature of $(\varphi_n)_{n\in\mathbb{N}}$. More specifically, at every round she either chooses the value of a variable, or the output value of a function at a given input value, or the Boolean value of a relation for a given pair of values. Only countably many pieces of information are needed to define the structure, and one can fix an order (independent of m) in which they are provided. Verifier wins if the structure she has defined is a model of $\wedge_{0\leq k\leq m}\varphi_k$. Let all histories of equal length be $\sim$-equivalent. Compactness says that if each $\wedge_{0\leq k\leq m}\varphi_k$ has a model, so does $\wedge_{0\leq k}\varphi_k$. Said otherwise, if Verifier has a winning strategy, she has a winning $\sim$-strategy, i.e. independent of Spoiler's first move. This $\sim$ satisfies all the conditions of Theorem 1.1 but weak W-closedness: the premise $(\forall n\in\mathbb{N},\boldsymbol{\rho}_{\leq n}\sim\boldsymbol{\rho}'_{\leq n})$ holds by universality, but the conclusion $(\boldsymbol{\rho}\in W\Leftrightarrow\boldsymbol{\rho}'\in W)$ is false since a model for $\wedge_{0\leq k\leq m}\varphi_k$ need not be a model for $\wedge_{0\leq k\leq m+1}\varphi_k$.

Acknowledgment. This article benefited from a useful conversation with François Schwarzentruber and Stéphane Demri, and from careful reading and comments by several anonymous referees.

References

1. Aghajohari, M., Avni, G., Henzinger, T.A.: Determinacy in discrete-bidding infinite-duration games. CoRR abs/1905.03588 (2019). http://arxiv.org/abs/1905.03588

2. Berthon, R., Maubert, B., Murano, A., Rubin, S., Vardi, M.Y.: Strategy logic with imperfect information. In: 32nd Annual ACM/IEEE Symposium on Logic in Computer Science, LICS 2017, Reykjavik, Iceland, 20–23 June 2017, pp. 1–12 (2017)

3. Berwanger, D., Doyen, L.: Observation and distinction. representing information in infinite games. CoRR abs/1809.05978 (2018). http://arxiv.org/abs/1809.05978

4. Chatterjee, K., Henzinger, T.A., Piterman, N.: Strategy logic. Inf. Comput. **208**(6), 677–693 (2010). https://doi.org/10.1016/j.ic.2009.07.004

5. Emerson, A., Jutla, C.: Tree automata, mu-calculus and determinacy. In: 32nd IEEE Symposium on Foundations of Computer Science, pp. 368–377 (1991)

6. Gödel, K.: Die vollständigkeit der axiome des logischen funktionenkalküls. Monatshefte für Mathematik und Physik **37**(1), 349–360 (1930)

7. Gurevich, Y., Harrington, L.: Trees, automata, and games. In: STOC 1982, pp. 60–65. ACM Press (1982)

8. Hintikka, J.: Knowledge and belief: an introduction to the logic of the two notions. Stud. Logica **16**, 119–122 (1962)

9. Kuhn, H.: Extensive games. Proc. Nat. Acad. Sci. **36**, 570–576 (1950)

10. Le Roux, S.: Concurrent games and semi-random determinacy. In: 43rd International Symposium on Mathematical Foundations of Computer Science, MFCS 2018, Liverpool, UK, 27–31 August 2018, pp. 40:1–40:15 (2018). https://doi.org/10.4230/LIPIcs.MFCS.2018.40

11. Le Roux, S.: Time-aware uniformization of winning strategies. CoRR abs/1907.05128 (2019). http://arxiv.org/abs/1907.05128

12. Martin, D.A.: Borel determinacy. Ann. Math. **102**, 363–371 (1975)

13. Maubert, B., Pinchinat, S.: A general notion of uniform strategies. Int. Game Theory Rev. **16**(01), 1440004 (2014). https://doi.org/10.1142/S0219198914400040

14. Maubert, B., Pinchinat, S., Schwarzentruber, F.: Reachability games in dynamic epistemic logic. CoRR abs/1905.12422 (2019). http://arxiv.org/abs/1905.12422

15. Mostowski, A.: Games with forbidden positions. Research report 78, University of Gdansk (1991)

16. Neumann, J., Morgenstern, O.: Theory of Games and Economic Behavior. Princeton University Press, Princeton (1944)

17. Rabin, M.O., Scott, D.: Finite automata and their decision problems. IBM J. Res. Dev. **3**(2), 114–125 (1959). https://doi.org/10.1147/rd.32.0114

18. Reif, J.H.: The complexity of two-player games of incomplete information. J. Comput. Syst. Sci. **29**(2), 274–301 (1984)

19. Roy, S., Ellis, C., Shiva, S.G., Dasgupta, D., Shandilya, V., Wu, Q.: A survey of game theory as applied to network security. In: 43rd Hawaii International International Conference on Systems Science (HICSS-43 2010), Proceedings, Koloa, Kauai, HI, USA, 5–8 January 2010, pp. 1–10 (2010). https://doi.org/10.1109/HICSS.2010.35

20. Van Benthem, J.: Games in dynamic-epistemic logic. Bull. Econ. Res. **53**(4), 219–248 (2001). https://doi.org/10.1111/1467-8586.00133

Quantitative Coding and Complexity Theory of Compact Metric Spaces

Donghyun Lim and Martin Ziegler[(✉)]

School of Computing, KAIST, Daejeon, Republic of Korea
ziegler@kaist.ac.kr

Abstract. Specifying a computational problem requires fixing encodings for input and output: encoding graphs as adjacency matrices, characters as integers, integers as bit strings, and vice versa. For such discrete data, the actual encoding is usually straightforward and/or complexity-theoretically inessential (up to polynomial time, say); but concerning continuous data, already real numbers naturally suggest various encodings with very different computational properties. With respect to qualitative computability, Kreitz and Weihrauch (1985) had identified *admissibility* as crucial property for "reasonable" encodings over the Cantor space of infinite binary sequences, so-called representations. For (precisely) these does the Kreitz-Weihrauch representation (aka *Main*) Theorem apply, characterizing continuity of functions in terms of continuous realizers. We similarly identify refined criteria for representations suitable for quantitative complexity investigations. Higher type complexity is captured by replacing Cantor's as ground space with more general compact metric spaces, similar to equilogical spaces in computability.

1 Introduction

Machine models formalize computation: they specify means of input, operations, and output of elements from some fixed set Γ; as well as measures of cost and of input/output 'size'; such that Complexity Theory can investigate the dependence of the former on the latter. Problems over spaces X other than Γ are treated by encoding its elements/instances over Γ.

Example 1. a) Recall the Turing machine model operating on the set Γ of finite (e.g. decimal or binary) sequences, and consider the space X of graphs: encoded for example as adjacency matrices' binary entries. Operations amount to local transformations of, and in local dependence of, the tape

Supported by the National Research Foundation of Korea (grant NRF-2017R1E1A1A03071032) and the International Research & Development Program of the Korean Ministry of Science and ICT (grant NRF-2016K1A3A7A03950702). This extended abstract builds on preprints arXiv:2002.04005 and arXiv:1809.08695 and on discussions with Akitoshi Kawamura, Sewon Park, Matthias Schröder, and Florian Steinberg. We thank the organizers for the opportunity of this invited contribution.

M. Anselmo et al. (Eds.): CiE 2020, LNCS 12098, pp. 205–214, 2020.
https://doi.org/10.1007/978-3-030-51466-2_18

contents. Size here is an integer: n commonly denotes the number of nodes of the graph, or the binary length of the encoded matrix, both polynomially related to each other.

b) Consider the space $X = \mathbb{N}$ of natural numbers, either encoded in binary or in unary: their lengths are computably but *not* polynomially related, and induce computably equivalent but significantly different notions of computational complexity.

c) Recall the type-2 machine model [Wei00, §2.1] operating on the Cantor space $\mathcal{C} := \{0,1\}^{\mathbb{N}}$ of infinite binary sequences; and the real unit interval $X = [0;1]$, equipped with various so-called *representations* [Wei00, §4.1]: surjective partial mappings from $\mathcal{C}$ onto X that formalize (sequences of) approximations up to any given absolute error bound $1/2^n$, $n \in \mathbb{N}$. Different representations of X may induce non-/equivalent notions of computability [Wei00, §4.2].

d) Computational cost of a type-2 computation is commonly gauged in dependence of the index position n within the binary input/output sequence, that is, the length of the finite initial segment read/written so far [Wei00, §7.1]. For $X = [0;1]$ and for some of the representations, this notion of 'size' is polynomially (and for some even linearly) related to n occurring in the error bound $1/2^n$ [Wei00, §7.1]; for other computably equivalent representations it is not [Wei00, Examples 7.2.1+7.2.3].

e) Recall the Turing machine model with 'variable' oracles [KC12, §3], operating on a certain subset $\mathcal{B}$ of string functions

$$\{0,1\}^{**} := \{\varphi : \{0,1\}^* \to \{0,1\}^*\} \ .$$

The 'size' of $\varphi \in \mathcal{B}$ here is captured by an integer function $\ell : \mathbb{N} \ni n \mapsto \max\{|\varphi(\vec{x})| : |\vec{x}| \leq n\} \in \mathbb{N}$ [KC96]; and polynomial complexity means bounded by a *second-order* polynomial in $\ell \in \mathbb{N}^{\mathbb{N}}$ and in $n \in \mathbb{N}$ [Meh76].

f) Equip the space $X = \mathcal{C}[0;1]$ of continuous functions $f : [0;1] \to \mathbb{R}$ with the surjective partial mapping $\delta_{\square} :\subseteq \mathcal{B} \twoheadrightarrow X$ from [KC12, §4.3]. Then, up to a second-order polynomial, the 'size' $\ell = \ell(\varphi)$ from (e) is related to a modulus of continuity (cmp. Subsect. 3.1 below) of $f = \delta_{\square}(\varphi) \in \mathcal{C}[0;1]$ and to the computational complexity of the application operator $(f, r) \mapsto f(r)$ [KS17, KS20, NS20].

g) Spaces X of continuum cardinality beyond real numbers are also commonly encoded over Cantor space [Wei00, §3], or over 'Baire' space $\{0,1\}^{**}$ [KC12, §3]. Matthias Schröder has recommended the Hilbert Cube as domain for partial surjections onto suitable X. Also *equilogical* spaces serve as such domains [BBS04].

To summarize, computation on various spaces is commonly formalized by various models of computation (Turing machine, type-2 machine, oracle machine) using encodings over various domains (Cantor space, 'Baire' space, Hilbert Cube, etc.) with various notions of 'size' and of polynomial time.

Question 2. Fix two mathematical structures X and Y, expansions over topological spaces. What machine models, what encodings, what notions of size and polynomial time, are suitable to formalize computation of (multi)functions f from X to Y?

In the sequel we will focus on the part of the question concerned with encoding continuous data. Section 2 recalls classical criteria and notions: qualitative *admissibility* of computably 'reasonable' representations for the Kreitz-Weihrauch *Main Theorem* (Subsect. 2.1), and complexity parameters for a quantitative *Main Theorem* in the real case (Subsect. 2.2). Section 4 combines both towards generic quantitative admissibility and an intrinsic complexity-theoretic *Main Theorem*. The key is to consider metric properties of the *inverse* of a representation, which is inherently multivalued a 'function'. To this end Sect. 3 adopts from [PZ13] a notion of quantitative (uniform) continuity multifunctions (Subsect. 3.1) and establishes important properties (Subsect. 3.2), including closure under a generalized conception of *restriction*. We close with applications to higher-type complexity.

2   Coding Theory of Continuous Data

Common models of computation naturally operate on some particular domain Γ (e.g., in/finite binary sequences, string functions, etc.); processing data from another domain X (graphs, real numbers, continuous functions) requires agreeing on some way of encoding (the elements x of) X over Γ.

Formally, a *representation* is a surjective partial mapping $\xi :\subseteq \Gamma \twoheadrightarrow X$; any $\gamma \in \mathrm{dom}(\xi)$ is called a *name* of $x = \xi(\gamma) \in X$; and for another representation υ of Y, *computing* a total function $f : X \to Y$ means to compute some (ξ, υ)-*realizer*: a transformation $F : \mathrm{dom}(\xi) \to \mathrm{dom}(\upsilon)$ on names such that $f \circ \xi = \upsilon \circ F$.

Some representations are computably 'unsuitable' [Tur37], including the binary expansion $\Gamma = \{0, 1\}^{\mathbb{N}} \ni \bar{b} \mapsto \sum_{n=0}^{\infty} b_n 2^{-n-1} \in [0; 1]$; cmp. [Wei00, Exercise 7.2.7]. Others are suitable for computability investigations [Wei00, Theorem 4.3.2], but not for complexity purposes [Wei00, Examples 7.2.1+7.2.3].

Example 3. The *signed digit representation* of $[0; 1]$ is the partial map

$$\sigma :\subseteq \{00, 01, 10\}^{\mathbb{N}} \subseteq \mathcal{C} \ni \bar{b} \mapsto \tfrac{1}{2} + \sum_{m=0}^{\infty} (2b_{2m} + b_{2m+1} - 1) \cdot 2^{-m-2} \in [0; 1]$$

Already for the case $X = [0; 1]$ of real numbers, it thus takes particular care to arrive at a complexity-theoretically 'reasonable' representation [Wei00, Theorem 7.3.1]; and even more so for continuous real functions [KC12], not to mention for more involved spaces [Ste17].

2.1   Qualitative Admissibility and Computability

Regarding computability on a large class of topological spaces X, an important criterion for a representation is *admissibility* [KW85, Sch02]:

Definition 4. *Call* $\xi :\subseteq \Gamma \twoheadrightarrow X$ *admissible iff it is (i) continuous and satisfies (ii):*
(ii) To every continuous $\zeta :\subseteq \Gamma \twoheadrightarrow X$ *there exists a continuous mapping* $G :$ $\mathrm{dom}(\zeta) \to \mathrm{dom}(\xi)$ *with* $\zeta = \xi \circ G$; *see* [Wei00, Theorem 3.2.9.2].

Admissible representations exist (at least) for T_0 spaces; they are Cartesian closed; and yield the Kreitz-Weihrauch (aka *Main*) Theorem [Wei00, Theorem 3.2.11]:

Fact 5. *Let* $\xi :\subseteq \Gamma \twoheadrightarrow X$ *and* $\upsilon :\subseteq \Gamma \twoheadrightarrow Y$ *be admissible. Then* $f : X \to Y$ *is continuous iff it admits a continuous* (ξ, υ)-realizer $F : \mathrm{dom}(\xi) \to \mathrm{dom}(\upsilon)$.

In particular *dis*continuous functions are *in*computable.

2.2   Real Quantitative Admissibility

The search for quantitative versions of admissibility and the Main Theorem is guided by above notion of qualitative admissibility. It revolves around quantitative metric versions of qualitative topological properties, such as continuity and compactness, obtained via *Skolemization*. Further guidance comes from reviewing the real case.

Recall that a *modulus of continuity* of a function $f : X \to Y$ between compact metric spaces (X, d) and (Y, e) is a strictly increasing mapping $\mu : \mathbb{N} \to \mathbb{N}$ such that

$$d(x, x') \leq 2^{-\mu(n)} \quad \Rightarrow \quad e\big(f(x), f(x')\big) \leq 2^{-n} . \tag{1}$$

In this case one says that f is μ-continuous. Actually we shall occasionally slightly weaken this notion and require Condition (1) only for all sufficiently large n.

Example 6. The signed digit representation $\sigma :\subseteq \mathcal{C} \twoheadrightarrow [0; 1]$ from Example 3 has modulus of continuity $\kappa(n) = 2n$.

Proposition 11d) below provides a converse. Together with Theorem 13 and Lemma 10 below, they yield the following quantitative strengthening of Fact 5 aka qualitative *Main Theorem*, where $\mathcal{O}()$ of refers to the asymptotic Landau symbol:

Theorem 7. *Fix strictly increasing* $\mu : \mathbb{N} \to \mathbb{N}$. *A function* $f : [0; 1] \to [0; 1]$ *has modulus of continuity* $\mathcal{O}\big(\mu(\mathcal{O}(n))\big)$ *iff it has a* (σ, σ)-realizer with modulus of continuity $\mathcal{O}\big(\mu(\mathcal{O}(n))\big)$.

In particular functions f with (only) 'large' modulus of continuity are inherently 'hard' to compute; cmp. [Ko91, Theorem 2.19]. This suggests gauging the efficiency of some actual computation of f *relative* to it modulus of continuity, rather than absolutely [KC12, KS17, KS20, NS20]:

Definition 8. *Function* $f : [0;1] \rightarrow [0;1]$ *is* polynomial-time computable *iff it can be computed in time bounded by a (first or second order) polynomial in the output precision parameter* n *and in* f*'s modulus of continuity.*

In the sequel we consider continuous total (multi)functions whose domains are compact: The latter condition ensures them to have a modulus of (uniform) continuity. Moreover computable functions with compact domains admit complexity bounds depending only on the output precision parameter n; cmp. [Ko91, Theorem 2.19] or [Wei00, Theorems 7.1.5+7.2.7] or [Sch03].

3   Multifunctions

Multifunctions are unavoidable in real computation [Luc77]. Their introduction simplifies several considerations; for example, every function $f : X \rightarrow Y$ has a (possibly multivalued) inverse $f^{-1} : Y \rightrightarrows X$.

Formally, a partial multivalued function (multifunction) F between sets X, Y is a relation $F \subseteq X \times Y$ that models a computational *search* problem: Given (any name of) $x \in X$, return some (name of some) $y \in Y$ with $(x, y) \in F$. One may identify the relation f with the single-valued total function $F : X \ni x \mapsto \{y \in Y \mid (x, y) \in F\}$ from X to the powerset 2^Y; but we prefer the notation $f :\subseteq X \rightrightarrows Y$ to emphasize that not every $y \in F(x)$ needs to occur as output. Letting the answer y depend on the code of x means dropping the requirement for ordinary functions to be extensional; hence, in spite of the oxymoron, such F is also called a *non*-extensional function. Note that no output is feasible in case $F(x) = \emptyset$.

Definition 9. *Abbreviate with* $\mathrm{dom}(F) := \{x \mid F(x) \neq \emptyset\}$ *for the domain of* F*; and* $\mathrm{range}(F) := \{y \mid \exists x : (x, y) \in F\}$*.* F *is total in case* $\mathrm{dom}(F) = X$*; surjective in case* $\mathrm{range}(F) = Y$*. The composition of multifunctions* $F :\subseteq X \rightrightarrows Y$ *and* $G :\subseteq Y \rightrightarrows Z$ *is* $G \circ F =$

$$\big\{(x, z) \mid x \in X, z \in Z, F(x) \subseteq \mathrm{dom}(G), \exists y \in Y : (x, y) \in F \wedge (y, z) \in G\big\}$$

Call F pointwise compact *if* $F(x) \subseteq Y$ *is compact for every* $x \in \mathrm{dom}(F)$*.*

Note that every (single-valued) function is pointwise compact. A computational problem, considered as total single-valued function $f : X \rightarrow Y$, becomes 'easier' when *restricting* arguments to $x \in X' \subseteq X$, that is, when proceeding to $f' = f|_{X'}$ for some $X' \subseteq X$. A search problem, considered as total multifunction $F : X \rightrightarrows Y$, additionally becomes 'easier' when proceeding to any $F' \subseteq X \rightrightarrows Y$ satisfying the following: $F'(x) \supseteq F(x)$ for every $x \in \mathrm{dom}(F')$. We call such F' also a *restriction* of F, and write $F' \sqsubseteq F$. A single-valued function $f : \mathrm{dom}(F) \rightarrow Y$ is a *selection* of $F :\subseteq X \rightrightarrows Y$ if F is a restriction of f.

Lemma 10. *Fix partial multifunctions* $F :\subseteq X \rightrightarrows Y$ *and* $G :\subseteq Y \rightrightarrows Z$*.*

a) If both F *and* G *are pointwise compact, then so is their composition* $G \circ F$*.*

b) *The composition of restrictions* $F' \sqsubseteq F$ *and* $G' \sqsubseteq G$, *is again a restriction* $G' \circ F' \sqsubseteq G \circ F$.

c) *It holds* $F^{-1} \circ F \sqsubseteq \mathrm{id}_X : X \to X$. *Single-valued surjective partial* $g :\subseteq X \twoheadrightarrow Y$ *furthermore satisfy* $g \circ g^{-1} = \mathrm{id}_Y$.

d) *For representations* ξ *of* X *and* υ *of* Y, *the following are equivalent: (i)* $f \circ \xi$ *is a restriction of* $\upsilon \circ F$ *(ii)* f *is a restriction of* $\upsilon \circ F \circ \xi^{-1}$ *(iii)* $\upsilon^{-1} \circ f \circ \xi$ *is a restriction of* F.

3.1   Quantitative Continuity for Multifunctions

Every restriction f' of a single-valued continuous function f is again continuous. This is not true for multifunctions with respect to *hemi*continuity. Instead Definition 12 below adapts, and quantitatively refines, a notion of continuity for multifunctions from [PZ13] such as to satisfy the following properties:

Proposition 11. *a) A single-valued function is* μ-*continuous iff it is* μ-*continuous when considered as a multifunction.*

b) *Suppose that* $F :\subseteq X \rightrightarrows Y$ *is* μ-*continuous. Then every restriction* $F' \sqsubseteq F$ *is again* μ-*continuous.*

c) *If additionally* $G :\subseteq Y \rightrightarrows Z$ *is* ν-*continuous, then* $G \circ F$ *is* $\nu \circ \mu$-*continuous*

d) *The multivalued inverse of the signed digit representation* σ^{-1} *is* $\mathcal{O}(n)$-*continuous.*

e) *For every* $\varepsilon > 0$, *the soft Heaviside 'function'* h_ε *is* id-*continuous, but not for* $\varepsilon = 0$:

$$h_\varepsilon(t) \ := \ \begin{cases} 0 : t \le \varepsilon \\ 1 : t \ge -\varepsilon \end{cases}$$

Our notion of quantitative (uniform) continuity is inspired by [BH94] and [PZ13, §4+§6]:

Definition 12. *Fix metric spaces* (X, d) *and* (Y, e) *and strictly increasing* $\mu :$ $\mathbb{N} \to \mathbb{N}$. *A total multifunction* $F : X \rightrightarrows Y$ *is called* μ-continuous *if there exists some* $n_0 \in \mathbb{N}$, *and to every* $x_0 \in X$ *there exists some* $y_0 \in F(x_0)$, *such that the following holds for every* $k \in \mathbb{N}$:

$$\forall n_1 \ge n_0 \ \forall x_1 \in \overline{\mathrm{B}}_{\mu(n_1)}(x_0) \ \exists y_1 \in F(x_1) \cap \overline{\mathrm{B}}_{n_1}(y_0)$$
$$\forall n_2 \ge n_1 + n_0 \ \forall x_2 \in \overline{\mathrm{B}}_{\mu(n_2)}(x_1) \ \exists y_2 \in F(x_2) \cap \overline{\mathrm{B}}_{n_2}(y_1) \qquad \cdots$$
$$\forall n_{k+1} \ge n_k + n_0 \ \forall x_{k+1} \in \overline{\mathrm{B}}_{\mu(n_{k+1})}(x_k) \exists y_{k+1} \in F(x_{k+1}) \cap \overline{\mathrm{B}}_{n_{k+1}}(y_k) \ .$$

The parameter n_0 is introduced for the purpose of Proposition 11d+e). Recall that also in the single-valued case we sometimes understand Eq. (1) to hold only for all $n \ge n_0$.

A continuous multifunction on Cantor space, unlike one for example on the reals [PZ13, Fig. 5], does admit a continuous selection, and even a bound on the modulus:

Theorem 13. *Suppose (Y, e) is compact of diameter* $\mathrm{diam}(Y) \leq 1$ *and satisfies the* strong *triangle inequality*

$$e(x, z) \leq \max\{e(x, y), e(y, z)\} \leq e(x, y) + e(y, z). \tag{2}$$

If $G :\subseteq \mathcal{C} \rightrightarrows Y$ *is* μ-continuous *and pointwise compact with compact domain* $\mathrm{dom}(G) \subseteq \mathcal{C}$, *then* G *admits a* $\mu(n + \mathcal{O}(1))$-*continuous selection.*

3.2 Generic Quantitative *Main Theorem*

Generalizing both Fact 5 and Theorem 7, Lemma 10 and Proposition 11 and Theorem 13 together in fact yields the following quantitative counterpart to the qualitative *Main Theorem* for generic compact metric spaces:

Theorem 14. *Fix compact metric spaces* (X, d) *and* (Y, e) *of* $\mathrm{diam}(X)$, $\mathrm{diam}(Y) \leq 1$. *Consider representations* $\xi :\subseteq \mathcal{C} \twoheadrightarrow X$ *and* $\upsilon :\subseteq \mathcal{C} \twoheadrightarrow Y$.

Let $\mu, \mu', \nu, \nu', \kappa, K : \mathbb{N} \to \mathbb{N}$ *be strictly increasing such that* ξ *is* μ-*continuous with compact domain and* μ'-*continuous multivalued inverse* $\xi^{-1} : X \rightrightarrows \mathcal{C}$; υ *is* ν-*continuous with compact domain and* ν'-*continuous multivalued inverse* $\upsilon^{-1} : Y \rightrightarrows \mathcal{C}$.

a) *If total multifunction* $g : X \rightrightarrows Y$ *has a* K-*continuous* (ξ, υ)-*realizer* G, *then* g *is* $(\nu \circ K \circ \mu')$-*continuous.*

b) *If total multifunction* $g : X \rightrightarrows Y$ *is* κ-*continuous and pointwise compact, then it has a* $\nu' \circ \kappa \circ \mu(n + \mathcal{O}(1))$-*continuous* (ξ, υ)-*realizer* G.

Following up on Definition 8, this suggests gauging the efficiency of some actual computation of g *relative* to both it modulus of continuity and moduli of continuity of the representations (and their multivalued inverses) involved.

4 Generic Quantitative Admissibility

According to Theorem 14, quantitative continuity of a (multi)function g is connected to that of a (single-valued) realizer G, subject to properties of the representations ξ, υ under consideration.

A 'true' quantitative *Main Theorem* should replace these extrinsic parameters with ones intrinsic to the co/domains X, Y: by imposing suitable conditions on the representations as quantitative variant of qualitative admissibility [Lim19].

Definition 15. *The* entropy *of a compact metric space* (X, d) *is the mapping* $\eta = \eta_X : \mathbb{N} \to \mathbb{N}$ *such that* X *can be covered by* $2^{\eta(n)}$ *closed balls* $\overline{\mathrm{B}}_n(x)$ *of radius* 2^{-n}, *but not by* $2^{\eta(n)-1}$.

Introduced by Kolmogorov [KT59], η thus quantitatively captures total boundedness [Koh08, Definition 18.52]. Its connections to computational complexity are well-known [Wei03, KSZ16].

Example 16. a) The d-dimensional real unit cube $X = [0; 1]^d$ has linear entropy
$\eta(n) = \Theta(dn)$. Cantor space $\mathcal{C} = \{0, 1\}^{\mathbb{N}}$, equipped with the metric $d(\bar{x}, \bar{y}) = 2^{-\min\{n: x_n \neq y_n\}}$, has linear entropy $\eta(n) = \Theta(n)$.

b) The space $[0; 1]'$ of non-expansive (aka 1-Lipschitz) functions $f : [0; 1] \to [0; 1]$ is compact when equipped with the supremum norm and has entropy $\eta(n) = \Theta(2^n)$.

c) More generally fix a connected compact metric space (X, d) of diameter $\mathrm{diam}(X) := \sup\{(d(x, x') : x, x' \in X\}$ with entropy η. Then the space X' of non-expansive functionals $\Lambda : X \to [0; 1]$ is compact when equipped with the supremum norm and has entropy $\eta'(n) = 2^{\eta(n \pm \mathcal{O}(1))}$.

Items (b) and (c) are relevant for higher-type complexity theory.

Since computational efficiency is connected to quantitative continuity (Subsect. 2.2), in Theorem 14 one prefers ξ and ξ^{-1} with 'small' moduli; similarly for υ and υ^{-1}. A simple but important constraint has been identified in [Ste16, Lemma 3.1.13]—originally for single-valued functions, but its proof immediately extends to multifunctions.

Lemma 17. *If surjective (multi)function $g : X \rightrightarrows Y$ is μ-continuous, then it holds $\eta_Y(n) \leq \eta_X \circ \mu(n)$ (for all sufficiently large n).*

This suggests the following tentative definition:

Definition 18. *Fix some compact metric space Γ, and recall Example 1g).*

a) A representation of compact metric space (X, d) is a continuous partial surjective (single-valued) mapping $\xi :\subseteq \Gamma \twoheadrightarrow X$.

b) Fix another compact metric space Δ and representation $\upsilon :\subseteq \Delta \twoheadrightarrow Y$. A (ξ, υ)-realizer of a total (multi)function $f : X \rightrightarrows Y$ is a (single-valued) function $F : \mathrm{dom}(\xi) \to \mathrm{dom}(\upsilon)$ satisfying any/all conditions of Lemma 10d).

c) Representation $\xi :\subseteq \Gamma \twoheadrightarrow X$ is polynomially admissible if (i) It has a modulus of continuity μ such that $\eta_\Gamma \circ \mu$ is bounded by a (first or second order) polynomial in the precision parameter n and in the entropy η of X. (ii) Its multivalued inverse ξ^{-1} has polynomial modulus of continuity μ'.

d) Call total (multi)function $f : X \rightrightarrows Y$ polynomial-time computable iff it can be computed in time bounded by a (first or second order) polynomial in the output precision parameter n and in the entropy η of X.

In view of Lemma 10c+d) we deliberately consider only single-valued representations [Wei05]. Item d) includes Definition 8 as well as higher types, such as Example 16b) and c). Note that Item (c i) indeed quantitatively strengthens Definition 4i). And Item (c ii) quantitatively strengthens Definition 4ii): For υ-continuous ζ, Theorem 13 yields a $\upsilon \circ \mu'$-continuous selection G of $\xi^{-1} \circ \zeta$, that is, with $\zeta = \xi \circ G$ according to Lemma 10.

References

[BBS04] Bauer, A., Birkedal, L., Scott, D.S.: Equilogical spaces. Theoret. Comput. Sci. **315**, 35–59 (2004)

[BH94] Brattka, V., Hertling, P.: Continuity and computability of relations (1994)

[KC96] Kapron, B.M., Cook, S.A.: A new characterization of type-2 feasibility. SIAM J. Comput. **25**(1), 117–132 (1996)

[KC12] Kawamura, A., Cook, S.: Complexity theory for operators in analysis. ACM Trans. Comput. Theory **4**(2), 1–24 (2012)

[Ko91] Ko, K.-I.: Complexity Theory of Real Functions. Progress in Theoretical Computer Science. Birkhäuser, Boston (1991)

[Koh08] Kohlenbach, U.: Applied Proof Theory. Springer, Heidelberg (2008). https://doi.org/10.1007/978-3-540-77533-1

[KS17] Kawamura, A., Steinberg, F.: Polynomial running times for polynomial-time oracle machines. In: 2nd International Conference on Formal Structures for Computation and Deduction, FSCD 2017, 3–9 September, 2017, Oxford, UK, pp. 23:1–23:18 (2017)

[KS20] Kapron, B.M., Steinberg, F.: Type-two polynomial-time and restricted lookahead. Theor. Comput. Sci. **813**, 1–19 (2020)

[KSZ16] Kawamura, A., Steinberg, F., Ziegler, M.: Complexity theory of (functions on) compact metric spaces. In: Proceedings of the 31st Annual ACM/IEEE Symposium on Logic in Computer Science, LICS 2016, New York, NY, USA, July 5–8 2016, pp. 837–846 (2016)

[KT59] Kolmogorov, A.N., Tikhomirov, V.M.: $\mathcal{E}$-entropy and $\mathcal{E}$-capacity of sets infunctional spaces. Uspekhi Mat. Nauk **14**(2), 3–86 (1959)

[KW85] Kreitz, C., Weihrauch, K.: Theory of representations. Theoret. Comput. Sci. **38**, 35–53 (1985)

[Lim19] Lim, D.: Representations of totally bounded metric spaces and their computational complexity. Master's thesis, KAIST, School of Computing, August 2019

[Luc77] Luckhardt, H.: A fundamental effect in computations on real numbers. Theoret. Comput. Sci. **5**(3), 321–324 (1977)

[Meh76] Mehlhorn, K.: Polynomial and abstract subrecursive classes. J. Comput. Syst. Sci. **12**(2), 147–178 (1976)

[NS20] Neumann, E., Steinberg, F.: Parametrised second-order complexity theory with applications to the study of interval computation. Theoret. Comput. Sci. **806**, 281–304 (2020)

[PZ13] Pauly, A., Ziegler, M.: Relative computability and uniform continuity of relations. J. Logic Anal. **5**(7), 1–39 (2013)

[Sch02] Schröder, M.: Effectivity in spaces with admissible multi representations. Math. Logic Q. **48**(Suppl 1), 78–90 (2002)

[Sch03] Schröder, M.: Spaces allowing type-2 complexity theory revisited. In: Brattka, V., Schröder, M., Weihrauch, K., Zhong, N. (eds.) Computability and Complexity in Analysis, Informatik Berichte, FernUniversität in Hagen, August 2003. International Conference, CCA 2003, Cincinnati, USA, 28–30 August 2003, vol. 302, pp. 345–361 (2003)

[Ste16] Steinberg, F.: Computational complexity theory for advanced function spaces in analysis. Ph.D. thesis, TU Darmstadt, Darmstadt (2016). http://tuprints.ulb.tu-darmstadt.de/6096/

[Ste17] Steinberg, F.: Complexity theory for spaces of integrable functions. Log. Methods Comput. Sci. **13**(3) (2017)

[Tur37] Turing, A.M.: On computable numbers, with an application to the "Entscheidungsproblem". Proc. Lond. Math. Soc. **42**(2), 230–265 (1937)

[Wei00] Weihrauch, K.: Computable Analysis. Springer, Berlin (2000). https://doi.org/10.1007/978-3-642-56999-9

[Wei03] Weihrauch, K.: Computational complexity on computable metric spaces. Math. Logic Q. **49**(1), 3–21 (2003)

[Wei05] Weihrauch, K.: Multi-functions on multi-represented sets are closed under flowchart programming. In: Grubba, T., Hertling, P., Tsuiki, H., Weihrauch, K. (eds.) Computability and Complexity in Analysis, Informatik Berichte, Fern Universität in Hagen, July 2005. Proceedings, Second International Conference, CCA 2005, Kyoto, Japan, 25–29 August 2005, vol. 326, pp. 267–300 (2005)

Functions of Baire Class One
over a Bishop Topology

Iosif Petrakis[✉][iD]

University of Munich, Munich, Germany
petrakis@math.lmu.de

Abstract. If $\mathcal{T}$ is a topology of open sets on a set X, a real-valued function on X is of Baire class one over $\mathcal{T}$, if it is the pointwise limit of a sequence of functions in the corresponding ring of continuous functions $C(X)$. If F is a Bishop topology of functions on X, a constructive and function-theoretic alternative to $\mathcal{T}$ introduced by Bishop, we define a real-valued function on X to be of Baire class one over F, if it is the pointwise limit of a sequence of functions in F. We show that the set $B_1(F)$ of functions of Baire class one over a given Bishop topology F on a set X is a Bishop topology on X. Consequently, notions and results from the general theory of Bishop spaces are naturally translated to the study of Baire class one-functions. We work within Bishop's informal system of constructive mathematics BISH*, that is BISH extended with inductive definitions with rules of countably many premises.

1 Introduction

If $\mathcal{T}$ is a topology of open sets on a set X, a function $f : X \to \mathbb{R}$ is of *Baire class one over* $\mathcal{T}$, if it is the pointwise limit of a sequence of functions in the corresponding ring of continuous functions $C(X)$. Such functions, which may no longer be in $C(X)$, were introduced by Baire in [2], suggesting the use of functions, instead of sets, to tackle problems of real analysis. If $B_0(X) = C(X)$, and if $B_1(X)$ is the set of all Baire class one-functions, one defines for every ordinal $\alpha \le \Omega$, where Ω is the first uncountable ordinal Ω, the set

$$B_\alpha(X) := \mathrm{Lim}_p\left(\bigcup_{\beta < \alpha} B_\beta(X) \right),$$

where, if $\mathbb{F}(X)$ is the set of real-valued functions on X, $\Phi \subseteq \mathbb{F}(X)$, and $f_n \xrightarrow{p} f$ denotes that f is the pointwise limit of $(f_n)_{n=1}^\infty$, we set

$$\mathrm{Lim}_p(\Phi) := \{ f \in \mathbb{F}(X) \mid \exists_{(f_n)_{n=1}^\infty \subseteq \Phi}\left(f_n \xrightarrow{p} f\right) \}.$$

The theory of Baire class-functions is a function-theoretic version of the theory of Baire sets i.e., of sets the characteristic function of which is in some Baire class[1].

---

[1] For that see the Lebesgue-Hausdorff theorem in [15], p. 393, and Lorch's comment in [16], p. 751, on the "coextension" of the two theories.

© Springer Nature Switzerland AG 2020
M. Anselmo et al. (Eds.): CiE 2020, LNCS 12098, pp. 215–227, 2020.
https://doi.org/10.1007/978-3-030-51466-2_19

Generalisations of Baire class functions between metrizable spaces are central objects of study in descriptive set theory (see e.g., [13,14]), with Baire class one-functions having applications to the theory of Banach spaces (see e.g., [9]).

The theory of Bishop spaces (TBS) is a function-theoretic approach to constructive topology within Bishop's informal system of constructive mathematics BISH. The fundamental notion of a *function space*, here called a Bishop space, was only introduced by Bishop in [3], p. 71. The subject was revived much later by Bridges in [5], where the notion of a Bishop morphism was also defined, and by Ishihara in [11]. In [18–27] we try to develop TBS.

A Bishop topology of functions F on a set X is a set of real-valued functions defined on X that satisfies the main properties of the set of all Bishop continuous functions from $\mathbb{R}$ to $\mathbb{R}$. A function $\phi : \mathbb{R} \to \mathbb{R}$ is called (*Bishop*) *continuous*, if it is uniformly continuous on every bounded subset B of $\mathbb{R}$ i.e., if for every bounded subset[2] B of $\mathbb{R}$ and for every $\varepsilon > 0$ there exists $\omega_{\phi,B}(\varepsilon) > 0$ such that

$$\forall_{a,b \in B}\big(|a - b| \leq \omega_{\phi,B}(\varepsilon) \Rightarrow |\phi(a) - \phi(b)| \leq \varepsilon\big),$$

where the function $\omega_{\phi,B} : \mathbb{R}^+ \to \mathbb{R}^+$, $\varepsilon \mapsto \omega_{\phi,B}(\varepsilon)$, is called a *modulus of continuity* for ϕ on B. Their set is denoted by $\mathrm{Bic}(\mathbb{R})$, and two functions $\phi_1, \phi_2 \in \mathrm{Bic}(\mathbb{R})$ are *equal*, if $\phi_1(a) = \phi_2(a)$, for every $a \in \mathbb{R}$. The restriction of this notion of continuity to a compact interval $[a, b]$ of $\mathbb{R}$ is equivalent to uniform continuity. By using this stronger notion of continuity, rather than the standard pointwise continuity, Bishop managed to avoid the use of fan theorem in the proof of the uniform continuity theorem and to remain "neutral" with respect to classical mathematics (CLASS), intuitionistic mathematics (INT), and intuitionistic computable mathematics (RUSS).

Extending our work [22], where the Baire sets over a Bishop topology F are studied, here we give an introduction to the constructive theory of Baire class one-functions over a Bishop topology. In analogy to the classical concept, if F is a Bishop topology on a set X, we define a function $f : X \to \mathbb{R}$ to be of Baire class one over F, if it is the pointwise limit of a sequence of functions in F. Our constructive translation of the fundamentals of the classical theory of Baire class one-functions (see e.g., [10]) within TBS is summarized by Theorem 1, according to which the set $B_1(F)$ of Baire class one-functions over F is a Bishop topology on X that includes F. As we explain in Sect. 5, and based on the examples of Baire class one-functions included in Sect. 4, this result offers a way to study constructively classically discontinuous functions.

We work within BISH*, that is BISH extended with inductive definitions with rules of countably many premises. A formal system for BISH* is Myhill's system CST*, developed in [17], or CZF with dependence choice[3] (see [6], p. 12), and some very weak form of Aczel's regular extension axiom (see [1]).

---

[2] It suffices to say that ϕ is uniformly continuous on every interval $[-n, n]$, and the quantification over the powerset of $\mathbb{R}$ is replaced by quantification over $\mathbb{N}$.

[3] Here we use the principle of dependent choice in the proof of Lemma 6.

2   Fundamentals of Bishop Spaces

In this section we include all definitions and facts necessary to the rest of the paper. All proofs not given given here are found in [18].

If $a, b \in \mathbb{R}$, let $a \vee b := \max\{a, b\}$ and $a \wedge b := \min\{a, b\}$. Hence, $|a| = a \vee (-a)$. If $f, g \in \mathbb{F}(X)$, let $f =_{\mathbb{F}(X)} g :\Leftrightarrow \forall_{x \in X}\big(f(x) =_{\mathbb{R}} f(y)\big)$, where for all definitions related to $\mathbb{R}$ see [4], chapter 2. If $f \in \mathbb{F}(X)$ and $(f_n)_{n=1}^\infty \subseteq \mathbb{F}(X)$, the *pointwise convergence* $(f_n) \xrightarrow{p} f$ and the *uniform convergence* $(f_n) \xrightarrow{u} f$ on $A \subseteq X$ are defined, respectively, by

$$(f_n) \xrightarrow{p} f :\Leftrightarrow \forall_{x \in A}\forall_{\epsilon > 0}\exists_{n_0 \in \mathbb{N}}\forall_{n \geq n_0}\big(|f_n(x) - f(x)| < \varepsilon\big),$$

$$(f_n) \xrightarrow{u} f :\Leftrightarrow \forall_{\epsilon > 0}\exists_{n_0 \in \mathbb{N}}\forall_{n \geq n_0}\big(U(A; f, f_n, \varepsilon)\big),$$

$$U(A; f, f_n, \varepsilon) :\Leftrightarrow \forall_{x \in A}\big(|f_n(x) - f(x)| < \varepsilon\big).$$

A set X is *inhabited*, if it has an element. We denote by $\bar{a}$, or simply by a, the constant function on X with value $a \in \mathbb{R}$, and by $\mathrm{Const}(X)$ their set.

Definition 1. *A Bishop space is a pair $\mathcal{F} := (X, F)$, where X is an inhabited set and F is an extensional subset of $\mathbb{F}(X)$ i.e., $[f \in F$ & $g =_{\mathbb{F}(X)} f] \Rightarrow g \in F$, such that the following conditions hold:*

(BS$_1$) $\mathrm{Const}(X) \subseteq F$.

(BS$_2$) *If $f, g \in F$, then $f + g \in F$.*

(BS$_3$) *If $f \in F$ and $\phi \in \mathrm{Bic}(\mathbb{R})$, then $\phi \circ f \in F$.*

(BS$_4$) *If $f \in \mathbb{F}(X)$ and $(f_n)_{n=1}^\infty \subseteq F$ such that $(f_n) \xrightarrow{u} f$ on X, then $f \in F$.*

We call F a Bishop topology on X. If $\mathcal{G} := (Y, G)$ is a Bishop space, a Bishop morphism from $\mathcal{F}$ to $\mathcal{G}$ is a function $h : X \to Y$ such that $\forall_{g \in G}\big(g \circ h \in F\big)$. We denote by $\mathrm{Mor}(\mathcal{F}, \mathcal{G})$ the set of Bishop morphisms from $\mathcal{F}$ to $\mathcal{G}$. If $h \in \mathrm{Mor}(\mathcal{F}, \mathcal{G})$, we say that h is open, if $\forall_{f \in F}\exists_{g \in G}\big(f = g \circ h\big)$.

A Bishop morphism $h \in \mathrm{Mor}(\mathcal{F}, \mathcal{G})$ is a "continuous" function from $\mathcal{F}$ to $\mathcal{G}$. If $h \in \mathrm{Mor}(\mathcal{F}, \mathcal{G})$ is a bijection, then $h^{-1} \in \mathrm{Mor}(\mathcal{G}, \mathcal{F})$ i.e., h is a *Bishop isomorphism*, if and only if h is open. Let $\mathcal{R}$ be the *Bishop space of reals* $(\mathbb{R}, \mathrm{Bic}(\mathbb{R}))$. It is easy to show that if F is a topology on X, then $F = \mathrm{Mor}(\mathcal{F}, \mathcal{R})$ i.e., an element of F is a real-valued "continuous" function on X. A Bishop topology F on X is an algebra and a lattice, where $f \vee g$ and $f \wedge g$ are defined pointwise, and $\mathrm{Const}(X) \subseteq F \subseteq \mathbb{F}(X)$. If $\mathbb{F}^*(X)$ denotes the bounded elements of $\mathbb{F}(X)$, then $F^* := F \cap \mathbb{F}^*(X)$ is a Bishop topology on X. If $x =_X y$ is the given equality on X, a Bishop topology F on X *separates the points* of X, or F is *completely regular* (see [19] for their importance in the theory of Bishop spaces), if

$$\forall_{x, y \in X}\big[\forall_{f \in F}\big(f(x) =_{\mathbb{R}} f(y)\big) \Rightarrow x =_X y\big].$$

In Proposition 5.1.3. of [18] it is shown that F separates the points of X if and only if the induced by F apartness relation on X

$$x \neq_F y :\Leftrightarrow \exists_{f \in F}\big(f(x) \neq_{\mathbb{R}} f(y)\big)$$

is *tight* i.e., $\neg(x \neq_F y) \Rightarrow x =_X y$. We use the last result in the proof of Proposition 1(iv). An apartness relation on X is a positively defined inequality on X. E.g., if $a, b \in \mathbb{R}$, then $a \neq_{\mathbb{R}} b :\Leftrightarrow |a - b| > 0$. In Proposition 5.1.2. of [18] we show that $a \neq_{\mathbb{R}} b \Leftrightarrow a \neq_{\text{Bic}(\mathbb{R})} b$.

Definition 2. *Turning the definitional clauses* $(\text{BS}_1) - (\text{BS}_4)$ *into inductive rules, the least topology* $\bigvee F_0$ *generated by a set* $F_0 \subseteq \mathbb{F}(X)$, *called a subbase of* $\bigvee F_0$, *is defined by the following inductive rules:*

$$\frac{f_0 \in F_0}{f_0 \in \bigvee F_0}, \quad \frac{f \in \bigvee F_0, \; g \in \mathbb{F}(X), \; g =_{\mathbb{F}(X)} f}{g \in \bigvee F_0}, \quad \frac{a \in \mathbb{R}}{\overline{a} \in \bigvee F_0}, \quad \frac{f, g \in \bigvee F_0}{f + g \in \bigvee F_0},$$

$$\frac{f \in \bigvee F_0, \; \phi \in \text{Bic}(\mathbb{R})}{\phi \circ f \in \bigvee F_0}, \quad \frac{f \in \mathbb{F}(X), \; \big(g \in \bigvee F_0, \; \text{U}(X; f, g, \varepsilon)\big)_{\varepsilon > 0}}{f \in \bigvee F_0},$$

where the last rule is reduced to the following rule with countably many premisses

$$\frac{f \in \mathbb{F}(X), \; g_1 \in \bigvee F_0, \; \text{U}(X; f, g_1, \tfrac{1}{2}), \; g_2 \in \bigvee F_0, \; \text{U}(X; f, g_2, \tfrac{1}{4}), \ldots}{f \in \bigvee F_0}.$$

The above rules induce the corresponding induction principle $\text{Ind}_{\bigvee F_0}$ *on* $\bigvee F_0$. *If* $A \subseteq X$, *the relative topology* $F_{|A}$ *on* A *has the set* $\{f_{|A} \mid f \in F\}$ *as a subbase. Unless otherwise stated, from now on,* X, Y *are inhabited sets, and* F, G *are Bishop topologies on* X *and* Y, *respectively.*

3   The Bishop Topology of Baire Class One-Functions

Definition 3. *A function* $g \in \mathbb{F}(X)$ *is called of Baire class one over* F, *or simply of Baire class one when* F *is clear from the context, if there is a sequence* $(f_n)_{n=1}^{\infty} \subseteq F$ *such that* $(f_n) \xrightarrow{p} g$ *on* X. *We denote their set by* $B_1(F)$.

Lemma 1. *Let* $(f_n)_{n=1}^{\infty} \subseteq \mathbb{F}(X)$ *and* $g \in \mathbb{F}(X)$ *with* $(f_n) \xrightarrow{p} g$. *If* $x \in X$, *there is* $M_x > 0$, *such that* $\{f_n(x) \mid n \geq 1\} \cup \{g(x)\} \subseteq [-M_x, M_x]$.

Proof. Let $n_0 \geq 1$ such that if $n \geq n_0$, then $|f_n(x) - g(x)| \leq 1$, hence $|f_n(x)| \leq |f_n(x) - g(x)| + |g(x)| \leq 1 + |g(x)|$. If $M_x := \max\{1 + |g(x)|, |f_1(x)|, \ldots, |f_{n_0-1}(x)|\}$, then $|f_n(x)| \leq M_x$, for every $n \geq 1$, and $|g(x)| \leq M_x$.

Lemma 2. *If* $g \in B_1(F)$ *and* $\phi \in \text{Bic}(\mathbb{R})$, *then* $\phi \circ g \in B_1(F)$.

Proof. Let $(f_n)_{n=1}^{\infty} \subseteq F$ such that $(f_n) \xrightarrow{p} g$. If $x \in X$ and $\varepsilon > 0$ are fixed, there is $n_0\big(\omega_{\phi, [-M_x, M_x]}(\varepsilon)\big)$ such that for every $n \geq n_0\big(\omega_{\phi, [-M_x, M_x]}(\varepsilon)\big)$ we have that $|f_n(x) - g(x)| \leq \omega_{\phi, [-M_x, M_x]}(\varepsilon)$. Since $f_n(x) \in [-M_x, M_x]$, for every $n \geq 1$, and $g(x) \in [-M_x, M_x]$, by the uniform continuity of ϕ we have that

$$|f_n(x) - g(x)| \leq \omega_{\phi, [-M_x, M_x]}(\varepsilon) \Rightarrow |\phi(f_n(x)) - \phi(g(x))| \leq \varepsilon.$$

Hence, for every $n \geq m_0(\varepsilon) := n_0\big(\omega_{\phi, [-M_x, M_x]}(\varepsilon)\big)$ we have that $|(\phi \circ f_n)(x) - (\phi \circ g)(x)| \leq \varepsilon$. Since $\varepsilon > 0$ is arbitrary, we get $(\phi \circ f_n)(x) \xrightarrow{n} (\phi \circ g)(x)$. Since $x \in X$ is arbitrary, we get $\phi \circ f_n \xrightarrow{p} \phi \circ g$.

Note that $(\mathbb{R}, \vee, \wedge)$ is not a distributive lattice, since not even $(\mathbb{Q}, \vee, \wedge)$ is one. For the properties of $a \wedge b$ and $a \vee c$ used in the next proof see [7], p. 52.

Lemma 3. *If $a, b, c, M \in \mathbb{R}$ and $M > 0$, the following hold.*
(i) If $a \leq b$, then $a \vee c \leq b \vee c$ and $a \wedge c \leq b \wedge c$.
(ii) $[a \vee (-M)] \wedge M = [a \wedge M] \vee (-M)$.

Proof. (i) Since $b \leq b \vee c$, we get $a \leq b \vee c$. Since also $c \leq b \vee c$, we get $a \vee c \leq b \vee c$. Since $a \wedge c \leq a \leq b$, and since also $a \wedge c \leq c$, we get $a \wedge c \leq b \wedge c$.
(ii) By Corollary 2.17 in [4], p. 26, $a > -M$ or $a < M$. If $a > -M$, then $[a \vee (-M)] \wedge M = a \wedge M$ and, since $a, M > -M$ we also get $a \wedge M > -M$, hence $[a \wedge M] \vee (-M) = a \wedge M$. If $a < M$, and since also $-M < M$, we get $a \vee (-M) \vee M$, hence $[a \vee (-M)] \wedge M = a \vee (-M)$. Moreover, $[a \wedge M] \vee (-M) = a \vee (-M)$ and the required equality holds.

Lemma 4. *If $g \in B_1(F)$ is bounded by some $M > 0$, there is a sequence $(h_n)_{n=1}^{\infty} \subseteq F$ such that $(h_n) \xrightarrow{p} g$ and h_n is bounded by M, for every $n \geq 1$.*

Proof. If $(f_n)_{n=1}^{\infty} \subseteq F$ such that $(f_n) \xrightarrow{p} g$, let $h_n := [f_n \vee (-M)] \wedge M \in F$, for every $n \geq 1$. We show that $(h_n) \xrightarrow{p} g$. Let $x \in X$, $\varepsilon > 0$ and $n_0(\varepsilon) \geq 1$, such that for every $n \geq n_0(\varepsilon)$ we have that $|f_n(x) - g(x)| \leq \varepsilon$, or equivalently $g(x) - \varepsilon \leq f_n(x) \leq g(x) + \varepsilon$. By Lemma 3(i), and since $-M \leq g(x) \leq M$, we get

$$f_n(x) \vee (-M) \leq [g(x) + \varepsilon] \vee (-M) = g(x) + \varepsilon.$$

Hence

$$[f_n(x) \vee (-M)] \wedge M \leq f_n(x) \vee (-M) \leq [g(x) + \varepsilon] \vee (-M) = g(x) + \varepsilon$$

i.e., $h_n(x) - g(x) \leq \varepsilon$. Since $g(x) - \varepsilon \leq f_n(x)$, by Lemma 3(i) we get $[g(x) - \varepsilon] \wedge M \leq f_n(x) \wedge M$. Since $g(x) - \varepsilon \leq g(x) \leq M$, we get $g(x) - \varepsilon = [g(x) - \varepsilon] \wedge M \leq f_n(x) \wedge M$, hence by Lemma 3(ii) we get

$$g(x) - \varepsilon \leq f_n(x) \wedge M \leq [f_n(x) \wedge M] \vee (-M) = [f_n(x) \vee (-M)] \wedge M$$

i.e., $g(x) - \varepsilon \leq h_n(x)$, which implies $g(x) - h_n(x) \leq \varepsilon$. Since we have already shown that $h_n(x) - g(x) \leq \varepsilon$, by the definition of $|h_n(x) - g(x)|$ we conclude that $|h_n(x) - g(x)| \leq \varepsilon$, for every $n \geq n_0(\varepsilon)$. Of course, $|h_n| \leq M$, for every $n \geq 1$.

The proofs of the following two lemmas for $B_1(X)$ (see [8]) are constructive.

Lemma 5. *Let $(g_k)_{k=1}^{\infty} \subseteq B_1(F)$ and $(M_k)_{k=1}^{\infty} \subseteq \mathbb{R}$ with $M_k > 0$, for every $k \geq 1$, and $\sum_{k=1}^{\infty} M_k \in \mathbb{R}$. If $|g_k| \leq M_k$, for every $k \geq 1$, then $g = \sum_{k=1}^{\infty} g_k \in B_1(F)$.*

Proof. Since g_k is bounded by M_k, for every $k \geq 1$, by Lemma 4 there is $(f_m^k)_{m=1}^{\infty} \subseteq F$ with $f_m^k \xrightarrow{p} g_k$ and $|f_m^k| \leq M_k$. If $n \geq 1$, let $h_n := \sum_{k=1}^{n} f_n^k = f_n^1 + f_n^2 + \ldots + f_n^n \in F$. Let $\varepsilon > 0$. Since $\sum_{k=1}^{\infty} M_k \in \mathbb{R}$, there is $N \geq 1$ with

$\sum_{k=N+1}^{\infty} M_k \leq \frac{\varepsilon}{3}$. If $x \in X$, there is $n_0 \geq N$, such that for every $n \geq n_0$ we have that $|g_k(x) - f_n^k(x)| \leq \frac{\varepsilon}{3N}$, for every $k \in \{1, \dots, N\}$. If $n \geq n_0$, then

$$
|g(x) - h_n(x)| := \left| \sum_{k=1}^{\infty} g_k(x) - \sum_{k=1}^{n} f_n^k \right|
$$

$$
\leq \left| \sum_{k=1}^{n} g_k(x) - f_n^k \right| + \left| \sum_{k=n+1}^{\infty} g_k \right|
$$

$$
\leq \sum_{k=1}^{n} |g_k(x) - f_n^k| + \sum_{k=n+1}^{\infty} |g_k|
$$

$$
= \sum_{k=1}^{N} |g_k(x) - f_n^k| + \sum_{k=N+1}^{n} |g_k(x) - f_n^k| + \sum_{k=n+1}^{\infty} |g_k|
$$

$$
\leq \sum_{k=1}^{N} |g_k(x) - f_n^k| + \sum_{k=N+1}^{n} |f_n^k| + \sum_{k=N+1}^{n} |g_k(x)| + \sum_{k=n+1}^{\infty} |g_k|
$$

$$
= \sum_{k=1}^{N} |g_k(x) - f_n^k| + \sum_{k=N+1}^{n} |f_n^k| + \sum_{k=N+1}^{\infty} |g_k|
$$

$$
\leq \sum_{k=1}^{N} \frac{\varepsilon}{3N} + \sum_{k=N+1}^{n} M_k + \sum_{k=N+1}^{\infty} M_k
$$

$$
\leq N \left(\frac{\varepsilon}{3N} \right) + \frac{\varepsilon}{3} + \frac{\varepsilon}{3} = \varepsilon.
$$

Lemma 6. *If $(g_n)_{n=1}^{\infty} \subseteq B_1(F)$ and $g \in \mathbb{F}(X)$ with $(g_n) \xrightarrow{u} g$, then $g \in B_1(F)$.*

Proof. Using dependent choice there is a subsequence $(g_{n_k})_{k=1}^{\infty}$ of $(g_n)_{n=1}^{\infty}$ with $U\left(X; g, g_{n_k}, \frac{1}{2^k}\right)$, for every $k \geq 1$. Let $h_k := g_{n_{k+1}} - g_{n_k} \in B_1(F)$. If $x \in X$, then

$$
|h_k(x)| \leq |g_{n_{k+1}} - g(x)| + |g(x) - g_{n_k}(x)| \leq \frac{1}{2^{k+1}} + \frac{1}{2^k} = \frac{3}{2} \frac{1}{2^k} =: M_k.
$$

By Lemma 5 we have that $h := \sum_{k=1}^{\infty} h_k \in B_1(F)$. Since

$$
h(x) = \lim_{N \to \infty} \sum_{k=1}^{N} \left(g_{n_{k+1}}(x) - g_{n_k}(x) \right)
$$

$$
= \lim_{N \to \infty} \left[\left(g_{n_2}(x) - g_{n_1}(x) \right) + \dots + \left(g_{n_{N+1}}(x) - g_{n_N}(x) \right) \right]
$$

$$
= \lim_{N \to \infty} \left(g_{n_{N+1}}(x) - g_{n_1}(x) \right)
$$

$$
= \lim_{N \to \infty} \left(g_{n_{N+1}}(x) \right) - g_{n_1}(x)
$$

$$
= g(x) - g_{n_1}(x),
$$

we get $g = h + g_{n_1} \in B_1(F)$, as $B_1(F)$ is trivially closed under addition.

Theorem 1. $B_1(F)$ *is a Bishop topology on* X *that includes* F.

Proof. $B_1(F)$ is an extensional subset of $\mathbb{F}(X)$, since if $g \in \mathbb{F}(X)$ and $(f_n)_{n=1}^{\infty} \subseteq F$ such that $(f_n) \xrightarrow{p} g$ on X, then if $g^* =_{\mathbb{F}(X)} g$, we also get $(f_n) \xrightarrow{p} g^*$ on X. Clearly, $F \subseteq B_1(F)$, and hence $\mathrm{Const}(X) \subseteq B_1(F)$. Moreover, $B_1(F)$ is closed under addition. By Lemma 2 $B_1(F)$ is closed under composition with elements of $\mathrm{Bic}(\mathbb{R})$, and by Lemma 6 $B_1(F)$ is closed under uniform limits.

By Theorem 1, if $g_1, g_2 \in B_1(F)$, then $g_1 \vee g_2, g_1 \wedge g_2, g_1 \cdot g_2$, and $|g_1|$ are in $B_1(F)$. These facts also follow trivially by the definition of $B_1(F)$. The importance of Theorem 1 though, is revealed by the use of the general theory of Bishop spaces in the proof of non-trivial properties of $B_1(F)$ that, consequently, depend only on the Bishop space-structure of $B_1(F)$.

Corollary 1. (*i*) $B_1(F)^* := B_1(F) \cap \mathbb{F}^*(X)$ *is a Bishop topology on* X.

(*ii*) *If* $g \in B_1(F)$ *such that* $g \geq \bar{c}$, *for some* $c \in \mathbb{R}$ *with* $c > 0$, *then* $\frac{1}{g} \in B_1(F)$.

(*iii*) *If* $g \in B_1(F)$ *such that* $g \geq \bar{0}$, *then* $\sqrt{g} \in B_1(F)$.

(*iv*) *The collection* $Z(B_1(F)) = \{\zeta(g) \mid g \in B_1(F)\}$ *of zero sets of* $B_1(F)$, *where* $\zeta(g) := \{x \in X \mid g(x) = 0\}$, *is closed under countable intersections.*

(*v*) [*Urysohn lemma for* $B_1(F)$-*zero sets*] *If* $A, B \subseteq X$, *then there is* $h \in B_1(F)$ *with* $h(A) = 0$ *and* $h(B) = 1$ *if and only if there are* $g_1, g_2 \in B_1(F)$, *and* $c > 0$. *such that* $A \subseteq \zeta(g_1)$, $B \subseteq \zeta(g_2)$, *and* $|g_1| + |g_2| \geq \bar{c}$.

(*vi*) [*Urysohn extension theorem for* $B_1(F)$] *Let* $Y \subseteq X$ *such that* $f_{|Y} \in G$, *for every* $f \in F$. *If for every* $A, B \subseteq Y$, *whenever* A, B *are separated by some function in* $B_1(G)^*$, *then* A, B *are separated by some function in* $B_1(F)^*$, *then every* $g^* \in B_1(G)^*$ *is the restriction of some* $f^* \in B_1(F)^*$.

Proof. These facts follow from the corresponding facts on general Bishop spaces. See [18], p. 41, for (i), Theorem 5.4.8. in [18] for (ii), [26] for a proof of (iii), Proposition 5.3.3.(ii) in [18] for (iv), Theorem 5.4.9. in [18] for (v), and the Urysohn extension theorem for general Bishop spaces in [20] for (vi).

Corollary 1, except from case (iii), are classically shown in [8] *specifically* for $B_1(X)$. Notice that in [20] we avoid quantification over the powerset of Y in the formulation of the Urysohn extension theorem, formulating it predicatively.

Proposition 1. *Let* $x, y \in X$.

(*i*) *If* $g \in B_1(F)$ *with* $g(x) \neq_{\mathbb{R}} g(y)$, *there is* $f \in F$ *such that* $f(x) \neq_{\mathbb{R}} f(y)$.

(*ii*) $x \neq_{B_1(F)} y \Leftrightarrow x \neq_F y$.

(*iii*) *The apartness* $\neq_{B_1(F)}$ *is tight if and only if the apartness* $\neq_F$ *is tight.*

(*iv*) $B_1(F)$ *separates the points of* X *if and only if* F *separates them.*

(*v*) $B_1(F)$ *separates the points of* X *if and only if* $B_1(F)^*$ *separates them.*

Proof. (i) Since $|g(x) - g(y)| > 0$, let the well-defined function

$$g^*(z) := \frac{1}{g(x) - g(y)} g(z) - g(y), \quad z \in X.$$

g^* is in $B_1(F)$, $g^*(x) = 1$ and $g^*(y) = 0$. If $(f_n)_{n=1}^\infty \subseteq F$ with $(f_n) \xrightarrow{p} g^*$, then

$$\exists_{n_0^x(\frac{1}{2}) \in \mathbb{N}} \forall_{n \geq n_0^x(\frac{1}{2})} \left(|f_n(x) - 1| < \frac{1}{2} \right) \quad \& \quad \exists_{n_0^y(\frac{1}{2}) \in \mathbb{N}} \forall_{n \geq n_0^y(\frac{1}{2})} \left(|f_n(y)| < \frac{1}{2} \right).$$

If $m := \max \left\{ n_0^x(\frac{1}{2}), n_0^y(\frac{1}{2}) \right\}$, then $f_m \in F$ with $f_m(x) \in (\frac{1}{2}, \frac{3}{2})$ and $f_m(y) \in (-\frac{1}{2}, \frac{1}{2})$, hence $f_m(x) \neq_\mathbb{R} f_m(y)$.

(ii) If $x \neq_{B_1(F)} y$, there is $g \in B_1(F)$ such that $g(x) \neq_\mathbb{R} g(y)$. By (i) we get $x \neq_F y$. Conversely, if $x \neq_F y$, there is $f \in F$ with $f(x) \neq_\mathbb{R} f(y)$. Since f is also in $B_1(F)$, we get $x \neq_{B_1(F)} y$.

(iii) Let $\neq_{B_1(F)}$ be tight. If $\neg(x \neq_F y)$, then by (ii) we get $\neg(x \neq_{B_1(F)} y)$, hence $x =_X y$. The converse implication is shown similarly.

(iv) It follows from (iii) and the result mentioned in Sect. 2 that a Bishop topology separates the points if and only if its induced apartness is tight.

(v) It follows from the general fact that F separates the points if and only if F^* separates them (see Proposition 5.7.2. in [18]).

Proposition 2. Let $\mathcal{F}_1 := (X, B_1(F))$ and $\mathcal{G}_1 := (Y, B_1(G))$.

(i) If $h \in \mathrm{Mor}(\mathcal{F}, \mathcal{G})$, then $h \in \mathrm{Mor}(\mathcal{F}_1, \mathcal{G}_1)$.

(ii) Let $h : X \to Y$ be a surjection with $\sigma : Y \to X$ a modulus of surjectivity[4] for h i.e., $h \circ \sigma = \mathrm{id}_Y$. If $h \in \mathrm{Mor}(\mathcal{F}, \mathcal{G})$ is open, then $h \in \mathrm{Mor}(\mathcal{F}_1, \mathcal{G}_1)$ is open.

Proof. (i) We need to show that $\forall_{g \in B_1(G)} (g \circ h \in B_1(F))$. If we fix $g \in B_1(G)$, let $(g_n)_{n=1}^\infty \subseteq G$ such that $(g_n) \xrightarrow{p} g$. Then, we get $(g_n \circ h) \xrightarrow{p} g \circ h$. Since $h \in \mathrm{Mor}(\mathcal{F}, \mathcal{G})$, we have that $g_n \circ h \in F$, for every $n \geq 1$, and hence $g \circ h \in B_1(F)$.

(ii) By case (i) $h \in \mathrm{Mor}(\mathcal{F}_1, \mathcal{G}_1)$. By Definition 1, if $\forall_{f \in F} \exists_{g \in G} (f = g \circ h)$, we prove $\forall_{f^* \in B_1(F)} \exists_{g^* \in B_1(G)} (f^* = g^* \circ h)$. Let $f^* \in B_1(F)$ and $(f_n)_{n=1}^\infty \subseteq F$ with $(f_n) \xrightarrow{p} f^*$ on X. By the principle of countable choice (see [6], p. 12) there is $(g_n)_{n=1}^\infty \subseteq G$ such that $f_n = g_n \circ h$, for every $n \geq 1$. Let $g^* : Y \to \mathbb{R}$, defined by $g^* := f^* \circ \sigma$. First we show that $g^* \circ h =_{\mathbb{F}(X)} f^*$. If $x \in X$, we show that

$$(g^* \circ h)(x) := g^*(h(x)) := f^*(\sigma(h(x))) = f^*(x).$$

Since $h(\sigma(h(x))) := (h \circ \sigma)(h(x)) =_Y \mathrm{id}_Y(h(x)) = h(x)$, we get

$$(g_n \circ h)(\sigma(h(x))) := g_n(h(\sigma(h(x)))) = g_n(h(x)) := (g_n \circ h)(x)$$

i.e., $f_n(\sigma(h(x))) = f_n(x)$, for every $n \geq 1$. Since $f_n(\sigma(h(x))) \xrightarrow{n} f^*(\sigma(h(x)))$ and $f_n(x) \xrightarrow{n} f^*(x)$, we get $f^*(\sigma(h(x))) = f^*(x)$. Since $B_1(F)$ is an extensional subset of $\mathbb{F}(X)$ and $g^* \circ h =_{\mathbb{F}(X)} f^* \in B_1(F)$, we conclude that $g^* \circ h \in B_1(F)$ too. To prove $g^* \in B_1(G)$, we show that $(g_n) \xrightarrow{p} g^*$. If $y \in Y$, then

$$g_n(y) = g_n(h(\sigma(y))) := (g_n \circ h)(\sigma(y)) = f_n(\sigma(y)).$$

Since $f_n(\sigma(y)) \xrightarrow{n} f^*(\sigma(y)) := g^*(y)$, we conclude that $g_n(y) \xrightarrow{n} g^*(y)$.

---

[4] We use σ in order to avoid the general axiom of choice in the proof.

4   Examples of Functions of Baire Class One over F

First we find an unbounded Baire class one-function over some Bishop topology. If $n \geq 1$, let $f_n : \{0\} \cup (0,1] \to \mathbb{R}$ defined by

$$f_n(x) := \begin{cases} 0 & , \ x = 0 \\ (n^2 x \wedge n) \wedge \frac{1}{x} & , \ x \in (0,1]. \end{cases}$$

Clearly, $f_n \leq n$, for every $n \geq 1$. If $0 < x < \frac{1}{n}$, then $0 < n^2 x < n$ and $n < \frac{1}{x}$, hence $(n^2 x \wedge n) \wedge \frac{1}{x} = (n^2 x) \wedge \frac{1}{x} = n^2 x$. If $\frac{1}{n} \leq x \leq 1$, then $n \leq n^2 x \leq n^2$ and $\frac{1}{x} \leq n$, hence $(n^2 x \wedge n) \wedge \frac{1}{x} = n \wedge \frac{1}{x} = \frac{1}{x}$. Hence,

$$f_n(x) = \begin{cases} n^2 x & , \ x \in \{0\} \cup (0, \frac{1}{n}) \\ \frac{1}{x} & , \ x \in [\frac{1}{n}, 1]. \end{cases}$$

If $F_0 := \{f_n \mid n \geq 1\}$, we consider the Bishop topology $\bigvee F_0$ on $X := \{0\} \cup (0,1]$. Let the function $g : \{0\} \cup (0,1] \to \mathbb{R}$, defined by

$$g(x) := \begin{cases} 0 & , \ x = 0 \\ \frac{1}{x} & , \ x \in (0,1]. \end{cases}$$

Clearly, g is unbounded on its domain. We show that $f_n \xrightarrow{p} g$, hence $g \in B_1(\bigvee F_0)$. If $x = 0$, then $0 = f_n(0) \xrightarrow{n} g(0) = 0$. Let $x \in (0,1]$. We fix some $\varepsilon > 0$, and we find $n_0 \geq 1$ such that $\frac{1}{n_0} < x$. Hence, if $n \geq n_0$, then $\frac{1}{n} < x$ too. Since then $n < n^2 x$ and $\frac{1}{x} < n$, we have that

$$|f_n(x) - g(x)| = \left| \left[(n^2 x \wedge n) \wedge \frac{1}{x} \right] - \frac{1}{x} \right| = \left| \left[(n \wedge \frac{1}{x} \right] - \frac{1}{x} \right| = \left| \frac{1}{x} - \frac{1}{x} \right| = 0 \leq \varepsilon.$$

A *pseudo-compact* Bishop topology is a topology all the elements of which are bounded functions. Since boundedness is a "liftable" property from F_0 to F i.e., if every $f_0 \in F_0$ is bounded, then every $f \in \bigvee F_0$ is bounded (see Proposition 3.4.4 in [18], p. 46), the topology $\bigvee F_0$ of the previous example is pseudo-compact, and hence the above construction is also an example of an unbounded Baire-class one function over a pseudo-compact Bishop topology!

It is immediate to show that $B_1(\mathbb{F}(X)) = \mathbb{F}(X)$ and $B_1(\mathrm{Const}(X)) = \mathrm{Const}(X)$. Next we find a Baire class one-function over some F that is not in F. Let $Y := [0,1) \cup \{1\}$ be equipped with the relative Bishop topology $C_u([0,1])_{|Y} = \mathrm{Bic}([0,1])_{|Y}$, where $C_u([0,1])$ is the Bishop topology of uniformly continuous functions on $[0,1]$, and $C_u([0,1]) = \mathrm{Bic}([0,1])$, with $\mathrm{Bic}([0,1])$ being defined similarly to $\mathrm{Bic}(\mathbb{R})$. Let $f_n : Y \to \mathbb{R}$, where $f_n := \mathrm{id}_{|Y}^n$, for every $n \geq 1$. By the definition of relative Bishop topology (see Definition 2) we have that $f_n \in \mathrm{Bic}([0,1])_{|Y}$, for every $n \geq 1$, and $(f_n) \xrightarrow{p} g$, where $g : Y \to \mathbb{R}$ is given by

$$g(x) := \begin{cases} 0 & , \ x \in [0,1) \\ 1 & , \ x \in \{1\}. \end{cases}$$

Since Y is dense in $[0,1]$, g is not in $\mathrm{Bic}([0,1])_{|Y}$; if it was, by Proposition 4.7.15. in [18] we get $g = h_{|Y}$ with $h \in C_u([0,1])$, which is impossible.

A similar example is the following. Let $Z := (-\infty, 1) \cup \{1\} \cup (1, +\infty)$ be equipped with the relative topology $\mathrm{Bic}(\mathbb{R})_{|Z}$. If $n \geq 1$, let $\phi_n = \mathrm{nid}_{\mathbb{R}} + (1-n) \in \mathrm{Bic}(\mathbb{R})$ and $\theta_n := -\mathrm{nid}_{\mathbb{R}} + (1+n) \in \mathrm{Bic}(\mathbb{R})$. If $\psi_n := (\phi_n \vee 0) \wedge (\theta_n \vee 0) \in \mathrm{Bic}(\mathbb{R})$,

$$\psi_n(x) := \begin{cases} 0 & , \ x < \frac{n-1}{n} \\ nx + 1 - n & , \ x \in [\frac{n-1}{n}, 1] \\ -nx + 1 + n & , \ x \in (1, \frac{n+1}{n}] \\ 0 & , \ x > \frac{n+1}{n}. \end{cases}$$

Let ψ_n^* be the restriction of ψ_n to Z, for every $n \geq 1$. Clearly, $\psi_n^* \xrightarrow{p} h$, where

$$h(x) := \begin{cases} 0 \ , \ x \in (-\infty, 1) \\ 1 \ , \ x \in \{1\} \\ 0 \ , \ x \in (1 + \infty). \end{cases}$$

Since Z is dense in $\mathbb{R}$ (see Lemma 2.2.8. of [18]), and arguing as in the previous example, h cannot be in the specified Bishop topology on Z.

As in the classical case, all derivatives of differentiable functions in $\mathbb{F}(\mathbb{R})$ are Baire class one-functions over $\mathrm{Bic}(\mathbb{R})$. We reformulate the definition in [4], p. 44, as follows.

Definition 4. *Let $a < b$, $f, f' : [a,b] \to \mathbb{R}$ (uniformly) continuous on $[a,b]$, and $\delta_{f,[a,b]} : \mathbb{R}^+ \to \mathbb{R}^+$. We say that f is differentiable on $[a,b]$ with derivative f' and modulus of differentiability $\delta_{f,[a,b]}$, in symbols $\mathrm{Dif}(f, f', \delta_{f,[a,b]})$, if*

$$\forall_{\varepsilon > 0} \forall_{x,y \in [a,b]} \left(|y - x| \leq \delta_{f,[a,b]}(\varepsilon) \Rightarrow |f(y) - f(x) - f'(x)(y-x)| \leq \varepsilon |y - x| \right).$$

If $\phi, \phi' \in \mathrm{Bic}(\mathbb{R})$, we say that ϕ is differentiable with derivative ϕ', in symbols $\mathrm{Dif}(\phi, \phi')$, if for every $n \geq 1$ we have that $\mathrm{Dif}(\phi_{|[-n,n]}, \phi'_{|[-n,n]}, \delta_{\phi_{|[-n,n]},[-n,n]})$.

Proposition 3. *If $\phi, \phi' \in \mathrm{Bic}(\mathbb{R})$ such that $\mathrm{Dif}(\phi, \phi')$, then $\phi' \in B_1(\mathrm{Bic}(\mathbb{R}))$.*

Proof. If $n \geq 1$, let $\phi_n := n[\phi \circ (\mathrm{id}_{\mathbb{R}} + \frac{1}{n}) - \phi] \in \mathrm{Bic}(\mathbb{R})$. We show that $(\phi_n) \xrightarrow{p} \phi'$. Let $x \in \mathbb{R}$ and $\varepsilon > 0$. Let $N \geq 1$ with $x \in [-N, N]$. Since $(x + \frac{1}{n}) \xrightarrow{n} x$ and $\delta_{\phi,[-N,N]}(\varepsilon) > 0$, there is $n_0 \geq 1$ such that for every $n \geq n_0$ we have that $x + \frac{1}{n} \in [-N, N]$ and $\frac{1}{n} \leq \delta_{\phi,[-N,N]}(\varepsilon)$, hence $\frac{1}{n} = |(x + \frac{1}{n} - x| \leq \delta_{\phi,[-N,N]}(\varepsilon)$, and by Definition 4 we have that

$$\left| \phi\left(x + \frac{1}{n}\right) - \phi(x) - \phi'(x)\left(x + \frac{1}{n} - x\right) \right| \leq \varepsilon \left| x + \frac{1}{n} - x \right| \Leftrightarrow$$

$$\left| \phi\left(x + \frac{1}{n}\right) - \phi(x) - \phi'(x)\frac{1}{n} \right| \leq \varepsilon \frac{1}{n} \Rightarrow$$

$$\left| n\phi\left(x + \frac{1}{n}\right) - n\phi(x) - \phi'(x) \right| \leq \varepsilon \Leftrightarrow$$

$$|\phi_n(x) - \phi'(x)| \leq \varepsilon.$$

5   Concluding Comments

In this paper we introduced the notion of a function of Baire class one over a Bishop topology F, translating a fundamental notion of classical real analysis and topology into the constructive topology of Bishop spaces. Our central result, that the set $B_1(F)$ of Baire class one-functions over F is a Bishop topology that includes F, is used to apply concepts and results from the general theory of Bishop spaces to the theory of functions of Baire class one over a Bishop topology. These first applications suggest that the structure of Bishop space, treated classically, would also be useful to the classical study of function spaces like $B_1(X)$.

For constructive topology, the fact that $B_1(F)$ is a Bishop topology provides a second way, within the theory of Bishop spaces, to treat classically discontinuous, real-valued functions as "continuous" i.e., as Bishop morphisms. The first way is to consider such discontinuous functions as elements of a subbase F_0. Since by definition $F_0 \subseteq \bigvee F_0$, the elements of F_0 are Bishop morphisms from the resulting least Bishop space $\mathcal{F}$ to the Bishop space $\mathcal{R}$ of reals. In [27], and based on a notion of convergence of test functions introduced by Ishihara, we follow this way to make the Dirac delta function δ and the Heaviside step function H "continuous". We consider a certain set $D_0(\mathbb{R})$ of linear maps on the test functions on $\mathbb{R}$, where $\delta, H \in D_0(\mathbb{R})$, and the Bishop topology $\bigvee D_0(\mathbb{R})$ is used to define the set of distributions on $\mathbb{R}$. The second way, is to start from a Bishop topology F and find elements of $B_1(F)$ i.e., Bishop morphisms from $\mathcal{F}_1$ to $\mathcal{R}$, that are pointwise discontinuous, as the functions g and h in the last two example before Definition 4. This second way is sort of a constructive analogue to the classical result that the points of pointwise continuity of some $f \in B_1(\mathbb{R})$ is dense in $\mathbb{R}$, hence f is almost everywhere continuous.

There are numerous interesting questions stemming from this introductory work. Can we prove constructively that the characteristic function of a (complemented) Baire set $\boldsymbol{B} = (B^1, B^0)$ over a Bishop topology F (see [22]) is a Baire class-one function over the relative topology $F_{|B^1 \cup B^0}$? Can we show constructively other classical characterisations of $B_1(X)$, like for example through F_σ-sets? What is the exact relation between $B_1(F)_{|A}$ and $B_1(F_{|A})$, or between $B_1(F \times G)$ and the product Bishop topology (see [18], Sect. 4.1 for its definition) $B_1(F) \times B_1(G)$? How far can we go constructively with the study of Baire class two-functions?

A *base* of a Bishop topology F is a subset B of F such that every $f \in F$ is the uniform limit of a sequence in B. If B is a base of F, it follows easily that

$$\mathrm{Lim}_p(B) = B_1(F),$$

hence for the uniform closure $\overline{\mathrm{Lim}_p(B)}$ of B in $\mathbb{F}(X)$ we get

$$\overline{\mathrm{Lim}_p(B)} = \overline{B_1(F)} = B_1(F)$$

i.e., $\mathrm{Lim}_p(B)$ is a base of $B_1(F)$. If $F_0 \subseteq F$ is a subbase of F i.e., $F = \bigvee F_0$, we have that $\mathrm{Lim}_p(F_0) \subseteq \mathrm{Lim}_p(\bigvee F_0) = B_1(F)$, hence $\bigvee \mathrm{Lim}_p(F_0) \subseteq B_1(F)$. When does the inverse inclusion also hold?

We hope to address some of these questions in a future work.

References

1. Aczel, P., Rathjen, M.: Constructive Set Theory, Book Draft (2010)
2. Baire, R.-L.: Sur les fonctions des variable réelles. Annali. Mat. **3**, 1–123 (1899)
3. Bishop, E.: Foundations of Constructive Analysis. McGraw-Hill, New York City (1967)
4. Bishop, E., Bridges, D.S.: Constructive Analysis, Grundlehren der math. Wissenschaften, vol. 279, Springer, Heidelberg (1985)
5. Bridges, D.S.: Reflections on function spaces. Ann. Pure Appl Log. **163**, 101–110 (2012)
6. Bridges, D.S., Richman, F.: Varieties of Constructive Mathematics. Cambridge University Press, Cambridge (1987)
7. Bridges, D.S., Vîţă, L.S.: Techniques of Constructive Analysis. Universitext, Springer, New York (2006)
8. Deb Ray, A., Mondal, A.: On rings of Baire one functions. Appl. Gen. Topol. **20**(1), 237–249 (2019)
9. Haydon, R., Odell, E., Rosenthal, H.P.: Certain subsclasses of Baire-1 functions with Banach space applications. In: Longhorn Notes, University of Texas at Austin Functional Analysis Seminar (1987–1989)
10. Hu, J.: Baire one functions, preprint
11. Ishihara, H.: Relating Bishop's function spaces to neighborhood spaces. Ann. Pure Appl. Log. **164**, 482–490 (2013)
12. Jayne, J.E.: Space of Baire functions. I, Annales de l'institut Fourier, pp. 47–76 (1974)
13. Kechris, A.S., Louveau, A.: A classification of Baire class 1 functions. Trans. Am. Math. Soc. **318**(1), 209–236 (1990)
14. Kechris, A.S.: Classical Descriptive Set Theory. Springer, Heidelberg (1995)
15. Kuratowski, K.: Topology, vol. I. Academic Press, New York (1966)
16. Lorch, E.R.: Continuity and Baire functions. Am. Math. Monthly **78**(7), 748–762 (1971)
17. Myhill, J.: Constructive set theory. J. Symb. Log. **40**, 347–382 (1975)
18. Petrakis, I.: Constructive Topology of Bishop Spaces, Ph.D. Thesis, Ludwig-Maximilians-Universität, München (2015)
19. Petrakis, I.: Completely regular Bishop spaces. In: Beckmann, A., Mitrana, V., Soskova, M. (eds.) CiE 2015. LNCS, vol. 9136, pp. 302–312. Springer, Cham (2015). https://doi.org/10.1007/978-3-319-20028-6_31
20. Petrakis, I.: The Urysohn extension theorem for Bishop spaces. In: Artemov, S., Nerode, A. (eds.) LFCS 2016. LNCS, vol. 9537, pp. 299–316. Springer, Cham (2016). https://doi.org/10.1007/978-3-319-27683-0_21
21. Petrakis, I.: A constructive function-theoretic approach to topological compactness. In: LICS, pp. 605–614 (2016)
22. Petrakis, I.: Borel and Baire sets in Bishop spaces. In: Manea, F., Martin, B., Paulusma, D., Primiero, G. (eds.) CiE 2019. LNCS, vol. 11558, pp. 240–252. Springer, Cham (2019). https://doi.org/10.1007/978-3-030-22996-2_21
23. Petrakis, I.: Constructive uniformities of pseudometrics and Bishop topologies. J. Log. Anal. **11**(FT2), 1–44 (2019)
24. Petrakis, I.: Direct spectra of Bishop spaces and their limits. https://arxiv.org/abs/1907.03273 (2019)

25. Petrakis, I.: Embeddings of Bishop spaces. J. Logi. Comput. **30**(1), 349–379 (2020). 10.1093/logcom/exaa015
26. Petrakis, I.: A constructive theory of $C^*(X)$, preprint (2020)
27. Petrakis, I.: Towards a constructive approach to the theory of distributions, preprint (2020)

Combinatorial Properties of Degree Sequences of 3-Uniform Hypergraphs Arising from Saind Arrays

A. Frosini[1], G. Palma[2], and S. Rinaldi[2(✉)]

[1] Dipartimento di Matematica e Informatica "U. Dini", Università di Firenze, Florence, Italy
andrea.frosini@unifi.it
[2] Dipartimento di Ingegneria dell'Informazione e Scienze Matematiche, Università di Siena, Siena, Italy
giulia.palma@phd.unipi.it, rinaldi@unisi.it

Abstract. The characterization of k-uniform hypergraphs by their degree sequences, say k-sequences, has been a longstanding open problem for $k \geq 3$. Very recently its decision version was proved to be *NP*-complete in [3]. In this paper, we consider Saind arrays S_n of length $3n-1$, i.e. arrays $(n, n-1, n-2, \ldots, 2-2n)$, and we compute the related 3-uniform hypergraphs incidence matrices M_n as in [3], where, for any M_n, the array of column sums, $\pi(n)$ turns out to be the degree sequence of the corresponding 3-uniform hypergraph. We show that, for a generic $n \geq 2$, $\pi(n)$ and $\pi(n+1)$ share the same entries starting from an index on. Furthermore, increasing n, these common entries give rise to the integer sequence A002620 in [15]. We prove this statement introducing the notion of *queue-triad* of size n and pointer k. Sequence A002620 is known to enumerate several combinatorial structures, including symmetric Dyck paths with three peaks, some families of integers partitions in two parts, bracelets with beads in three colours satisfying certain constraints, and special kind of genotype frequency vectors. We define bijections between queue triads and the above mentioned combinatorial families, thus showing an innovative approach to the study of 3-hypergraphic sequences which should provide subclasses of 3-uniform hypergraphs polynomially reconstructable from their degree sequences.

1 Introduction

A fundamental and widely investigated notion related both to graphs and to hypergraphs is the characterization of their degree sequences (i.e. the array of their vertex degrees), indeed they are of relevance to model and gather information about a wide range of statistics of complex systems (see the book by Berge [14]). The characterization of the degree sequences of k-uniform simple hypergraphs, i.e., those hypergraphs whose hyperedges have the same cardinality k and such that no loops and no parallel hyperedges are present, called k-hypergraphic sequences, has been a long-standing open problem for the case

© Springer Nature Switzerland AG 2020
M. Anselmo et al. (Eds.): CiE 2020, LNCS 12098, pp. 228–238, 2020.
https://doi.org/10.1007/978-3-030-51466-2_20

$k > 2$, until very recently has been proved to be *NP*-complete [3]. Formally, that is: Given $\pi = (d_1, d_2, \ldots, d_n)$ a sequence of positive integers, can π be the degree sequence of a *k*-uniform simple hypergraph?

The degree sequences for $k = 2$, that is for simple graphs, have been studied by many authors, including the celebrated work of Erdös and Gallai [2], which effectively characterizes them. A polyonomial time algorithm to reconstruct the adjacency matrix of a graph G having π as degree sequence (if G exists) has been defined by Havel and Hakimi [16].

In this article, we push further the study of degree sequences of simple *k*-uniform hypergraphs, with $k \geq 3$: the NP-completeness result of their character-ization led our interest to find some subclasses that are polynomially tractable in order to restrict the NP-hard core of the problem. Some literature on recent developments on this subject can be found in [3,5,7,10,12]. In particular, the present study aims at studying degree sequences that generalize those used as gadget for the NP-completeness proof in [3] and that can be computed starting from a generic integer vector, as shown in the next section. We mainly focus on vectors of the form $(n, n-1, n-2, \ldots, 2-2n)$, that we call *Saind arrays*, and we analyze the combinatorial properties of the derived degree sequences in order to characterize them and to gather information about the associated 3−uniform hypergraphs. The results are obtained by borrowing some mathematical tools from recent research areas involving Discrete Mathematics: Discrete Tomogra-phy, Enumerative Combinatorics and Combinatorics on words. The next section is devoted to definitions and results about graphs and hypergraphs that are useful for our study.

In Sect. 3 we first introduce the notion of Saind array and we restrict our investigation to 3−uniform hypergraphs and Saind sequences. We compute the related incidence matrices M_n as in [3], where, for any M_n, the array of column sums, $\pi(n)$ turns out to be the degree sequences of the corresponding 3-uniform hypergraph. We show that, for a generic $n \geq 2$, $\pi(n)$ and $\pi(n+1)$ share the same entries starting from an index on. Furthermore, increasing n, these com-mon entries give rise to the integer sequence A002620 in [15], that we call the *Saind sequence*. Then, in Sect. 4 we analyze the combinatorial properties of the computed degree sequences and then we are able to describe the Saind sequences by introducing the notion of queue triads of given size and fixed pointer.

Then, we show their connections with other families of combinatorial struc-tures known in the literature. Precisely, we show bijections between queue triads and integer partitions in two parts; queue triads and symmetric Dyck paths with three peaks.

2   Definitions and Previous Results

The seminal books by Berge [14] will give to the reader the formal definitions and vocabulary, some results with the related proofs, and more about applications of hypergraphs. In the following we recall the main concepts.

The notion of hypergraph generalizes that of graph, in the sense that each hyperedge is a non-void subset of the set of vertices, without constraints on its cardinality. Formally, a *hypergraph* $\mathcal{H}$ is a pair (V, E), where $V = \{v_1, \ldots, v_n\}$ is a finite set of vertices, and $E \subset 2^{|V|} \setminus \emptyset$ is a set of hyperedges, i.e. a collection of subsets of V. A hypergraph is *simple* if none of its hyperedges is a singleton and there are no two hyperedges one included in (or equal to) another. From now on we will only consider simple hypergraphs.

The *degree* of a vertex is the number of hyperedges that contain it. The degree sequence $\pi = (d_1, d_2, \ldots, d_n)$ of a simple hypergraph $\mathcal{H}$ is the sequence of the degrees of its vertices, usually arranged in non increasing order. When $\mathcal{H}$ is k-uniform (i.e. each hyperedge contains exactly k vertices) the sequence π is called k-hypergraphic.

The study of k-hypergraphic sequences started with the simplest case of $k = 2$, i.e. the case of *graphs*. A 2-graphic sequence is simply called *graphic*. Observe that a simple graph is then a graph without loops or parallel edges. The problem of characterizing graphic sequences of simple graphs was solved by Erdös and Gallai [2]:

Theorem 1. *A sequence* $\pi = (d_1, d_2, \ldots, d_n)$, *where* $d_1 \geq d_2 \geq \cdots \geq d_n$ *is graphic if and only if* $\sum_{i=1}^{n} d_i$ *is even and*

$$\sum_{i=1}^{k} d_i \leq k(k-1) + \sum_{i=k+1}^{n} \min\{k, d_i\}, 1 \leq k \leq n.$$

Let us denote with k-Seq the problem of deciding if an integer sequence π is a k-sequence. The problem k-Seq for $k \geq 3$ was raised by Colbourn et al. in [9] and only recently proved to be NP-complete by Deza et al. in [3]. The proof consists in a reduction of the NP-complete problem 3-Partition into 3-Seq and it is based, in an intermediate step, on the construction of a 3-uniform hypergraph $\mathcal{H}_S$ from an integer sequence S related to an instance of 3-Partition.

We provide a generalized version of this construction: let $S = (s_1, \ldots, s_k)$ be an array of integers. We define a binary matrix M_S of dimension $k' \times k$ collecting all the distinct rows (arranged in lexicographical order) that satisfy the following constraint: for every index i, the i-th row of M_S has all elements equal to zero except three entries in positions j_1, j_2 and j_3 such that $s_{j_1} + s_{j_2} + s_{j_3} > 0$. The number of rows k' is bounded by $\binom{k}{3}$. For instance, the matrix M_S with $S = (5, 2, 2, -1, -4, -4)$ is

5	2	2	−1	−4	−4
1	1	1	0	0	0
1	1	0	1	0	0
1	1	0	0	1	0
1	1	0	0	0	1
1	0	1	1	0	0
1	0	1	0	1	0
1	0	1	0	0	1
0	1	1	1	0	0
7	5	5	3	2	2

where S is depicted in red, and π_S in blue.

The matrix M_S is **incidence matrix** of a (simple) 3-uniform hypergraph $\mathcal{H}_S = (V, E)$ such that the element $M_S(i, j) = 1$ if and only if the hyperedge $e_i \in E$ contains the vertex v_j. Let $\pi_S = (p_1, \ldots, p_k)$ denote the degree sequence of $\mathcal{H}_S$. It holds $\sum_{i=1}^{k'} M_S(i,j) = p_j$. In [3], the authors underline the remarkable property that $\mathcal{H}_S$ is the only 3-uniform hypergraph (up to isomorphism) having degree sequence π_S. Moving the spot on M_S, it is the only binary matrix having distinct rows, 3-constant row sums and π_S column sums. The following problems can be addressed:

Problem 1: determine the computational complexity of 3-Seq restricted to the class of the instances π_S.

Problem 2: provide a combinatorial characterization of the 3-sequences of 3-uniform hypergraphs which are unique up to isomorphism. Determine the computational complexity of 3-Seq restricted to that class of instances.

The present study constitutes a step ahead in the solution of Problem 1: in the next section we consider a family of non decreasing integer sequences and we establish a connection between the 3-uniform hypergraphs constructed from these sequences to several different combinatorial objects in order to find some common properties that will provide a useful starting point for their characterization and for the reconstruction of the associated hypergraphs.

3   Saind Arrays and Their Incidence Matrices

In our analysis of the k-sequences, we restrict the investigation to those π_{S_n} obtained when S_n is a *Saind arrays*, i.e. a sequence defined, for any $n \geq 2$, as $S_n = (n, n-1, n-2, \ldots, 2-2n)$. For the sake of simplicity we will often refer to the elements of the array S_n as $(s_1, \ldots, s_{3n-1})$, where $s_i = n - i + 1$, and to the related degree sequence π_{S_n} as $\pi(n)$. For every $n \geq 2$, according to [3] we associate to S_n its (unique) incidence matrix M_{S_n} (briefly, M_n), obtained as described in the previous section.

So for example, $S_2 = (2, 1, 0, -1, -2)$, $S_3 = (3, 2, 1, 0, -1, -2, -3, -4)$, and their incidence matrices, M_2 and M_3, respectively, are depicted in Fig. 1.

By definition, for any $n \geq 2$, $\pi(n) = (\pi_1, \ldots, \pi_{3n-1})$ is such that $\pi_1 \geq \pi_2 \geq \ldots \geq \pi_{3n-1} = 1$. For small values of $n \geq 2$, the vectors $\pi(n)$ are reported below:

$$\pi(2) = (4, 3, 2, \mathbf{2}, \ \mathbf{1})$$
$$\pi(3) = (12, 10, 8, 6, 5, 4, \ \mathbf{2}, \ \mathbf{1})$$
$$\pi(4) = (25, 21, 18, 15, 12, 10, 9, \ \mathbf{6}, \ \mathbf{4}, \ \mathbf{2}, \ \mathbf{1})$$
$$\pi(5) = (42, 37, 32, 28, 24, 20, 17, 15, \mathbf{12}, \ \mathbf{9}, \ \mathbf{6}, \ \mathbf{4}, \ \mathbf{2}, \ \mathbf{1}).$$

We observe that, for every $n \geq 2$ a final sequence of elements at the end of a vector $\pi(n)$ is repeated at the end of the vector $\pi(n+1)$ (in the list above, these elements are in boldface). We refer to this array of elements as the *queue* $Q(n)$ of $\pi(n)$. To avoid problems with the indices we often consider the *reverse* $\tilde{Q}(n)$ of $Q(n)$, i.e. the vector obtained reading the entries of $Q(n)$ from right to left. So, we have for instance:

$$\tilde{Q}(2) = (\mathbf{1}, \ \mathbf{2}); \ \ \tilde{Q}(3) = (1, 2, 4); \ \ \tilde{Q}(4) = (1, 2, 4, \mathbf{6,9}); \ \ \tilde{Q}(5) = (1, 2, 4, 6, 9, \mathbf{12}).$$

$$
M_3 = \begin{bmatrix}
3 & 2 & 1 & 0 & -1 & -2 & -3 & -4 \\
\hline
1 & 1 & 1 & 0 & 0 & 0 & 0 & 0 \\
1 & 1 & 0 & 1 & 0 & 0 & 0 & 0 \\
1 & 1 & 0 & 0 & 1 & 0 & 0 & 0 \\
1 & 1 & 0 & 0 & 0 & \boxed{1} & 0 & 0 \\
1 & 1 & 0 & 0 & 0 & 0 & 1 & 0 \\
1 & 1 & 0 & 0 & 0 & 0 & 0 & 1 \\
1 & 0 & 1 & 1 & 0 & 0 & 0 & 0 \\
1 & 0 & 1 & 0 & 1 & 0 & 0 & 0 \\
1 & 0 & 1 & 0 & 0 & \boxed{1} & 0 & 0 \\
1 & 0 & 1 & 0 & 0 & 0 & 1 & 0 \\
1 & 0 & 0 & 1 & 1 & 0 & 0 & 0 \\
1 & 0 & 0 & 1 & 0 & \boxed{1} & 0 & 0 \\
\hline
0 & 1 & 1 & 1 & 0 & 0 & 0 & 0 \\
0 & 1 & 1 & 0 & 1 & 0 & 0 & 0 \\
0 & 1 & 1 & 0 & 0 & \boxed{1} & 0 & 0 \\
0 & 1 & 0 & 1 & 1 & 0 & 0 & 0
\end{bmatrix}
$$

$$
M_2 = \begin{bmatrix}
2 & 1 & 0 & -1 & -2 \\
\hline
1 & 1 & 1 & 0 & 0 \\
1 & 1 & 0 & \boxed{1} & 0 \\
1 & 1 & 0 & 0 & \boxed{1} \\
1 & 0 & 1 & \boxed{1} & 0
\end{bmatrix}
$$

Fig. 1. The matrices M_2 (left) and M_3 (right). Queue triads are in boldface and the pointer of each triad has been put in box.

An element that belongs to $\tilde{Q}(n+1)$ but not to $\tilde{Q}(n)$ is said a *last entry* of $\tilde{Q}(n+1)$ (they are the boldface elements in the above list). By extension, we speak of *last entry* of $Q(n+1)$ (and of $\pi(n)$).

Remark 1. A neat inspection shows that the queue of $\pi(n)$ has two last entries if n is even, and one last entry otherwise (red elements in $\pi(n)$). The reason for this fact will be made clear in the sequel.

As n increases, the entries of $\tilde{Q}(n)$ (and of the queue of $\pi(n)$) give rise to an infinite sequence, that we call the *Saind sequence* $(w_n)_{n \geq 1}$. The first few terms of w_n are:

$$1, 2, 4, 6, 9, 12, 16, 20, 25, 30, 36, 42, 49, 56, 64, 72, 81, 90, 100, 110, 121, 132, \ldots$$

In the following part of this section we prove some properties of the rows of M_n. First of all, given $n \geq 2$, the generic row r of the incidence matrix M_n associated with S_n is uniquely described by the triad of indices $t_r = (i_r, j_r, k_r)$ of the entries in r that are equal to 1. Each of these triads contributes to increase by one three entries of $\pi(n)$, precisely the entries in the position specified by the indices i, j, and k. By abuse of notation, with n fixed, we will sometimes refer to a generic triad (i, j, k) of M_n by means of the corresponding elements of the Saind array, i.e. (s_i, s_j, s_k), where clearly, for any h, the two triples are related by $s_h = n - h + 1$.

Let us introduce further notation. With $1 \leq i < j$, we denote by $B_n(i)$ (briefly, $B(i)$) the submatrix of M_n comprising rows that have the leftmost 1 in position i, and by $B_n(i, j)$ (briefly, $B(i, j)$) the submatrix of $B(i)$ where the rows have the second occurrence of 1 in position j. For instance, the matrix $B_6(2,6)$ is the following

$$\begin{bmatrix} 6 & 5 & 4 & 3 & 2 & 1 & 0 & -1 & -2 & -3 & -4 & -5 & -6 & -7 & -8 & -9 & -10 \\ 0 & 1 & 0 & 0 & 0 & 1 & 1 & 0 & 0 & 0 & 0 & 0 & 0 & 0 & 0 & 0 & 0 \\ 0 & 1 & 0 & 0 & 0 & 1 & 0 & 1 & 0 & 0 & 0 & 0 & 0 & 0 & 0 & 0 & 0 \\ 0 & 1 & 0 & 0 & 0 & 1 & 0 & 0 & 1 & 0 & 0 & 0 & 0 & 0 & 0 & 0 & 0 \\ 0 & 1 & 0 & 0 & 0 & 1 & 0 & 0 & 0 & 1 & 0 & 0 & 0 & 0 & 0 & 0 & 0 \\ 0 & 1 & 0 & 0 & 0 & 1 & 0 & 0 & 0 & 0 & 1 & 0 & 0 & 0 & 0 & 0 & 0 \\ 0 & 1 & 0 & 0 & 0 & 1 & 0 & 0 & 0 & 0 & 0 & 1 & 0 & 0 & 0 & 0 & 0 \end{bmatrix}$$

We observe that $B_n(n-1)$ is the bottom block of M_n, and its last row, for any n, is obtained by considering $s_1 = 2$, $s_2 = 0$, $s_3 = -1$. Therefore, $M_n = \cup_{i=1}^{n-1} B_n(i)$.

Lemma 1. *Let $n \geq 2$ and $1 \leq i < j \leq 3n - 1$. Then we have:*

1. $|B_n(i,j)| > 0$ *if and only if* $j < \frac{3n-i+2}{2}$;
2. *For any* $j < \frac{3n-i+2}{2}$, *we have* $|B_n(i,j)| = 3n - i - 2j + 2$;
3. $|B_n(i)| = \lfloor \frac{n-i+1}{2} \rfloor \cdot \lceil \frac{n-i+1}{2} \rceil$;
4. $|M_n| = \sum_{i=1}^{n-1} \lfloor \frac{n-i+1}{2} \rfloor \cdot \lceil \frac{n-i+1}{2} \rceil$.

Proof. 1. $B_n(i,j)$ contains at least a row if and only if $s_i + s_j + s_{j+1} > 0$. This holds if and only if $j < \frac{3n-i+2}{2}$.

2. With i,j fixed, $B_n(i,j)$ contains all the triads of the form (i,j,k), so that $s_i + s_j + s_k > 0$. This is satisfied by the values of $k = j+1, \ldots, 3n - (i+j)+2$. So the number of rows of $B_n(i,j)$ is $(3n-(i+j)+2)-(j+1)+1 = 3n-i-2j+2$.

3. It is clear that for any i, n, $|B_n(i)| = |B_{n-i+1}(1)|$. So it is sufficient to compute the cardinality of $|B_m(1)|$, for a generic $m \geq 2$. So, using the formulas in 1. and 2. , and omitting the computation, we obtain that:

$$|B_m(1)| = \begin{cases} 1+3+5+\ldots+3(m-1) & \text{with } m \text{ even} \\ 2+4+6+\ldots+3(m-1) & \text{with } m \text{ odd}. \end{cases}$$

Using standard methods, both the expressions above can be written in a more compact way, i.e.

$$|B_m(1)| = \left\lfloor \frac{m}{2} \right\rfloor \cdot \left\lceil \frac{m}{2} \right\rceil .$$

Then, we have that:

$$|M_n| = |B_n(1)| + |B_{n-1}(2)| + \ldots + |B_2(n-1)| = \sum_{i=1}^{n-1} \left\lfloor \frac{n-i+1}{2} \right\rfloor \cdot \left\lceil \frac{n-i+1}{2} \right\rceil .$$

$\square$

4 Queue Triads and the Saind Sequence

As mentioned above, for every n, there are triads of indices $(i_k, j_k, k(n))$ (briefly, $(i,j;k)$) that contribute to the appearance of the new entry(entries) of the queue, in correspondence with the index k, and these triads are called *queue triads* of size n and *pointer* k. These triads are depicted in bold in the matrices in Fig. 1, whereas, for any triad, the pointer is put in a box. The following gives account of the existence of the queue of the arrays $\pi(n)$ (hence of the Saind sequence).

Theorem 2. *For any given $h \geq 1$ there is a positive integer $\tilde{n}(h)$ such that, with $n \geq \tilde{n}(h)$, the h-th entry of $\tilde{\pi}(n)$ (the reverse of $\pi(n)$) is equal to h-th entry of $\tilde{\pi}(n+1)$.*

Proof. We observe that the h-th element of $\tilde{\pi}(n)$ corresponds to the entry $2 - 2n + h - 1$ in S_n. A generic triad that contributes to the value of the h-th element of the sequence $\tilde{\pi}(n)$ will contain two other nonnegative distinct integers k_1, k_2, such that the corresponding entries in S_n are $n - k_1$ and $n - k_2$, with and $k_1 < k_2$. It holds, by construction, $1 - 2n + h > 0$, so $k_1 + k_2 < h + 1$. Hence, setting $\tilde{n}(h) = \sum_{i=1}^{h-2} p(i)$, with $p(i) = (\lceil i/2 \rceil - 1)$ being the number of pairs of different elements that sum to i (see also Lemma 1, 3.), we get the result. $\square$

Using the simple argument above we provide a characterization of queue triads of size n.

Proposition 1. *Queue triads (i, j, k) of size n can be generated as follows:*

Step 1: We determine the pointers which can be associated with n:

$$\begin{cases} k_o = 3 \cdot \frac{n+1}{2} & \text{if } n \text{ is odd;} \\ k_e = \frac{3n+2}{2} + 1, \ k'_e = \frac{3n+2}{2} & \text{if } n \text{ is even.} \end{cases}$$

Step 2: We calculate the values of i for the pointers determined in Step 1:

$-$ n odd:

$$\begin{cases} 1 \leq i \leq \frac{3 \cdot n - k_o}{2} & \text{if } k_o \text{ odd;} \\ 1 \leq i \leq \frac{3 \cdot n - k_o + 1}{2} & \text{if } k_o \text{ even.} \end{cases}$$

$-$ n even, and $k \in \{k_e, k'_e\}$: $\begin{cases} 1 \leq i \leq \frac{3 \cdot n - k + 1}{2} & k \text{ odd;} \\ 1 \leq i \leq \frac{3 \cdot n - k}{2} & k \text{ even.} \end{cases}$

Step 3: We calculate j for each of the values of i, $k \in \{k_o, k_e, k'_e\}$ obtained in Steps 1,2: $i + 1 \leq j \leq 3 \cdot n - k - (i - 2)$.

Proof. The proof is obtained using technical arguments similar to those used for Theorem 2. $\square$

Let us denote by $\mathcal{Q}_n$ the set of queue triads of size n generated using Proposition 1. As an example, we determine the queue triads for $n = 4$:

Step 1: $k'_e = 7, k_e = 8$;

Step 2: if $k_e = 8$, then $1 \leq i \leq 2$; if $k'_e = 7$, then $1 \leq i \leq 3$;

Step 3: If $k_e = 8$: with $i = 1$, $2 \leq j \leq 6$; with $i = 2$, $3 \leq j \leq 5$; with $i = 3$, $4 \geq j \geq 4$. Otherwise, if $k'_e = 7$: with $i = 1$: $2 \leq j \leq 5$; with $i = 2$: $3 \leq j \leq 4$. Therefore, the queue triads of size 4 are:

$$(1, 2, 8), (1, 3, 8), (1, 4, 8), (1, 5, 8), (2, 3, 8), (2, 4, 8)$$

with pointer 8, and

$$(1, 2, 7), (1, 3, 7), (1, 4, 7), (1, 5, 7), (1, 6, 7), (2, 3, 7), (2, 4, 7), (2, 5, 7), (3, 4, 7)$$

with pointer 7, giving $w_4 = 6$, and $w_5 = 9$.

We would like to point out that the relation between queue triads of given size and the Saind sequence can be described as follows:

1. if n is odd then the pointer k_o is unique and the number of queue triads of size n gives the term w_{3n-k_o};
2. if n is even, then we have two pointers: k_e and k'_e, then the number of queue triads of size n and pointer k_e (resp. k'_e) gives the term w_{3n-k_e} (resp. $w_{3n-k'_e}$).

Summarizing, the number of queue triads of size n and pointer k is the $m = (3n - k)$-th term of the Saind sequence w_m. This clearly explains the observation in Remark 1. Moreover, using the arguments above, we can obtain a closed formula for m-th entry of the Saind sequence:

Theorem 3. *For any $m \geq 1$, we have $w_m = \left\lfloor \frac{m+1}{2} \right\rfloor \cdot \left\lceil \frac{m+1}{2} \right\rceil$.*

We point out that the nth term of the Saind sequence w_n coincides with the $(n + 1)$th term of sequence $A002620$ in the On-line Encyclopedia of Integer Sequences, [15]. This sequence has several combinatorial interpretations. Therefore, rather than giving an analytical proof of Theorem 3, we will prove it bijectively, in the next section, by establishing bijections between the queue triads of a given size and pointer, and other combinatorial objects counted by sequence $A002620$.

5 Bijections Between Queue Triads and Other Combinatorial Objects

In this section we provide bijections between queue triads and other combinatorial objects counted by $A002620$, precisely: symmetric Dyck paths with 3 peaks and integer partitions in two parts. These bijections provide a combinatorial proof for the formula of Theorem 3.

Queue Triads and Integer Partitions in Two Parts. A *partition* of a positive integer n in k parts is a sequence of positive integers $(\lambda_1, \lambda_2, \ldots, \lambda_k)$, such that $\lambda_1 \geq \lambda_2 \geq \cdots \geq \lambda_k$ and $\lambda_1 + \lambda_2 + \cdots + \lambda_k = n$. Let $P(i, 2)$ denote the number of integer partitions of i into 2 parts, it is known that $P(i, 2)$ is $\lfloor \frac{i}{2} \rfloor$, and $a_n = \sum_{i=2}^{n} P(i, 2)$. It is known that a_n is the $n - th$ entry of sequence $A002620$ (see [15]).

Proposition 2. *For any $n \geq 2$, there is a bijection between queue triads of size n and pointer k and integer partitions in two parts of the integers $2, 3, \ldots, k - 2$.*

Proof. Using the characterization of the pointers associated with a given n, provided in Proposition 1, we have that $k - 2$ is equal to:

i) $3\left(\frac{n-1}{2}\right) + 1$, if n is odd;
ii) $3\left(\frac{n-2}{2}\right) + 2$ or $3\left(\frac{n-2}{2}\right) + 3$, if n is even.

We define the function f as follows: given a queue triad $t = (x, y, k)$, the corresponding integer partition $f(t) = (g, p)$ is obtained by setting $g = y - 1$ and $p = x$. The sum of the partition is then $x + y - 1$, and this value runs from 2 to $k - 2$.

Conversely, given an integer partitions of $2, 3, \ldots, i$ in two parts, we can find the corresponding queue triads in the following way: we calculate the size n of the Saind array, that is the only integer element of the set $N = \left\{ \frac{2 \cdot i + 1}{3}, \frac{2 \cdot (i+1)}{3}, \frac{2 \cdot i}{3} \right\}$.

Then, we define $f^{-1}(g, p)$ as the queue triad $(p, g + 1, k)$ of S_n, where, if n is odd, then $k = 3 \cdot \frac{n+1}{2}$, else if $n = \frac{2 \cdot (i+1)}{3}$ (resp. $n = \frac{2 \cdot i}{3}$), then $k = \frac{3 \cdot n + 2}{2} + 1$ (resp. $k = \frac{3 \cdot n + 2}{2}$). $\qquad \square$

Let us see an example with $n = 3$. The pointer associated with $n = 3$ is $k = 6$, and the queue triads are $(1, 2, 6)$, $(1, 3, 6)$, $(1, 4, 6)$ and $(2, 3, 6)$. They are in correspondence with the partitions of the integers $2, 3, \ldots, 3 \cdot \frac{3-1}{2} + 1 = 4$, precisely: $(1, 1)$, $(2, 1)$, $(3, 1)$, $(2, 2)$. According to Proposition 2 we have: $f(1, 2, 6) = (1, 1)$, $f(1, 3, 6) = (2, 1)$, $f(1, 4, 6) = (3, 1)$, and $f(2, 3, 6) = (2, 2)$.

Queue Triads and Symmetric Dyck Paths with Three Peaks. Recall that a Dyck path of semi-length n is a lattice path using up $U = (1, 1)$ and down $D = (1, -1)$ unit steps, running from $(0, 0)$ to $(2n, 0)$ and remaining weakly above the x-axis. Any occurrence of a UD factor in a Dyck path is called a *peak* of the path. Here, we consider Dyck paths that are *symmetric* with respect to the line which passes through the upper end of the $n - th$ step and it is parallel to the y-axis (see Fig. 2). Deutsch showed that the number of symmetric Dyck paths with three peaks and semi-length n is given by the $(n - 1)$th term of sequence $A002620$, [15].

Fig. 2. A symmetric Dyck path of length 8 and its axis of symmetry.

Proposition 3. *For any $n \geq 1$, the family of queue triads with size n and pointer k is in bijection with symmetric Dyck paths with exactly three peaks and semi-length $\ell = (3n - 1) - k + 3$.*

Proof. We observe that a symmetric Dyck path with 3 peaks of semi-length 2ℓ has the central peak lying on the symmetry line and is uniquely determined by its prefix of length ℓ, which ends with an U step. We denote this family of prefixes by $\mathcal{D}(\ell)$. The function g maps queue triads with size n and pointer k onto the paths of $\mathcal{D}(\ell)$, where $\ell = (3n - 1) - k + 3$. Precisely, the queue triad (x, y, k) with size n is mapped onto the unique path of $\mathcal{D}(\ell)$ obtained as follows: y gives the number of steps between the vertex of the peak and the axis of symmetry that passes through the upper end of the last step of the prefix, and x gives the number of D steps between the first peak and the first valley.

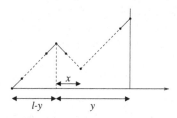

Conversely, given a path $G \in \mathcal{D}(\ell)$, first we calculate the size n of the queue triad, as the only integer element of the set: $N = \left\{ \frac{2\cdot\ell-1}{3}, \frac{2\cdot\ell}{3}, \frac{2\cdot(\ell-1)}{3} \right\}$.

Then the queue triad $g^{-1}(G) = (x, y, k)$, is obtained as follows: x is equal to the number of D steps immediately after the vertex of the first peak, y is equal to the number of steps between the vertex of the first peak and the axis of symmetry and k is equal to $k = 3 \cdot \frac{n+1}{2}$ if n is odd, or to $k = \ell + 2$ (resp. $k = \ell$) if $n = \frac{2\cdot\ell}{3}$ (resp. $n = \frac{2\cdot(\ell-1)}{3}$).   □

For example, the queue triads of size 3 and pointer 6 (i.e. $(1, 2, 6)$, $(1, 3, 6)$, $(1, 4, 6)$, $(2, 3, 6)$) are mapped onto the paths of $\mathcal{D}(5)$. The correspondence is shown in Fig. 3.

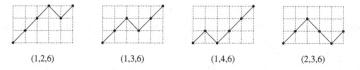

$(1,2,6)$ $(1,3,6)$ $(1,4,6)$ $(2,3,6)$

Fig. 3. The 4 paths in $\mathcal{D}(5)$ and the corresponding queue triads of size 3 and pointer 6.

6   Future Developments

Other than the bijections we showed in the previous section, in our study we also established bijections (not presented in this paper) between queue triads of size n and other families of objects as: (a) bracelets with $n + 3$ beads, two of which are red and one of which is blue; (b) distinct genotype frequency vectors possible for a sample of n diploid individuals at a biallelic genetic locus with a specified major allele [15].

The study concerning Saind arrays led us to consider significant the investigation of other arrays having regular shapes. The following table shows some examples that we have obtained experimentally with an approach similar to that for the Saind sequence. We point out that each term of the sequence shown in the first (resp. second) row is repeated twice (resp. three times). The second column provides the sequence reference according to [15].

Array	Number sequence	First terms
$(n, n, n-1, n-1, \ldots, 1-2n, 1-2n)$	A035608	$1, 5, 10, 18, 27, 39, 52, 68, 85, \ldots$
$(n, n, n, \ldots, -n, -n, -n)$	A079079	$3, 6, 12, 24, 42, 63, 90, 120 \ldots$

This study seems extremely relevant both from a combinatorial point of view, since it would lead to new combinatorial interpretations of the computed sequences, and from a graph theoretical perspective, due to their close connection with hypergraphic degree sequences.

References

1. Aigner, M., Triesch, E.: Realizability and uniqueness in graphs. Discrete Math. **136**(1–3), 3–20 (1994)
2. Erdös, P., Gallai, T.: Graphs with prescribed degrees of vertices (Hungarian). Mat. Lapok **11**, 264–274 (1960)
3. Deza, A., Levin, A., Meesum, S.M., Onn, S.: Optimization over degree sequences. SIAM J. Disc. Math. **32**(3), 2067–2079 (2018)
4. Barrus, M., Harttke, S., Jao, K., West, D.: Length thresholds for graphic lists given fixed largest and smallest entries and bounded gaps. Discrete Math. **312**(9), 1494–1501 (2012)
5. Dewdney, A.: Degree sequences in complexed and hypergraphs. Proc. Am. Math. Soc. **53**(2), 535–540 (1975)
6. Yina, J.-H., Lib, J.-S.: Two sufficient conditions for a graphic sequence to have a realization with prescribed clique size. Discrete Math. **301**, 218–227 (2005)
7. Behrens, S., et al.: New results on degree sequences of uniform hypergraphs. Electron. J. Comb. **20**(4), paper 14, 18 (2013)
8. Brlek, S., Frosini, A.: A tomographical interpretation of a sufficient condition on h-graphical sequences. In: Normand, N., Guédon, J., Autrusseau, F. (eds.) DGCI 2016. LNCS, vol. 9647, pp. 95–104. Springer, Cham (2016). https://doi.org/10.1007/978-3-319-32360-2_7
9. Colbourn, C.J., Kocay, W.L., Stinson, D.R.: Some NP-complete problems for hyper- graph degree sequences. Discrete Appl. Math. **14**, 239–254 (1986)
10. Frosini, A., Picouleau, C., Rinaldi, S.: New sufficient conditions on the degree sequences of uniform hypergraphs (submitted)
11. Frosini, A., Palma, G., Rinaldi, S.: On step-two degree sequences of k-uniform hypergraphs (submitted)
12. Kocay, W., Li, P.C.: On 3-hypergraphs with equal degree sequences. Ars Comb. **82**, 145–157 (2006)
13. Palma, G.: Combinatorial properties of degree sequences of $k-$uniform hypergraphs, Master thesis (2019)
14. Berge, C.: Hypergraphs Combinatorics of Finite Sets, North-Holland (1989)
15. Sloane, N.J.A.: The On-line Encyclopedia of Integer Sequences (2011)
16. West, D.: Introduction to Graph Theory. Prentice Hall, Upper Saddle River (2001)

The Computational Significance of Hausdorff's Maximal Chain Principle

Peter Schuster$^{(\boxtimes)}$ ⓘ and Daniel Wessel$^{(\boxtimes)}$ ⓘ

Dipartimento di Informatica, Università degli Studi di Verona,
Strada le Grazie 15, 37134 Verona, Italy
{peter.schuster,daniel.wessel}@univr.it

Abstract. As a fairly frequent form of the Axiom of Choice about relatively simple structures (posets), Hausdorff's Maximal Chain Principle appears to be little amenable to computational interpretation. This received view, however, requires revision. When attempting to convert Hausdorff's principle into a conservation theorem, we have indeed found out that maximal chains are more reminiscent of maximal ideals than it might seem at first glance. The latter live in richer algebraic structures (rings), and thus are readier to be put under computational scrutiny. Exploiting the newly discovered analogy between maximal chains and ideals, we can carry over the concept of Jacobson radical from a ring to an arbitrary set with an irreflexive symmetric relation. This achievement enables us to present a generalisation of Hausdorff's principle first as a semantic and then as a syntactical conservation theorem. We obtain the latter, which is nothing but the desired computational core of Hausdorff's principle, by passing from maximal chains to paths of finite binary trees of an adequate inductively generated class. In addition to Hausdorff's principle, applications include the Maximal Clique Principle for undirected graphs. Throughout the paper we work within constructive set theory.

Keywords: Axiom of Choice · Maximal chain · Maximal ideal · Maximal clique · Jacobson radical · Proof-theoretic conservation · Computational content · Constructive set theory · Finite binary tree · Inductive generation

1 Introduction

Hausdorff's maximal chain principle asserts that every totally ordered subset of a partially ordered set S is contained in a maximal one. Equivalently, this can be put as a completeness criterion in first-order terms: a chain C is maximal precisely when, for every $x \in S$, if $C \cup \{a\}$ is a chain, then $a \in C$. So a chain C is maximal if and only if, for every $a \in S$, either $a \in C$ or a is incomparable with at least one $b \in C$, i.e.,

$$a \in C \lor (\exists b \in C)(a \not\leq b \land b \not\leq a). \tag{1}$$

© Springer Nature Switzerland AG 2020
M. Anselmo et al. (Eds.): CiE 2020, LNCS 12098, pp. 239–250, 2020.
https://doi.org/10.1007/978-3-030-51466-2_21

This is somewhat reminiscent of the characterisation of maximal ideals in commutative ring theory [21]. In this setting an ideal J of a commutative unital ring takes the place of C, and the respective right-hand disjunct of (1) expresses that the ring element a is invertible modulo J. Moreover, it is possible to describe the common part of all maximal ideals in first-order terms. This encodes Krull's Maximal Ideal Theorem as an intersection principle, and yields a notion of Jacobson radical suitable for constructive algebra [21,31,38].

By analogy, we can define the Jacobson radical $\mathrm{Jac}(C)$ of a chain C, and prove (assuming the Axiom of Choice AC) that $\mathrm{Jac}(C)$ coincides with the intersection of all maximal chains containing C. Hence Hausdorff's principle too can be recast as an intersection principle. All this will even be done in a slightly more general fashion. The main point to be stressed is that a simple *constructive* interpretation is possible, whence the purpose of this paper is twofold: we communicate a new choice principle, and describe its constructive underpinning.

We proceed as follows. In Sect. 2, alongside the analogy with ring theory, we describe our concepts of coalition and Jacobson radical. In Sect. 3 we briefly relate this to past work [25–27] on the interplay of single- and multi-conclusion entailment relations [9,35]. In Sect. 4 we give a constructive account of complete coalitions by means of a suitable inductively generated class of binary trees. In Sect. 5 we briefly discuss two applications: maximal chains of partially ordered sets, and maximal cliques of undirected graphs. The main results are Proposition 1 and its constructive companion Proposition 3.

Foundations

The content of this paper is elementary and can be formalised in a suitable fragment of constructive set theory **CZF** [2,3]. Due to the choice of this setting, sometimes certain assumptions have to be made explicit which otherwise would be trivial in classical set theory. For instance, a subset T of a set S is *detachable* if, for every $a \in S$, either $a \in T$ or $a \notin T$. A set S is *finitely enumerable* if there is $n \geqslant 0$ and a surjective function $f : \{1, \ldots, n\} \to S$. We write $\mathrm{Fin}(S)$ for the set of finitely enumerable subsets of S. To pin down a rather general, classical intersection principle, and to point out certain of its incarnations, requires some classical logic and the Axiom of Choice (AC) in its classically equivalent form of Zorn's Lemma (ZL) [40]. For simplicity we switch in such a case to classical set theory **ZFC**, signalling this appropriately.

2   Coalitions

Throughout, let S be a set, and let R be an irreflexive symmetric relation on S. We say that a subset C of S be a *coalition*[1] (with respect to R) if $\neg aRb$ for all $a, b \in C$. This is the same as demanding that C be $\overline{R}$-*connected*, which is to say

---

[1] Incidentally, the term "coalition", which we use here for sake of intuition, is standard terminology in game theory to denote a group of agents [39].

that $a \in C$ only if $a\overline{R}b$ for every $b \in C$, where $\overline{R}$ denotes the complementary relation. For instance, the empty subset is a coalition, as is every singleton subset of S, by the irreflexivity of R. Notice that coalitions are closed under directed union. A coalition C is called *complete* if, for every $a \in S$,

$$a \in C \vee (\exists b \in C)\, aRb. \tag{2}$$

It is perhaps instructive to read aRb as "a opposes b" (and *vice versa*, to account for symmetry), under which reading it makes sense to require irreflexivity. A coalition is then a subset of S in which no two members oppose one another. A complete coalition C is such that, given any $a \in S$, this a either belongs to C, or else C exhibits a witness b which opposes a.

Lemma 1. *Every complete coalition is detachable and maximal (with respect to set inclusion) among coalitions. Conversely, with classical logic every maximal coalition is complete.*

Proof. Let C be a complete coalition. Since $a\overline{R}b$ for all $a, b \in C$, the second alternative of completeness (2) entails that $a \notin C$; whence C is detachable. As regards C being maximal, let D be a coalition such that $C \subseteq D$ and let $a \in D$. By completeness, either $a \in C$ right away, or else there is $b \in C$ such that aRb, but the latter case is impossible as D is a coalition. As regards the converse, if C is a complete coalition and $a \notin C$, then $C' = C \cup \{a\}$ cannot, due to maximality of C, in turn be a coalition. With classical logic, the latter statement is witnessed by a certain element $b \in C$. This yields completeness. $\square$

If C is a coalition, let us write

$$\mathrm{Comp}/C$$

for the collection of all complete coalitions that contain C, with the special case $\mathrm{Comp} = \mathrm{Comp}/\emptyset$. Since every complete coalition is detachable (Lemma 1), these collections are sets due to the presence in **CZF** of the Exponentiation Axiom [2,3].

All this is fairly reminiscent of the characteristics of maximal ideals in ring theory [21]. Given a commutative ring A with 1, recall from [12,21] that the Jacobson radical [20] of an ideal J of A can be defined as

$$\mathrm{Jac}(J) = \{\, a \in A \mid (\forall b \in A)(1 \in \langle a, b \rangle \rightarrow (\exists c \in J)\, 1 \in \langle b, c \rangle)\,\}, \tag{3}$$

where sharp brackets denote generated ideals. By plain analogy with the ring-theoretic setting, let us then define the *Jacobson radical* of an arbitrary subset C of S, of course with respect to our default, irreflexive symmetric relation R:

$$\mathrm{Jac}(C) = \{\, a \in S \mid (\forall b \in S)(aRb \rightarrow (\exists c \in C)\, bRc)\,\}. \tag{4}$$

In particular, the Jacobson radical of the empty coalition is

$$\mathrm{Jac}(\emptyset) = \{\, a \in S \mid (\forall b \in S)\, a\overline{R}b\,\}.$$

Thus, we substitute the property of mutual opposition in (4) for the one of comaximality in (3), i.e., for the property of two ring elements to generate the unit ideal. Assuming AC, the Jacobson radical of an ideal J is the intersection of all maximal ideals that contain J [21]. Similarly, and still with AC, the Jacobson radical of a coalition C turns out to be the intersection of all complete coalitions containing C (Proposition 1).

Lemma 2. *The Jacobson radical defines a closure operator on S which restricts to a mapping on coalitions, i.e., if C is a coalition, then so is $\mathrm{Jac}(C)$.*

Proof. As for the first statement we only show idempotency, i.e., $\mathrm{Jac}(\mathrm{Jac}(C)) \subseteq \mathrm{Jac}(C)$, where $C \subseteq S$. In fact, if $a \in \mathrm{Jac}(\mathrm{Jac}(C))$ and $b \in S$ is such that aRb, then there is $c \in \mathrm{Jac}(C)$ with cRb. It follows that there is $c' \in C$ such that bRc', and so $a \in \mathrm{Jac}(C)$.

As regards the second statement, suppose that $C \subseteq S$ is a coalition, and let $a_0, a_1 \in \mathrm{Jac}(C)$. Assuming that $a_1 R a_0$, since $a_1 \in \mathrm{Jac}(C)$, there is $c_0 \in C$ such that $a_0 R c_0$. Since $a_0 \in \mathrm{Jac}(C)$ too, there is $c_1 \in C$ such that $c_0 R c_1$, which is in conflict with C being a coalition. □

Proposition 1 (ZFC). *If C is a coalition, then*

$$\mathrm{Jac}(C) = \bigcap \mathrm{Comp}/C.$$

Proof. Let $a \in \mathrm{Jac}(C)$ and suppose that D is a complete coalition which contains C. By completeness, either $a \in D$ right away, or else there is $b \in D$ such that aRb. But since $a \in \mathrm{Jac}(C)$, the latter case would imply that there were $c \in C \subseteq D$ with bRc, by way of which D would fail to be a coalition after all.

For the right-to-left inclusion we concentrate on the contrapositive. Thus, suppose that $a \notin \mathrm{Jac}(C)$. Accordingly, there is b such that aRb and $C' := C \cup \{b\}$ is a coalition. ZL yields a coalition D which is maximal among those containing C'. This D is complete by way of being maximal, and it must avoid a, because if $a \in D$, then D were not a coalition since $b \in C' \subseteq D$ and aRb. □

Remark 1. The argument in the right-to-left part of the proof of Proposition 1 can also be used in a more affirmative manner. ZL, which is said to be constructively neutral [4],[2] directly implies that

$$\bigcap \mathrm{Max}/C \subseteq \{\, a \in S \mid (\forall b \in S)(aRb \to \neg(\forall c \in C)\, b\overline{R}c)\,\},$$

where Max/C denotes the collection of all maximal coalitions over C. The crucial direction of Proposition 1 can also be proved in a more direct manner by using Open Induction [6,11,23] in place of Zorn's Lemma. For similar cases see [10, 24,29,30].

---

[2] Forms of ZL have been considered over classical [14], intuitionistic [5] as well as constructive set theory [1,32].

By a *radical* coalition C we understand one which is closed with respect to Jac, i.e., which is such that $\mathrm{Jac}(C) = C$. Clearly, every complete coalition is radical, and by Lemma 2 so is the intersection of an inhabited family of complete coalitions. By Proposition 1, in **ZFC** the radical coalitions are precisely the intersections of complete coalitions; so in particular

$$\{\, a \in S \mid (\forall b \in S)\, a\overline{R}b \,\} = \mathrm{Jac}(\emptyset) = \bigcap \mathrm{Comp}.$$

With Proposition 3 we will give a constructive version of Proposition 1 in Sect. 4, to which end Proposition 2 below will be crucial.

In the following, we write

$$R(x) = \{\, y \in S \mid xRy \,\}$$

for the image of x under R, and use $\mathrm{Jac}(C, x)$ as a shorthand for $\mathrm{Jac}(C \cup \{x\})$.

Proposition 2. *The following is provable for the Jacobson radical:*

$$\frac{a \in \mathrm{Jac}(C, x) \quad (\forall y \in R(x))\, a \in \mathrm{Jac}(C, y)}{a \in \mathrm{Jac}(C)}$$

where $a, x \in S$ and C is an arbitrary subset of S.

Proof. Given the displayed premises, to check that $a \in \mathrm{Jac}(C)$, consider $b \in S$ such that aRb. We need to find $c \in C$ such that bRc. The left-hand premise yields $c' \in C \cup \{x\}$ such that bRc'. If $c' \in C$, then $c = c'$ is as required. In case of $c' = x$, the right-hand premise for $y = b$ yields $a \in \mathrm{Jac}(C, b)$. Again with aRb it follows that there is $c \in C \cup \{b\}$ such that bRc, whence in fact $c \in C$ since R is irreflexive. □

Remark 2. Given a binary relation R on S, an R-*clique* is a subset C such that, for every $a \in S$,

$$a \in C \Leftrightarrow (\forall b \in C)\, aRb.$$

Bell's *Clique Property* asserts that, for any reflexive symmetric relation R on S, an R-clique exists. This is in fact an intuitionistic equivalent of ZL [5]. Classically, given an irreflexive symmetric relation R, every $\overline{R}$-clique is a complete R-coalition. Conversely, and constructively, every complete R-coalition is an $\overline{R}$-clique. More precisely, a subset C of S is an $\overline{R}$-clique if and only if it is $\overline{R}$-connected as well as $\overline{R}$-*saturated*, the latter of which is to say that $a \in C$ already if $a\overline{R}b$ for all $b \in C$.

3   Entailment for Completeness

Consider on S the relation $\rhd \subseteq \mathrm{Fin}(S) \times S$ which is defined by the Jacobson radical, i.e., stipulate

$$U \rhd a \equiv a \in \mathrm{Jac}(U).$$

Lemma 2 tells us that this $\rhd$ is a *single-conclusion entailment relation*, which is to say that it is *reflexive*, *monotone*, and *transitive* in the following sense:

$$\frac{U \ni a}{U \rhd a}(\text{R}) \qquad \frac{U \rhd a}{U, V \rhd a}(\text{M}) \qquad \frac{U \rhd b \quad U, b \rhd a}{U \rhd a}(\text{T})$$

where the usual shorthand notation is at work with $U, V \equiv U \cup V$ and $U, b \equiv U \cup \{b\}$. In **ZFC**, the consequences with respect to $\rhd$ of a coalition $U \in \text{Fin}(S)$ are semantically determined by the complete coalitions over U, i.e.,

$$(\forall C \in \text{Comp})(C \supseteq U \implies a \in C) \implies U \rhd a.$$

Proposition 2 implies that the following is provable:

$$\frac{U, x \rhd a \quad (\forall y \in R(x)) \, U, y \rhd a}{U \rhd a}$$

This is to say that the infinitary axiom of completeness (2), which in the present context can be put in the form

$$\vdash x, R(x)$$

is in fact *conservative* [25, 26] over $\rhd$. To make this precise requires extending the results of [25, 26] to an infinitary setting [36], but upon which those results go through verbatim. We do not require such a development here; an elementary constructive interpretation of Proposition 1 will be given in the following section using instead a suitable inductively generated collection of finite binary trees. For related uses of conservativity see also, e.g., [16, 27, 28].

4   Binary Trees for Complete Coalitions

In this section we carry over the approach recently followed in [34] for prime ideals of commutative rings, so as to accommodate complete coalitions. Readers familiar with dynamical algebra [13, 21, 38] will draw a connection between the tree methods of [13] and the one employed here.

Let again S be a set. For every $a \in S$ we first introduce a corresponding letter X_a. Let

$$\mathcal{S} = (S \cup \{X_a \mid a \in S\})^*$$

be the set of finite sequences of elements of S and such letters, with the usual provisos on notation, concatenation, etc. Next, we generate inductively a class $\mathcal{T}$ of finite rooted binary trees $T \subseteq \mathcal{S}$ as follows:

$$\frac{}{\{[\,]\} \in \mathcal{T}}(\text{root}) \qquad \frac{T \in \mathcal{T} \quad u \in \text{Leaf}(T) \quad a \in S}{T \cup \{ua, uX_a\} \in \mathcal{T}}(\text{branch}) \qquad (5)$$

As usual, by a *leaf* we understand a sequence $u \in T$ without immediate successor in T. The second rule is to say that, given $T \in \mathcal{T}$, if u is a leaf of T, then each element a of S gives rise to a new member of $\mathcal{T}$ by way of an additional branching at u. More precisely, u gives birth to two children ua and uX_a. Here is a possible instance, where $a, b \in S$:

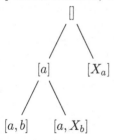

As an auxiliary tool, we further need a sorting function $\text{sort} : \mathcal{S} \to \mathcal{S}$ which gathers all occurring letters X_a at the tail of a finite sequence. As the resulting order of the entries won't matter later on, this function may be defined recursively in the simplest manner, as follows:

$$\text{sort}([]) = []$$
$$\text{sort}(ua) = a\,\text{sort}(u)$$
$$\text{sort}(uX_a) = \text{sort}(u)X_a$$

Last but not least, given a subset C of S, we introduce a relation $\Vdash_C$ between elements of S and sorted finite sequences in $\mathcal{S}$ by defining

$$c \Vdash_C [a_1, \dots, a_k, X_{b_1}, \dots, X_{b_\ell}] \equiv$$
$$(\forall x_1, \dots, x_\ell \in S)\Big(\bigwedge_{j=1}^{\ell} x_j R b_j \to c \in \text{Jac}(C, a_1, \dots, a_k, x_1, \dots, x_\ell)\Big),$$

where we drop the quantifier in case of $\ell = 0$. In particular,

$$c \Vdash_C [] \Leftrightarrow c \in \text{Jac}(C). \tag{6}$$

Keeping in mind Proposition 1, with AC the semantics of this relation is that if $u = [a_1, \dots, a_k, X_{b_1}, \dots, X_{b_\ell}]$ as above, then $c \Vdash_C u$ precisely when, for every simultaneous instantiation of respective opponents $x_1, \dots, x_\ell$ of $b_1, \dots, b_\ell$, this c is a member of every complete coalition over C that further contains $a_1, \dots, a_k, x_1, \dots, x_\ell$. The case in which this holds with respect to *every* leaf of a certain tree $T \in \mathcal{T}$ will later be of particular interest.

With the relation $\Vdash_C$ in place, we can now rephrase Proposition 2 as follows.

Lemma 3. *Let $a, c \in S$ and let $u \in \mathcal{S}$ be sorted. If $c \Vdash_C au$ and $c \Vdash_C uX_a$, then $c \Vdash_C u$.*

Proof. Consider $u = [a_1, \dots, a_k, X_{b_1}, \dots, X_{b_\ell}]$ and suppose that (i) $c \Vdash_C au$ and (ii) $c \Vdash_C uX_a$. To show that $c \Vdash_C u$, let $x_1 \in R(b_1), \dots, x_\ell \in R(b_\ell)$. We write $C' = C \cup \{a_1, \dots, a_k, x_1, \dots, x_\ell\}$ and need to check that $c \in \text{Jac}(C')$. With $x_1, \dots, x_\ell$ fixed, premise (i) yields $c \in \text{Jac}(C', a)$, while (ii) implies that, for every $x \in R(a)$, $c \in \text{Jac}(C', x)$. Now Proposition 2 implies $a \in \text{Jac}(C')$. □

Given a subset C and an element c of S, let us say that a tree $T \in \mathcal{T}$ *terminates for C in c* if $c \Vdash_C \text{sort}(u)$ for *every* leaf u of T. Intuitively, this is to say that, along every path of T, no matter how we instantiate indeterminates X_b that we might encounter with a corresponding opponent x of b, if C' is a complete coalition over C and contains the elements we will have collected at the leaf, then c is a member of C'. The idea is now to fold up branchings by inductive application of Lemma 3, to capture termination by way of the Jacobson radical, and thus to resolve indeterminacy in the spirit of [33].

The following is the constructive counterpart of Proposition 1 and does not require that C be a coalition to start with.

Proposition 3. *Let C be a subset and c an element of S. The following are equivalent.*

1. $c \in \text{Jac}(C)$.
2. There is $T \in \mathcal{T}$ which terminates for C in c.

Proof. If $c \in \text{Jac}(C)$, then $c \Vdash_C []$ by (6), which is to say that $[]$ terminates for C in c. Conversely, suppose that $T \in \mathcal{T}$ is such that $c \Vdash_C \text{sort}(u)$ for every leaf u of T. We argue by induction on T to show that $c \in \text{Jac}(C)$. The case $T = []$ is trivial (6). Suppose that T is the result of a branching at a certain leaf u of an immediate subtree T', and suppose further that $c \Vdash_C \text{sort}(ua) = a \, \text{sort}(u)$ as well as $c \Vdash_C \text{sort}(uX_a) = \text{sort}(u)X_a$ for a certain $a \in S$. Lemma 3 implies that $c \Vdash_C \text{sort}(u)$, whence we reduce to T', to which the induction hypothesis applies. □

Membership in a radical coalition C is thus tantamount to termination.

Remark 3. Very much in the spirit of dynamical algebra [13,21,37,38], every tree $T \in \mathcal{T}$ represents the course of a dynamic argument *as if* a given coalition were complete. Note that every complete coalition C_m of S gives rise to a path through a given tree $T \in \mathcal{T}$. In fact, at each branching, corresponding to an element a of S, by way of completeness this a either belongs to C_m or else the latter assigns a value to X_a in the sense of exhibiting a witness $b \in C_m$ for which aRb. The entries in the terminal node of this path, with values assigned appropriately, then belong to C_m. In particular, if T terminates in c for a certain subset $C \subseteq C_m$, then $c \in C_m$ because $c \in \text{Jac}(C) \subseteq \text{Jac}(C_m) = C_m$ by Proposition 3 and the fact that every complete coalition is radical.

Remark 4. In general it cannot be decided effectively, i.e., without using some excluded middle, whether, given $c \in S$ and $C \subseteq S$, there is a tree $T \in \mathcal{T}$ which terminates for C in c.[3] This is due to the constructive character of Proposition 3 and the following Brouwer–style counterexample. Let φ be a bounded formula.[4] Let $S = \{0, 1\}$ and put

$$R_\varphi = \{ (0,1) \mid \varphi \} \cup \{ (1,0) \mid \varphi \}.$$

---

[3] One of the anonymous referees has kindly drawn our attention to this.
[4] A set-theoretic formula φ is *bounded* if only set-bounded quantifiers $\forall x \in y$ and $\exists x \in y$ occur in φ.

By definition, this relation clearly is irreflexive and symmetric. Consider now the corresponding Jacobson radical $\mathrm{Jac}(\emptyset)$. It is easy to see that

$$0 \in \mathrm{Jac}(\emptyset) \quad \Leftrightarrow \quad \neg\varphi.$$

Therefore, if $\mathrm{Jac}(\emptyset)$ is detachable, then

$$\neg\varphi \vee \neg\neg\varphi.$$

This is to say that the *Weak Restricted Law of Excluded Middle* (WREM) holds.

5   Applications

We will now briefly discuss two instantiations of Proposition 1, concerning maximal chains of partially ordered sets and maximal cliques in undirected graphs. In both cases Proposition 3 provides the corresponding constructive underpinning, which we leave to the reader to spell out in detail. Incidentally, the trick is to start with a relation R of which only the complement $\overline{R}$ is the relation one actually one wants to consider. This clearly fits the concept of coalition we are employing.

Hausdorff's Principle

Let $(S, \leqslant)$ be a partially ordered set. On S we consider the binary relation R of *incomparability*, which is

$$aRb \equiv a \not\leqslant b \wedge b \not\leqslant a,$$

and for which $\overline{R}$ means comparability. Classically, a coalition for R is nothing but a *chain*, i.e., a totally ordered subset of S, and the complete coalitions are the maximal chains. As regards the Jacobson radical in this setting, Proposition 1 applied to the empty chain yields that

$$\{\, a \in S \mid (\forall b \in S)(a \leqslant b \vee b \leqslant a)\,\} = \bigcap \mathrm{Comp}. \tag{7}$$

This is a way to rephrase Hausdorff's maximal chain principle [17]. In fact, if S is not totally ordered by $\leqslant$, as witnessed by a certain element a of S incomparable to some $b \in S$, then by (7) and classical logic there is a maximal chain that avoids a. Incidentally, this application helps to calibrate Proposition 1, which over classical set theory **ZF** thus turns out equivalent to AC through Hausdorff's principle [18, 19, 22].

Maximal Cliques

Let $G = (V, E)$ be an undirected graph, V being its set of vertices, E its set of edges, i.e., E is a set of unordered pairs of elements of V. On the set of vertices we consider the binary relation R of *nonadjacency*, which is

$$aRb \equiv a \neq b \wedge \{a, b\} \notin E.$$

In this setting, classically, a coalition for R is nothing but a *clique*[5] [7], i.e., a subset of V every two distinct elements of which are adjacent, and the complete coalitions are the maximal cliques. Concerning the Jacobson radical, Proposition 1 implies that

$$\{\, a \in V \mid (\forall b \in V)(a \neq b \rightarrow \{\, a, b \,\} \in E)\,\} = \bigcap \mathrm{Comp}.$$

Similar to the preceding application, this yields a solution to the problem of finding a maximal clique with AC.[6]

6   Conclusion

Hausdorff's Maximal Chain Principle, a forerunner of Zorn's Lemma [8,40], is presumably one of the most well-known order-theoretic forms of the Axiom of Choice. We have seen that the property of a chain to be maximal can be put as a completeness criterion, reminiscent of the case in commutative ring theory for maximal ideals. By analogy with Krull's Theorem for maximal ideals, employing a suitably adapted form of Jacobson radical, it has become possible to put a new variant of Hausdorff's Principle in terms of a universal statement. This has paved the way to a constructive, purely syntactic rereading by means of an inductively defined class of finite binary trees which encode computations along generic maximal chains. It remains to be seen, however, to what extent in a concrete setting our method allows to bypass invocations of Hausdorff's Principle.

Along similar lines, we have carried over the concept of Jacobson radical from commutative rings to the setting of universal algebra and thus to broaden considerably the range of applications that our approach has opened up so far [33,34]. In fact, every single-conclusion entailment relation is accompanied by a Jacobson radical which in turn encodes a corresponding maximality principle. In particular, this encompasses the Jacobson radical for distributive lattices [12], commutative rings [31], as well as for propositional theories [15,16]. We keep for future research to put all this under computational scrutiny, and to compare with ours the related methods employed in dynamical algebra [13].

Acknowledgements. The present study was carried out within the projects "A New Dawn of Intuitionism: Mathematical and Philosophical Advances" (ID 60842) funded by the John Templeton Foundation, and "Reducing complexity in algebra, logic, combinatorics - REDCOM" belonging to the programme "Ricerca Scientifica di Eccellenza 2018" of the Fondazione Cariverona. The authors are members of the Gruppo Nazionale per le Strutture Algebriche, Geometriche e le loro Applicazioni (GNSAGA) within the Italian Istituto Nazionale di Alta Matematica (INdAM) (The opinions expressed in this paper are those of the authors and do not necessarily reflect the views of those foundations.). The anonymous referees' careful readings of the manuscript and several helpful remarks and suggestions are gratefully acknowledged.

---

[5] A caveat on terminology: this notion is *a priori* different from the one used in Bell's Clique Property (cf. Remark 2) but which carries over to graph theory.

[6] Clique problems, e.g., the problem of finding a *maximum* clique and that of listing all maximal cliques are prominent in finite graph theory and computational complexity theory [7].

References

1. Aczel, P.: Zorn's Lemma in CZF (2002). Unpublished note
2. Aczel, P., Rathjen, M.: Notes on constructive set theory. Technical report, Institut Mittag-Leffler (2000). Report No. 40
3. Aczel, P., Rathjen, M.: Constructive set theory (2010). https://www1.maths.leeds. ac.uk/~rathjen/book.pdf, book draft
4. Bell, J.L.: Zorn's lemma and complete Boolean algebras in intuitionistic type theories. J. Symb. Log. **62**(4), 1265–1279 (1997)
5. Bell, J.L.: Some new intuitionistic equivalents of Zorn's lemma. Arch. Math. Logic **42**(8), 811–814 (2003)
6. Berger, U.: A computational interpretation of open induction. In: Titsworth, F. (ed.) Proceedings of the Ninetenth Annual IEEE Symposium on Logic in Computer Science, pp. 326–334. IEEE Computer Society (2004)
7. Bondy, J., Murty, U.: Graph Theory, Graduate Texts in Mathematics, vol. 244. Springer, London (2008)
8. Campbell, P.J.: The origin of "Zorn's lemma". Historia Math. **5**, 77–89 (1978)
9. Cederquist, J., Coquand, T.: Entailment relations and distributive lattices. In: Buss, S.R., Hájek, P., Pudlák, P. (eds.) Logic Colloquium 1998. Proceedings of the Annual European Summer Meeting of the Association for Symbolic Logic, Prague, Czech Republic, 9–15 August 1998. Lect. Notes Logic, vol. 13, pp. 127–139. A. K. Peters, Natick, MA (2000)
10. Ciraulo, F., Rinaldi, D., Schuster, P.: Lindenbaum's Lemma via Open Induction. In: Kahle, R., Strahm, T., Studer, T. (eds.) Advances in Proof Theory. PCSAL, vol. 28, pp. 65–77. Springer, Cham (2016). https://doi.org/10.1007/978-3-319-29198-7_3
11. Coquand, T.: A note on the open induction principle. Technical report, Göteborg University (1997). www.cse.chalmers.se/~coquand/open.ps
12. Coquand, T., Lombardi, H., Quitté, C.: Dimension de Heitmann des treillis distributifs et des anneaux commutatifs. Publications Mathématiques de Besançon. Algèbre et Théorie des Nombres, pp. 57–100 (2006)
13. Coste, M., Lombardi, H., Roy, M.F.: Dynamical method in algebra: effective Nullstellensätze. Ann. Pure Appl. Logic **111**(3), 203–256 (2001)
14. Felgner, U.: Untersuchungen über das Zornsche Lemma. Compos. Math. **18**, 170–180 (1967)
15. Fellin, G.: The Jacobson radical: from algebra to logic. Master's thesis. Università di Verona, Dipartimento di Informatica (2018)
16. Fellin, G., Schuster, P., Wessel, D.: The Jacobson radical of a propositional theory. In: Piecha, T., Schroeder-Heister, P. (eds.) Proof-Theoretic Semantics: Assessment and Future Perspectives. Proceedings of the Third Tübingen Conference on Proof-Theoretic Semantics, 27–30 March 2019, pp. 287–299. University of Tübingen (2019). http://dx.doi.org/10.15496/publikation-35319
17. Hausdorff, F.: Grundzüge der Mengenlehre. Verlag von Veit & Comp, Leipzig (1914)
18. Herrlich, H.: Axiom of Choice, Lecture Notes in Mathematics, vol. 1876. Springer, Berlin (2006). https://doi.org/10.1007/11601562
19. Howard, P., Rubin, J.: Consequences of the Axiom of Choice. American Mathematical Society, Providence (1998)
20. Jacobson, N.: The radical and semi-simplicity for arbitrary rings. Am. J. Math. **67**(2), 300–320 (1945). https://doi.org/10.2307/2371731

21. Lombardi, H., Quitté, C.: Commutative Algebra: Constructive Methods. Finite Projective Modules, Algebra and Applications, vol. 20. Springer, Netherlands (2015). https://doi.org/10.1007/978-94-017-9944-7
22. Moschovakis, Y.: Notes on Set Teory. Undergraduate Texts in Mathematics, Second edn. Springer, New York (2006)
23. Raoult, J.C.: Proving open properties by induction. Inform. Process. Lett. **29**(1), 19–23 (1988)
24. Rinaldi, D., Schuster, P.: A universal Krull-Lindenbaum theorem. J. Pure Appl. Algebra **220**, 3207–3232 (2016)
25. Rinaldi, D., Schuster, P., Wessel, D.: Eliminating disjunctions by disjunction elimination. Bull. Symb. Logic **23**(2), 181–200 (2017)
26. Rinaldi, D., Schuster, P., Wessel, D.: Eliminating disjunctions by disjunction elimination. Indag. Math. (N.S.) **29**(1), 226–259 (2018)
27. Rinaldi, D., Wessel, D.: Cut elimination for entailment relations. Arch. Math. Logic **58**(5–6), 605–625 (2019)
28. Schlagbauer, K., Schuster, P., Wessel, D.: Der Satz von Hahn-Banach per Disjunktionselimination. Confluentes Math. **11**(1), 79–93 (2019)
29. Schuster, P.: Induction in algebra: a first case study. In: 2012 27th Annual ACM/IEEE Symposium on Logic in Computer Science, LICS 2012. Dubrovnik, Croatia, pp. 581–585. IEEE Computer Society Publications (2012)
30. Schuster, P.: Induction in algebra: a first case study. Log. Methods Comput. Sci. **9**(3), 20 (2013)
31. Schuster, P., Wessel, D.: Syntax for semantics: Krull's maximal ideal theorem. In: Heinzmann, G., Wolters, G. (eds.) Paul Lorenzen: Mathematician and Logician. Springer, forthcoming
32. Schuster, P., Wessel, D.: A general extension theorem for directed-complete partial orders. Rep. Math. Logic **53**, 79–96 (2018)
33. Schuster, P., Wessel, D.: Resolving finite indeterminacy: a definitive constructive universal prime ideal theorem. In: Proceedings of the 35th Annual ACM/IEEE Symposium on Logic in Computer Science (LICS 2020), 8–11 July 2020, Saarbrücken, Germany. ACM, New York (2020)
34. Schuster, P., Wessel, D., Yengui, I.: Dynamic evaluation of integrity and the computational content of Krull's lemma (2019, preprint)
35. Scott, D.: Completeness and axiomatizability in many-valued logic. In: Henkin, L., Addison, J., Chang, C., Craig, W., Scott, D., Vaught, R. (eds.) Proceedings of the Tarski Symposium (Proceedings of Symposia in Pure Mathematics, vol. XXV, University of California, Berkeley, California (1971). pp. 411–435. American Mathematical Society, Providence (1974)
36. Wessel, D.: Points, ideals, and geometric sequents. Technical report, University of Verona (2018)
37. Yengui, I.: Making the use of maximal ideals constructive. Theoret. Comput. Sci. **392**, 174–178 (2008)
38. Yengui, I.: Constructive Commutative Algebra. Projective Modules Over Polynomial Rings and Dynamical Gröbner Bases, Lecture Notes in Mathematics, vol. 2138. Springer, Cham (2015). https://doi.org/10.1007/978-3-319-19494-3
39. Young, H.P., Zamir, S. (eds.): Handbook of Game Theory, Volume 4. Handbooks in Economics, 1st edn. North-Holland, Amsterdam (2015)
40. Zorn, M.: A remark on method in transfinite algebra. Bull. Am. Math. Soc. **41**, 667–670 (1935)

Number of Prefixes in Trace Monoids: Clique Polynomials and Dependency Graphs

Cyril Banderier[1]([⊠]) and Massimiliano Goldwurm[2]

[1] Laboratoire d'Informatique de Paris-Nord (UMR CNRS 7030),
Université Paris-Nord, Villetaneuse, France
`cb@lipn.fr`
[2] Dipartimento di Matematica "Federigo Enriques",
Università degli Studi di Milano, Milan, Italy

Abstract. We present some asymptotic properties on the average number of prefixes in trace languages. Such languages are characterized by an alphabet and a set of commutation rules, also called concurrent alphabet, which can be encoded by an independency graph or by its complement, called dependency graph. One key technical result, which has its own interest, concerns general properties of graphs and states that "if an undirected graph admits a transitive orientation, then the multiplicity of the root of minimum modulus of its clique polynomial is smaller or equal to the number of connected components of its complement graph". As a consequence, under the same hypothesis of transitive orientation of the independency graph, one obtains the relation $E[T_n] = O(E[W_n])$, where the random variables T_n and W_n represent the number of prefixes in traces of length n under two different fundamental probabilistic models:

– the uniform distribution among traces of length n (for T_n),
– the uniform distribution among words of length n (for W_n).

These two quantities are related to the time complexity of algorithms for solving classical membership problems on trace languages.

Keywords: Trace monoids · Clique polynomials · Möbius functions · Automata theory · Analytic combinatorics · Patterns in words

1 Introduction

In computer science, trace monoids have been introduced by Mazurkiewicz [22] as a model of concurrent events, describing which action can permute or not with another action (we give a formal definition of traces and trace monoids in Sect. 2, see also [14] for a treatise on the subject). In combinatorics, they are related to the fundamental studies of the "monoïde partiellement commutatif" introduced by Cartier and Foata in [10], and to its convenient geometrical view as heap of pieces proposed by Viennot in [25].

© Springer Nature Switzerland AG 2020
M. Anselmo et al. (Eds.): CiE 2020, LNCS 12098, pp. 251–263, 2020.
https://doi.org/10.1007/978-3-030-51466-2_22

Several classical problems in language theory (recognition of rational and context-free trace languages, determination of the number of representative words of a given trace, computing the finite state automaton recognizing these words) can be solved by algorithms that work in time and space proportional to (or strictly depending on) the number of prefixes of the input trace [3,6–8,15,23]. This is due to the fact that prefixes represent the possible decompositions of a trace in two parts and hence they are natural indexes for computations on traces.

This motivates the analysis of the number of prefixes of a trace of given length both in the worst and in the average case. In the average case analysis, two natural sequences of random variables play a key role:

- $\{T_n\}_{n\in\mathbb{N}}$, the number of prefixes of traces of length n generated at random under the equidistribution of traces of given size;
- $\{W_n\}_{n\in\mathbb{N}}$, the number of prefixes of traces of length n generated at random under the equidistribution of representative words of given size.

For some families of trace monoids, the asymptotic average, variance, and limit distributions of $\{T_n\}$ and $\{W_n\}$ are known [6,7,19–21]. It is interesting that they rely on the structural properties of an underlying graph (the independency graph, defined in Sect. 2). For example, it is known that, for every trace monoid $\mathcal{M}$, the maximum number of prefixes of a trace of length n is of the order $\Theta(n^\alpha)$, where α is the size of the largest clique in the concurrent alphabet defining $\mathcal{M}$ [8]. We summarize further such results in Sect. 3. In analytic combinatorics (see [17] for an introduction to this field), it remains a nice challenge to get a more universal description of the possible asymptotics of T_n and W_n.

In this work we prove that, if the concurrent alphabet $(\Sigma, \mathcal{C})$ admits a transitive orientation, then

$$E[T_n] = O(E[W_n]).$$

This is obtained by showing a general property of undirected graphs, which in our context is applied to the concurrent alphabet (Σ, C) and its complement (Σ, C^c). Such a property states that, for any undirected graph $\mathcal{G}$ admitting a transitive orientation of its edges, the number of connected components of its complement is greater or equal to the multiplicity of the root of smallest modulus in the clique polynomial of $\mathcal{G}$. The interest for the present discussion mainly relies on the use of finite state automata and on classical tools of formal languages to study properties of integer random variables in particular the asymptotic behaviour of their moments.

The paper is organized as follows: in Sect. 2 we recall the basic definitions on trace monoids; in Sect. 3 we summarize some asymptotic results on the random variables T_n and W_n; in Sects. 4 and 5, we present our main results on cross-sections of trace monoids, clique polynomials, and a new bound relating the asymptotic behaviour of T_n and W_n; we then conclude with possible future extensions of our work.

2   Notation and Preliminary Notions

For the reader not already familiar with the terminology of trace languages, we present in this section the key notions used in this article (see e.g. [14] for more details on all these notions).

Given a finite alphabet Σ, as usual Σ^* denotes the free monoid of all words over Σ, ε is the empty word and $|w|$ is the length of a word w for every $w \in \Sigma^*$. We recall that, for any $w \in \Sigma^*$, a prefix of w is a word $u \in \Sigma^*$ such that $w = uv$, for some $v \in \Sigma^*$. Also, for any finite set $\mathcal{S}$, we denote by $\#\mathcal{S}$ the cardinality of $\mathcal{S}$.

A **concurrent alphabet** is then a pair $(\Sigma, \mathcal{C})$, where $\mathcal{C} \subseteq \Sigma \times \Sigma$ is a symmetric and irreflexive relation over Σ. Such a pair can alternatively be defined by an undirected graph, which we call **independency graph**, where Σ is the set of nodes and $\{\{a, b\} \mid (a, b) \in \mathcal{C}\}$ is the set of edges. Its complement $(\Sigma, \mathcal{C}^c)$ is called **dependency graph**. As the notions of concurrent alphabet and independency graph are equivalent, in the sequel we indifferently refer to either of them. Informally, a concurrent alphabet lists the pairs of letters which can commute.

The **trace monoid** generated by a concurrent alphabet $(\Sigma, \mathcal{C})$ is defined as the quotient monoid $\Sigma^* / \equiv_{\mathcal{C}}$, where $\equiv_{\mathcal{C}}$ is the smallest congruence extending the equations $\{ab = ba : (a, b) \in \mathcal{C}\}$, and is denoted by $\mathcal{M}(\Sigma, \mathcal{C})$ or simply by $\mathcal{M}$. Its elements are called **traces** and its subsets are named **trace languages**. In other words, a trace is an equivalence class of words with respect to the relation $\equiv_{\mathcal{C}}$ given by the reflexive and transitive closure of the binary relation $\sim_{\mathcal{C}}$ over Σ^* such that $uabv \sim_{\mathcal{C}} ubav$ for every $(a, b) \in \mathcal{C}$ and every $u, v \in \Sigma^*$. For any $w \in \Sigma^*$, we denote by $[w]$ the trace represented by w; in particular $[\varepsilon]$ is the empty trace, i.e. the unit of $\mathcal{M}$. Note that the product of two traces $r, s \in \mathcal{M}$, where $r = [x]$ and $s = [y]$, is the trace $t = [xy]$, which does not depend on the **representative words** $x, y \in \Sigma^*$ and we denote the product by $t = s \cdot r$. The length of a trace $t \in \mathcal{M}$, denoted by $|t|$, is the length of any representative word. For any $n \in \mathbb{N}$, let $\mathcal{M}_n := \{t \in \mathcal{M} : |t| = n\}$ and $m_n := \#\mathcal{M}_n$.

Note that if $\mathcal{C} = \emptyset$ then $\mathcal{M}$ reduces to Σ^*, while if $\mathcal{C} = \{(a, b) \in \Sigma \times \Sigma \mid a \neq b\}$ then $\mathcal{M}$ is the commutative monoid of all monomials with letters in Σ.

Any trace $t \in \mathcal{M}$ can be represented by a **partial order** over the multiset of letters of t, denoted by $\mathrm{PO}(t)$. It works as follows: first, consider a word w satisfying $t = [w]$. Then, for any pair of letters (a, b) of w, let a_i be the i-th occurrence of the letter a and b_j the j-th occurrence of the letter b. The partial order is then defined as $a_i < b_j$ whenever a_i precedes b_j in all representative words of $[w]$. (See Example 1 hereafter.)

A **prefix** of a trace $t \in \mathcal{M}$ is a trace p such that $t = p \cdot s$ for some $s \in \mathcal{M}$. Clearly, any prefix of t is a trace $p = [u]$ where u is a prefix of a representative of t. It is easy to see that if p is a prefix of t then the $\mathrm{PO}(u)$ is an order ideal of $\mathrm{PO}(t)$ and can be represented by the corresponding antichain. We recall that an **antichain** of a partial order set $(\mathcal{S}, \leq)$ is a subset $A \subseteq \mathcal{S}$ such that $a \leq b$ does not hold for any pair of distinct elements $a, b \in A$, while an order ideal in $(\mathcal{S}, \leq)$ is a subset $\{a \in \mathcal{S} \mid \exists b \in A \text{ such that } a \leq b\}$ for some antichain A of $(\mathcal{S}, \leq)$. For every $t \in \mathcal{M}$, we denote by $\mathrm{Pref}(t)$ the set of all prefixes of t.

Example 1. Let $\mathcal{M}$ be the trace monoid characterized by the following independency graph:

$$\textcircled{a}\!\!-\!\!\textcircled{b}\!\!-\!\!\textcircled{c}\!\!-\!\!\textcircled{d}$$

That is, one has $ab = ba, bc = cb, cd = dc$. Then, the trace $[bacda]$ (i.e., the equivalence class of the word $bacda$) is the set of words $\{bacda, badca, abdca, abcda, acbda\}$. The corresponding partially ordered set is given by the following diagram

$$\mathrm{PO}([bacda]) \;=\; \begin{array}{c} a_1 \longrightarrow c_1 \searrow \\ \searrow \qquad\quad a_2 \\ b_1 \longrightarrow d_1 \nearrow \end{array}$$

where an arrow from x_i to y_j means that x_i always precedes y_j and where we omitted the arrows implied by transitivity. The set of prefixes is given by

$$\mathrm{Pref}([bacda]) \;=\; \{[\varepsilon], [a], [b], [ab], [ac], [abc], [abd], [abcd], [abcda]\}.$$

In this set, we now overline the letters belonging to the antichain of each prefix:
$\{[\varepsilon], [\overline{a}], [\overline{b}], [\overline{ab}], [a\overline{c}], [a\overline{bc}], [ab\overline{d}], [ab\overline{cd}], [abcd\overline{a}]\}$.   ∎

Recognizable, rational and context-free trace languages are well defined by means of linearization and closure operations over traditional string languages; their properties and in particular the complexity of their membership problems are widely studied in the literature (see for instance [8, 14, 15, 23]).

For any alphabet Σ and trace monoid $\mathcal{M}$, we denote by $\mathbb{Z}\langle\!\langle\Sigma\rangle\!\rangle$ the set of formal series on words (they are thus series in noncommutative variables) and by $\mathbb{Z}\langle\!\langle\mathcal{M}\rangle\!\rangle$ the set of formal series on traces (they are thus series in partially commutative variables), and $\mathbb{Z}[\![z]\!]$ stands for ring of classical power series in the variable z. These three distinct rings (with the operations of sum and Cauchy product, see [5, 14, 24]) will be used in Sects. 4 and 5.

3   Asymptotic Results for the Number of Prefixes

Several algorithms are presented in the literature for the recognition of rational and context-free trace languages, or for other problems like computing the number of representative words of a trace, that take a trace t as input and then carry out some operations on all prefixes of t [3, 6–9, 15, 23]. Thus, their time and space complexity strictly depend on the number of prefixes of t and in many cases they work just in time $\Theta(\#\,\mathrm{Pref}(t))$. Now, it follows from [8] that

$$\max\{\#\,\mathrm{Pref}(t) : t \in \mathcal{M}_n\} = \Theta(n^\alpha), \tag{1}$$

where α is the size of the largest clique in the independency graph of $\mathcal{M}$. It is thus essential to get a more refined analysis of the asymptotic behaviour of $\#\,\mathrm{Pref}(t)$ under natural distribution models in order to obtain a better understanding of the average complexity of all these algorithms.

In this section, we recall the main results on the number of prefixes of a random trace, under two different probabilistic models.

3.1 Probabilistic Analysis on Equiprobable Words

A main goal of the present contribution is to compare the random variables T_n and W_n, defined by

$$T_n = \# \mathrm{Pref}(t) \quad \text{and} \quad W_n = \# \mathrm{Pref}([w]), \tag{2}$$

where t is uniformly distributed over $\mathcal{M}_n$, while w is uniformly distributed over Σ^n. Clearly the properties of T_n and W_n immediately yield results on time complexity of the algorithms described in [3,6,7] assuming, respectively, equiprobable input traces of length n and equiprobable representative words of length n. Since every trace of length n has at least $n + 1$ prefixes, a first crude asymptotic bound is

$$n + 1 \leq T_n \leq dn^\alpha, \quad n + 1 \leq W_n \leq dn^\alpha \qquad (\forall\, n \in \mathbb{N}),$$

for a suitable constant $d > 0$, where α is defined as in (1). More precise results on the moments of W_n are studied in [6,7,20]:

$$\mathrm{E}[W_n^j] = \Theta(n^{jk}) \quad \forall\, j \in \mathbb{N}, \tag{3}$$

where k is the number of connected components of the dependency graph of $\mathcal{M}$. This relation is obtained by constructing suitable bijections between each moment of W_n and the set of words of length n in a regular language [6]. These bijections also allow proving a first order cancellation of the variance, i.e. $\mathrm{var}(W_n) = O(n^{2k-1})$ [20]. Further, when the dependency graph is transitive, this leads to two different limit laws, either chi-squared or Gaussian, according whether all the connected components of (Σ, C^c) have the same size or not [19].

3.2 Probabilistic Analysis on Equiprobable Traces

Now, in order to analyse T_n (the number of prefixes of a random trace of size n), it is useful to introduce the **generating function** of the trace monoid $\mathcal{M}$:

$$M(z) := \sum_{n \in \mathbb{N}} m_n z^n, \quad \text{with } m_n := \#\mathcal{M}_n = \#\{t \in \mathcal{M} : |t| = n\}.$$

The **Möbius function** of $\mathcal{M}$ is defined as $\mu_{\mathcal{M}} := \sum_{t \in \mathcal{M}} \mu_{\mathcal{M}}(t)\, t$, where

$$\mu_{\mathcal{M}}(t) = \begin{cases} 1 & \text{if } t = [\varepsilon], \\ (-1)^n & \text{if } t = [a_1 a_2 \cdots a_n], \\ & \text{where all } a_i \in \Sigma \text{ are distinct and } (a_i, a_j) \in C \text{ for any } i \neq j, \\ 0 & \text{otherwise.} \end{cases}$$

It is in fact a polynomial belonging to $\mathbb{Z}\langle\!\langle \mathcal{M} \rangle\!\rangle$. As established by Cartier and Foata in [10], an important property of $\mu_{\mathcal{M}}$ is that

$$\xi_{\mathcal{M}} \cdot \mu_{\mathcal{M}} = \mu_{\mathcal{M}} \cdot \xi_{\mathcal{M}} = 1, \tag{4}$$

where $\xi_{\mathcal{M}} = \sum_{t \in \mathcal{M}} t$ is the characteristic series of $\mathcal{M}$. Here, $\xi_{\mathcal{M}}$ can be seen as a partially commutative analogue of $M(z)$.

Now, let $p_\mathcal{M} \in \mathbb{Z}[z]$ be the commutative analogue of $\mu_\mathcal{M}$. It then follows that

$$p_\mathcal{M}(z) = 1 - c_1 z + c_2 z^2 - \cdots + (-1)^\alpha c_\alpha z^\alpha, \tag{5}$$

where c_i is the number of cliques of size i in the independency graph of $\mathcal{M}$. For this reason, we call $p_\mathcal{M}$ the **clique polynomial** of the independency graph (Σ, C). Its properties are studied in several papers (see for instance [18,21]). In particular, the commutative analogue of Eq. (4) is then

$$M(z) \cdot p_\mathcal{M}(z) = p_\mathcal{M}(z) \cdot M(z) = 1. \tag{6}$$

This entails that $M(z) = (p_\mathcal{M}(z))^{-1}$, a fundamental identity which can also be derived by an inclusion-exclusion principle.

As it is known from [21] that $p_\mathcal{M}$ has a unique root ρ of smallest modulus (and clearly $\rho > 0$ via Pringsheim's theorem, see [17]), one gets $m_n = \#\mathcal{M}_n = c\rho^{-n} n^{\ell-1} + O\left(\rho^{-n} n^{\ell-2}\right)$, where $c > 0$ is a constant and ℓ is the multiplicity of ρ in $p_\mathcal{M}(z)$. We observe that the existence of a unique root of smallest modulus for $p_\mathcal{M}(z)$ is not a consequence of the strict monotonicity of the sequence $\{m_n\}$. Indeed, if one considers $M(z) = \frac{1}{(1-z^3)(1-z)^2}$, one has $m_{n+3} = ((n+5)m_n + 2m_{n+1} + 2m_{n+2})/(n+3)$ so the sequence $\{m_n\}$ is strictly increasing; however, the polynomial $(1-z^3)(1-z)^2$ has 3 distinct roots of smallest modulus. Therefore, such a $M(z)$ *cannot* be the generating function of a trace monoid.

In our context, clique polynomials are particularly relevant as they are related to the average value of the number of prefixes of traces [7,21]. In fact, for any trace monoid $\mathcal{M}$, we have $E[T_n] = \frac{P_n}{m_n}$, where $P_n = \sum_{t \in \mathcal{M}_n} \# \text{Pref}(t)$. Since $\xi_\mathcal{M}^2 = \sum_{t \in \mathcal{M}} \# \text{Pref}(t)t$, from (4) and (6) its commutative analogue becomes $\sum_n P_n z^n = p_\mathcal{M}(z)^{-2}$ and hence $P_n = \Theta(\rho^{-n} n^{2\ell-1})$, which proves

$$E[T_n] = \Theta(n^\ell), \tag{7}$$

where ℓ is the multiplicity of the smallest root of $p_\mathcal{M}(z)$.

4   Cross-Sections of Trace Monoids

Cross-sections are standard tools to study the properties of trace monoids by lifting the analysis at the level of free monoids. Intuitively, a cross-section of a trace monoid $\mathcal{M}$ is a language $\mathcal{L}$ having exactly one representative string for each trace in $\mathcal{M}$. Thus, the generating function of $\mathcal{L}$ coincides with $M(z)$ and hence it satisfies equality (6). As a consequence, by choosing an appropriate regular cross-section $\mathcal{L}$, one can use the property of a finite state automaton recognizing $\mathcal{L}$ to study the singularities of $M(z)$, i.e. the roots of $p_\mathcal{M}(z)$.

Formally, a **cross-section** of a trace monoid $\mathcal{M}$ over a concurrent alphabet (Σ, C) is a language $\mathcal{L} \subseteq \Sigma^*$ such that

- for each trace $t \in \mathcal{M}$, there exists a word $w \in \mathcal{L}$ such that $t = [w]$,
- for each pair of words $x, y \in \mathcal{L}$, if $[x] = [y]$ then $x = y$.

Among all cross-sections of $\mathcal{M}$, it is convenient to consider a canonical one. A natural one is based on a normal form using the lexicographic order [1]. Alternatively, one can see it as based on the orientations of edges in the independency graph of $\mathcal{M}$, as used in [12,13] to study properties of Möbius functions in trace monoids. It works as follows. Let $\leq$ be any total order on the alphabet Σ and let $\leq^*$ be the lexicographic linear order induced by $\leq$ over Σ^*. We denote by $<_{\mathcal{C}}$ the binary relation over Σ such that $a <_{\mathcal{C}} b$ if $(a,b) \in \mathcal{C}$ and $a \leq b$. Thus, $<_{\mathcal{C}}$ is an orientation of the independency graph of $\mathcal{M}$. We now consider the following cross-section of $\mathcal{M}$: the language $\mathcal{L}_\leq$ of all minimal lexicographic representatives of traces in $\mathcal{M}$, i.e. $\mathcal{L}_\leq = \{w \in \Sigma^* \mid w \leq^* y \text{ for every } y \in [w]\}$. Moreover, $\mathcal{L}_\leq$ is regular, as it satisfies the equality

$$\mathcal{L}_\leq = \Sigma^* \setminus \bigcup_{\substack{(a,b) \in \mathcal{C} \\ a <_{\mathcal{C}} b}} \Sigma^* b \, \mathcal{C}_a^* \, a \, \Sigma^*, \tag{8}$$

where $\mathcal{C}_a := \{c \in \Sigma \mid (a,c) \in \mathcal{C}\}$ is the set of letters allowed to commute with a. Thus, $\mathcal{L}_\leq$ is the set of all words in Σ^* that do not contain any factor of the form bva where $a <_{\mathcal{C}} b$ and $v \in \mathcal{C}_a^*$. Then, for any $w \in \Sigma^*$, in order to verify whether $w \in \mathcal{L}_\leq$, one can read the letters of w in their order, updating at each step the family of letters $a \in \Sigma$ forming a "forbidden" factor of the form bva, with $a <_{\mathcal{C}} b$, $v \in \mathcal{C}_a^*$. If one of these letters is met then w is rejected, otherwise it is accepted.

To formalize the definition, for each $b \in \Sigma$, the predecessors of b are $\mathrm{Pred}(b) = \{a \in \Sigma \mid a <_{\mathcal{C}} b\}$. Define the finite state automaton $\mathcal{A}$ as the 4-tuple $(2^\Sigma, \emptyset, \delta, F)$, where the set of states is 2^Σ, i.e. the power set of Σ, the initial state is the empty set $\emptyset$, $F = \{\mathcal{S} \in 2^\Sigma \mid \mathcal{S} \neq \Sigma\}$ is the family of final states and the transition function $\delta : (2^\Sigma \times \Sigma) \to 2^\Sigma$ is given by

$$\delta(\mathcal{S}, b) = \begin{cases} \Sigma & \text{if } b \in \mathcal{S} \\ \mathrm{Pred}(b) \cup (\mathcal{S} \cap \mathcal{C}_b) & \text{otherwise} \end{cases} \qquad (\forall \, \mathcal{S} \subseteq \Sigma, \, \forall \, b \in \Sigma).$$

Note that, during a computation, the current state $\mathcal{S}$ represents the set of forbidden letters. At the beginning, all input letters are allowed, as $\emptyset$ is the initial state, while Σ is a trap state, where all letters are forbidden. In a general step, if $\mathcal{S} \subseteq \Sigma$ is the current state and $b \notin \mathcal{S}$ is an input letter, the new set of forbidden letters must be obtained from $\mathcal{S} \cup \mathrm{Pred}(b)$ by removing those elements that do not commute with b. This justifies the above definition of δ and it is clear that $\mathcal{A}$ recognizes $\mathcal{L}_\leq$.

Moreover, the state set of the above automaton can be reduced to the states $\mathcal{S} \subsetneq \Sigma$ reachable from $\emptyset$. Setting

$$Q = \{\mathcal{S} \subseteq \Sigma \mid \mathcal{S} \neq \Sigma, \exists \, w \in \Sigma^* : \delta(\emptyset, w) = \mathcal{S}\},$$

the entries of the transition matrix $\widetilde{A}$ of the automaton $\mathcal{A}$ are given by:

$$\widetilde{A}_{\mathcal{S}, \mathcal{S}'} = \sum_{b \in \Sigma : \delta(\mathcal{S}, b) = \mathcal{S}'} b \qquad (\forall \, \mathcal{S}, \mathcal{S}' \in Q).$$

The commutative analogue in $\mathbb{N}[[z]]$ of this transition matrix has therefore all its entries which are monomials of degree one in z. Factorizing by z, this commutative analogue can thus be written zA, for a matrix A we call the *adjacency matrix* of $\mathcal{A}$. Note that A strictly depends on both the concurrent alphabet (Σ, C) and the total order $\leq$ over Σ.

As a consequence, since $\mathcal{A}$ recognizes a cross-section of $\mathcal{M}$, denoting by π and η, respectively, the characteristic (column) vectors of $\emptyset$ and Q, the generating function $M(z)$ is given by

$$M(z) = \sum_{n=0}^{+\infty} \pi' A^n \eta z^n = \pi'(I - zA)^{-1}\eta, \tag{9}$$

where I is the identity matrix of size $\#Q \times \#Q$ and π' is the transposed of π. This identity, together with relation (6) proves the following proposition.

Proposition 1 (Factorisation property). *For any trace monoid $\mathcal{M}$ with a concurrent alphabet (Σ, C), let $\leq$ be a total order on Σ, let A be the adjacency matrix of the automaton $\mathcal{A}$ recognizing the cross-section $\mathcal{L}_\leq$ of $\mathcal{M}$, and assume I, π and η defined as in (9). Then, $M(z)$ and $p_\mathcal{M}(z)$ satisfy the identities*

$$M(z) = \pi'(I - zA)^{-1}\eta = \frac{\pi' \operatorname{adj}(I - zA)\eta}{\det(I - zA)} \quad, \quad p_\mathcal{M}(z) = \frac{\det(I - zA)}{\pi' \operatorname{adj}(I - zA)\eta}. \tag{10}$$

Example 2. Consider the concurrent alphabet (Σ, C) defined by the graph

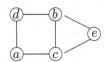

Then, the clique polynomial and the generating function of $\mathcal{M}$ are given by

$$p_\mathcal{M}(z) = 1 - 5z + 6z^2 - z^3, \quad M(z) = \sum_{n=0}^{+\infty} m_n z^n = \frac{1}{1 - 5z + 6z^2 - z^3}.$$

The standard ordering (a, b, c, d, e) on Σ induces the following (non-transitive) orientation $<_C$ over the independency graph

$$<_C = \quad$$

Thus the predecessors of each letter are given by $\operatorname{Pred}(a) = \operatorname{Pred}(b) = \emptyset$, $\operatorname{Pred}(c) = \operatorname{Pred}(d) = \{a, b\}$, $\operatorname{Pred}(e) = \{b, c\}$ and the transition matrix of $\mathcal{A}$ is defined by the following table, where rows and columns are labelled by the states of $\mathcal{A}$:

	$\emptyset$	$\{a,b\}$	$\{b,c\}$	$\{c\}$
$\emptyset$	$a+b$	$c+d$	e	0
$\{a,b\}$	0	$c+d$	e	0
$\{b,c\}$	0	d	e	a
$\{c\}$	0	d	e	$a+b$

$\widetilde{A} =$ (the table above)

From that $I - zA$ is easily computed (where A is the adjacency matrix of $\mathcal{A}$):

$$I - zA = \begin{bmatrix} 1 - 2z & -2z & -z & 0 \\ 0 & 1 - 2z & -z & 0 \\ 0 & -z & 1 - z & -z \\ 0 & -z & -z & 1 - 2z \end{bmatrix}$$

and, accordingly, $\det(I - zA) = 1 - 7z + 16z^2 - 13z^3 + 2z^4 = (1 - 2z)p_{\mathcal{M}}(z)$. ∎

Proposition 2. *For any trace monoid $\mathcal{M}$ over a concurrent alphabet (Σ, C) and any total order $\leq$ on Σ, all roots of the clique polynomial $p_{\mathcal{M}}(z)$ are reciprocals of eigenvalues of the corresponding adjacency matrix A. More precisely, the clique polynomial of any independency graph (Σ, C) is of the form*

$$p_{\mathcal{M}}(z) = \prod_{i=1}^{\alpha}(1 - x_i z)$$

where α is the size of the maximum clique in (Σ, C) and all x_i's are eigenvalues of a adjacency matrix A.

Proof (sketch). The result follows from Proposition 1 by refining equalities (10) and recalling that all roots of clique polynomials are different from 0. □

We observe that the reverse property does not hold in general, i.e. it may occur that an eigenvalue of A is not the reciprocal of a root of $p_{\mathcal{M}}(z)$. However, as shown in the following section, such a reverse sentence is true whenever the graph (Σ, C) admits a transitive orientation.

5 Concurrent Alphabets with Transitive Orientation

Now let us consider a trace monoid $\mathcal{M}$ such that its independency graph (Σ, C) admits a transitive orientation. Then, we may fix a total order $\leq$ on Σ such that $<_C$ is transitive. In this case, the definition of cross-section $\mathcal{L}_\leq$ and of the automaton $\mathcal{A}$ can be simplified, since the set of "forbidden" factors of the form bwa, with $a <_C b$ and $w \in C_a^*$, can be reduced to the simple set of words $S = \{\tau\sigma \in \Sigma^2 \mid \sigma <_C \tau\}$. To prove this property, consider a forbidden factor of the above form bwa, with $a <_C b$ and $w \in C_a^*$; thus any symbol c occurring in w must verify $(a, c) \in C$. As a consequence, either $a <_C c$ or $c <_C a$: in the first case

ca belongs to $\mathcal{S}$ while, in the second case, by transitivity of $<_{\mathcal{C}}$ we have $c <_{\mathcal{C}} b$ and hence bc is in $\mathcal{S}$.

Thus, identity (8) can be simplified as $\mathcal{L}_{\leq} = \Sigma^* \backslash \bigcup_{a <_{\mathcal{C}} b} \Sigma^* ba\Sigma^*$. Moreover, the state set of the automaton $\mathcal{A}$ can be reduced to $Q = \{\operatorname{Pred}(a) \mid a \in \Sigma\}$ and the transition function now assumes values $\delta(\mathcal{S}, b) = \operatorname{Pred}(b)$, for every $\mathcal{S} \in Q$ and every $b \in \Sigma \backslash \mathcal{S}$.

Proposition 3. *Let $(\Sigma, \mathcal{C})$ be a concurrent alphabet with an associated independency graph admitting a transitive orientation $<_{\mathcal{C}}$. Let $\leq$ be a total order on Σ extending $<_{\mathcal{C}}$. Also assume that the dependency graph $(\Sigma, \mathcal{C}^c)$ is connected. Then the adjacency matrix A is primitive.*

Proof (sketch). Under these hypotheses, by the simplifications above, it turns out that the state diagram of the automaton $\mathcal{A}$ (defined by $\leq$) is strongly connected and has at least one loop. □

The hypothesis of transitivity for $<_{\mathcal{C}}$ cannot be avoided to guarantee that A is primitive. For instance, in Example 2 the dependency graph $(\Sigma, \mathcal{C}^c)$ is connected but the orientation $<_{\mathcal{C}}$ of $(\Sigma, \mathcal{C})$ is not transitive, and in fact observe that the corresponding transition matrix is not irreducible and hence A is not primitive. Nevertheless, the smallest root of $p_{\mathcal{M}}(z)$ is simple and then the same concurrent alphabet satisfies the following theorem.

Theorem 4. *Let $(\Sigma, \mathcal{C})$ be a concurrent alphabet. If its independency graph admits a transitive orientation $<_{\mathcal{C}}$, then one has $\ell \leq k$, where ℓ and k denote, respectively, the multiplicity of the smallest root of $p_{\mathcal{M}}(z)$ and the number of connected components of the dependency graph $(\Sigma, \mathcal{C}^c)$.*

Proof (sketch). First, it is well-known [18,21] that $p_{\mathcal{M}}(z)$ is always the product of the clique polynomials of all independency subalphabets given by the connected components of $(\Sigma, \mathcal{C}^c)$. Then, each of these clique polynomials (using the additional condition that one has a transitive orientation) has a smallest root of multiplicity 1: this follows from Proposition 3 and a commutative analogue of a result in [11] stating that, when $(\Sigma, \mathcal{C})$ has a transitive orientation, its clique polynomial equals $\det(I - zA)$. □

Applying the previous theorem to relations (3) and (7), one gets the following.

Theorem 5. *Let $(\Sigma, \mathcal{C})$ be a concurrent alphabet. If its independency graph admits a transitive orientation $<_{\mathcal{C}}$, then the random variables counting the number of prefixes in traces (as defined in (2)) satisfy $\mathrm{E}[T_n] = O(\mathrm{E}[W_n])$.*

Example 3. Consider the concurrent alphabet $(\Sigma, \mathcal{C})$ and the orientation $<_{\mathcal{C}}$ of Example 2. Note that $(\Sigma, \mathcal{C})$ is connected but $<_{\mathcal{C}}$ is not transitive and in fact A is not primitive. However, $(\Sigma, \mathcal{C})$ admits a (different) orientation that is transitive, given by

A total order extending the previous orientation is $c < d < a < e < b$. Computing matrix A with respect to this total order we obtain

$$I - zA = \begin{bmatrix} 1-2z & -z & -z & -z \\ 0 & 1-z & -z & -z \\ -z & -z & 1-z & -z \\ 0 & -z & 0 & 1-z \end{bmatrix},$$

and hence $\det(I - zA) = 1 - 5z + 6z^2 - z^3 = p_{\mathcal{M}}(z)$.   ∎

The following example considers an independency graph of $\mathcal{M}$ that does not admit any transitive orientation. In this case $p_{\mathcal{M}}(z)$ is a proper factor of $\det(I - zA)$, but its smallest root is again simple and hence $\ell \leq k$ is still true even if the hypothesis of Theorem 4 is not satisfied.

Example 4. Consider the concurrent alphabet corresponding to the following independency graph $\mathcal{G}$, associated to the following partial order $<_C$:

Thus the transition matrix, defined according to Sect. 4, is given by the following table:

		$\emptyset$	$\{a,b\}$	$\{a\}$	$\{b,d\}$	$\{d\}$
$\tilde{A} =$	$\emptyset$	$a+b$	c	d	e	0
	$\{a,b\}$	0	c	d	e	0
	$\{a\}$	b	c	d	e	0
	$\{b,d\}$	0	c	0	e	a
	$\{d\}$	b	c	0	e	a

Accordingly, one has $\det(I - zA) = 1 - 6z + 10z^2 - 5z^3 = (1-z)p_{\mathcal{M}}(z)$.   ∎

6   Conclusion

We have investigated the fundamental role played by the clique polynomial in asymptotic studies of trace monoids. Building on the factorization property (stated in Proposition 1), we got a link between the multiplicity of its smallest root and the number of connected components of some associated graph (Theorem 4). This, in turn, is the key for a new asymptotic relation between the number of prefixes in traces of length n: $E[T_n] = O(E[W_n])$ (Theorem 5), where T_n and W_n correspond to two natural models (uniform distribution over traces and over words). In the long version of this article, we plan to extend these analyses to more general cases (including concurrent alphabets without transitive orientation).

Several other problems remain open in our context and could be at the centre of future investigations. The first one concerns the adjacency matrix A defined in Sect. 4, which does not seem to be studied too much in the previous literature; in particular, in all our examples $\det(I - zA)$ is a clique polynomial, even when the concurrent alphabet $(\Sigma, \mathcal{C})$ does not admit any transitive orientation. For this purpose, similarly to the approach used in [11] and in our proof of Theorem 4, it is possible to adapt a noncommutative approach building on links to words with forbidden patterns (see [2]). We plan to use these links to tackle the asymptotic behaviour of the variance and higher moments of $\{T_n\}$, and the limit distributions of both $\{T_n\}$ and $\{W_n\}$ for all trace monoids.

In conclusion, all these studies are further illustration of the nice interplay between complex analysis (analytic combinatorics) and the structural properties of formal languages, as also illustrated e.g. in [4,5,16,17,19,20].

References

1. Anisimov, A.V., Knuth, D.E.: Inhomogeneous sorting. Int. J. Comput. Inf. Sci. **8**, 255–260 (1979). https://doi.org/10.1007/BF00993053
2. Asinowski, A., Bacher, A., Banderier, C., Gittenberger, B.: Analytic combinatorics of lattice paths with forbidden patterns, the vectorial kernel method, and generating functions for pushdown automata. Algorithmica **82**, 1–43 (2020). https://doi.org/10.1007/s00453-019-00623-3
3. Avellone, A., Goldwurm, M.: Analysis of algorithms for the recognition of rational and context-free trace languages. RAIRO Theoret. Inform. Appl. **32**, 141–152 (1998)
4. Banderier, C., Drmota, M.: Formulae and asymptotics for coefficients of algebraic functions. Comb. Probab. Comput. **24**, 1–53 (2015)
5. Berstel, J., Reutenauer, C.: Rational Series and Their Languages. Springer, Heidelberg (1988)
6. Bertoni, A., Goldwurm, M.: On the prefixes of a random trace and the membership problem for context-free trace languages. In: Huguet, L., Poli, A. (eds.) AAECC 1987. LNCS, vol. 356, pp. 35–59. Springer, Heidelberg (1989). https://doi.org/10.1007/3-540-51082-6_68
7. Bertoni, A., Goldwurm, M., Sabadini, N.: Analysis of a class of algorithms for problems on trace languages. In: Beth, T., Clausen, M. (eds.) AAECC 1986. LNCS, vol. 307, pp. 202–214. Springer, Heidelberg (1988). https://doi.org/10.1007/BFb0039193
8. Bertoni, A., Mauri, G., Sabadini, N.: Equivalence and membership problems for regular trace languages. In: Nielsen, M., Schmidt, E.M. (eds.) ICALP 1982. LNCS, vol. 140, pp. 61–71. Springer, Heidelberg (1982). https://doi.org/10.1007/BFb0012757
9. Breveglieri, L., Crespi Reghizzi, S., Goldwurm, M.: Efficient recognition of trace languages defined by repeat until loops. Inf. Comput. **208**, 969–981 (2010)
10. Cartier, P., Foata, D.: Problèmes combinatoire de commutation et réarrangements. Lecture Notes in Mathematics, vol. 85. Springer, Heidelberg (1969). https://doi.org/10.1007/BFb0079468

11. Choffrut, C., Goldwurm, M.: Determinants and Möbius functions in trace monoids. Discrete Math. **194**, 239–247 (1999)
12. Diekert, V.: Transitive orientations, Möbius functions, and complete semi-thue systems for free partially commutative monoids. In: Lepistö, T., Salomaa, A. (eds.) ICALP 1988. LNCS, vol. 317, pp. 176–187. Springer, Heidelberg (1988). https://doi.org/10.1007/3-540-19488-6_115
13. Diekert, V.: Möbius functions and confluent semi-commutations. Theor. Comput. Sci. **108**, 25–43 (1993)
14. Diekert, V., Rozenberg, G. (eds.): The Book of Traces. World Scientific, Singapore (1995)
15. Duboc, C.: Commutations dans les monoïdes libres: un cadre théorique pour l'étude du parallélisme. Thèse, Faculté des Sciences de l'Université de Rouen (1986)
16. Flajolet, P.: Analytic models and ambiguity of context-free languages. Theor. Comput. Sci. **49**, 283–309 (1987)
17. Flajolet, P., Sedgewick, R.: Analytic Combinatorics. Cambridge University Press, Cambridge (2009)
18. Fisher, D., Solow, A.: Dependence polynomials. Discrete Math. **82**, 251–258 (1990)
19. Goldwurm, M.: Some limit distributions in analysis of algorithms for problems on trace languages. Int. J. Found. Comput. Sci. **1**(3), 265–276 (1990)
20. Goldwurm, M.: Probabilistic estimation of the number of prefixes of a trace. Theor. Comput. Sci **92**, 249–268 (1992)
21. Goldwurm, M., Santini, M.: Clique polynomials have a unique root of smallest modulus. Inf. Process. Lett. **75**, 127–132 (2000)
22. Mazurkiewicz, A.: Concurrent program schemes and their interpretations, DAIMI Rep. PB 78, Aarhus University, Aarhus (1977)
23. Rytter, W.: Some properties of trace languages. Fund. Inform. **7**, 117–127 (1984)
24. Salomaa, A., Soittola, M.: Automata-Theoretic Aspects of Formal Power Series. Springer, New York (1978). https://doi.org/10.1007/978-1-4612-6264-0
25. Viennot, G.X.: Heaps of pieces, I : Basic definitions and combinatorial lemmas. In: Labelle, G., Leroux, P. (eds.) Combinatoire énumérative. LNM, vol. 1234, pp. 321–350. Springer, Heidelberg (1986). https://doi.org/10.1007/BFb0072524

Repetitions in Toeplitz Words
and the Thue Threshold

Antonio Boccuto[ID] and Arturo Carpi[(✉)][ID]

Dipartimento di Matematica e Informatica, University of Perugia, via Vanvitelli 1,
06123 Perugia, Italy
{antonio.boccuto,arturo.carpi}@unipg.it

Abstract. A (finite or infinite) word is said to be k-th power-free if it
does not contain k consecutive equal blocks. A colouring of the integer
lattice points in the n-dimensional Euclidean space is power-free if there
exists a positive integer k such that the sequence of colours of consecutive
points on any straight line is a k-th power-free word. The Thue threshold
of $\mathbb{Z}^n$ is the least number of colours $t(n)$ allowing a power-free colouring
of the integer lattice points in the n-dimensional Euclidean space.

Answering a question of Grytczuk (2008), we prove that $t(2) = t(3) =$
2. Moreover, we show the existence of a 2-colouring of the integer lattice
points in the Euclidean plane such that the sequence of colours of consecutive points on any straight line does not contain squares of length
larger than 26.

In order to obtain these results, we study repetitions in Toeplitz words.
We show that the Toeplitz word generated by any sequence of primitive
partial words of maximal length k is k-th power-free. Moreover, adding
a suitable hypothesis on the positions of the holes in the generating
sequence, we obtain that also the subwords occurring in the considered
Toeplitz word according to an arithmetic progression of suitable difference, are k-th power-free words.

Keywords: Power-free word · Toeplitz word · Partial word · Word
with bounded square · Thue threshold · Arithmetic subword

1 Introduction

The study of repetitions in words has been one of the main fields of interest
in Combinatorics on Words since its origins [6,20,21]. This subject has several
applications in other fields such as Algebra, Symbolic Dynamics, Game Theory.

Let k be a positive integer. We recall that a word is said to be k-th power-
free if it does not contain k consecutive equal blocks. For instance, the word
"barbarian" is not square-free, since it contains two consecutive occurrences of
the block "bar", while it is cube-free. It is known that there exist cube-free
infinite words over a two-letter alphabet and square-free infinite words over a
three-letter alphabet [20].

© Springer Nature Switzerland AG 2020
M. Anselmo et al. (Eds.): CiE 2020, LNCS 12098, pp. 264–276, 2020.
https://doi.org/10.1007/978-3-030-51466-2_23

In [8], one of the authors considered the following multidimensional extension of k-th power-freeness, suggested by J. Berstel. Consider a colouring of the integer lattice points in the n-dimensional Euclidean space, with finitely many colours. Such a colouring will be called k-th power-free if the sequence of colours of consecutive points on any straight line is an infinite k-th power-free word. The main result of [8] is the existence of a square-free colouring of the integer lattice points in the n-dimensional Euclidean space, for all positive integer n.

The minimum number of colours $C(n)$ required for such a colouring is actually unknown. While the construction of [8] uses 4^n colours, recently [16] it has been shown that $C(n) \geq 9 \cdot 2^{n-2}$, for all $n \geq 2$. Kao et al. [15] used a smaller number of colours to build k-th power-free colourings of the integer lattice points of the plane, with $k > 2$. In this context, an interesting notion is the Thue threshold, introduced by Grytczuk [13]. A colouring of the integer lattice points in the n-dimensional Euclidean space is power-free if it is k-th power-free for some positive k. The Thue threshold of $\mathbb{Z}^n$ is the minimum number of colours $t(n)$ needed for such a colouring. One of the problems proposed by Grytczuk asks for the value of $t(n)$, at least in the case $n = 2$. The existence of infinite cube-free words on a binary alphabet shows that $t(1) = 2$. The results of [15] quoted above imply, in particular, that $t(2) \leq 4$ and, more generally, $t(n) \leq 2^n$. On the other side, it is clear that $t(n) \geq 2$ for all n. In this paper, we will show that $t(2) = t(3) = 2$. More precisely, we construct a 9-th power-free 2-colouring of the 3-dimensional Euclidean space and an 8-th power-free 2-colouring of the Euclidean plane. In view of these results, we conjecture that $t(n) = 2$ for all positive n.

Another notion very close to power-freeness is that of bounded repetition. An infinite word is said to have bounded repetition if the length of the squares occurring in it is upperbounded by a constant. An infinite word with bounded repetition over a binary alphabet has been given in [10]. Further examples can be found, for instance, in [1,11,19]. Similarly to the Thue threshold, one can introduce the bounded repetition threshold as follows. A colouring of the integer lattice points in the n-dimensional Euclidean space is said to have bounded repetition if the length of the squares occurring in it is upperbounded by a constant. The minimum number of colours needed for such a colouring will be called the bounded repetition threshold of $\mathbb{Z}^n$. We will show that the bounded repetition threshold of $\mathbb{Z}^2$ is 2.

Now we describe some of the tools used for the construction of the power-free colourings considered above. According to [4], we call arithmetic subsequence (of difference d) of an infinite word the subsequence obtained by extracting the letters which are positioned according to an arithmetic progression of difference d. The study of infinite words whose arithmetic subsequences, for some prescribed difference, are power-free has been started in [8]. Further results on this subject can be found in [9,15]. In [8] it is shown that there exists an infinite word on a 4-letter alphabet such that all its arithmetic subsequences of odd difference are square-free. As noticed in [15] this word is related to paperfolding words. There is a large literature on paperfolding words and on the more general class of Toeplitz words (see, e.g., [1,3,14,18,19] and the references therein). Avgustinovich *et al.* [4] showed that the factors of arithmetic subsequences of odd difference of paperfolding words are themselves factors of paperfolding words. Since all paperfolding words are 4-th power-free [1], one obtains that arithmetic subsequences of odd difference of paperfolding words are 4-th power-free words.

In this paper we study in more details power-freeness of Toeplitz words and of their arithmetic subwords, obtaining some extension of the results above. In order to describe our results, we recall some further notion. Partial words were introduced by Berstel and Boasson [5] in order to study some extension of the Fine and Wilf Periodicity Theorem. Roughly speaking, a partial word is a word such that the letters in some positions (called holes) are unknown. Primitive partial words have been studied by Blanchet-Sadri [7]. Given a sequence of partial words $s^{(n)}$, $n \geq 1$, one can construct an infinite word as follows: first concatenate infinitely many copies of $s^{(1)}$, thus obtaining an infinite word with holes; next, ordinately replace the holes by the letters of the infinite word obtained by concatenating infinitely many copies of $s^{(2)}$, thus obtaining a new infinite word with holes; and so on... If there are infinitely many n such that $s^{(n)}$ does not start with a hole, then the previous construction converges to an infinite word with no hole, which is called the Toeplitz word generated by the sequence $s^{(n)}$. We will show that, if the words $s^{(n)}$, $n \geq 1$, are primitive partial words of maximal length k, then the generated Toeplitz word is k-th power-free. Concerning the arithmetic subsequences of Toeplitz words, we prove a useful structural property. If the positions of the holes in the words of the generating sequence of a Toeplitz word U satisfy a suitable hypothesis and d is an integer relatively prime with their lengths, then any arithmetic subsequence of difference d of U is a Toeplitz word and its generating sequence can be obtained by rearranging the letters of the generating sequence according to a particular rule. We notice that some results related to this one can be found in [4,12].

As a consequence of the two previous results, we obtain that if U is a Toeplitz word generated by a sequence of primitive partial words of maximal length k satisfying the condition on the position of the holes considered above, if q is the least common multiple of the lengths of the generating words, and if d is any integer which is not a multiple of q, then all arithmetic subsequences of U of difference d are k-th power-free. This last result gives us a powerful tool to construct words with power-free arithmetic subsequences.

The paper is organized as follows. In the next section we recall some basic definitions and properties needed in the sequel. In Sect. 3 we establish the result on power-freeness of Toeplitz words generated by primitive partial words. Arithmetic subsequences of Toeplitz words are studied in Sect. 4. In Sect. 5 we prove that the Thue threshold of $\mathbb{Z}^3$ is 2 and in Sect. 6 we prove that the bounded square threshold of $\mathbb{Z}^2$ is 2. Finally, in Sect. 7 we discuss some open problems.

2   Preliminaries

Let A be a finite nonempty set, or *alphabet*, and A^* be the free monoid generated by A. The elements of A are usually called *letters* and those of A^* *words*. The identity element of A^* is called *empty word* and denoted by ε. We set $A^+ = A^* \setminus \{\varepsilon\}$. A word $w \in A^+$ can be written uniquely as a sequence of letters as $w = w_1 w_2 \cdots w_n$, with $w_i \in A$, $1 \leq i \leq n$, $n > 0$. The integer n is called the *length* of w and denoted by $|w|$. All words $w_i w_{i+1} \cdots w_j$ with $1 \leq i \leq j \leq n$ are called *factors* of w. A *period* of the word w is any positive integer p such that $w_i = w_{i+p}$ for all $i = 1, 2, \ldots, n - p$. For all $n \geq 0$, the set of all words of length n on the alphabet A is denoted by A^n.

Let k be a positive integer. Any word of the form u^k, with $u \neq \epsilon$, is called a *k-th power*. In particular, 2-nd and 3-rd powers are usually called *squares* and *cubes*. As is known, a word w is a k-th power if and only if w has a period p such that $|w| = kp$. A word is *primitive* if it is not a k-th power, for all $k \geq 2$.

An *infinite word* U on the alphabet A is any unending sequence of letters. For all $n \geq 1$, we let U_n denote the n-th letter of the infinite word U. Thus, $U = U_1 U_2 \cdots U_n \cdots$. The set of infinite words on the alphabet A is denoted by A^ω. A *bi-infinite word* on the alphabet A is any map $V : \mathbb{Z} \to A$, where $\mathbb{Z}$ denotes the semiring of relative integers. The image of any $n \in \mathbb{Z}$ by V is denoted by V_n. The *factors* of an infinite (resp., bi-infinite) word U are the words $U_i U_{i+1} \cdots U_j$ with $1 \leq i \leq j$ (resp., $i, j \in \mathbb{Z}$ and $i \leq j$). A *period* of an infinite (resp., bi-infinite) word U is any positive integer p such that $U_i = U_{i+p}$ for all $i \geq 1$ (resp., $i \in \mathbb{Z}$). If s is a finite word, we let s^ω denote the infinite word obtained by concatenating infinitely many copies of s.

A (finite or infinite or bi-infinite) word is said to be *k-th power-free* if none of its factors is a k-th power.

Let U be an infinite word, and i and d be two positive integers. The infinite word $V = U_i U_{i+d} U_{i+2d} \cdots U_{i+nd} \cdots$ is called an *arithmetic subsequence* of U of *difference* d. Any factor of any arithmetic subsequence of U of difference d is said to be an *arithmetic subword* of U of *difference* d. Arithmetic subsequences and subwords of bi-infinite words can be defined similarly.

Let A be a k-letter alphabet and n be a positive integer. Any map $\alpha : \mathbb{Z}^n \to A$ will be called a *k-colouring of the lattice points of the n-dimensional space*, or, briefly, a $\mathbb{Z}^n$-*word*. A bi-infinite word V is a *line* of α if there exist integers $j_1, \ldots, j_n, m_1, \ldots, m_n \in \mathbb{Z}$ such that $\gcd(m_1, \ldots, m_n) = 1$ and

$$V_q = \alpha(j_1 + qm_1, \ldots, j_n + qm_n) \quad \text{for all } q \in \mathbb{Z}.$$

This definition [8] is motivated by the fact that, as one can easily verify, the lines of α are the sequences of letters corresponding to the integer lattice points of the n-dimensional space $\mathbb{R}^n$ which lie on a same straight line. For instance, if α is the 'chessboard coloring' of the plane, that is the 2-coloring of $\mathbb{Z}^2$ defined by $\alpha(x, y) = x + y \mod 2$, then its lines are the bi-infinite words

$$\cdots 01010101 \cdots, \quad \cdots 00000000 \cdots \quad \text{and} \quad \cdots 11111111 \cdots$$

Indeed, the first one corresponds to all straight lines with directive numbers (m_1, m_2) with $m_1 + m_2$ odd, while the other two correspond to the straight lines with directive numbers (m_1, m_2) where both m_1 and m_2 are odd and, consequently, $m_1 + m_2$ is even.

3   Repetitions in Toeplitz Words

Partial words were introduced by Berstel and Boasson [5]. We call *partial word* over the alphabet A any word over the alphabet $A \cup \{?\}$, where ? is a distinguished letter, not belonging to A. The occurrences of ? in a partial word are usually called *holes*. Let x and y be partial words. We say that x is *contained* in y and we write $x \subset y$, if the partial word y can be obtained from x replacing some hole by letters of A. A partial word x is *primitive* if there do not exist another partial word z and an integer $n \geq 2$ such that $x \subset z^n$. For instance, if $x = ab??ab?a$ and $y = ab?bab?a$, then one has $x \subset y$. The word x is not primitive, since $x \subset (abaa)^2$, while the word y is primitive.

The following proposition is a slight modification of a result of [5]. The proof is identical.

Proposition 1. *Let x and y be partial words. One has $x^k \subset y^\ell$ for some integers k, ℓ if and only if $x \subset z^n$ and $z^m \subset y$ for some partial word z and integers n, m.*

The following is a straightforward consequence of the previous proposition.

Corollary 1. *Let x be a primitive partial word. If one has $x^k \subset y^\ell$ for some partial word y and integers k, ℓ, then $x^m \subset y$ for some integer m.*

The following proposition [7] will be useful in the sequel.

Proposition 2. *Let x and y be partial words. If xy is primitive, then yx is primitive.*

Now, we recall the notion of Toeplitz word [14]. Let U be an infinite word over the alphabet $A \cup \{?\}$ containing infinitely many holes and s be a partial word over the alphabet $A \cup \{?\}$. We let $T_s(w)$ denote the infinite word obtained by ordinately replacing in U the holes by the letters of s^ω. For instance, if $U = (aa?)^\omega$ and $s = b?c?$, then $T_s(U) = (aabaa?aacaa?)^\omega$.

Now, let $s^{(n)}$, $n \geq 1$ be a sequence of non-empty partial words over the alphabet A. We suppose that each word $s^{(n)}$ contains at least one hole and that there are infinitely many n such that the initial letter of $s^{(n)}$ is not a hole. We define a sequence of periodic words as follows:

$$U^{(0)} = ?^\omega, \quad U^{(n)} = T_{s^{(n)}}(U^{(n-1)}), \; n \geq 1. \tag{1}$$

Our assumption on the initial letters of $s^{(n)}$ ensures that the sequence $U^{(n)}$ converges to an infinite word $U \in A^\omega$. It will be called the *Toeplitz word* generated by the sequence $s^{(n)}$. The following lemma, whose proof is left to the reader, will be useful in the sequel.

Lemma 1. *Let U be the Toeplitz word generated by a sequence $s^{(n)}$, $n \geq 1$ and V be the Toeplitz word generated by the sequence $s^{(n)}$, $n \geq 2$. Then U can be obtained by ordinately replacing in $(s^{(1)})^\omega$ the holes by the letters of V.*

Now, we are ready to state the main result of this section.

Theorem 1. *Let U be the Toeplitz word generated by a sequence $s^{(n)}$, $n \geq 1$. If all $s^{(n)}$ are primitive partial words of maximal length k, then U is k-th power-free.*

Proof. We suppose, by contradiction, that U contains a k-th power u^k, with $u \in A^+$. With no loss of generality, we assume that u^k is the shortest k-th power occurring in any Toeplitz word satisfying the statement.

The proof requires several steps. The first one consists in showing that $|u|$ is a multiple of $|s^{(1)}|$.

Indeed, set $s = s^{(1)}$, $p = |s|$, $q = |u|$. By the construction of U, there is a factor v of $U^{(1)} = s^{\omega}$ such that $v \subset u^p$. Since $|v| = pq$, one derives that $s = xy$ and $v = (yx)^q$ for some partial words x, y. Hence, in view of Proposition 2 and Corollary 1, one has $(yx)^m \subset u$, for a suitable integer m. Replacing the words by their lengths, one has $mp = q$, which proves our claim.

As a second step, we show that the word V generated by the sequence $s^{(n)}$, $n \geq 2$, also contains a k-th power.

Let h be the integer such that $u^k = U_h U_{h+1} \cdots U_{h+kq-1}$ and $i_1, i_2, \ldots, i_t$ be the positions of the holes of s^{ω} in the interval $[h, h+kq-1]$. In view of Lemma 1, the word $w = U_{i_1} U_{i_2} \cdots U_{i_t}$ is a factor of V. We shall verify that w is a k-th power. Indeed, let ℓ denote the number of holes in s^m. Then each factor of s^{ω} of length $q = mp$ contains exactly ℓ holes. One derives, in particular, $t = k\ell$. Moreover, one has

$$i_{j+\ell} = i_j + q, \quad j = 1, 2, \ldots, t - q.$$

Taking into account that q is a period of the word $U_h U_{h+1} \cdots U_{h+kq-1} = u^k$, one obtains $U_{i_j} = U_{i_{j+\ell}}$, $j = 1, 2, \ldots, t - q$. This proves that w has period ℓ. Since $|w| = t = k\ell$, we conclude that $w = z^k$, for some word z of length ℓ.

By the minimality of u, one derives $\ell \geq q = |s^m|$. But s can contain at most $p - 1$ holes, so that $\ell \leq m(p-1) < q$. This leads to a contradiction. $\qquad\square$

Example 1. A Toeplitz word generated by a sequence of words $s^{(n)} \in \{0?1?, 1?0?\}$, $n \geq 1$, is called a *paperfolding word*. By Theorem 1, one derives the known fact [2] that paperfolding words are 4-th power-free.

Let $s^{(n)} = 01?$ for all $n \geq 1$. The Toeplitz word generated by the sequence $s^{(n)}$ is cube-free. More generally, if $p > 2$ and $r^{(n)} = 0^{p-2}1?$ for all $n \geq 1$, then the Toeplitz word generated by the sequence $r^{(n)}$ is p-th power-free.

4   Arithmetic Subsequences of Toeplitz Words

A permutation σ of the set $\{1, 2, \ldots, n\}$ will be called *arithmetic* if there exists an integer d such that

$$\sigma(i+1) \equiv \sigma(i) + d \pmod{n}, \quad i = 1, 2, \ldots, n - 1.$$

One easily verifies that, in such a case, one has $\sigma(j) - \sigma(i) \equiv (j - i)d \pmod{n}$, for all $i, j \in \{1, 2, \ldots, n\}$. Moreover, the composition of arithmetic permutations

270 A. Boccuto and A. Carpi

is an arithmetic permutation. Thus, the set of all arithmetic permutations is a subgroup of the symmetric group on n objects.

Now, let x and y be two words of length n. We say that x and y are *arithmetically conjugate* and we write $x \approx y$ if there exists an arithmetic permutation σ of $\{1, 2, \ldots, n\}$ such that

$$x = x_1 x_2 \cdots x_n, \quad y = x_{\sigma(1)} x_{\sigma(2)} \cdots x_{\sigma(n)}, \tag{2}$$

with $x_1, x_2, \ldots, x_n$ letters. Roughly speaking, x and y are arithmetically conjugate if y is obtained by rearranging the letters of x according to an arithmetic permutation. Taking into account that by composing (or inverting) arithmetic permutations one obtains an arithmetic permutation, one easily checks that the relation $\approx$ is an equivalence relation.

The following statement extends Proposition 2 to the case of arithmetic conjugacy. For the sake of brevity, the proof is omitted .

Proposition 3. *Let $x, y \in A^*$ be arithmetically conjugate partial words. Then x is primitive if and only if y is primitive.*

The following lemma shows a useful application of arithmetic conjugacy to the analysis of arithmetic subsequences of periodic words.

Lemma 2. *Let s be a word of length p and V be an arithmetic subsequence of difference d of the word s^ω. If d and p are coprime, then there exists a word r such that*

$$r \approx s \quad and \quad V = r^\omega.$$

We will study Toeplitz words with a generating sequence whose elements lie in the set

$$P = \bigcup_{0 \leq i \leq \ell} A^i ?(A^\ell ?)^* A^{\ell - i}. \tag{3}$$

Arithmetic subwords of Toeplitz words of this type have been studied in [4, 12] as they are prototypes of uniformly recurrent words with linear arithmetic complexity. In this section, we will investigate in more details the relationship between a Toeplitz word with a generating sequence whose elements are in P and its arithmetic subsequences.

Let U be an infinite word over the alphabet $A \cup \{?\}$. The sequence of the integers i such that $U_i = ?$, in increasing order, will be called the *hole sequence* of U. One easily verifies that one has $s \in P$ if and only if the hole sequence of s^ω is an arithmetic progression whose difference divides $|s^{(n)}|$. The following lemma shows that, in the case of our interest, the hole sequences of the words approximating a Toeplitz word are, in fact, arithmetic progressions.

Lemma 3. *For all $n \geq 1$, let $s^{(n)} \in P$. Then, for all $n \geq 0$, the hole sequence of the infinite word $U^{(n)}$ defined by (1) is an arithmetic progression whose difference divides the number $\prod_{i=1}^{n} |s^{(i)}|$.*

Now we are ready to prove the main result of this section.

Theorem 2. *Let U be the Toeplitz word generated by a sequence of words $s^{(n)} \in P$, $n \geq 1$ and V be an arithmetic subsequence of U of difference d. If one has $\gcd(d, |s^{(n)}|) = 1$ for all $n \geq 1$, then V is a Toeplitz word with a generating sequence $r^{(n)}$ such that for all $n \geq 1$, $r^{(n)} \approx s^{(n)}$.*

Proof. Let $(i_m)_{m \geq 1}$ be the arithmetic progression of difference d such that

$$V = U_{i_1} U_{i_2} \cdots U_{i_m} \cdots .$$

Moreover, for all $n \geq 0$, let $U^{(n)}$ be the infinite word defined by (1) and $V^{(n)}$ be the arithmetic subsequence $V^{(n)} = U^{(n)}_{i_1} U^{(n)}_{i_2} \cdots U^{(n)}_{i_m} \cdots$. Fix $n \geq 1$, let $(h_m)_{m \geq 1}$ be the hole sequence of $U^{(n-1)}$ and $(h_{j_m})_{m \geq 1}$ be the subsequence of the elements occurring also in the arithmetic progression $(i_m)_{m \geq 1}$. In view of Lemma 3, $(h_m)_{m \geq 1}$ is an arithmetic progression whose difference ℓ is coprime with d. From this fact, one easily derives that $(h_{j_m})_{m \geq 1}$ is an arithmetic progression of difference $d\ell$. This implies that $(j_m)_{m \geq 1}$ is an arithmetic progression of difference d.

Let W be the arithmetic subsequence of $(s^{(n)})^{\omega}$ obtained by taking the letters of position j_m, $m \geq 1$. By Lemma 2, there is a partial word $r^{(n)}$ such that $r^{(n)} \approx s^{(n)}$ and $W = (r^{(n)})^{\omega}$. Since the word $U^{(n)}$ is obtained by ordinately replacing the holes of $U^{(n-1)}$ by the letters of $(s^{(n)})^{\omega}$, the word $V^{(n)}$ will be obtained by ordinately replacing the holes of $V^{(n-1)}$ by the letters of W, that is $V^{(n)} = T_{r^{(n)}}(V^{(n-1)})$.

One easily verifies that the sequence $V^{(n)}$ converges to V, so that V is the Toeplitz word with generating sequence $r^{(n)}$, $n \geq 1$. $\qquad\square$

From Theorems 1 and 2 and Proposition 3, one easily derives the following

Corollary 2. *Let U be the Toeplitz word generated by a sequence of primitive partial words $s^{(n)} \in P$, $n \geq 1$. If $k = \max\{|s^{(n)}|, n \geq 1\}$, then every arithmetic subsequence of U whose difference d is coprime with $\mathrm{lcm}\{|s^{(n)}|, n \geq 1\}$ is a k-th power-free infinite word.*

Example 2. From the corollary above, one obtains that all arithmetic subsequences of odd difference of a paperfolding word are 4-th power-free. However, this fact has been directly proved in [15] using a result of [4].

Let U be the Toeplitz word generated by the (constant) sequence $s^{(n)} = 01?$, $n \geq 1$. By the corollary above, all arithmetic subsequences of U whose difference is not a multiple of 3 are cube-free. More generally, if $p > 2$ and V is the Toeplitz word generated by the (constant) sequence $s^{(n)} = 0^{p-2}1?$, $n \geq 1$, then all arithmetic subsequences of U whose difference is not a multiple of p are p-th power-free.

5   Thue Threshold

In this section we shall prove that the 3-dimensional Thue threshold is equal to 2. More precisely, we will exhibit a 2-colouring of the lattice points of the 3-dimensional space such that any line is a 9-th power-free word. As a consequence, also the 2-dimensional Thue threshold is equal to 2.

We let $\mathbb{Z}_3$ denote the field of the integers modulo 3. The following lemma gives a useful combinatorial property of the vector space $\mathbb{Z}_3^3$.

Lemma 4. *There is a partition (A_1, A_2, A_3) of $\mathbb{Z}_3^3$ such that for all $\boldsymbol{j}, \boldsymbol{m} \in \mathbb{Z}_3^3$ with $\boldsymbol{m} \neq \boldsymbol{0}$, there exist $k \in \mathbb{Z}_3$ and $i \in \{1, 2, 3\}$ such that*

$$\boldsymbol{j} + k\boldsymbol{m} \in A_i, \quad m_i \neq 0.$$

We limit ourselves to list the elements of the classes A_1, A_2, A_3, as the verification of Lemma 4 merely requires a finite but long and tedious check.

$$A_1 = \{\boldsymbol{x} + (k, k, k) \mid \boldsymbol{x} = (0, 0, 0), (0, 1, 0), (0, 0, 1), \; k = 0, 1, 2\}$$
$$A_2 = \{(i_3 + 1, i_1, i_2) \mid (i_1, i_2, i_3) \in A_1\}, \quad A_3 = \{(i_3 + 1, i_1, i_2) \mid (i_1, i_2, i_3) \in A_2\}.$$

Remark 1. The previous lemma has an interesting interpretation in terms of Galois geometry: let $AG(3, 3)$ be the 3-dimensional affine space over the 3-element field $\mathbb{Z}_3$. Then there is a partition (A_1, A_2, A_3) of the points of $AG(3, 3)$ with the following property: for every line r there is an axis x_i, $i = 1, 2, 3$ such that r intersects A_i and is not orthogonal to x_i. We notice that an analogous property holds for the 2-dimensional affine space over the 2-element field $\mathbb{Z}_2$, taking the partition (A_1, A_2) with $A_1 = \{(0, 0), (1, 1)\}$ and $A_2 = \{(0, 1), (1, 0)\}$.

In Example 2, we have shown an infinite binary word such that all arithmetic subsequences whose difference is not a multiple of 3 are cube-free. A bi-infinite word with the same property can be easily obtained via Koenig's Lemma (see, e.g., [17]). Now, we define a 2-colouring of the lattice points of the 3-dimensional space whose lines are 9-th power-free. Let (A_1, A_2, A_3) be the partition of $\mathbb{Z}_3^3$ given by Lemma 4 and let U be a binary bi-infinite word such that all arithmetic subsequences whose difference is not a multiple of 3 are cube-free. We let π denote the natural projection $\pi \colon \mathbb{Z}^3 \to \mathbb{Z}_3^3$ and define the map $\alpha \colon \mathbb{Z}^3 \to \{0, 1\}$ as follows:

$$\text{if } \pi(j_1, j_2, j_3) \in A_i, \text{ then } \alpha(j_1, j_2, j_3) = U_{\lfloor j_i/3 \rfloor},$$

$j_1, j_2, j_3 \in \mathbb{Z}$, $i \in \{1, 2, 3\}$.

Proposition 4. *All lines of α are 9-th power-free.*

Proof. Actually, it suffices to verify that no line of α contains a cube u^3 with $|u|$ multiple of 3. Indeed, any 9-th power v^9 can be written as $v^9 = u^3$, with $u = v^3$ and $|u| = 3|v|$.

Let V be a line of α. Then, there are $j_1, j_2, j_3, m_1, m_2, m_3 \in \mathbb{Z}$ such that $\gcd(m_1, m_2, m_3) = 1$ and

$$V_n = \alpha(j_1 + nm_1, j_2 + nm_2, j_3 + nm_3) \text{ for all } n \in \mathbb{Z}. \tag{4}$$

By Lemma 4, there are $i \in \{1, 2, 3\}$ and $k \in \mathbb{Z}$ such that

$$\pi(j_1 + km_1, j_2 + km_2, j_3 + km_3) \in A_i, \quad m_i \not\equiv 0 \pmod 3.$$

From (4) and the definition of α, it follows that for all $n \in \mathbb{Z}$, one has

$$V_{k+3n} = \alpha(j_1 + km_1 + 3nm_1, j_2 + km_2 + 3nm_2, j_3 + km_3 + 3nm_3) = U_{j'+nm_i},$$

where $j' = \lfloor(j_i + km_i)/3\rfloor$. This equation shows that there is a bi-infinite word W which is simultaneously an arithmetic subsequence of difference m_i of U and an arithmetic subsequence of difference 3 of V. Since $m_i \not\equiv 0 \pmod 3$, the word W is cube-free. Now suppose, by contradiction, that a cube u^3 with $|u| = 3\ell$ occurs in V, $\ell \geq 1$. Then the subword made by the letters of this occurrence whose position is congruent to $k \pmod 3$ would have the form v^3 with $|v| = \ell$ and would be a factor of W. This yields a contradiction, since W is cube-free.

Thus, no line of α contains a factor of the form u^3 with $|u|$ multiple of 3. This implies, in particular, that all lines of α are 9-th power-free.   □

As a corollary of the proposition above one obtains

Theorem 3. *The Thue threshold of $\mathbb{Z}^3$ is equal to 2. Thus, the Thue threshold of $\mathbb{Z}^2$ is also equal to 2.*

6   Avoiding Long Squares

In this section we will produce a 2-colouring of the lattice points of the plane such that any line contains only squares of length not larger than 26. In this case, the results of Sects. 3 and 4 do not seem to help, so that we need an ad-hoc construction based on paperfolding words. For the sake of brevity, we limit ourselves to outline the construction.

As a consequence of Koenig's Lemma, there exists a bi-infinite binary word U whose factors are all finite paperfolding words. We define the map $\beta \colon \mathbb{Z}^2 \to \{0, 1\}$ as

$$\beta(j_1, j_2) = \begin{cases} U_{\lfloor j_1/2\rfloor} & \text{if } j_1 + j_2 \text{ is odd,} \\ U_{\lfloor j_2/2\rfloor} & \text{if } j_1 + j_2 \text{ is even,} \end{cases} \tag{5}$$

$j_1, j_2 \in \mathbb{Z}$.

Finite factors of paperfolding words are usually called *finite paperfolding words*. By exploiting some known properties of paperfolding words we obtain the following description of the factors of the lines of β.

Lemma 5. *Let x be a factor of even length of any line of β. Then one can write*

$$x = b_1 c_1 b_2 c_2 \cdots b_k c_k,$$

with $b_1, b_2, \ldots, b_k, c_1, c_2, \ldots, c_k \in \{0, 1\}$, $k \geq 0$, and the words $y = b_1 b_2 \cdots b_k$ and $z = c_1 c_2 \cdots c_k$ are either finite paperfolding words or words of period 2. Moreover, at least one of the words y and z is a paperfolding word.

We need also the following combinatorial property, which we state without proof, for the sake of brevity.

Proposition 5. *Let $x, y \in \{0,1\}^*$ be words such that both xy and yx are finite paperfolding words. If $|x|, |y| > 6$, then $|xy|$ is even.*

Now we are ready to establish the main result of this section.

Proposition 6. *Let β be the 2-colouring of the integer lattice points of the plane defined by (5). No line of β contains a square u^2 with $|u| > 13$.*

Proof (outline). Suppose that u^2 is a factor of some line of β. Then $x = u^2$ can be factorized as in Lemma 5. If $k = |u|$ is even, then both y and z are squares. Since one of them is a paperfolding word and no paperfolding word of length larger than 10 is a square [1], one derives that $|u| = k \leq 10$.

Now suppose that $k = |u|$ is odd, say $k = 2\ell + 1$ and $\ell > 6$. Since x is a square, with some computation, one obtains

$$y = b_1 b_2 \cdots b_k = c_{\ell+1} c_{\ell+2} \cdots c_k c_1 c_2 \cdots c_\ell. \tag{6}$$

If y and z are both paperfolding words, then from (6) and Proposition 5 one derives that k is even, which is a contradiction. If, on the contrary, one of the words y and z has period 2, then from (6) one derives that both y and z have a factor of period 2 and length 4, that is one of the words 0^4, 1^4, 0101, or 1010. Since none of them is a paperfolding word, we conclude that neither y nor z are paperfolding words, obtaining a contradiction. We conclude that $\ell \leq 6$, so that $k \leq 13$. □

Remark 2. A careful analysis of the proof of the previous proposition shows that if u^2 is a factor of a line of β, then either $|u|$ is an odd integer not larger than 13 or $|u| \in \{2, 6, 10\}$. In particular, no line of β contains a square whose length is a multiple of 8. Since any 8-th power is also a square whose length is a multiple of 8, we conclude that all lines of β are 8-th power-free words.

7   Concluding Remarks

We have proved that the Thue threshold $t(n)$ satisfies the equality $t(1) = t(2) = t(3) = 2$ and that the bounded square threshold of $\mathbb{Z}^2$ is 2. It would be interesting to evaluate the Thue threshold and the bounded square threshold of $\mathbb{Z}^n$ for all positive integer n. In [15] it is shown that $t(n) \leq 2^n$. Combining the results of this paper with the construction of [8], one could obtain the tighter bound $t(3n) \leq 2^n$. We conjecture that $t(n) = 2$ for all $n \geq 1$. Actually, since $t(n)$ is a non-decreasing sequence, it would suffice to verify that $t(n) = 2$ for arbitrarily large n. We notice that our proof of the equality $t(n) = 2$ for $n = 3$ could be easily generalized to the case $n = p$ for any prime p provide one is able to establish, for the affine Galois space $AG(p, p)$, a property analogous to that of Remark 1. Moreover, we have treated the case of Toeplitz words such that the elements of the generating sequence belong to the set P, where P is as in (3). Other interesting unanswered questions are the following:

1. What is the least integer k such that there exists a 2-colouring of $\mathbb{Z}^2$ (resp., $\mathbb{Z}^3$) whose lines are k-th power-free?
2. What is the least integer k such that there exists a 2-colouring of $\mathbb{Z}^2$ whose lines do not contain squares of length larger than k?

We have built a 2-colouring of $\mathbb{Z}^2$ such that all lines are 8-th power-free and do not contain squares of length larger than 26 and a 2-colouring of $\mathbb{Z}^3$ such that all lines are 9-th power-free. However, there is no evidence that these bounds are optimal.

References

1. Allouche, J.-P.: Suites infinies à répétitions bornées, Séminaire de Théorie des Nombres de Bordeaux (1983–1984) Exposé no. 20, 11 p
2. Allouche, J.-P., Bousquet-Mélou, M.: Facteurs des suites de Rudin-Shapiro généralisées. Bull. Belg. Math. Soc. **1**, 145–164 (1994)
3. Allouche, J.-P., Bousquet-Mélou, M.: Canonical positions for the factors in paper-folding sequences. Theoret. Comput. Sci. **129**, 263–278 (1994)
4. Avgustinovich, S.V., Fon-Der-Flaas, D.G., Frid, A.E.: Arithmetical complexity of infinite words. In: Ito, M., Imaoka, T. (eds.) Words, Languages & Combinatorics III, pp. 51–62. World Scientific, Singapore (2003)
5. Berstel, J., Boasson, L.: Partial words and a theorem of Fine and Wilf. Theoret. Comput. Sci. **218**, 135–141 (1999)
6. Berstel, J., Perrin, D.: The origins of combinatorics on words. Eur. J. Comb. **28**, 996–1022 (2007)
7. Blanchet-Sadri, F.: Primitive partial words. Discret. Appl. Math. **148**, 195–213 (2005)
8. Carpi, A.: Multidimensional unrepetitive configurations. Theoret. Comput. Sci. **56**, 233–241 (1988)
9. Currie, J., Simpson, J.: Non-repetitive tilings. Electron. J. Comb. **9**, #R28 (2002)
10. Etringer, R.C., Jackson, D.E., Schatz, J.A.: On nonrepetitive sequences. J. Comb. Theory (A) **16**, 159–164 (1974)
11. Fraenkel, A.S., Simpson, J.: How many squares can a string contain? J. Comb. Theory (A) **82**, 112–120 (1998)
12. Frid, A.E.: Sequences of linear arithmetical complexity. Theoret. Comput. Sci. **339**, 68–87 (2005)
13. Grytczuk, J.: Thue type problems for graphs, points, and numbers. Discret. Math. **308**, 4419–4429 (2008)
14. Jacobs, K., Keane, M.: 0-1-sequences of Toeplitz type. Z. Wahr. verw. Geb. **13**(2), 123–131 (1969). https://doi.org/10.1007/BF00537017
15. Kao, J.-Y., Rampersad, N., Shallit, J., Silva, M.: Words avoiding repetitions in arithmetic progressions. Theoret. Comput. Sci. **391**, 126–137 (2008)
16. Kenkireth, B.G., Singh, M.: On the minimal alphabet size in multidimensional unrepetitive configurations. Discret. Appl. Math. **255**, 258–266 (2019)
17. Lothaire, M.: Combinatorics on Words. Encyclopaedia of Mathematics and its Applications, vol. 17. Addison-Wesley, Reading (1983)
18. Mendès France, M., van der Poorten, A.J.: Arithmetic and analytic properties of paperfolding sequences. Bull. Austr. Math. Soc. **24**, 123–131 (1981)

19. Prodinger, H., Urbanek, F.J.: 0–1 sequences without long adjacent identical blocks. Discret. Math. **28**, 277–289 (1979)
20. Thue, A.: Über unendliche Zeichenreihen, Norske vid. Selsk. Skr. Mat. Nat. Kl. **7**, 1–22 (1906). Reprinted in T. Nagell (ed.), Selected mathematical papers of Axel Thue, Universitetsforlaget (Oslo, 1977) pp. 139–158
21. Thue, A.: Über die gegenseitige Lage gleicher Teile gewisser Zeichenreihen, Norske vid. Selsk. Skr. Mat. Nat. Kl. **1**, 1–67 (1912). Reprinted in T. Nagell (ed.), Selected mathematical papers of Axel Thue, Universitetsforlaget (Oslo, 1977) pp. 413–478

On Simulation in Automata Networks

Florian Bridoux[1], Maximilien Gadouleau[2], and Guillaume Theyssier[3(✉)]

[1] Université Aix-Marseille, CNRS, LIS, Marseille, France
`florian.bridoux@lis-lab.fr`
[2] Department of Computer Science, Durham University, Durham, UK
`m.r.gadouleau@durham.ac.uk`
[3] Université Aix-Marseille, CNRS, I2M, Marseille, France
`guillaume.theyssier@cnrs.fr`

Abstract. An automata network is a finite graph where each node holds
a state from some finite alphabet and is equipped with an update function that changes its state according to the configuration of neighboring
states. More concisely, it is given by a finite map $f : Q^n \to Q^n$. In this
paper we study how some (sets of) automata networks can be simulated by some other (set of) automata networks with prescribed update
mode or interaction graph. Our contributions are the following. For non-
Boolean alphabets and for any network size, there are intrinsically non-
sequential transformations (i.e. that can not be obtained as composition
of sequential updates of some network). Moreover there is no universal automaton network that can produce all non-bijective functions via
compositions of asynchronous updates. On the other hand, we show that
there are universal automata networks for sequential updates if one is
allowed to use a larger alphabet and then use either projection onto or
restriction to the original alphabet. We also characterize the set of functions that are generated by non-bijective sequential updates. Following
Tchuente, we characterize the interaction graphs D whose semigroup of
transformations is the full semigroup of transformations on Q^n, and we
show that they are the same if we force either sequential updates only,
or all asynchronous updates.

1 Introduction

An automata network is a network of entities each equipped with a local update
function that changes its state according to the states of neighboring entities.
Automata networks have been used to model different kind of networks: gene networks, neural networks, social networks, or network coding (see [9] and references
therein). They can also be considered as a model of distributed computation with
various specialized definitions [18,19]. The architecture of an automata network
can be represented via its interaction graph, which indicates which update functions depend on which variables. An important stream of research is to determine

This work was partially funded by the CNRS and Royal Society joint research project
PRC1861 and the French ANR project FANs ANR-18-CE40-0002 and the ECOS
project C16E01.

M. Anselmo et al. (Eds.): CiE 2020, LNCS 12098, pp. 277–288, 2020.
https://doi.org/10.1007/978-3-030-51466-2_24

how the interaction graph affects different properties of the network or to design networks with a prescribed interaction graph and with a specific dynamical property (see [8] for a review of known results). On the other hand, automata networks are usually associated with an update mode describing how local update functions of each entity are applied at each step of the evolution. In particular, three categories of update modes can be distinguished: sequential (one node update at a time), asynchronous (any subset of nodes at a time) or synchronous (all nodes simultaneously). Studying how changing the update mode affects the properties of an automata network with fixed local update functions is another major trend in this field [12,13,15]. Comparing the computational power of sequential and parallel machines is of course at the heart of computer science, but the questioning on update modes is also meaningful for applications of automata networks in modeling of natural systems where the synchronous update mode is often considered unrealistic.

For both parameters (interaction graphs and update modes), the set of properties that could be potentially affected is unlimited. In this paper, instead of choosing a set of properties to analyze, we adopt an intrinsic approach: we study how some (sets of) automata networks can be simulated by some other (set of) automata networks with prescribed update mode or interaction graph.

Notations. We will always consider alphabets of the form $[\![q]\!] = \{0,\ldots,q-1\}$ for some q and usually denote by n the number of nodes of the network which are identified by integers in the interval $[1,n]$. An **automata network** is a map $f : [\![q]\!]^n \to [\![q]\!]^n$. An element $x \in [\![q]\!]^n$ is a configuration and x_v denotes the state of node v in configuration x. By extension f_v denotes the map $x \mapsto f(x)_v$. The rank of f is the size of its image. For any set of coordinates $V \subseteq [1,n]$, $f^{(V)} : [\![q]\!]^n \to [\![q]\!]^n$ denotes the following map:

$$f^{(V)}(x)_i = \begin{cases} f(x)_i & \text{if } i \in V \\ x_i & \text{else.} \end{cases}$$

The notation is extended to words of subsets $w = (w_1,\ldots,w_k)$ as follows: $f^{(w)} = f^{(w_k)} \circ \cdots \circ f^{(w_1)}$. For $v \in [1,n]$ we overload this notation by $f^{(v)} = f^{(\{v\})}$.

We will often consider semigroups of functions under compositions: $\langle X \rangle$ where X is a set of functions that denotes the semigroup generated by compositions of elements of X. We denote the fact that S_1 is a sub-semigroup of S_2 by $S_1 \leq S_2$. For any set X, $\mathrm{Sym}(X)$ is the set of permutations on X. We denote the set of all networks $f : [\![q]\!]^n \to [\![q]\!]^n$ as $\mathrm{F}(n,q)$. We denote by $\mathrm{Sym}(n,q)$ the set of $f \in \mathrm{F}(n,q)$ which are bijective and by $\mathrm{Sing}(n,q)$ the set of $f \in \mathrm{F}(n,q)$ which are non-bijective. For any set F of functions in $\mathrm{F}(n,q)$, what they can simulate (asynchronously, sequentially, synchronously) is denoted as follows:

$$\langle F \rangle_{\mathrm{Asy}} := \left\langle \left\{ f^{(V)} : f \in F, V \subseteq [1,n] \right\} \right\rangle,$$

$$\langle F \rangle_{\mathrm{Seq}} := \left\langle \left\{ f^{(v)} : f \in F, v \in [1,n] \right\} \right\rangle,$$

$$\langle F \rangle_{\mathrm{Syn}} = \langle F \rangle.$$

Then we say that F **simulates** $g \in F(n,q)$ asynchronously (sequentially, synchronously, respectively) if $g \in \langle F \rangle_{Asy}$ ($\langle F \rangle_{Seq}$, $\langle F \rangle_{Syn}$, respectively). When $F = \{f\}$ we use notations $\langle f \rangle_{Asy}$, $\langle f \rangle_{Seq}$, $\langle f \rangle_{Syn}$, respectively.

Previous Works. Simulation of automata networks is the topic of two main strands of work. The first stream investigates what a single network can simulate. The main observation, made in [2], is that there is no sequentially complete network for $F(n,q)$, i.e. for all $f \in F(n,q)$, $\langle f \rangle_{Seq} \neq F(n,q)$. This was refined in several ways. Firstly, there is no sequentially complete network for singular (*i.e.* non-permutation) transformations: for all $f \in F(n,q)$, $\mathrm{Sing}(n,q) \not\subseteq \langle f \rangle_{Seq}$ [2]. Secondly, for all $n \geq 2$ and $q \geq 2$ (unless $n = q = 2$), there exists a sequentially complete network for permutations: there exists $f \in F(n,q)$ such that $\langle f \rangle_{Seq} = \mathrm{Sym}(n,q)$ [7]. These results illustrate a clear dichotomy between permutations and non-permutations. Thirdly, the simulation model was extended in [2] to include situations whereby a large network $f \in F(m,q)$ could simulate a smaller network $g \in F(n,q)$ for $n \leq m$; notably, there always exists a complete network of size $m = n + 1$ which can sequentially simulate any $g \in F(n,q)$.

Another strand of work considers simulation by (possibly large) sets of networks. Firstly, Tchuente [16] investigated what networks with a prescribed reflexive interaction graph D could simulate synchronously. The main result is that this set of networks $F(D,q)$ is complete, i.e. $\langle F(D,q) \rangle_{Syn} = F(n,q)$, if and only if D is strongly connected and has a vertex of in-degree n. Secondly, in the context of in-situ computation (a.k.a. memoryless computation), Burckel proved that any network could be sequentially simulated, if we allow the updates to differ at each time step; in our language: for all n and q, $\langle F(n,q) \rangle_{Seq} = F(n,q)$ [4]. This seminal result was subsequently refined (see [6,10]); notably linear bounds on the shortest word required to simulate a transformation were obtained in [5,6].

Our Contributions. In this paper, we are further developing both strands of the theory of simulation of automata networks. We make the following contributions. We first consider simulation by a single network. Firstly, we show that for any $q \geq 3$ and any $n \geq 2$, there exists a network $g \in F(n,q)$ which is not sequentially simulatable. Secondly, we consider asynchronous simulation, and we show that there is no asynchronously complete network: for all $f \in F(n,q)$, $\mathrm{Sing}(n,q) \not\subseteq \langle f \rangle_{Asy}$. This is a clear strengthening of the result in [2] for sequential simulation. Thirdly, we extend the framework to let a network over a large alphabet $f \in F(n,q')$ simulate a network $g \in F(n,q)$ over a smaller alphabet. We consider two ways to extend the alphabet, and for each we prove the existence of sequentially complete networks for $q' = 2q$ and $q' = q + 1$, respectively. We then consider simulation by large sets of networks. The seminal result in [4] shows that instructions (updates of the form $f^{(v)}$ for some $v \in [1,n]$) can simulate any network; in this paper, we determine what singular instructions can simulate (and even idempotent instructions for $q \geq 3$). We finally strengthen the main result in [16] by showing that it also holds when considering sequential and asynchronous updates as well.

Proof. Complete proofs of all lemmas, propositions and theorems can be found in [3].

2   Sequential Simulation

We say $g \in F(n, q)$ is **sequentially simulatable** if $g \in \langle f \rangle_{\text{Seq}}$ for some $f \in F(n, q)$. Recall that unless $n = q = 2$ any $g \in \text{Sym}(n, q)$ is sequentially simulatable since there is a universal $f \in F(n, q)$ such that $\langle f \rangle_{\text{Seq}} = \text{Sym}(n, q)$ [7]. Concerning non-bijective maps, the situation is radically different for non-Boolean alphabets as shown in the following theorem. For any function $\phi \in F(n, q)$, we denote by $O(()\phi)$ the set of its orphans: $O(()\phi) = \{c \in [\![q]\!]^n : \phi^{-1}(c) = \emptyset\}$. The analysis of oprhans configurations under sequential updates is the key behind the following theorem.

Theorem 1. *For any $n \geq 2$ and $q \geq 3$, there exists $h \in F(n, q)$ which is not sequentially simulatable.*

The functions which are not sequentially simulatable produced in the proof of Theorem 1 have two configurations a and b in $[\![q]\!]^n$ with the same image and another d which is an orphan with the following property: for each coordinate i where a_i and b_i differ, d_i is different from both a_i and b_i. Note that this situation is impossible in the Boolean case since if $a_i \neq b_i$ then necessarily $d_i \in \{a_i, b_i\}$.

F. Bridoux did an exhaustive search in $F(n, 2)$ with $n = 2$ and $n = 3$ to test which one are sequentially simulatable [1]. It turns out that all $f \in F(3, 2)$ are sequentially simulatable. However, some functions in $F(2, 2)$ are not and one example is the circular permutation $00 \rightarrow 01 \rightarrow 11 \rightarrow 10 \rightarrow 00$ [1, Proposition 12]. More details (including the code of the test program) are available at http://theyssier.org/san2020.

3   Asynchronous Simulation

In this section, we consider asynchronous simulation, where at each step we allow any update $f^{(T)}$ for $T \subseteq [1, n]$. We then refine the result in [2] that there is no network that can sequentially simulate all singular networks.

We say that a function $h : B \rightarrow C$, where B and C are finite sets, is **balanced** if for any $c, c' \in C$, $|h^{-1}(c)| = |h^{-1}(c')|$. In particular, if $f \in F(n, q)$ is bijective, then all its coordinate functions $f_v : [\![q]\!]^n \rightarrow [\![q]\!]$ must be balanced.

Theorem 2. *For all $f \in F(n, q)$, $\text{Sing}(n, q) \not\leq \langle f \rangle_{\text{Asy}}$.*

Proof. Suppose, for the sake of contradiction, that $\text{Sing}(n, q) \leq \langle f \rangle_{\text{Asy}}$. We first show that not all coordinate functions of f are balanced. There exists $S \subseteq [1, n]$ such that $f^{(S)}$ has rank $q^n - 1$. (Otherwise, no function in $\langle f \rangle_{\text{Asy}}$ has rank $q^n - 1$.) Then there exist $a, b \in [\![q]\!]^n$ such that

$$\left| \left(f^{(S)} \right)^{-1} (x) \right| = \begin{cases} 2 & \text{if } x = a \\ 0 & \text{if } x = b \\ 1 & \text{otherwise.} \end{cases}$$

Then let $v \in S$ such that $a_v \neq b_v$. We have

$$|f_v^{-1}(a_v)| = \sum_{x : x_v = a_v} \left|\left(f^{(S)}\right)^{-1}(x)\right| = 2 + \sum_{x : x_v = a_v, x \neq a} 1 = q^{n-1} + 1,$$

thus f_v is not balanced.

Thus, suppose f_v is not balanced, and let $q_0 \in [\![q]\!]$ such that $|f_v^{-1}(q_0)| < q^{n-1}$. Say a network $h \in \mathrm{F}(n,q)$ is defective if $h^{-1}(x) = \emptyset$ for some x with $x_v = q_0$. Let $g \in \mathrm{Sing}(n,q)$ not be deficient, and have a nontrivial g_v; and suppose $g = f^{(w_1 \cdots w_k)}$. Let $i = \max\{1 \leq j \leq k : v \in w_j\}$, then $f^{(w_i)}$ is defective, and so is $f^{(w_1 \cdots w_i)}$. Since $f^{(w_{i+1} \cdots w_k)}$ fixes the coordinate v, $f^{(w_1 \cdots w_k)} = g$ is also deficient, which is the desired contradiction. $\qquad \square$

Similarly to Theorem 1, the obstacle in Theorem 2 was found in the set of maps of rank $q^n - 1$. We now show that maps of rank $q^n - 2$ form another obstruction to having complete simulation in the asynchronous case. Let $T(n,q)$ be the set of networks in $\mathrm{F}(n,q)$ whose rank is not equal to $q^n - 1$. It is clear that $T(n,q)$ is a semigroup, generated by maps of rank q^n or $q^n - 2$.

Proposition 1. *For all $f \in \mathrm{F}(n,q)$, $T(n,q) \not\leq \langle f \rangle_{\mathrm{Asy}}$.*

Proof. Suppose, for the sake of contradiction, that $T(n,q) \leq \langle f \rangle_{\mathrm{Asy}}$. Firstly, all the coordinate functions of f are balanced. Indeed, let $g(x) = x + (1, \ldots, 1)$ and express $g = f^{(w_1 \cdots w_k)}$. Then $f^{(w_i)}$ is bijective and hence f_v is balanced for all $v \in w_i$; since $\bigcup_{i=1}^{k} w_i = [1, n]$, we obtain that f_v is balanced for all $v \in [1, n]$. Secondly, the proof of Theorem 2 showed that there is no $f^{(S)}$ of rank $q^n - 1$.

Now, there are two types of networks with rank $q^n - 2$:

- Say g is of type I if there exists $a \in [\![q]\!]^n$ such that $|g^{-1}(a)| = 3$ (and hence any other $x \neq a$ has $|g^{-1}(x)| \leq 1$).
- Say h is of type II if there exist $a, b \in [\![q]\!]^n$ such that $|h^{-1}(a)| = |h^{-1}(b)| = 2$ (and hence any other $x \notin \{a, b\}$ has $|h^{-1}(x)| \leq 1$).

By an argument similar to the proof of Theorem 2, there is no $S \subseteq [1, n]$ such that $f^{(S)}$ is of type I. Let g be of type I and let us express it as $g = f^{(w_1 \cdots w_k)}$. Each $f^{(w_l)}$ has rank at least $q^n - 2$, and there exists $1 \leq i \leq k$ such that $f^{(w_i)}$ is singular. By the argument above, $f^{(w_i)}$ is of type II and so is $h := f^{(w_1 \cdots w_i)}$, say $|h^{-1}(a)| = |h^{-1}(b)| = 2$. Denote $g = h' \circ h$ for $h' := f^{(w_{i+1} \cdots w_k)}$. If $h'(a) = h'(b)$, then g has rank at most $q^n - 3$; otherwise $|g^{-1}(h'(a))| = |g^{-1}(h'(b))| = 2$ and hence g is of type II, which is the desired contradiction. $\qquad \square$

4   Simulation Using Larger Alphabets

As said earlier, there is no universal automata network in $\mathrm{F}(n,q)$ able to sequentially simulate all functions of $\mathrm{F}(n,q)$ (actually Theorem 2 gives a stronger negative result). In this section, we revisit this problem when the simulator is allowed

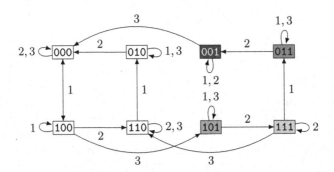

Fig. 1. Definition and sequential behavior of $\rho : [\![2]\!]^3 \to [\![2]\!]^3$ from Theorem 3. Label on arcs represent the coordinate updated.

to use a larger alphabet. In this case we can consider two natural types of simulations: one requires the simulation to work on any initial configuration of the simulator and uses a projection onto configurations of the simulated functions; the other does not use projection, but works only on initial configurations using the alphabet of the simulated function.

Definition 1. *Let* $n \in \mathbb{N}$, $2 \leq q < q'$ *and consider* $f \in F(n, q')$. *We say that* f *is* (n,q)-*universal by factor if there is a surjection* $\pi : [\![q']\!] \to [\![q]\!]$ *such that for any* $h \in F(n, q)$ *there is a word* $w \in [1, n]^*$ *such that*

$$\forall x \in [\![q']\!]^n, \overline{\pi} \circ f^{(w)}(x) = h \circ \overline{\pi}(x)$$

where $\overline{\pi}(x_1, \ldots, x_n) = (\pi(x_1), \ldots, \pi(x_n))$. f *is said* (n,q)-*universal by initialization if for any* $h \in F(n, q)$ *there is a word* $w \in [1, n]^*$ *such that*

$$\forall x \in [\![q]\!]^n, f^{(w)}(x) = h(x).$$

We are going to show that universality can be achieved for each kind of simulation. In both cases, the larger alphabet allows us to encode more information than the configuration of the simulated function. This additional information is used as a global controlling state that commands transformations applied on the simulated configuration and evolves according to a finite automaton. In the case of simulation by factor, the encoding is straightforward but the global controlling state is uninitialized. The key is to use a control automaton with a synchronizing word (see Fig. 1). In the case of simulation by initialization, the difficulty lies in the encoding.

The following theorems were obtained by F. Bridoux during his PhD thesis [1].

Theorem 3. *For any* $q \geq 2$ *and* $n \geq 3$, *there exists* $f \in F(n, 2q)$ *which is* (n, q)-*universal by factor.*

Proof. We can see any configuration of $[2q]^n$ as a pair made of a configuration of $[q]^n$ and a Boolean configuration, so we can as well describe f as a function acting on $[q]^n \times [2]^n$ to simplify notations and use the surjective map $\pi : [q]^n \times [2]^n \to [q]^n$ that projects onto the first component. We will actually choose f which is the identity map on the coordinates 4 to n on the Boolean component. So, to simplify even further, we will define a function $f : [q]^n \times [2]^3 \to [q]^n \times [2]^3$.

Consider first the function $\rho : [2]^3 \to [2]^3$ defined by Fig. 1 and consider the map $\Psi : [q]^n \times [2]^3 \to [q]^n$ defined by:

$$\Psi(x,y)_1 = \begin{cases} x_1 + 1 \bmod q & \text{if } y = 101, \\ 1 & \text{if } x = (0)^n \text{ and } (y = 011 \text{ or } y = 001), \\ 0 & \text{if } x = 1(0)^{n-1} \text{ and } y = 011, \\ x_1 & \text{otherwise,} \end{cases}$$

$$\Psi(x,y)_2 = \begin{cases} x_2 + 1 \bmod q & \text{if } x_1 = 0 \text{ and } y = 111, \\ x_2 & \text{otherwise,} \end{cases}$$

$$\Psi(x,y)_3 = \begin{cases} x_3 + 1 \bmod q & \text{if } x_1 = x_2 = 0 \text{ and } y = 011, \\ x_3 & \text{otherwise,} \end{cases}$$

$$\forall i \in [4, n], \ \Psi(x,y)_i = \begin{cases} x_i + 1 \bmod q & \text{if } x_1 = x_2 = \cdots = x_{i-1} = 0, \\ x_i & \text{otherwise.} \end{cases}$$

Then we define f by $f(x, y) = \big(\Psi(x, y), \rho(y)\big)$. We now prove properties about f implying that it is (n, q)-universal by factor.

Claim 1. For any $(x, y) \in [q]^n \times [2]^3$ it holds $f^{((3)^q, 2, 3, 1, 1, 2, 1, 3)}(x, y) = (x, 101)$.

Proof. First, let us remark that updating q times coordinate 3 starting from (x, y), there are two cases:

- $y \neq 011$ or $x_1 \neq 0$ or $x_2 \neq 0$ and then the component x is not modified;
- $y = 011$ and $x_1 = x_2 = 0$, and then the modification $x_3 \leftarrow x_3 + 1$ is applied q times.

Therefore we have $f^{((3)^q)}(x, y) = (x', y')$ with

$$x' = (x_1, x_2, x_3 + q, \dots) = (x_1, x_2, x_3, \dots) = x.$$

To show that the update sequence $((3)^q, 2, 3, 1, 1, 2, 1, 3)$ does not modify the component x, it is sufficient to verify the following:

- coordinate 1 is not updated when $y \in \{101, 011, 001\}$;
- coordinate 2 is not updated when $y = 111$;
- when coordinate 3 is updated and $y = 011$, it is updated q times.

By definition of $f^{((3)^q,2,3,1,1,2,1,3)}$, we obtain:

$$x000 \xrightarrow{(3)^q} x000 \xrightarrow{2} x000 \xrightarrow{3} x000 \xrightarrow{1} x100 \xrightarrow{1} x000 \xrightarrow{2} x000 \xrightarrow{1} x100 \xrightarrow{3} x101,$$
$$x100 \xrightarrow{(3)^q} x101 \xrightarrow{2} x111 \xrightarrow{3} x110 \xrightarrow{1} x010 \xrightarrow{1} x010 \xrightarrow{2} x000 \xrightarrow{1} x100 \xrightarrow{3} x101,$$
$$x010 \xrightarrow{(3)^q} x010 \xrightarrow{2} x000 \xrightarrow{3} x000 \xrightarrow{1} x100 \xrightarrow{1} x000 \xrightarrow{2} x000 \xrightarrow{1} x100 \xrightarrow{3} x101,$$
$$x110 \xrightarrow{(3)^q} x110 \xrightarrow{2} x110 \xrightarrow{3} x110 \xrightarrow{1} x010 \xrightarrow{1} x010 \xrightarrow{2} x000 \xrightarrow{1} x100 \xrightarrow{3} x101,$$
$$x001 \xrightarrow{(3)^q} x000 \xrightarrow{2} x000 \xrightarrow{3} x000 \xrightarrow{1} x100 \xrightarrow{1} x000 \xrightarrow{2} x000 \xrightarrow{1} x100 \xrightarrow{3} x101,$$
$$x011 \xrightarrow{(3)^q} x011 \xrightarrow{2} x001 \xrightarrow{3} x000 \xrightarrow{1} x100 \xrightarrow{1} x000 \xrightarrow{2} x000 \xrightarrow{1} x100 \xrightarrow{3} x101,$$
$$x101 \xrightarrow{(3)^q} x101 \xrightarrow{2} x111 \xrightarrow{3} x110 \xrightarrow{1} x010 \xrightarrow{1} x010 \xrightarrow{2} x000 \xrightarrow{1} x100 \xrightarrow{3} x101,$$
$$x111 \xrightarrow{(3)^q} x110 \xrightarrow{2} x110 \xrightarrow{3} x110 \xrightarrow{1} x010 \xrightarrow{1} x010 \xrightarrow{2} x000 \xrightarrow{1} x100 \xrightarrow{3} x101.$$

$\qquad\qquad\qquad\qquad\qquad\qquad\qquad\qquad\qquad\qquad\qquad\qquad\qquad\square$

Let us now show that, starting from $(x, 101)$, f can realize three kinds of transformations on x that will turn out to be sufficient to generate all $F(n, q)$.

– Let $c \in \mathrm{Sym}(n, q)$ be the following circular permutation:

$$c : ((0)^n \to 1(0)^{n-1} \to \cdots \to (q-1)(0)^{n-1} \to 01(0)^{n-2} \to \dots).$$

then for any $x \in [\![q]\!]^n$ we have $f^{(1,2,2,1,(3,4,\dots,n))}(x, 101) = (c(x), 011)$ because:

$$x101 \xrightarrow{1} c^{(1)}(x)101 \xrightarrow{2} c^{(1)}(x)111 \xrightarrow{2} c^{([1,2])}(x)111 \xrightarrow{1} c^{([1,2])}(x)011$$
$$\xrightarrow{3} c^{([1,3])}(x)011 \xrightarrow{4} c^{([1,4])}(x)011 \xrightarrow{5} \dots \xrightarrow{n} c(x)011.$$

– Consider the transposition $k = ((0)^n \leftrightarrow 1(0)^{n-1})$, then we have, for any $x \in [\![q]\!]^n$, $f^{(2,1,1)}(x, 101) = (k(x), 011)$ because:

$$x101 \xrightarrow{2} x111 \xrightarrow{1} x011 \xrightarrow{1} k(x)011.$$

– Finally, consider the assignment $d = ((0)^n \to 1(0)^{n-1})$, then for any $x \in [\![q]\!]^n$ it holds $f^{(2,1,2,1)}(x, 101) = (d(x), 001)$ because:

$$x101 \xrightarrow{2} x111 \xrightarrow{1} x011 \xrightarrow{2} x001 \xrightarrow{1} d(x)001.$$

Since functions c, k and d generate $F(n, q)$ (see [14] or [11]), the theorem follows. $\square$

Theorem 4. *For any $q \geq 2$ and $n \geq 3q$, there is $f \in F(n, q+1)$ which is (n, q)-universal by initialization.*

5   Simulation by Sets of Networks

So far we studied what a single function can simulate. We know shift our interest to semigroups generated by some sets of functions.

5.1   Singular Instructions

An instruction is any $f^{(v)}$ for some $f \in F(n, q)$ and some $v \in [1, n]$. Burckel showed that any network is the composition of instructions: $\langle F(n, q) \rangle_{\text{Seq}} = \langle \{ f^{(v)} : f \in F(n, q), v \in [1, n] \} \rangle = F(n, q)$. As an immediate consequence, any permutation in $\text{Sym}(n, q)$ is the composition of permutation instructions: $\text{Sym}(n, q)$ is exactly $\langle \{ f^{(v)} \in \text{Sym}(n, q) : f \in F(n, q), v \in [1, n] \} \rangle$. We now determine what singular instructions generate: let

$$S(n, q) := \left\langle \left\{ f^{(v)} \in \text{Sing}(n, q) : f \in F(n, q), v \in [1, n] \right\} \right\rangle.$$

Proposition 2. *The semigroup $S(n, q)$ generated by singular instructions consists of all networks f such that there exist $a, b \in [\![q]\!]^n$ with $f(a) = f(b)$ and $d_{\text{H}}(a, b) = 1$.*

Any network f can be seen as a vertex colouring of the Hamming graph $H(n, q)$ (x colored by $f(x)$). From the proposition above, networks in $S(n, q)$ correspond to improper colouring. Since the chromatic number of $H(n, q)$ is equal to q, we deduce that any network with rank at most $q - 1$ can be generated by singular instructions. However, the network $f(x) = (x_1 + \ldots + x_n, 0, \ldots, 0)$ cannot be generated by singular instructions, since it generates a proper colouring of the Hamming graph.

A network f is idempotent if $f^2 = f$. Idempotents are pivotal in the theory of semigroups, for they are the identity elements of the subgroups of a given semigroup. In particular it is interesting to know whether a semigroup S is generated by its set of idempotents, because then any element $s \in S$ can be expressed as a product of consecutively distinct idempotents: $s = e_1 e_2 \ldots e_k$. We remark that if $f \in S(n, q)$ is idempotent and has rank $q^n - 1$, then it must be an assignment instruction.

Theorem 5. *$S(n, q)$ is generated by assignment instructions for $q \geq 3$.*

The previous result could be proved using the so-called fifteen-puzzle. In the original puzzle, an image is cut into a four-by-four grid of tiles; one of the tiles is removed, thus creating a hole; the remaining fifteen tiles are scrambled by sliding a tile into the hole. The player is then given the scrambled image, and has to reconstruct it by repeatedly sliding a tile in the hole.

Clearly, this game can be played on any simple graph D, where a hole is created at a vertex (say h), and one can "slide" one vertex into the hole, the hole thus moving to that vertex. If the hole goes back to its original place h, then we have created a permutation of $V(D) \setminus h$. The set of all possible permutations is closed under composition and hence it forms a group, called the **puzzle group** $G(D, h)$. Wilson [17] fully characterised that group for 2-connected simple graphs; we give a simpler version of the theorem below.

Theorem 6. (Wilson's fifteen-puzzle theorem). *Let D be a 2-connected simple graph, then $G(D, h) \cong G(D, h')$ for all vertices $h, h' \in V(D)$. Moreover, if D is the undirected cycle, then $G(D, h)$ is trivial. Otherwise, the following hold.*

1. *If D is not bipartite and has at least eight vertices, then $G(D, h) = \text{Sym}(V(D) \setminus h)$.*
2. *If D is bipartite, then $G(D, h) = \text{Alt}(V(D) \setminus h)$.*

Using assignment instructions $(a \to b)$ to simulate a network f of rank $q^n - 1$ can be viewed as playing the fifteen-puzzle on the Hamming graph $H(n, q)$: the first $(a^1 \to b^1)$ places a hole in vertex a^1 and any subsequent $(a^k \to b^k)$ slides the vertex a^k into the hole b^k (and the hole moves to a^k instead). Since $H(n, q)$ is not bipartite for $q \geq 3$ (and it has at least nine vertices for $n \geq 2$), we can apply Wilson's theorem and, after a bit more work, prove Theorem 5 that way. However, the hypercube $H(n, 2)$ is bipartite, then the puzzle group is only the alternating group. Thus, $S(n, 2)$ is not generated by assignment instructions, and in particular $f = (010 \cdots 0 \leftrightarrow 110 \cdots 0) \circ (000 \cdots 0 \to 100 \cdots 0)$ cannot be generated by assignment instructions.

5.2 Simulation by Graphs

The **interaction graph** of $f \in F(n, q)$ is the (directed graph) which has vertex set $V = [1, n]$ and has an arc from u to v if and only if f_v depends essentially on u, i.e. there exists $a, b \in [\![q]\!]^n$ such that $a_{V \setminus u} = b_{V \setminus u}$ and $f_v(a) \neq f_v(b)$. For any graph D with n nodes, we denote the set of networks in $F(n, q)$ whose interaction graph is a subgraph of D as $F(D, q)$.

A graph is reflexive if for any vertex v, (v, v) is an arc in D. Note that for any reflexive graph D it holds $\langle F(D, q) \rangle_{\text{Seq}} \subseteq \langle F(D, q) \rangle_{\text{Asy}} = \langle F(D, q) \rangle_{\text{Syn}}$. The first inclusion is trivial; the equality follows from the fact that for any $f \in F(D, q)$ and any $S \subseteq [1, n]$, $f^{(S)}$ belongs to $F(D, q)$ as well. Moreover, it is clear that if $\langle F(H, q) \rangle_{\text{Seq}} = F(n, q)$, then H is reflexive (otherwise, $\langle F(H, q) \rangle_{\text{Seq}}$ would not contain any permutation). The reflexive graphs which can simulate the whole of $F(n, q)$ synchronously were classified by Tchuente in [16]. In fact, the same graphs can simulate the whole of $F(n, q)$ asynchronously or sequentially.

Theorem 7. *Let D be a reflexive graph on n vertices. Then the following are equivalent.*

1. *$\langle F(D, q) \rangle_{\text{Seq}} = F(n, q)$.*
2. *$\langle F(D, q) \rangle_{\text{Asy}} = F(n, q)$.*
3. *$\langle F(D, q) \rangle_{\text{Syn}} = F(n, q)$.*
4. *D is strongly connected and it has a vertex of in-degree n.*

A permutation of variables is any network $f := \bar{\phi}$ defined by $f_i(x) = x_{\phi(i)}$ for some $\phi \in \text{Sym}([1, n])$. We first show that we can permute variables freely if the graph is strongly connected (and is reflexive for the sequential case).

Lemma 1. *The following are equivalent for a reflexive graph D.*

1. *$\langle F(D, q) \rangle_{\text{Seq}}$ contains all permutations of variables of $F(n, q)$.*
2. *$\langle F(D, q) \rangle_{\text{Asy}}$ contains all permutations of variables of $F(n, q)$.*
3. *D is strong.*

Proof (Proof of Theorem 7). Clearly, 1 implies 2, which in turn is equivalent to 3. We prove 2 implies 4. Let D such that $\langle F(D, q) \rangle_{\text{Asy}} = F(n, q)$. By Lemma 1, D is strong. We now prove that D has a vertex of in-degree n. Otherwise, let $f \in F(D, q)$ of rank $q^n - 1$. Let $a \in O(()f)$ and b with $|f^{-1}(b)| = 2$ (and hence $|f^{-1}(x)| = 1$ for any other x). We then have

$$\sum_{x \in [\![q]\!]^n} f(x) \bmod q^n = b - a \neq 0.$$

On the other hand, it is easily seen that for any $y \in [\![q]\!]$, $|f_v^{-1}(y)|$ is a multiple of q^{n-d_v} where d_v is the in-degree of v in D, hence

$$\sum_{x \in [\![q]\!]^n} f_v(x) \bmod q = \sum_{y \in [\![q]\!]} |f_v^{-1}(y)| y \bmod q = 0.$$

Doing this componentwise for all v, we obtain $\sum_{x \in [\![q]\!]^n} f(x) = 0$, which is the desired contradiction.

We prove 4 implies 1. We only need to show that all instructions in $F(n, q)$ belong to $\langle F(D, q) \rangle_{\text{Seq}}$. Let u be a vertex of in-degree n, then we already have any instruction updating u. Let v be another vertex, and g be an instruction updating v, then $g = \overline{(u \leftrightarrow v)} \circ h \circ \overline{(u \leftrightarrow v)}$, where h is the instruction updating u such that $h_u = g_v \circ (u \leftrightarrow v)$. Then $(u \leftrightarrow v) \in \langle F(D, q) \rangle_{\text{Seq}}$ according to Lemma 1. Thus, any instruction can be generated. □

6 Future Work

The contrast between the complete sequential simulator for $\text{Sym}(n, q)$ and the existence of non-bijective functions that are not sequentially simulatable in the non-Boolean case is striking. We would like first to settle the Boolean case: we conjecture that all functions of $F(n, 2)$ are sequentially simulatable for large enough n. For $q \geq 3$, in order to better understand the set of sequentially simulatable networks, one could for instance analyze how much synchronism is required to simulate them (how large are the sets V in the asynchronous updates $f^{(V)}$ used to simulate them). In particular, one may ask whether, for all n, there exists some network with n entities that require a synchronous update $f^{([1,n])}$ in order to be simulated asynchronously. Besides, the networks considered in Sects. 2, 3 and 4 have an unconstrained interaction graph. The situation could be very different when restricting all networks to particular a family of interaction graphs (bounded degree, bounded tree-width, etc.). Finally, still concerning interaction graphs, the characterization of Theorem 7 is about reflexive graphs. We would like to extend it to any graph (not necessarily reflexive).

References

1. Bridoux, F.: Intrinsic simulations and complexities in automata networks. Theses, Aix-Marseile Université, July 2019. https://hal.archives-ouvertes.fr/tel-02436228

2. Bridoux, F., Castillo-Ramirez, A., Gadouleau, M.: Complete simulation of automata networks. J. Comput. Syst. Sci. **109**, 1–21 (2020)
3. Bridoux, F., Gadouleau, M., Theyssier, G.: Simulation of automata networks. CoRR abs/2001.09198 (2020). https://arxiv.org/abs/2001.09198
4. Burckel, S.: Closed iterative calculus. Theor. Comput. Sci. **158**, 371–378 (1996)
5. Burckel, S., Gioan, E., Thomé, E.: Mapping computation with no memory. In: Proceedings of International Conference on Unconventional Computation, pp. 85–97. Ponta Delgada, Portugal, September 2009
6. Burckel, S., Gioan, E., Thomé, E.: Computation with no memory, and rearrangeable multicast networks. Discrete Math. Theor. Comput. Sci. **16**, 121–142 (2014)
7. Cameron, P.J., Fairbairn, B., Gadouleau, M.: Computing in permutation groups without memory. Chicago J. Theor. Comput. Sci. **2014**(07), 1–20 (2014)
8. Gadouleau, M.: On the influence of the interaction graph on a finite dynamical system. Nat. Comput. **19**(1), 15–28 (2019). https://doi.org/10.1007/s11047-019-09732-y
9. Gadouleau, M., Richard, A.: Simple dynamics on graphs. Theor. Comput. Sci. **628**, 62–77 (2016)
10. Gadouleau, M., Riis, S.: Memoryless computation: new results, constructions, and extensions. Theor. Comput. Sci. **562**, 129–145 (2015)
11. Ganyushkin, O., Mazorchuk, V.: Classical Finite Transformation Semigroups: An Introduction, Algebra and Applications, vol. 9. Springer, London (2009)
12. Goles, E., Noual, M.: Disjunctive networks and update schedules. Adv. Appl. Math. **48**(5), 646–662 (2012)
13. Goles, E., Montealegre, P., Salo, V., Törmä, I.: PSPACE-completeness of majority automata networks. Theor. Comput. Sci. **609**, 118–128 (2016). https://doi.org/10.1016/j.tcs.2015.09.014
14. Howie, J.M.: Fundamentals of Semigroup Theory. Oxford Science Publications (1995)
15. Noual, M., Sené, S.: Synchronism versus asynchronism in monotonic Boolean automata networks. Nat. Comput. **17**(2), 393–402 (2017). https://doi.org/10.1007/s11047-016-9608-8
16. Tchuente, M.: Computation on binary tree-networks. Discrete Appl. Math. **14**, 295–310 (1986)
17. Wilson, R.M.: Graph puzzles, homotopy, and the alternating group. J. Comb. Theory B **16**, 86–96 (1974)
18. Wu, A., Rosenfeld, A.: Cellular graph automata. I. basic concepts, graph property measurement, closure properties. Inf. Control **42**(3), 305–329 (1979)
19. Wu, A., Rosenfeld, A.: Cellular graph automata. II. graph and subgraph isomorphism, graph structure recognition. Inf. Control **42**, 330–353 (1979). https://doi.org/10.1016/S0019-9958(79)90296-1

Theoretical and Implementational Aspects of the Formal Language Server (LaSer)

Stavros Konstantinidis[(✉)] [iD]

Mathematics and Computing Science, Saint Mary's University,
923 Robie St., Halifax, NS B3H 3C3, Canada
`s.konstantinidis@smu.ca`

Abstract. LaSer, the formal language server, allows a user to enter a question about an independent language and provides an answer either in real time or by generating a program that can be executed at the user's site. Typical examples of independent languages are codes in the classic sense, such as prefix codes and error-detecting languages, or DNA-computing related codes. Typical questions about independent languages are the satisfaction, maximality and construction questions. We present some theoretical and implementational aspects of LaSer, as well as some ongoing progress and research plans.

Keywords: Independent languages · Regular languages · Codes · DNA codes · Property satisfaction · Maximality · Implementation

1 Introduction

LaSer, [18], the formal language server, allows a user to enter a question about an independent language and provides an answer either in real time or by generating a program that can be executed at the user's site. Typical examples of independent languages are codes in the classic sense, such as prefix codes and error-detecting languages, or DNA-computing related codes. Typical questions about independent languages are the satisfaction, maximality and construction questions. For example, the satisfaction question is to decide, given an independence $\mathcal{I}$ and a regular language L, whether L is independent with respect to $\mathcal{I}$.

We present some theoretical and implementational aspects of LaSer, as well as some ongoing progress and research plans.

2 Transducer Independences Allowed in LaSer

Let R be a binary relation, that is, a subset of $\Sigma^* \times \Sigma^*$, where Σ is an alphabet. A language L is R-independent, [21,22], if

$$(x,y) \in R \text{ and } x,y \in L \text{ implies } x = y. \tag{1}$$

Research supported by NSERC, Canada.

M. Anselmo et al. (Eds.): CiE 2020, LNCS 12098, pp. 289–295, 2020.
https://doi.org/10.1007/978-3-030-51466-2_25

Examples of R-independent languages are error-detecting languages for various error combinations, variable length codes, such as prefix codes and suffix codes, as well as DNA-related languages [2,5,6,10,12–14,23]. LaSer allows users to represent rational relations via transducers[1], and regular languages via NFAs[2]. The satisfaction question is whether $L(a)$ is $R(t)$-independent, given an NFA a and a transducer t. If the answer is NO, a witness word pair (x, y) is computed such that $x \neq y$ and one of (x, y) and (y, x) is in $R(t)$. The maximality question is whether $L(a)$ is a maximal $R(t)$-independent language knowing that $L(a)$ is $R(t)$-independent. If the answer is NO, a witness word $x \notin L(a)$ is computed such that $L(a) \cup \{x\}$ is $R(t)$-independent. The construction question is to make an n-element language (if possible) that is $R(t)$-independent, given transducer t, integer $n > 0$ and the size of the alphabet.

If t is a transducer then the following language class

$$\mathcal{P}_t = \{L \mid L \text{ satisfies (1) for } R = R(t)\}$$

is called the independence, or property, described by t. In this case any language $L \in \mathcal{P}_t$ is said to satisfy $\mathcal{P}_t$. For example, the independence "prefix codes" is described by the transducer px and the independence "2-synchronization-error detecting languages" is described by the transducer id$_2$ (see Fig. 1).

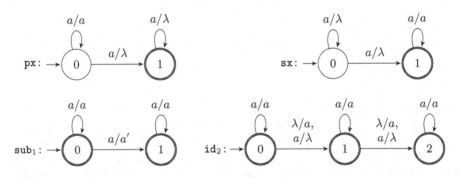

Fig. 1. Various transducers. An arrow with label a/a denotes multiple transitions: one with label a/a for each $a \in \Sigma$, and similarly for labels a/λ. An arrow with label a/a' denotes multiple transitions: one with label a/a' for all $a, a' \in \Sigma$ with $a \neq a'$. Let x be any word. We have: px$(x) =$ the set of proper prefixes of x, equivalently, $R(\text{px}) =$ the set of word pairs (x, y) such that y is a proper prefix of x; sx$(x) =$ set of proper suffixes of x; sub$_1(x) =$ set of words resulting by substituting at most one symbol in x with another one; id$_2(x) =$ set of words resulting by inserting and/or deleting at most 2 symbols in x.

Transducer independences are closed under intersection, that is, if $\mathcal{P}_{t_1}$ and $\mathcal{P}_{t_2}$ are transducer independences, then also $\mathcal{P}_{t_1} \cap \mathcal{P}_{t_2}$ is a transducer independence.

---

[1] See [1,20], for instance, for transducer concepts.
[2] NFA = Nondeterministic Finite Automaton.

This is useful when we are interested in languages L satisfying two or more properties, such as the combined property of being a prefix and 1-substitution detecting code. As prefix codes are described by the transducer px and 1-substitution detecting codes are described by the transducer sub_1 then also the combined property is described by a transducer. This implies that LaSer can answer the satisfaction, maximality and construction questions for the combined property.

LaSer's backend is based on the Python package FAdo [8], which implements automata, transducers, and independences described by transducer objects in the module codes.py [17]. The choice to use FAdo is based on the facts that its installation is very simple, it contains a rich set of easy to use methods, and is written in Python which in turn provides a rich availability of high level methods.

3   Rational Independence Expressions

Three examples of independences that are not of the form $\mathcal{P}_t$ are the following.

- The class of UD codes (uniquely decodable/decipherable codes). LaSer supports this class.
- The class of comma-free codes: that is, all languages L satisfying the equation

$$LL \cap \Sigma^+ L \Sigma^+ = \emptyset. \tag{2}$$

- The class of language pairs (L_1, L_2) satisfying the equation

$$L_1 \overset{\text{sdi}}{\Leftarrow} L_2 = \emptyset. \tag{3}$$

Here the site directed insertion operation $x \overset{\text{sdi}}{\Leftarrow} y$ between words x, y, introduced in [3], is such that $z \in x \overset{\text{sdi}}{\Leftarrow} y$ if $x = x_1 u v x_2$, $y = u w v$, $z = x_1 u w v x_2$, where u, v are nonempty. This operation models site-directed mutagenesis, an important technique for introducing a mutation into a DNA sequence.

That "UD codes" and "comma-free codes" are not transducer independences can be shown using dependence theory: every transducer independence is a 2-independence, but the "comma-free codes" independence, for instance, is a 3-independence and not a 2-independence—see [12,15]. In general, intersecting (combining) the above independences with a transducer independence results into a new independence that is not of the form $\mathcal{P}_t$ for some transducer t. LaSer does not handle non-transducer independences, with the exception of "UD codes" for which specific algorithms are employed. More specifically, for the satisfaction question LaSer uses the quadratic-time elegant algorithm of [9]. For the maximality question LaSer uses Schützenberger's theorem that maximality of L is equivalent to the condition that every word in Σ^* is subword (part) of some word of L^* [2]. A specific algorithm is also used in [23] for the satisfaction question of the combination of the UD code and two other DNA-related independences.

We now discuss possible approaches of representing non-transducer independences as finite objects in a way that these objects can be manipulated by
algorithms which can answer questions about the independences being represented. Some approaches are discussed in [15,19]. It is important to note that
there is a distinction between the terms "property" and "independence". A (language) property is simply a class (set) of languages. A (language) independence
is a property $\mathcal{P}$ for which the concept of maximality is defined, that is, $\mathcal{P}$ must
satisfy

> if $L \in \mathcal{P}$ then also $L' \in \mathcal{P}$ for all $L' \subseteq L$

(see [15] for details). A general type of independences can be defined via rational
language equations $\varphi(L) = \emptyset$, where $\varphi(L)$ is an independence expression involving the variable L. An independence expression φ is defined inductively as follows:
it is L or a language constant, or one of $\varphi_1\varphi_2, \varphi_1\cup\varphi_2, \varphi_1\cap\varphi_2, (\varphi_1)^*, t(\varphi_1), \theta(\varphi_1)$,
where φ_1, φ_2 are independence expressions, t is a transducer constant, and θ is
an antimorphic permutation[3] constant. A rational language equation is an expression of the form $\varphi(L) = \emptyset$, where $\varphi(L)$ is an independence expression containing
the variable L and the language constants occurring in $\varphi(L)$ represent regular
languages. A language L is φ-independent if it satisfies the equation $\varphi(L) = \emptyset$.
The independence $\mathcal{P}_\varphi$ described by φ is the set of φ-independent languages.

Example 1. Every transducer independence $\mathcal{P}_t$ such that t is an input-altering
transducer[4] is described by the rational language equation

$$t(L) \cap L = \emptyset, \qquad (4)$$

where we have used the standard notation $t(L) = \{y \mid (x, y) \in \mathrm{R}(t), x \in L\}$.
The independence "comma-free codes" is described by the rational language
equation (2). The independence "θ-free languages", [10], is described by the
rational language equation

$$LL \cap \Sigma^+\theta(L)\Sigma^+ = \emptyset.$$

$\square$

The satisfaction question for independences $\mathcal{P}_\varphi$ is implemented in [19], where
parsing of expressions $\varphi(L)$ is implemented using Python's `lark` library, and
evaluation of $\varphi(L)$ for given $L = \mathrm{L}(a)$ is implemented using the FAdo package.

4   Further Independences

What about independences described by equations like (3)? This is a non-
transducer independence. Various general types of independences are defined

---

[3] An antimorphic permutation θ maps the alphabet Σ onto Σ and extends to words
anti-morphically: $\theta(xy) = \theta(y)\theta(x)$. A typical example of this is the DNA involution
on the alphabet $\{a, c, g, t\}$ such that $\theta(a) = t$, $\theta(c) = g$, $\theta(g) = c$, $\theta(t) = a$. In this
case, $\theta(aac) = gtt$.

[4] This is a transducer t such that $w \notin t(w)$ for all words w. For example, px and sx
in Fig. 1 are input-altering transducers.

in [5,11,12,22]. Equation (3) however involves an independence expression with *two* language variables: L_1 and L_2. In analogy to transducer independences that are R-independences, for binary relations R, one can consider independences with respect to higher degree relations[5]. Consider the **ternary relation** $\text{SDI} = \{(x, y, z) \mid z \in x \overset{\text{sdi}}{\leftarrow} y\}$, which is realized by the transducer **sdi** in Fig. 2. A language pair (L_1, L_2) is SDI-independent if

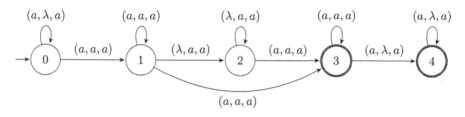

Fig. 2. The 3-tape transducer **sdi** realizing the set of all word triples (x, y, z) such that $x = x_1 uv x_2$, $y = uwv$ and $z = x_1 uwv x_2$, for some nonempty words u, v; that is, $z \in x \overset{\text{sdi}}{\leftarrow} y$.

$$\text{sdi}(L_1, L_2 : 3) = \emptyset.$$

Above we have made the following notation: for any k-tape transducer t, for any $i \in \{1, \ldots, k\}$, and for any list of $k - 1$ languages $L_1, \ldots, L_{k-1}$, the expression $t(L_1, \ldots, L_{k-1} : i)$ denotes the set of all words w that result if we consider the i-th tape as output tape and the rest $k - 1$ tapes as input tapes such that the input $k - 1$ words are from the $k - 1$ languages. Thus, we can talk about the independence described by the equation

$$t(L_1, \ldots, L_{k-1} : i) = \emptyset.$$

Given a k-tape transducer t and $k - 1$ NFAs accepting the languages $L_1, \ldots, L_{k-1}$, the satisfaction question can be decided if we construct the transducer resulting by intersecting t with the NFAs at the $k - 1$ positions other than i, and then testing whether the resulting transducer has a path from an initial to a final state.

5   Looking Ahead

We propose to investigate further the rational language equations defined here as well as in [15]. Topics of interest are maximality, embedding and expressibility[6] as well as enhancement and implementation of algorithms involved. Some of

---

[5] See references [4,7], for instance, for higher degree relations.

[6] What independences are and are not describable by the independence-describing method.

these topics could be complex. For example, the maximality question for transducer independences can be answered by a simple algorithm, but the question is PSPACE hard. The embedding question is to construct a maximal independent language containing the given independent language $L(a)$. For transducer independences described by Eq. (4), the embedding question is addressed in [16]. The maximality and embedding questions for non-transducer independences appear to be far more complex. For the satisfaction question of independences described by rational language equations, a useful question is to define and implement witnesses of non-satisfaction. For example, a witness of non-satisfaction for equation (2) would be a word triple (x, y, z) such that $x, y, z \in L$ and $xy = uzv$ for some nonempty words u and v.

Acknowledgement. We thank the anonymous referees for constructive comments.

References

1. Berstel, J.: Transductions and Context-Free Languages. B.G. Teubner, Stuttgart (1979)
2. Berstel, J., Perrin, D., Reutenauer, C.: Codes and Automata. Cambridge University Press, Cambridge (2009)
3. Cho, D.-J., Han, Y.-S., Salomaa, K., Smith, T.J.: Site-directed insertion: decision problems, maximality and minimality. In: Konstantinidis, S., Pighizzini, G. (eds.) DCFS 2018. LNCS, vol. 10952, pp. 49–61. Springer, Cham (2018). https://doi.org/10.1007/978-3-319-94631-3_5
4. Choffrut, C.: Relations over words and logic: a chronology. Bull. EATCS **89**, 159–163 (2006)
5. Domaratzki, M.: Trajectory-based codes. Acta Informatica **40**, 491–527 (2004). https://doi.org/10.1007/s00236-004-0140-4
6. Domaratzki, M.: Bond-free DNA language classes. Nat. Comput. **6**, 371–402 (2007). https://doi.org/10.1007/s11047-006-9022-8
7. Elgot, C.C., Mezei, J.E.: On relations defined by generalized finite automata. IBM J. Res. Dev. **9**(1), 47–68 (1965). https://doi.org/10.1147/rd.91.0047
8. FAdo: Tools for formal languages manipulation. http://fado.dcc.fc.up.pt/. Accessed Jan 2020
9. Head, T., Weber, A.: Deciding code related properties by means of finite transducers. In: Capocelli, R., de Santis, A., Vaccaro, U. (eds.) Sequences II, Methods in Communication, Security, and Computer Science, pp. 260–272. Springer, Berlin (1993). https://doi.org/10.1007/978-1-4613-9323-8_19
10. Hussini, S., Kari, L., Konstantinidis, S.: Coding properties of DNA languages. Theoret. Comput. Sci. **290**, 1557–1579 (2003). https://doi.org/10.1016/S0304-3975(02)00069-5
11. Jürgensen, H.: Syntactic monoids of codes. Acta Cybern. **14**, 117–133 (1999)
12. Jürgensen, H., Konstantinidis, S.: Codes. In: Rozenberg, G., Salomaa, A. (eds.) Handbook of Formal Languages, vol. 1, pp. 511–607. Springer, Berlin (1997). https://doi.org/10.1007/978-3-642-59136-5_8
13. Kari, L., Konstantinidis, S., Kopecki, S.: Transducer descriptions of DNA code properties and undecidability of antimorphic problems. Inf. Comput. **259**(2), 237–258 (2018). https://doi.org/10.1016/j.ic.2017.09.004

14. Konstantinidis, S.: An algebra of discrete channels that involve combinations of three basic error types. Inf. Comput. **167**(2), 120–131 (2001). https://doi.org/10.1006/inco.2001.3035

15. Konstantinidis, S.: Applications of transducers in independent languages, word distances, codes. In: Pighizzini, G., Câmpeanu, C. (eds.) DCFS 2017. LNCS, vol. 10316, pp. 45–62. Springer, Cham (2017). https://doi.org/10.1007/978-3-319-60252-3_4

16. Konstantinidis, S., Mastnak, M.: Embedding rationally independent languages into maximal ones. J. Automata Lang. Comb. **21**(4), 311–338 (2017). https://doi.org/10.25596/jalc-2016-311

17. Konstantinidis, S., Meijer, C., Moreira, N., Reis, R.: Symbolic manipulation of code properties. J. Automata Lang. Comb. **23**(1–3), 243–269 (2018). https://doi.org/10.25596/jalc-2018-243. (This is the full journal version of the paper "Implementation of Code Properties via Transducers" in the Proceedings of CIAA 2016, LNCS 9705, pp 1–13, edited by Yo-Sub Han and Kai Salomaa)

18. LaSer: Independent LAnguage SERver. http://laser.cs.smu.ca/independence/. Accessed Apr 2020

19. Rafuse, M.: Deciding rational property definitions, Honours Undergraduate Thesis, Department of Mathematics and Computing Science, Saint Mary's University, Halifax, NS, Canada (2019)

20. Sakarovitch, J.: Elements of Automata Theory. Cambridge University Press, Berlin (2009)

21. Shyr, H.J., Thierrin, G.: Codes and binary relations. In: Malliavin, M.P. (ed.) Séminaire d'Algèbre Paul Dubreil, Paris 1975–1976 (29ème Année). Lecture Notes in Mathematics, vol. 586, pp. 180–188. Springer, Heidelberg (1977). https://doi.org/10.1007/BFb0087133

22. Yu, S.S.: Languages and Codes. Tsang Hai Book Publishing, Taichung (2005)

23. Zaccagnino, R., Zizza, R., Zottoli, C.: Testing DNA code words properties of regular languages. Theoret. Comput. Sci. **608**, 84–97 (2015). https://doi.org/10.1016/j.tcs.2015.08.034

Balancing Straight-Line Programs
for Strings and Trees

Markus Lohrey[(✉)]

University of Siegen, Siegen, Germany
`lohrey@eti.uni-siegen.de`

Abstract. The talk will explain a recent balancing result according to which a context-free grammar in Chomsky normal form of size m that produces a single string w of length n (such a grammar is also called a straight-line program) can be transformed in linear time into a context-free grammar in Chomsky normal form for w of size $\mathcal{O}(m)$, whose unique derivation tree has depth $\mathcal{O}(\log n)$. This solves an open problem in the area of grammar-based compression, improves many results in this area and greatly simplifies many existing constructions. Similar balancing results can be formulated for various grammar-based tree compression formalism like top DAGs and forest straight-line programs. The talk is based on joint work with Moses Ganardi and Artur Jeż. An extended abstract appeared in [11]; a long version of the paper can be found in [12].

1 Grammar-Based String Compression

In *grammar-based compression* a combinatorial object (e.g. a string or tree) is compactly represented using a grammar of an appropriate type. In many grammar-based compression formalisms such a grammar can be exponentially smaller than the object itself. A well-studied example of this general idea is grammar-based string compression using context-free grammars that produce only one string each, which are also known as *straight-line programs*, SLPs for short. Formally, we define an SLP as a context-free grammar $\mathcal{G}$ in Chomsky normal form such that (i) $\mathcal{G}$ is acyclic and (ii) for every nonterminal A there is exactly one production, where A is the left-hand side. The *size* $|\mathcal{G}|$ of the SLP $\mathcal{G}$ is the number of nonterminals in $\mathcal{G}$ and the *depth* of $\mathcal{G}$ (depth($\mathcal{G}$) for short) is the height of the unique derivation tree of $\mathcal{G}$. Since we assume that $\mathcal{G}$ is in Chomsky normal form, we have depth($\mathcal{G}$) $\geq \log n$ if n is the length of the string produced by $\mathcal{G}$.

The goal of grammar-based string compression is to compute from a given string an SLP of small size. Grammar-based string compression is tightly related to dictionary-based compression: the famous LZ78 algorithm can be viewed as a particular grammar-based compressor. Moreover, the number of phrases in the LZ77-factorization is a lower bound for the smallest SLP for a string [24], and an LZ77-factorization of length m can be converted to an SLP of size $\mathcal{O}(m \cdot \log n)$

© Springer Nature Switzerland AG 2020
M. Anselmo et al. (Eds.): CiE 2020, LNCS 12098, pp. 296–300, 2020.
https://doi.org/10.1007/978-3-030-51466-2_26

where n is the length of the string [8,18,19,24]. For various other aspects of grammar-based string compression see [8,20].

2   Balancing String Straight-Line Programs

An advantage of grammar-based compression is that SLPs are well-suited for further algorithmic processing. There is an extensive body of literature on algorithms for SLP-compressed strings, see e.g. [20] for a survey. For many of these algorithms, the depth of the input SLP is an important parameter. Let us give a simple example: in the *random access problem* for an SLP-compressed string s an SLP $\mathcal{G}$ for s is given. Let n be the length of s. The goal is to produce from $\mathcal{G}$ a data structure that allows us to compute for a given position i ($1 \leq i \leq n$) the i-th symbol of s. As observed in [6] one can solve this problem in time $\mathcal{O}(\mathrm{depth}(\mathcal{G}))$ (assuming arithmetic operations on numbers from the interval $[0, n]$ need constant time): in the preprocessing phase one computes for every nonterminal A of $\mathcal{G}$ the length n_A of the string produced from A; this takes time $\mathcal{O}(\mathcal{G})$ and produces a data structure of size $\mathcal{O}(\mathcal{G})$ (assuming a number from $[0, n]$ fits into a memory location). In the query phase, one computes for a given position $i \in [1, n]$ the path from the root to the i-th leaf in the derivation tree of $\mathcal{G}$. Only the current nonterminal A and a relative position in the string produced from A has to be stored. Using the pre-computed numbers n_A the whole computation needs time $\mathcal{O}(\mathrm{depth}(\mathcal{G}))$. Recall that $\mathrm{depth}(\mathcal{G}) \geq \log n$. In [6] it is shown that one can compute from $\mathcal{G}$ a data structure of size $\mathcal{O}(|\mathcal{G}|)$ which allows to access every position in time $\mathcal{O}(\log n)$, irrespective of the depth of $\mathcal{G}$. The algorithm in [6] is quite complicated and used several sophisticated data structures. An alternative approach to obtain access time $\mathcal{O}(\log n)$ is to *balance* the input SLP $\mathcal{G}$ in the preprocessing phase, i.e., to reduce its depth. This is the approach that we will follow.

It is straightforward to show that any string s of length n can be produced by an SLP of size $\mathcal{O}(n)$ and depth $\mathcal{O}(\log n)$. A more difficult problem is to balance a given SLP: assume that the SLP $\mathcal{G}$ produces a string of length n. Several authors have shown that one can restructure $\mathcal{G}$ in time $\mathcal{O}(|\mathcal{G}| \cdot \log n)$ into an equivalent SLP $\mathcal{H}$ of size $\mathcal{O}(|\mathcal{G}| \cdot \log n)$ and depth $\mathcal{O}(\log n)$ [8,19,24]. Applied to the random access problem, these balancing procedures would yield access time $\mathcal{O}(\log n)$ at the cost of building a data structure of size $\mathcal{O}(|\mathcal{G}| \cdot \log n)$ during the preprocessing. Our main result shows that SLP balancing is in fact possible with a constant blow-up in SLP-size:

Theorem 1. *Given an SLP $\mathcal{G}$ producing a string of length n one can construct in linear time an equivalent SLP $\mathcal{H}$ of size $\mathcal{O}(|\mathcal{G}|)$ and depth $\mathcal{O}(\log n)$.*

As a corollary we obtain a very simple and clean algorithm for the random access problem with access time $\mathcal{O}(\log n)$ that uses a data structure of size $\mathcal{O}(m)$ (in words of bit length $\log n$). We can also obtain an algorithm for the random access problem with access time $\mathcal{O}(\log n / \log \log n)$ using a data structure with $\mathcal{O}(m \cdot \log^\epsilon n)$ words for any $\epsilon > 0$; previously this bound was only shown for

SLPs of height $\mathcal{O}(\log n)$. [1] The paper [11] contains a list of further applications of Theorem 1, which include the following problems on SLP-compressed strings: rank and select queries [1], subsequence matching [2], computing Karp-Rabin fingerprints [4], computing runs, squares, and palindromes [17], real-time traversal [14,23] and range-minimum queries [15]. In all these applications we either improve existing results or significantly simplify existing proofs by replacing depth($\mathcal{G}$) by $\mathcal{O}(\log n)$ in time/space bounds.

3   Proof Strategy

The proof of Theorem 1 consists of two main steps. Take an SLP $\mathcal{G}$ for the string s of length n and let m be the size of $\mathcal{G}$. We consider the derivation tree t for $\mathcal{G}$; it has size $\mathcal{O}(n)$. The SLP $\mathcal{G}$ can be viewed as a DAG for t of size m. We decompose this DAG into node-disjoint paths such that each path from the root to a leaf intersects $\mathcal{O}(\log n)$ paths from the decomposition. Each path from the decomposition is then viewed as a string of integer-weighted symbols, where the weights are the lengths of the strings derived from nodes that branch off from the path. For this weighted string we construct an SLP of linear size that produces all suffixes of the path in a certain weight-balanced way. Plugging these SLPs together yields the final balanced SLP.

Some of the ideas in our algorithm can be traced back to the area of parallel algorithms: the path decomposition for DAGs is related to the centroid path decomposition of trees [9], where it is the key technique in several parallel algorithms on trees. Moreover, the SLP of linear size that produces all suffixes of a weighted string can be seen as a weight-balanced version of the optimal prefix sum algorithm.

Our balancing procedure involves simple arithmetics on string positions, i.e., numbers of order n. Therefore we need machine words of bit-length $\Omega(\log n)$ in order to achieve a linear running time in Theorem 1; otherwise the running time increases by a multiplicative factor of order $\log n$. Note that such an assumption is realistic and standard in the field: machine words of bit length $\Omega(\log n)$ are needed, say, for indexing positions in the represented string. On the other hand, our procedure works in the pointer model regime.

4   Balancing Forest Straight-Line Programs and Top DAGs

Grammar-based compression has been generalized from strings to ordered node-labelled trees. In fact, the representation of a tree t by its smallest directed acyclic graph (DAG) is a form of grammar-based tree compression. This DAG is obtained by merging nodes where the same subtree of t is rooted. It can be seen as a regular tree grammar that produces only t. A drawback of DAG-compression is that the size of the DAG is lower-bounded by the height of the tree t. Hence, for deep narrow trees (like for instance caterpillar trees), the DAG-representation cannot achieve good compression. This can be overcome by representing a tree t by a linear context-free tree grammar that produces only t. Such grammars are also known as *tree straight-line programs* in the case of ranked trees (where the number of children of a node is uniquely determined by the node label) [7,21,22] and *forest straight-line programs* in the case of unranked trees [13]. The latter are tightly related to *top DAGs* [3,5,10,13,16], which are another tree compression formalism, also akin to grammars. Our balancing technique works similarly for these compression formalisms:

Theorem 2. *Given a top DAG/forest straight-line program/tree straight-line program $\mathcal{G}$ producing the tree t one can compute in time $\mathcal{O}(|\mathcal{G}|)$ a top DAG/forest straight-line program/tree straight-line program $\mathcal{H}$ for t of size $\mathcal{O}(|\mathcal{G}|)$ and depth $\mathcal{O}(\log|t|)$.*

For top DAGs, this solves an open problem from [5], where it was shown that from an unranked tree t of size n, whose minimal DAG has size m (measured in number of edges in the DAG), one can construct in linear time a top DAG for t of size $\mathcal{O}(m \cdot \log n)$ and depth $\mathcal{O}(\log n)$. It remained open whether the additional factor $\log n$ in the size bound can be avoided. A negative answer for the specific top DAG constructed in [5] was given in [3]. On the other hand, Theorem 2 yields another top DAG of size $\mathcal{O}(m)$ and depth $\mathcal{O}(\log n)$. To see this note that one can easily convert the minimal DAG of t into a top DAG of roughly the same size, which can then be balanced. This also gives an alternative proof of a result from [10], according to which one can construct in linear time a top DAG of size $\mathcal{O}(n/\log_\sigma n)$ and depth $\mathcal{O}(\log n)$ for a given tree of size n containing σ many different node labels.

Let us finally mention that Theorems 1 and 2 are instances of a general balancing result that applies to a large class of circuits over algebraic structures, see [12] for details.

References

1. Belazzougui, D., Cording, P.H., Puglisi, S.J., Tabei, Y.: Access, rank, and select in grammar-compressed strings. In: Bansal, N., Finocchi, I. (eds.) ESA 2015. LNCS, vol. 9294, pp. 142–154. Springer, Heidelberg (2015). https://doi.org/10.1007/978-3-662-48350-3_13

2. Bille, P., Cording, P.H., Gørtz, I.L.: Compressed subsequence matching and packed tree coloring. Algorithmica **77**(2), 336–348 (2017)

3. Bille, P., Fernström, F., Gørtz, I.L.: Tight bounds for top tree compression. In: Fici, G., Sciortino, M., Venturini, R. (eds.) SPIRE 2017. LNCS, vol. 10508, pp. 97–102. Springer, Cham (2017). https://doi.org/10.1007/978-3-319-67428-5_9

4. Bille, P., Gørtz, I.L., Cording, P.H., Sach, B., Vildhøj, H.W., Vind, S.: Fingerprints in compressed strings. J. Comput. Syst. Sci. **86**, 171–180 (2017)

5. Bille, P., Gørtz, I.L., Landau, G.M., Weimann, O.: Tree compression with top trees. Inf. Comput. **243**, 166–177 (2015)

6. Bille, P., Landau, G.M., Raman, R., Sadakane, K., Satti, S.R., Weimann, O.: Random access to grammar-compressed strings and trees. SIAM J. Comput. **44**(3), 513–539 (2015)

7. Busatto, G., Lohrey, M., Maneth, S.: Efficient memory representation of XML document trees. Inf. Syst. **33**(4–5), 456–474 (2008)
8. Charikar, M., Lehman, E., Liu, D., Panigrahy, R., Prabhakaran, M., Sahai, A., Shelat, A.: The smallest grammar problem. IEEE Trans. Inf. Theory **51**(7), 2554–2576 (2005)
9. Cole, R., Vishkin, U.: The accelerated centroid decomposition technique for optimal parallel tree evaluation in logarithmic time. Algorithmica **3**, 329–346 (1988)
10. Dudek, B., Gawrychowski, P.: Slowing down top trees for better worst-case compression. In: Proceedings of the Annual Symposium on Combinatorial Pattern Matching, CPM 2018, volume 105 of LIPIcs, pp. 16:1–16:8. Schloss Dagstuhl - Leibniz-Zentrum für Informatik (2018)
11. Ganardi, M., Jeż, A., Lohrey, M.: Balancing straight-line programs. In: Proceedings of the 60th IEEE Annual Symposium on Foundations of Computer Science, FOCS 2019, pp. 1169–1183. IEEE Computer Society (2019)
12. Ganardi, M., Jeż, A., Lohrey, M.: Balancing straight-line programs. CoRR, abs/1902.03568 (2019)
13. Gascón, A., Lohrey, M., Maneth, S., Reh, C.P., Sieber, K.: Grammar-based compression of unranked trees. In: Fomin, F.V., Podolskii, V.V. (eds.) CSR 2018. LNCS, vol. 10846, pp. 118–131. Springer, Cham (2018). https://doi.org/10.1007/978-3-319-90530-3_11
14. Gasieniec, L., Kolpakov, R.M., Potapov, I., Sant, P.: Real-time traversal in grammar-based compressed files. In: Proceedings of the 2005 Data Compression Conference, DCC 2005, p. 458. IEEE Computer Society (2005)
15. Gawrychowski, P., Jo, S., Mozes, S., Weimann, O.: Compressed range minimum queries. CoRR, abs/1902.04427 (2019)
16. Hübschle-Schneider, L., Raman, R.: Tree compression with top trees revisited. In: Bampis, E. (ed.) SEA 2015. LNCS, vol. 9125, pp. 15–27. Springer, Cham (2015). https://doi.org/10.1007/978-3-319-20086-6_2
17. Tomohiro, I., et al.: Detecting regularities on grammar-compressed strings. Inf. Comput. **240**, 74–89 (2015)
18. Jeż, A.: Approximation of grammar-based compression via recompression. Theor. Comput. Sci. **592**, 115–134 (2015)
19. Jeż, A.: A really simple approximation of smallest grammar. Theor. Comput. Sci. **616**, 141–150 (2016)
20. Lohrey, M.: Algorithmics on SLP-compressed strings: a survey. Groups Complex. Cryptol. 4(2), 241–299 (2012)
21. Lohrey, M.: Grammar-based tree compression. In: Potapov, I. (ed.) DLT 2015. LNCS, vol. 9168, pp. 46–57. Springer, Cham (2015). https://doi.org/10.1007/978-3-319-21500-6_3
22. Lohrey, M., Maneth, S., Mennicke, R.: XML tree structure compression using RePair. Inf. Syst. **38**(8), 1150–1167 (2013)
23. Lohrey, M., Maneth, S., Reh, C.P.: Constant-time tree traversal and subtree equality check for grammar-compressed trees. Algorithmica **80**(7), 2082–2105 (2018)
24. Rytter, W.: Application of Lempel-Ziv factorization to the approximation of grammar-based compression. Theor. Comput. Sci. **302**(1–3), 211–222 (2003)

Two-Dimensional Codes

Maria Madonia[✉]

Dipartimento di Matematica e Informatica, Università di Catania,
Viale Andrea Doria 6/a, 95125 Catania, Italy
madonia@dmi.unict.it

Extended Abstract

The theory of codes of strings takes its origin in the theory of information developed by Shannon in the 1950s. Since then, it has evolved in several different directions. Among them, we can mention the theory of entropy, a branch of probability theory, and the theory of error-correcting codes, more related to commutative algebra. Due to the nature of the involved concepts, the theory of codes exhibits both theoretical and practical features. The former are related to combinatorics on words, automata theory and formal languages (see [12] for some classical references), while the latter apply in finding efficient methods for transmitting and storing data.

In the 1960s, the interest in image processing, pattern recognition and pattern matching motivated the research on families of matrices with entries taken from a finite alphabet, as a two-dimensional counterpart of strings. Significant work has been done to transfer formalisms and results from formal language theory to a two-dimensional setting (see for example [2, 7, 13, 19–21, 30]). Several classes of two-dimensional objects have been introduced and investigated, namely polyominoes, labelled polyominoes, directed polyominoes, and rectangular labelled polyominoes. This paper focusses on this last kind of polyominoes, which we will refer to as *pictures*.

Recently, we are seeing a renewed interest in two-dimensional languages in different frameworks (see for example [15–18, 23, 28, 29, 31]). The motivations of this paper are mainly theoretical. Nevertheless, as formal language theory had very significant impact in several applications, we do not exclude that theoretical results on two-dimensional languages may be exploited for practical applications. Besides researchers who investigate open questions in the aforementioned fields of image processing, pattern recognition and pattern matching, researchers in other scientific areas are interested in the investigation on pictures. Please note that some families of picture languages are of particular meaning in physics; as a matter of fact, they represent the evolution of noteworthy discrete systems (see [25]). Moreover, the recognizability of picture languages by finite models is connected to the study of some properties of symbolic dynamical systems (see [24]).

Partially supported by INdAM-GNCS Project 2019 and CREAMS Project of University of Catania.

M. Anselmo et al. (Eds.): CiE 2020, LNCS 12098, pp. 301–305, 2020.
https://doi.org/10.1007/978-3-030-51466-2_27

Extending results from formal language theory to two dimensions is often a non-trivial task, and sometimes a very challenging one. The two-dimensional structure may give rise to new problems, even in some basic concepts (see [3,20]). As an example, the generalization of the classical operation of string concatenation to two dimensions, leads to the definition of two different operations between pictures, the horizontal and the vertical concatenations. Differently from the string case, these concatenations are partial operations and, as a very remarkable aspect, they do not induce a monoid structure on the set of all pictures over a given alphabet. Please note that the monoid structure of the set of all strings over a given alphabet has played an important role in the theory of formal languages. Another basic definition on strings, the one of prefix of a string, opens new scenarios when generalized to pictures. It can be extended to pictures in a natural way; a prefix of a picture is a rectangular portion in the top-left corner of the picture. Nevertheless, if a prefix is removed from a picture, the remaining part is no longer a picture. On the contrary, if a prefix is deleted from a string, the remaining part is still a string and this is a big advantage.

In the literature, we find several attempts to generalize the notion of code to two-dimensional objects. A set C of two-dimensional objects over a given alphabet Σ, is a code if every two-dimensional object over Σ can be tiled without holes or overlaps in at most one way with copies of elements of C. Most of the results show that we lose important properties when moving from one to two dimensions. A major result due to D. Beauquier and M. Nivat states that the problem whether a finite set of polyominoes is a code is undecidable; the same result also holds for dominoes (see [11]). Some particular cases have been studied in [1]. Codes of directed polyominoes with respect to some concatenation operation are considered in [22]; some special decidable cases are detected. Codes of labeled polyominoes, called bricks, as well as codes of directed figures are studied in [27] and in [26], respectively. In these papers, further undecidability results are proved. Doubly-ranked monoids are introduced in [14] with the aim of extending syntactic properties to two dimensions; in this framework a notion of picture code is introduced and studied. More recently, non-overlapping codes have been considered in [10].

In this paper, we consider the definition of *picture code* that has been introduced in [3]. Here, the pictures are composed using the operation of tiling star defined in [30]. The tiling star of a set of pictures X is the set X^{**} of all pictures that are tilable by elements of X, i.e. all pictures that can be covered by pictures of X without holes or overlapping. Then, X is a code if any picture in X^{**} is tilable in a unique way. Again, it is not decidable whether a finite language of pictures is a code. Consequently, it is a goal to find decidable subclasses of picture codes. Taking the families of codes of strings as starting point, many families of two-dimensional codes have been introduced (see [3–6,8,9]). The investigation mainly focussed on their combinatorial properties, their construction, their decidability, and the notions of maximality and completeness in this setting.

The generalization of the notion of prefix code of strings to two-dimensions leads to two different definitions of codes, namely the prefix and the strong prefix codes of pictures. In these definitions, pictures must be considered in relation to a given scanning direction; e.g. from the top-left corner toward the bottom-right one. The families of prefix and strong prefix picture codes inherit several properties from the family of prefix string codes. For example, the results in [5] present a recursive procedure to construct all finite maximal strong prefix codes of pictures, starting from the singleton pictures, i.e. those which contain only one alphabet symbol. This construction generalizes the tree representation of prefix codes of strings (cf. [12]). Subsequently, the codes of pictures with finite deciphering delay have been introduced, in analogy to the case of strings. More recently, three new classes of picture codes have been introduced in [8,9]: the comma-free, the cylindric and the toroidal codes. They extend the notions of comma-free and circular code of strings to two dimensions. Again, the generalization can be achieved in more than one way. Notably, these definitions share the property to be "non-oriented", in the sense that they do not require to set a specific scanning direction.

In this paper, we shall introduce some of the aforementioned families of picture codes and discuss their properties. Classes of two-dimensional codes will be compared each other and with their one-dimensional counterpart. The results highlight new scenarios which also help one in understanding some hidden features in the one-dimensional setting.

References

1. Aigrain, P., Beauquier, D.: Polyomino tilings, cellular automata and codicity. Theor. Comput. Sci. **147**, 165–180 (1995)
2. Anselmo, M., Giammarresi, D., Madonia, M.: Deterministic and unambiguous families within recognizable two-dimensional languages. Fundam. Inform. **98**(2–3), 143–166 (2010)
3. Anselmo, M., Giammarresi, D., Madonia, M.: Prefix picture codes: a decidable class of two-dimensional codes. Int. J. Found. Comput. Sci. **25–8**, 1017–1032 (2014)
4. Anselmo, M., Giammarresi, D., Madonia, M.: Picture codes and deciphering delay. Inf. Comput. **253**, 358–370 (2017)
5. Anselmo, M., Giammarresi, D., Madonia, M.: Structure and properties of strong prefix codes of pictures. Math. Struct. Comput. Sci. **27**, 123–142 (2017)
6. Anselmo, M., Giammarresi, D., Madonia, M.: Full sets of pictures to encode pictures. Theor. Comput. Sci. **777**, 55–68 (2019)
7. Anselmo, M., Giammarresi, D., Madonia, M., Restivo, A.: Unambiguous recognizable two-dimensional languages. RAIRO: Theor. Inform. Appl. **40**(2), 277–293 (2006)
8. Anselmo, M., Madonia, M.: Two-dimensional comma-free and cylindric codes. Theor. Comput. Sci. **658**, 4–17 (2017)
9. Anselmo, M., Madonia, M., Selmi, C.: Toroidal codes and conjugate pictures. In: Martín-Vide, C., Okhotin, A., Shapira, D. (eds.) LATA 2019. LNCS, vol. 11417, pp. 288–301. Springer, Cham (2019). https://doi.org/10.1007/978-3-030-13435-8_21
10. Barcucci, E., Bernini, A., Bilotta, S., Pinzani, R.: A 2D non-overlapping code over a q-ary alphabet. Crypt. Commun. **10**(4), 667–683 (2018)
11. Beauquier, D., Nivat, M.: A codicity undecidable problem in the plane. Theor. Comput. Sci. **303**, 417–430 (2003)

12. Berstel, J., Perrin, D., Reutenauer, C.: Codes and Automata. Cambridge University Press, Cambridge (2009)
13. Blum, M., Hewitt, C.: Automata on a two-dimensional tape. In: IEEE Symposium on Switching and Automata Theory, pp. 155–160 (1967)
14. Bozapalidis, S., Grammatikopoulou, A.: Picture codes. ITA **40**(4), 537–550 (2006)
15. Charlier, É., Puzynina, S., Vandomme, É.: Recurrence in multidimensional words. In: Martín-Vide, C., Okhotin, A., Shapira, D. (eds.) LATA 2019. LNCS, vol. 11417, pp. 397–408. Springer, Cham (2019). https://doi.org/10.1007/978-3-030-13435-8_29
16. Dennunzio, A., Formenti, E., Manzoni, L., Margara, L., Porreca, A.E.: On the dynamical behaviour of linear higher-order cellular automata and its decidability. Inf. Sci. **486**, 73–87 (2019)
17. Fernau, H., Paramasivan, M., Schmid, M.L., Thomas, D.G.: Simple picture processing based on finite automata and regular grammars. J. Comput. Syst. Sci. **95**, 232–258 (2018)
18. Gamard, G., Richomme, G., Shallit, J., Smith, T.J.: Periodicity in rectangular arrays. Inf. Process. Lett. **118**, 58–63 (2017)
19. Giammarresi, D., Restivo, A.: Recognizable picture languages. Int. J. Pattern Recogn. Artif. Intell. **6**(2–3), 241–256 (1992)
20. Giammarresi, D., Restivo, A.: Two-dimensional languages. In: Rozenberg, G., Salomaa, A. (eds.) Handbook of Formal Languages, vol. III, pp. 215–268. Springer, Heidelberg (1997). https://doi.org/10.1007/978-3-642-59126-6_4
21. Kari, J., Salo, V.: A survey on picture-walking automata. In: Kuich, W., Rahonis, G. (eds.) Algebraic Foundations in Computer Science. LNCS, vol. 7020, pp. 183–213. Springer, Heidelberg (2011). https://doi.org/10.1007/978-3-642-24897-9_9
22. Kolarz, M., Moczurad, W.: Multiset, set and numerically decipherable codes over directed figures. In: Arumugam, S., Smyth, W.F. (eds.) IWOCA 2012. LNCS, vol. 7643, pp. 224–235. Springer, Heidelberg (2012). https://doi.org/10.1007/978-3-642-35926-2_25
23. Kulkarni, M.S., Mahalingam, K.: Two-dimensional palindromes and their properties. In: Drewes, F., Martín-Vide, C., Truthe, B. (eds.) LATA 2017. LNCS, vol. 10168, pp. 155–167. Springer, Cham (2017). https://doi.org/10.1007/978-3-319-53733-7_11
24. Lind, D., Marcus, B.: An Introduction to Symbolic Dynamics and Coding. Cambridge University Press, Cambridge (1995)
25. Lindgren, K., Moore, C., Nordahl, M.: Complexity of two-dimensional patterns. J. Stat. Phys. **91**(5–6), 909–951 (1998)
26. Moczurad, W.: Decidability of multiset, set and numerically decipherable directed figure codes. Discrete Math. Theor. Comput. Sci. **19**(1) (2017). Article no 11
27. Moczurad, M., Moczurad, W.: Some open problems in decidability of brick (labelled polyomino) codes. In: Chwa, K.-Y., Munro, J.I.J. (eds.) COCOON 2004. LNCS, vol. 3106, pp. 72–81. Springer, Heidelberg (2004). https://doi.org/10.1007/978-3-540-27798-9_10
28. Mráz, F., Průša, D., Wehar, M.: Two-dimensional pattern matching against basic picture languages. In: Hospodár, M., Jirásková, G. (eds.) CIAA 2019. LNCS, vol. 11601, pp. 209–221. Springer, Cham (2019). https://doi.org/10.1007/978-3-030-23679-3_17
29. Otto, F., Mráz, F.: Deterministic ordered restarting automata for picture languages. Acta Inform. **52**, 593–623 (2015). https://doi.org/10.1007/s00236-015-0230-5

30. Simplot, D.: A characterization of recognizable picture languages by tilings by finite sets. Theor. Comput. Sci. **218**(2), 297–323 (1991)
31. Terrier, V.: Communication complexity tools on recognizable picture languages. Theor. Comput. Sci. **795**, 194–203 (2019)

Formal Languages in Information Extraction and Graph Databases

Wim Martens$^{(\boxtimes)}$

University of Bayreuth, Bayreuth, Germany
wim.martens@uni-bayreuth.de

This abstract covers two areas of data management research in which formal language theory plays a central role, namely in Information Extraction and Graph Databases.

Information Extraction. Automata-based foundations of Information Extraction (e.g., [9,16,34]) have become a popular research topic over the last years. One framework that has been studied in this context is that of *document spanners* [16]. Document spanners model information extraction tasks as functions that map input text documents to a relation of *spans*, i.e., intervals of start and end positions in the text. A particular interesting class of spanners is the class of regular spanners, which is based on regular languages with capture variables. This class satisfies a number of interesting complexity and expressiveness properties and therefore caused a revival of automata- and formal language techniques in database research. Examples of such work are on the enumeration of answers [1,18], expressiveness [20,21,33], complexity issues [22,29,32], integration of weights [15], and distributed evaluation [14]. That said, the document spanners framework is not the only one that is studied in this context, and there are other elegant frameworks that can express information extraction functions beyond the spanner framework, e.g., [9,34].

Graph Databases. Formal languages have played a central role in Graph Databases since the SIGMOD 1987 paper of Cruz et al. [12], which is one of the first and most influential papers on the topic. Indeed, this paper introduced regular expressions for querying paths (later named *regular path queries* or *RPQs*), which are still used in graph query languages today [13,19,24]. This early work on Graph Databases only allowed RPQs to match *simple paths* in graphs, i.e., paths without repeated nodes. However, after discovering that this restriction already makes simple queries difficult to evaluate [30], it was largely abandoned by the research community, and a huge body of research on fundamental problems followed, in which RPQs were allowed to match all paths. This line of work is too extensive to discuss here, but its state until 2013 is nicely surveyed by Barceló [7]. It still produces high-quality and exciting results today (e.g., [8,17]).

W. Martens—Supported by grant MA 4938/4–1 from the Deutsche Forschungsgemeinschaft and grant I-1502-407.6/2019 of the German-Israeli Foundation for Scientific Research and Development.

M. Anselmo et al. (Eds.): CiE 2020, LNCS 12098, pp. 306–309, 2020.
https://doi.org/10.1007/978-3-030-51466-2_28

Perhaps ironically, the *simple paths* and the similar *trail* restriction (which only allows paths without repeated edges) resurfaced on the systems side of graph databases. Indeed, an early incarnation[1] of SPARQL 1.1 [24] used (a variant of) the simple path restriction, whereas the default semantics of Cypher [13] uses the trail restriction. This new development on the practical side of graph query languages motivated several research groups to build a scientific basis that can be used to guide the design of RPQs in graph query languages [5,6,26,27]. Furthermore, it seems that, in order to understand RPQ evaluation in practical graph query languages, it is very useful to combine fundamental research with query log analysis [10,11,28].

To conclude, it seems that the research communities' efforts to connect theory and practice (which go far beyond what I've been able to mention here, see, e.g. [2–4,25,31,35]) are paying off, in the sense that we are now experiencing an increased interaction between researchers and practitioners in the *Graph Query Language (GQL) Standard* initiative [23] and the process that has led to it.[2] The GQL initiative was recently inaugurated as an official ISO project that aims at becoming an international standard for graph database querying. Furthermore, the story does not stop here at all—a large number of initiatives is currently brainstorming on next-generation logical foundations of graph databases and their query languages, schema languages for graphs, etc.

References

1. Amarilli, A., Bourhis, P., Mengel, S., Niewerth, M.: Constant-delay enumeration for nondeterministic document spanners. In: International Conference on Database Theory (ICDT), pp. 22:1–22:19 (2019)
2. Angles, R., et al.: G-CORE: a core for future graph query languages. In: International Conference on Management of Data (SIGMOD), pp. 1421–1432 (2018)
3. Angles, R., Arenas, M., Barceló, P., Hogan, A., Reutter, J.L., Vrgoc, D.: Foundations of modern query languages for graph databases. ACM Comput. Surv. **50**(5), 68:1–68:40 (2017)
4. Angles, R., et al.: The linked data benchmark council: a graph and RDF industry benchmarking effort. SIGMOD Rec. **43**(1), 27–31 (2014)
5. Arenas, M., Conca, S., Pérez, J.: Counting beyond a yottabyte, or how SPARQL 1.1 property paths will prevent adoption of the standard. In: Proceedings of the World Wide Web Conference (WWW), pp. 629–638 (2012)
6. Bagan, G., Bonifati, A., Groz, B.: A trichotomy for regular simple path queries on graphs. In: ACM Symposium on Principles of Database Systems (PODS), pp. 261–272 (2013)
7. Barceló, P.: Querying graph databases. In: ACM Symposium on Principles of Database Systems (PODS), pp. 175–188 (2013)
8. Barceló, P., Figueira, D., Romero, M.: Boundedness of conjunctive regular path queries. In: International Colloquium on Automata, Languages, and Programming (ICALP), pp. 104:1–104:15 (2019)

---

[1] https://www.w3.org/TR/2012/WD-sparql11-query-20120105, see the definition of *ZeroOrMorePath*.

[2] https://www.gqlstandards.org/existing-languages has an influence diagram.

9. Beedkar, K., Gemulla, R., Martens, W.: A unified framework for frequent sequence mining with subsequence constraints. ACM Trans. Database Syst. **44**(3), 11:1–11:42 (2019)

10. Bonifati, A., Martens, W., Timm, T.: Navigating the maze of Wikidata query logs. In: The World Wide Web Conference (WWW), pp. 127–138 (2019)

11. Bonifati, A., Martens, W., Timm, T.: An analytical study of large SPARQL query logs. VLDB J. (2020, to appear)

12. Cruz, I.F., Mendelzon, A.O., Wood, P.T.: A graphical query language supporting recursion. In: ACM Special Interest Group on Management of Data (SIGMOD), pp. 323–330 (1987)

13. Cypher Query Language. https://neo4j.com/cypher-graph-query-language/

14. Doleschal, J., Kimelfeld, B., Martens, W., Nahshon, Y., Neven, F.: Split-correctness in information extraction. In: ACM Symposium on Principles of Database Systems (PODS), pp. 149–163 (2019)

15. Doleschal, J., Kimelfeld, B., Martens, W., Peterfreund, L.: Weight annotation in information extraction. In: International Conference on Database Theory (ICDT) (2020, to appear)

16. Fagin, R., Kimelfeld, B., Reiss, F., Vansummeren, S.: Document spanners: a formal approach to information extraction. J. ACM **62**(2), 12:1–12:51 (2015)

17. Figueira, D.: Containment of UC2RPQ: the hard and easy cases. In: International Conference on Database Theory (ICDT) (2020, to appear)

18. Florenzano, F., Riveros, C., Ugarte, M., Vansummeren, S., Vrgoc, D.: Constant delay algorithms for regular document spanners. In: ACM Symposium on Principles of Database Systems (PODS), pp. 165–177 (2018)

19. Francis, N., et al.: Cypher: an evolving query language for property graphs. In: International Conference on Management of Data (SIGMOD), pp. 1433–1445 (2018)

20. Freydenberger, D.D.: A logic for document spanners. In: International Conference on Database Theory (ICDT), pp. 13:1–13:18 (2017)

21. Freydenberger, D.D., Holldack, M.: Document spanners: from expressive power to decision problems. In: International Conference on Database Theory (ICDT), pp. 17:1–17:17 (2016)

22. Freydenberger, D.D., Kimelfeld, B., Peterfreund, L.: Joining extractions of regular expressions. In: ACM Symposium on Principles of Database Systems (PODS), pp. 137–149 (2018)

23. GQL Standard. https://www.gqlstandards.org/

24. Harris, S., Seaborne, A.: SPARQL 1.1 Query Language (2013). https://www.w3.org/TR/sparql11-query

25. Libkin, L., Martens, W., Vrgoc, D.: Querying graphs with data. J. ACM **63**(2), 14:1–14:53 (2016)

26. Losemann, K., Martens, W.: The complexity of regular expressions and property paths in SPARQL. ACM Trans. Database Syst. **38**(4), 24:1–24:39 (2013)

27. Martens, W., Niewerth, M., Trautner, T.: A trichotomy for regular trail queries. In: Annual Symposium on Theoretical Aspects of Computer Science (STACS) (2020, to appear)

28. Martens, W., Trautner, T.: Dichotomies for evaluating simple regular path queries. ACM Trans. Database Syst. **44**(4), 16:1–16:46 (2019). Article 16

29. Maturana, F., Riveros, C., Vrgoc, D.: Document spanners for extracting incomplete information: expressiveness and complexity. In: ACM Symposium on Principles of Database Systems (PODS), pp. 125–136 (2018)

30. Mendelzon, A.O., Wood, P.T.: Finding regular simple paths in graph databases. SIAM J. Comput. **24**(6), 1235–1258 (1995)
31. Pérez, J., Arenas, M., Gutiérrez, C.: Semantics and complexity of SPARQL. ACM Trans. Database Syst. **34**(3), 16:1–16:45 (2009)
32. Peterfreund, L., Freydenberger, D.D., Kimelfeld, B., Kröll, M.: Complexity bounds for relational algebra over document spanners. In: ACM Symposium on Principles of Database Systems (PODS), pp. 320–334 (2019)
33. Peterfreund, L., ten Cate, B., Fagin, R., Kimelfeld, B.: Recursive programs for document spanners. In: International Conference on Database Theory (ICDT), pp. 13:1–13:18 (2019)
34. Renz-Wieland, A., Bertsch, M., Gemulla, R.: Scalable frequent sequence mining with flexible subsequence constraints. In: IEEE International Conference on Data Engineering (ICDE), pp. 1490–1501 (2019)
35. Reutter, J.L., Romero, M., Vardi, M.Y.: Regular queries on graph databases. In: International Conference on Database Theory (ICDT), pp. 177–194 (2015)

On the Perceptron's Compression

Shay Moran[1], Ido Nachum[2], Itai Panasoff[3], and Amir Yehudayoff[3(✉)]

[1] Google Brain, Princeton, NJ, USA
[2] EPFL, Lausanne, Switzerland
[3] Department of Mathematics, Technion-IIT, Haifa, Israel
amir.yehudayoff@gmail.com

Abstract. We study and provide exposition to several phenomena that are related to the perceptron's compression. One theme concerns modifications of the perceptron algorithm that yield better guarantees on the margin of the hyperplane it outputs. These modifications can be useful in training neural networks as well, and we demonstrate them with some experimental data. In a second theme, we deduce conclusions from the perceptron's compression in various contexts.

Keywords: Machine learning · Compression · Convex separation

1 Introduction

The perceptron is an abstraction of a biological neuron that was introduced in the 1950's by Rosenblatt [31], and has been extensively studied in many works (see e.g. the survey [27]). It receives as input a list of real numbers (various electrical signals in the biological case) and if the weighted sum of its input is greater than some threshold it outputs 1 and otherwise -1 (it fires or not in the biological case).

Formally, a perceptron computes a function of the form $\mathsf{sign}(w \cdot x - b)$ where $w \in \mathbb{R}^d$ is the weight vector, $b \in \mathbb{R}$ is the threshold, $\cdot$ is the standard inner product, and $\mathsf{sign} : \mathbb{R} \to \{\pm 1\}$ is 1 on the non-negative numbers. It is only capable of representing binary functions that are induced by partitions of $\mathbb{R}^d$ by hyperplanes.

Definition 1. *A map $Y : \mathcal{X} \to \{\pm 1\}$ over a finite set $\mathcal{X} \subset \mathbb{R}^d$ is (linearly)[1] separable if there exists $w \in \mathbb{R}^d$ such that $\mathsf{sign}(w \cdot x) = Y(x)$ for all*

---

[1] We focus on the linear case, when the threshold is 0. A standard lifting that adds a coordinate with 1 to every vector allows to translate the general (affine) case to the linear case. This lifting may significantly decrease the margin; e.g., the map Y on $\mathcal{X} = \{999, 1001\} \subset \mathbb{R}$ defined by $Y(999) = 1$ and $Y(1001) = -1$ has margin 1 in the affine sense, but the lift to $(999, 1)$ and $(1001, 1)$ in $\mathbb{R}^2$ yields very small margin in the linear sense. This solution may therefore cause an unnecessary increase in running time. This tax can be avoided, for example, if one has prior knowledge of $R = \max_{x \in \mathcal{X}} \|x\|$. In this case, setting the last coordinate to be R does not significantly decrease the margin. In fact, it can be avoided without any prior knowledge using the ideas in Algorithm 3 below.

A. Yehudayoff—Supported by ISF grant 1162/15.

M. Anselmo et al. (Eds.): CiE 2020, LNCS 12098, pp. 310–325, 2020.
https://doi.org/10.1007/978-3-030-51466-2_29

$x \in \mathcal{X}$. When the Euclidean norm of w is $\|w\| = 1$, the number $\mathsf{marg}(w, Y) = \min_{x \in \mathcal{X}} Y(x)w \cdot x$ is the margin of w with respect to Y. The number $\mathsf{marg}(Y) = \sup_{w \in \mathbb{R}^d : \|w\| = 1} \mathsf{marg}(w, Y)$ is the margin of Y. We call Y an ε-partition if its margin is at least ε.

Variants of the perceptron (neurons) are the basic building blocks of general neural networks. Typically, the sign function is replaced by some other activation function (e.g., sigmoid or rectified linear unit $\mathsf{ReLu}(z) = \max\{0, z\}$). Therefore, studying the perceptron and its variants may help in understanding neural networks, their design and their training process.

Overview. In this paper, we provide some insights into the perceptron's behavior, survey some of the related work, deduce some geometric applications, and discuss their usefulness in other learning contexts. Below is a summary of our results and a discussion of related work, partitioned to five parts numbered (i) to (v). Each of the results we describe highlights a different aspect of the perceptron's compression (the perceptron's output is a sum of small subset of examples). For more details, definitions and references, see the relevant sections.

(i) Variants of the perceptron (Sect. 2). The well-known perceptron algorithm (see Algorithm 1 below) is guaranteed to find a separating hyperplane in the linearly separable case. However, there is no guarantee on the hyperplane's margin compared to the optimal margin ε^*. This problem was already addressed in several works, as we now explain (see also references within). The authors of [9] and [23] defined a variant of the perceptron that yields a margin of the form $\Omega(\varepsilon^*/R^2)$; see Algorithm 2 below. The authors of [10] defined the passive-aggressive perceptron algorithms that allow e.g. to deal with noise, but provided no guarantee on the margin of the output. The authors of [22] defined a variant of the perceptron that yields provable margin under the assumption that a lower bound on the optimal margin is known. The author of [17] designed the ALMA algorithm and showed that it provides almost optimal margin under the assumption that the samples lie on the unit sphere. It is worth noting that normalizing the examples to be on the unit sphere may significantly alter the margin, and even change the optimal separating hyperplane. The author of [37] defined the minimal overlap algorithm which guarantees optimal margin but is not online since it knows the samples in advance. Finally, the authors of [36] analyzed gradient descent for a single neuron and showed convergence to the optimal separating hyperplane under certain assumptions (appropriate activation and loss functions). We provide two new ideas that improve the learning process. One that adaptively changes the "scale" of the problem and by doing so improves the guarantee on the margin of the output (Algorithm 3), and one that yields almost optimal margin (Algorithms 4).

(ii) Applications for neural networks (Sect. 3). Our variants of the perceptron algorithm are simple to implement, and can therefore be easily applied in the training process of general neural networks. We validate their benefits by training a basic neural network on the MNIST dataset.

(iii) Convex separation (Sect. 4). We use the perceptron's compression to prove a sparse separation lemma for convex bodies. This perspective also suggests a different proof of Novikoff's theorem on the perceptron's convergence [30]. In addition, we interpret this sparse separation lemma in the language of game theory as yielding sparse strategies in a related zero-sum game.

(iv) Generalization bounds (Sect. 5). An important aspect of a learning algorithm is its generalization capabilities; namely, its error on new examples that are independent of the training set (see the textbook [33] for background and definitions). We follow the theme of [18], and observe that even though the (original) perceptron algorithm does not yield an optimal hyperplane, it still generalizes.

(v) Robust concepts (Sect. 6). The robust concepts theme presented by Arriaga and Vempala [3] suggests focusing on well-separated data. We notice that the perceptron fits well into this framework; specifically, that its compression yields efficient dimension reductions. Similar dimension reductions were used in several previous works (e.g. [3–6,16,21]).

Summary. In parts (i)–(ii) we provide a couple of new ideas for improving the training process and explain their contribution in the context of previous work. In part (iii) we use the perceptron's compression as a tool for proving geometric theorems. We are not aware of previous works that studied this connection. Parts (iv)–(v) are mostly about presenting ideas from previous works in the context of the perceptron's compression. We think that parts (iv) and (v) help to understand the picture more fully.

2　Variants of the Perceptron

Deciding how to train a model from a list of input examples is a central consideration in any learning process. In the case of the perceptron algorithm the input examples $(x_1, y_1), (x_2, y_2), \ldots$ with $x_i \in \mathbb{R}^d$ and $y_i \in \{\pm 1\}$ are traversed while maintaining a hypothesis $w^{(t)}$ in a way that reduces the error on the current example:

> **initialize:** $w^{(0)} = \vec{0}$ and $t = 0$
> **while** $\exists i$ *with* $y_i w^{(t)} \cdot x_i \leq 0$ **do**
> $\quad \mid \quad w^{(t+1)} = w^{(t)} + y_i x_i$
> $\quad \mid \quad t = t + 1$
> **end**
> return $w^{(t)}$

Algorithm 1: The perceptron algorithm

The perceptron algorithm terminates whenever its input sample is linearly separable, in which case its output represents a separating hyperplane. Novikoff analyzed the number of steps T required for the perceptron to stop as a function of the margin of the input sample [30].

The standard analysis of the perceptron convergence properties uses the optimal separating hyperplane w^* (later in Sect. 4 we present an alternative analysis that does not use it):

$$w^* = \mathsf{argmax}_{w \in \mathbb{R}^d: \|w\|=1} \mathsf{marg}(w, S),$$

where we think of S as the map from $\{x_1, \ldots, x_m\}$ to $\{\pm 1\}$ defined by $x_i \mapsto y_i$.[2] Novikoff's analysis consists of the following two parts. Let $\varepsilon^* = \mathsf{marg}(w^*, S)$ and $R = \max_i \|x_i\|$.

Part I: The Projection Grows Linearly in Time. In each iteration, the projection of $w^{(t)}$ on w^* grows by at least ε^*, since $y_i x_i \cdot w^* \geq \varepsilon^*$. By induction, we get $w^{(t)} \cdot w^* \geq \varepsilon^* t$ for all $t \geq 0$.

Part II: The Norm Grows Sub-Linearly in Time. In each iteration,

$$\|w^{(t)}\|^2 = \|w^{(t-1)}\|^2 + 2y_i x_i \cdot w^{(t-1)} + \|x_i\|^2 \leq \|w^{(t-1)}\|^2 + R^2$$

(the term $2y_i x_i \cdot w^{(t-1)}$ is negative by choice). So by induction $\|w^{(t)}\| \leq R\sqrt{t}$ for all t.

Combining the two parts,

$$1 \geq \frac{w^{(t)} \cdot w^*}{\|w^{(t)}\|\|w^*\|} \geq \frac{\varepsilon^*}{R}\sqrt{t},$$

which implies that the number of iterations of the algorithm is at most $(R/\varepsilon^*)^2$.

As discussed in Sect. 1, Algorithm 1 has several drawbacks. Here we describe some simple ideas that allow to improve it. Below we describe three algorithms, each is followed by a theorem that summarizes its main properties.

In the following, $\mathcal{X} \subset \mathbb{R}^d$ is a finite set, Y is a linear partition, $\varepsilon^* = \mathsf{marg}(Y)$ is the optimal margin, and $R = \max_{x \in \mathcal{X}} \|x\|$ is the maximal norm of a point.

In the first variant that already appeared in [9,23], the suggestion is to replace the condition $y_i w^{(t)} \cdot x_i < 0$ by $y_i w^{(t)} \cdot x_i < \beta$ for some a priori chosen $\beta > 0$. that may change over time. As we will see, different choices of β yield different guarantees.

> **initialize:** $w^{(0)} = \vec{0}$ and $t = 0$
> **while** $\exists i$ *with* $y_i w^{(t)} \cdot x_i < \beta$ **do**
> $\quad \mid \quad w^{(t+1)} = w^{(t)} + y_i x_i$
> $\quad \mid \quad t = t + 1$
> **end**
> return $w^{(t)}$

Algorithm 2: The β-perceptron algorithm

---

[2] We assume that S is consistent with a function (does not contain identical points with opposite labels).

Theorem 1 ([9,23]). *The β-perceptron algorithm performs at most $\frac{2\beta+R^2}{(\varepsilon^*)^2}$ updates and achieves a margin of at least $\frac{\beta\varepsilon^*}{2\beta+R^2}$.*

Proof. We only replaced the ≤ 0 condition in the while loop by a $< \beta$ condition, for some $\beta > 0$. As before, by induction

$$\|w^{(t)}\|^2 = \|w^{(t-1)}\|^2 + 2y_i x_i \cdot w^{(t-1)} + \|x_i\|^2 \leq (2\beta + R^2)t$$

and

$$1 \geq \frac{w^{(t)} \cdot w^*}{\|w^{(t)}\|\|w^*\|} \geq \frac{\varepsilon^*}{\sqrt{2\beta + R^2}}\sqrt{t}$$

where $R = \max_i \|x_i\|$. The number of iterations is thus at most $\frac{2\beta+R^2}{(\varepsilon^*)^2}$. In addition, by choice, for all i,

$$y_i w^{(t)} \cdot x_i \geq \beta.$$

So, since

$$\|w^{(t)}\| \leq \sqrt{(2\beta + R^2)t} \leq \frac{2\beta + R^2}{\varepsilon^*},$$

we get

$$\mathsf{marg}(w^{(t)}, S) \geq \frac{\beta\varepsilon^*}{2\beta + R^2}. \qquad \square$$

To remove the dependence on R in the output's margin above, we propose to rescale β according to the observed examples.

initialize: $w^{(0)} = \vec{0}$ and $t = 0$ and $\beta = 0$
while $\exists i$ with $y_i w^{(t)} \cdot x_i \leq \beta$ **do**
  $\quad w^{(t+1)} = w^{(t)} + y_i x_i$
  $\quad t = t + 1$
  $\quad$**if** $\beta < \|x_i\|^2$ **then**
  $\quad\quad \beta = 4\|x_i\|^2$
  $\quad$**end**
end
return $w^{(t)}$

Algorithm 3: The R-independent perceptron algorithm

Theorem 2. *The R-independent perceptron algorithm performs at most $\frac{10R^2}{(\varepsilon^*)^2}$ updates and achieves a margin of at least $\frac{\varepsilon^*}{3}$.*

Proof. This version of the algorithm guarantees a margin of $\varepsilon^*/3$ coupled with a running time comparable to the original algorithm *without* knowing R. Indeed, to bound the running time, observe that before a change in β occurs, there could be at most $\frac{2\beta+R^2}{\varepsilon^2}$ errors (as before for the relevant β and R). The amount of changes in β is at most $\lceil \log(R/r) \rceil$, where $r = \min_i \|x_i\|$. The overall running time is at most

$$\sum_{k=1}^{\lceil \log(R/r) \rceil} \frac{2 \cdot 4 |x_{i_k}|^2 + (2 |x_{i_k}|)^2}{(\varepsilon^*)^2} \leq 2 \cdot \sum_{k=1}^{\lceil \log(R/r) \rceil} \frac{3 \cdot 4^k r^2}{(\varepsilon^*)^2}$$

$$\leq 6 \cdot 4/3 \cdot 4^{\lceil \log(R/r) \rceil} \frac{r^2}{(\varepsilon^*)^2} = O((R/\varepsilon^*)^2). \quad \Box$$

Finally, if one would like to improve upon the $\frac{\varepsilon^*}{3}$ guarantee, we suggest to change β with time. To run the algorithm, we should first decide how well do we want to approximate the optimal margin. To do so, we need to choose the parameter $\alpha \in (1, 2)$; the closer α is to 2, the better the approximation is (see Theorem 3).

initialize: $w^{(0)} = \vec{0}$ and $t = 0$ and $\beta = 0$ and $\alpha \in (1, 2)$
while $\exists i$ *with* $y_i w^{(t)} \cdot x_i \leq \beta$ **do**
$\quad | \quad w^{(t+1)} = w^{(t)} + y_i x_i$
$\quad | \quad t = t + 1$
$\quad | \quad \beta = 0.5((t+1)^\alpha - t^\alpha - 1)$
end
return $w^{(t)}$

Algorithm 4: The ∞-perceptron algorithm

Theorem 3. *If $R \leq 1$, the ∞-perceptron algorithm performs at most $(1/\varepsilon^*)^{2/(2-\alpha)}$ updates and achieves a margin of at least $\frac{\alpha \epsilon^*}{2}$.*

Proof. For simplicity, we assume here that $R = \max_i \|x_i\| = 1$. The idea is as follows. The analysis of the classical perceptron relies on the fact that $\|w^{(t)}\|^2 \leq t$ in each step. On the other hand, in an "extremely aggressive" version of the perceptron that always updates, one can only obtain a trivial bound $\|w^{(t)}\|^2 \leq t^2$ (as $w^{(t)}$ can be the sum of t unit vectors in the same direction). The update rule in the version below is tailored so that a bound of $\|w^{(t)}\|^2 \leq t^\alpha$ for $\alpha \in (1, 2)$ is maintained.

Here we use that for $t \geq 2$,

$$\|w^{(t)}\|^2 \leq \|w^{(t-1)}\|^2 + (t^\alpha - (t-1)^\alpha - 1) + \|x_i\|^2.$$

By induction, for all $t \geq 0$,

$$\|w^{(t)}\|^2 \leq t^\alpha.$$

This time

$$1 \geq \frac{w^{(t)} \cdot w^*}{\|w^{(t)}\| \|w^*\|} \geq \frac{\varepsilon^* t}{t^{\alpha/2}}.$$

So, the running time is at most $(1/\varepsilon^*)^{2/(2-\alpha)}$.

The output's margin is at least

$$\frac{0.5((t+1)^\alpha - t^\alpha - 1)}{t^{\alpha/2}}. \tag{1}$$

This is a decreasing function for $t > 0$, since its derivative is at most zero.

Since $(t+1)^\alpha - t^\alpha \geq \alpha t^{\alpha-1}$ for $t \geq 0$, the output's margin is at least

$$0.5\alpha \frac{(1/\varepsilon^*)^{2(\alpha-1)/(2-\alpha)} - 1}{(1/\varepsilon^*)^{\alpha/(2-\alpha)}} = 0.5\alpha\varepsilon^* - (\varepsilon^*)^{\alpha/(2-\alpha)}.$$

So we can get arbitrarily close to the true margin by setting $\alpha = 2(1-\delta)$ for some small $0 < \delta < 0.5$ of our choice. This gives margin

$$(1-\delta)\varepsilon^* - (\varepsilon^*)^{(2-\delta)/\delta} \geq \varepsilon^*\left(1 - \delta - (\varepsilon^*)^{1/\delta}\right).$$

The running time, however, becomes $(1/\varepsilon^*)^{1/\delta}$.

When ε^* is very close to 1, the lower bound on the margin above may not be meaningful. We claim that the margin of the output is still close to ε^* even in this case. To see this, let $\tilde{w}$ be a hyperplane with margin $\tilde{\varepsilon} = (1 - \delta \ln(1/\delta))\varepsilon^*$. We can carry the argument above with $\tilde{w}$ instead of w^*, and get that the margin is at least

$$\tilde{\varepsilon}\left(1 - \delta - (\tilde{\varepsilon})^{1/\delta}\right) > (1 - 2\delta - \delta\ln(1/\delta))\varepsilon^*.$$

So we can choose δ small enough, without knowing any information on ε^*, and get an almost optimal margin. □

Remark 1. *The bound on the running time is sharp, as the following example shows. The two points $(\sqrt{1-\varepsilon^2}, \varepsilon), (\sqrt{1-\varepsilon^2}, -\varepsilon)$ with labels $1, -1$ are linearly separated with margin $\Omega(\varepsilon)$. The algorithm stops after $\Omega\left((1/\varepsilon)^{2/(2-\alpha)}\right)$ iterations (if ε is small enough and α close enough to 2).*

Remark 2. *Algorithms 3 and 4 can be naturally combined to a single algorithm that arrives arbitrarily close to the optimal margin without assuming that $R \leq 1$.*

3   Application for Neural Networks

Our results explain some choices that are made in practice, and can potentially help to improve them. Observe that if one applies gradient descent on a neuron of the form $\mathsf{ReLu}(w \cdot x)$ with loss function of the form $\mathsf{ReLu}(\beta - y_x w \cdot x)$ with $\beta = 0$ then one gets the same update rule as in the perceptron algorithm. Choosing $\beta = 1$ corresponds to using the hinge loss to drive the learning process. The fact that $\beta = 1$ yields provable bounds on the output's margin of a single neuron suggests a formal evidence that supports the benefits of the hinge loss.

Moreover, in practice, β is treated as a hyper-parameter and tuning it is a common challenge that needs to be addressed in order to maximize performance. We proposed a couple of new options for choosing and updating β throughout the training process that may contribute towards a more systematic approach for setting β (see Algorithms 4 and 3). Theorems 2 and 3 explain the theoretical advantages of these options in the case of a single neuron.

We also provide some experimental data. Our experiments verify that our suggestions for choosing β can indeed yield better results. We used the MNIST database [24] of handwritten digits as a test case with no preprocessing. We used a simple and standard neural network with one hidden layer consisting of 800/300 neurons and 10 output neurons (the choice of 800 and 300 is the same as in Simard et al. [35] and Lecun et al. [24]). We trained the network by back-propagation (gradient descent). The loss function of each output neuron of the form $\mathsf{ReLu}(w \cdot G(x))$, where $G(x)$ is the output of the hidden layer, is $\mathsf{ReLu}(-y_x w \cdot G(x) + \beta)$ for different β's. This loss function is 0 if w provides a correct and confident (depending on β) classification of x and is linear in $G(x)$ otherwise. This choice updates the network even when the network classifies correctly but with less than β confidence. It has the added value of yielding simple and efficient calculations compared to other choices (like cross entropy or soft-max).[3]

We tested four values of β as shown in Fig. 1. In two tests, the value of β is fixed in time[4] to be 0 and 1. In two tests, β changes with the time t in a sublinear fashion. This choice can be better understood after reading the analysis of Algorithm 4. Roughly speaking, the analysis predicts that β should be of the form t^{1-c} for $c > 0$, and that the smaller c is, the smaller the error will be. This prediction is indeed verified in the experiments; it is evident that choosing β in a time-dependent manner yields better results. For comparison, the last row of the table shows the error of the two-layer MLP of the same size that is driven by the cross-entropy loss [35]. In fact, our network of 300 neurons performed better than all the general purpose networks with 300 neurons even with preprocessing of the data that appear in http://yann.lecun.com/exdb/mnist/ (Fig. 2).

	test error
$\beta = 0$	no convergence
$\beta = 1$	1.5 %
$\beta \approx t^{0.4}$	1.44 %
$\beta \approx t^{0.75}$	1.35 %
cross-entropy [35]	1.6 %

Fig. 1. One hidden layer with 800 neurons

	test error
$\beta \approx t^{0.75}$	1.49 %
mean square error [24]	4.7 %
MSE, [distortions] [24]	3.6 %
deskewing [24]	1.6 %

Fig. 2. One hidden layer with 300 neurons

Finally, a natural suggestion that emerges from our work is to add $\beta > 0$ as a parameter for each individual neuron in the network, and not just to the loss function. Namely, to translate the input to a ReLu neuron by β. The value

---

[3] An additional added value is that with this loss function there is a dichotomy, either an error occurred or not. This dichotomy can be helpful in making decisions throughout the learning process. For example, instead of choosing the batch-size to be of fixed size B, we can choose the batch-size in a dynamic but simple way: just wait until B errors occurred.

[4] Time is measured by the number of updates.

of β may change during the learning process. Figuratively, this can be thought of as "internal clocks" of the neurons. A neuron changes its behavior as time progresses. For example, the "older" the neuron is, the less it is inclined to change.

4   Convex Separation

Linear programming (LP) is a central paradigm in computer science and mathematics. LP duality is a key ingredient in many algorithms and proofs, and is deeply related to von Neumann's minimax theorem that is seminal in game theory [29]. Two related and fundamental geometric properties are Farkas' lemma [12], and the following separation theorem.

Theorem 4 (Convex separation theorem). *For every non empty convex sets $K, L \subset \mathbb{R}^d$, precisely one of the following holds: (i) $\mathrm{dist}(K, L) = \inf\{\|p-q\| : p \in K, q \in L\} = 0$, or (ii) there is a hyperplane separating K and L.*

We observe that the following stronger version of the separation theorem follows from the perceptron's compression (a similar version of Farkas' lemma can be deduced as well).

Lemma 1 (Sparse Separation). *For every non empty convex sets $K, L \subset \mathbb{R}^d$ so that $\sup\{\|p - q\| : p \in K, q \in L\} = 1$ and every $\varepsilon > 0$, one of the following holds:*

1. $\mathrm{dist}(K, L) < \varepsilon$.
2. *There is a hyperplane $H = \{x : w \cdot x = b\}$ separating K from L so that its normal vector is "sparse":*
   (i) $\frac{w \cdot p - b}{\|w\|} > \frac{\varepsilon}{30}$ *for all $p \in K$,*
   (ii) $\frac{w \cdot q - b}{\|w\|} < -\frac{\varepsilon}{30}$ *for all $q \in L$, and*
   (iii) *w is a sum of at most $(10/\varepsilon)^2$ points in K and $-L$.*

Proof. Let K, L be convex sets and $\varepsilon > 0$. For $x \in \mathbb{R}^d$, let $\tilde{x}$ in $\mathbb{R}^{d+1}$ be the same as x in the first d coordinates and 1 in the last (we have $\|\tilde{x}\| \leq \|x\| + 1$). We thus get two convex bodies $\tilde{K}$ and $\tilde{L}$ in $d+1$ dimensions (using the map $x \mapsto \tilde{x}$).

Run Algorithm 2 with $\beta = 1$ on inputs that positively label $\tilde{K}$ and negatively label $\tilde{L}$. This produces a sequence of vectors $w^{(0)}, w^{(1)}, \ldots$ so that $\|w^{(t)}\| \leq \sqrt{6t}$ for all t. For every $t > 0$, the vector $w^{(t)}$ is of the form $w^{(t)} = k^{(t)} - \ell^{(t)}$ where $k^{(t)}$ is a sum of t_1 elements of $\tilde{K}$ and $\ell^{(t)}$ is a sum of t_2 elements of $\tilde{L}$ so that $t_1 + t_2 = t$. In particular, we can write $\frac{1}{t}w^{(t)} = \alpha^{(t)}p^{(t)} - (1 - \alpha^{(t)})q^{(t)}$ for $\alpha^{(t)} \in [0, 1]$ where $p^{(t)} \in \tilde{K}$ and $q^{(t)} \in \tilde{L}$ (note that the last coordinate of $w^{(t)}$ equals $2\alpha^{(t)} - \frac{1}{2}$).

If the algorithm does not terminate after T steps for T satisfying $\sqrt{6/T} < \varepsilon/4$ then it follows that $\|\frac{1}{T}w^{(T)}\| < \varepsilon/4$. In particular, $|\alpha^{(T)} - 1/2| < \varepsilon/8$ and so

$$\frac{\varepsilon}{4} > \|\alpha^{(t)}p^{(t)} - (1 - \alpha^{(t)})q^{(t)}\| > \frac{\|p^{(t)} - q^{(t)}\|}{2} - \frac{\varepsilon}{4},$$

which implies that $\text{dist}(K, L) < \varepsilon$.

In the complementing case, the algorithm stops after $T < (10/\varepsilon)^2$ rounds. Let w be the first d coordinates of $w^{(T)}$ and b be its last coordinate. For all $p \in K$,

$$\frac{w \cdot p + b}{\|w\|} \geq \frac{1}{\|w^{(T)}\|} \geq \frac{1}{\sqrt{6T}} > \frac{\varepsilon}{30}.$$

Similarly, for all $q \in L$ we get $\frac{w \cdot q + b}{\|w\|} < -\frac{\varepsilon}{30}$. $\qquad\qquad\square$

The lemma is strictly stronger than the preceding separation theorem. Below, we also explain how this perspective yields an alternative proof of Novikoff's theorem on the convergence of the perceptron [30]. It is interesting to note that the usual proof of the separation theorem relies on a concrete construction of the separating hyperplane that is geometrically similar to hard-SVMs. The proof using the perceptron, however, does not include any "geometric construction" and yields a sparse and strong separator (it also holds in infinite dimensional Hilbert space, but it uses that the sets are bounded in norm).

Alternative Proof of the Perceptron's Convergence. Assume without loss of generality that all of examples are labelled positively (by replacing x by $-x$ if necessary). Also assume that $R = \max_i \|x_i\| = 1$. As in the proof above, let $w^{(0)}, w^{(1)}, \ldots$ be the sequence of vectors generated by the perceptron (Algorithm 1). Instead of arguing that the projection on w^* grows linearly with t, argue as follows. The vectors $v^{(1)}, v^{(2)}, \ldots$ defined by $v^{(t)} = \frac{1}{t} w^{(t)}$ are in the convex hull of the examples and have norm at most $\|v^{(t)}\| \leq \frac{1}{\sqrt{t}}$. Specifically, for every w of norm 1 we have $v^{(t)} \cdot w \leq \frac{1}{\sqrt{t}}$ and so there is an example x so that $x \cdot w \leq \frac{1}{\sqrt{t}}$. This implies that the running time T satisfies $\frac{1}{\sqrt{T}} \geq \varepsilon^*$ since for every example x we have $x \cdot w^* \geq \varepsilon^*$.

A Game Theoretic Perspective. The perspective of game theory turned out to be useful in several works in learning theory (e.g. [13,28]). The ideas above have a game theoretic interpretation as well. In the associated game there are two players. A Point player whose pure strategies are points v in some finite set $V \subset \mathbb{R}^d$ so that $\max\{\|v\| : v \in V\} = 1$, and a Hyperplane player whose pure strategies are w for $w \in \mathbb{R}^d$ with $\|w\| = 1$. For a given choice of v and w, the Hyperplane player's payoff is of $P(v, w) = v \cdot w$ coins (if this number is negative, then the Hyperplane player pays the Point player). The goal of the Point player is thus to minimize the amount of coins she pays. A mixed strategy of the Point player is a distribution μ on V, and of the Hyperplane player is a (finitely supported) distribution κ on $\{w : \|w\| = 1\}$. The expected gain is

$$P(\mu, \kappa) = \mathbb{E}_{(v,w) \sim \mu \times \kappa} P(v, w).$$

Claim (Sparse Strategies). Let ε^* be the minimax value of the game:

$$\varepsilon^* = \sup_\kappa \inf_\mu P(\mu, \kappa) \geq 0.$$

There is $T \geq \frac{1}{3(\varepsilon^*)^2}$ (if $\varepsilon^* = 0$ then $T = \infty$) and a sequence of mixed strategies $\mu_1, \mu_2, \ldots, \mu_T$ of the Point player so that for all $t \leq T$, the support size of μ_t is at most t and for every mixed strategy κ of the Hyperplane player,

$$P(\mu_t, \kappa) \leq \sqrt{3/t}.$$

Proof. Let $v^{(t)} = \frac{1}{t} w^{(t)}/t$ be as in the proof of Lemma 1 above, when we replace K by V and L by $\emptyset$. We can interpret $v^{(t)}$ as a mixed-strategy μ_t of the Point player (the uniform distribution over some multi-subset of V of size t). Specifically, for every κ and $t > 0$,

$$P(\mu_t, \kappa) = \mathbb{E}_{w \sim \kappa} v^{(t)} \cdot w \leq \|v^{(t)}\| \leq \sqrt{3/t}.$$

Denote by T the stopping time. If $T = \infty$ then indeed $P(\mu_t, \kappa)$ tends to zero as $t \to \infty$. If $T < \infty$, we have $v \cdot v^{(T)} \geq \frac{1}{T}$ for all $v \in V$. We can interpret $v^{(T)}$ as a non trivial strategy for the Hyperplane player: let

$$\tilde{w} = \frac{v^{(T)}}{\|v^{(T)}\|}.$$

Thus, for every μ,

$$P(\mu, \tilde{w}) \geq \frac{1}{T\|v^{(T)}\|} \geq \frac{1}{\sqrt{3T}}.$$

In particular, $\varepsilon^* \geq \frac{1}{\sqrt{3T}}$ and so

$$T \geq \frac{1}{3(\varepsilon^*)^2}. \qed$$

The last strategy in the sequence $\mu_1, \mu_2, \ldots$ guarantees the Point player a loss of at most $3\varepsilon^*$. This sequence is naturally and efficiently generated by the perceptron algorithm and produces a strategy for the Point player that is optimal up to a constant factor. The ideas presented in Sect. 2 allow to reduce the constant 3 to as close to 1 as we want, by paying in running time (see Algorithm 4).

5   Generalization Bounds

Generalization is one of the key concepts in learning theory. One typically formalizes it by assuming that the input sample consists of i.i.d. examples drawn from an unknown distribution D on $\mathbb{R}^d$ that are labelled by some unknown function $c : \mathbb{R}^d \to \{\pm 1\}$. The algorithm is said to generalize if it outputs an hypothesis $h : \mathbb{R}^d \to \{\pm 1\}$ so that $\Pr_D[h \neq c]$ is as small as possible.

We focus on the case that c is linearly separable. A natural choice for h in this case is given by hard-SVM; namely, the halfspace with maximum margin on the input sample. It is known that if D is supported on points that are γ-far from some hyperplane then the hard-SVM choice generalizes well (see Theorem 15.4 in [33]). The proof of this property of hard-SVMs uses Rademacher complexity.

We suggest that using the perceptron algorithm, instead the hard-SVM solution, yields a more general statement with a simpler proof. The reason is that the perceptron can be interpreted as a sample compression scheme.

Theorem 5 (similar to [18]). *Let D be a distribution on $\mathbb{R}^d$. Let $c : \mathbb{R}^d \to \{\pm 1\}$. Let $x_1, \ldots, x_m$ be i.i.d. samples from D. Let $S = ((x_1, c(x_1)), \ldots, (x_m, c(x_m)))$. If*

$$\Pr_S [\mathsf{marg}(S) < \varepsilon] < \delta/2 \tag{2}$$

for some $\varepsilon, \delta > 0$, then

$$\Pr_S \left[\mathbb{P}_D[\pi(S) \neq c] \leq 50 \frac{\log\left(\varepsilon^2 m\right) + \log(2/\delta)}{\varepsilon^2 m} \right] \geq 1 - \delta$$

where π is the perceptron algorithm.

The theorem can also be interpreted of as a local-to-global statement in the following sense. Assume that we know nothing of c, but we get a list of m samples that are linearly separable with significant margin (this is a local condition that we can empirically verify). Then we can deduce that c is close to being linearly separable. The perceptron's compression allows to deduce more general local-to-global statements, like bounding the global margin via the local/empirical margins (this is related to [32]).

Condition (2) holds when the expected value of one over the margin is bounded from above (and may hold when c is not linearly separable). This assumption is weaker than the assumption in [33] on the behavior of hard-SVMs (that the margin is always bounded from below).

For the proof of Theorem 5 we will need the following.

Definition 2 (Selection schemes). *A selection scheme of size d consists of a compression map κ and a reconstruction map ρ such that for every input sample S:*

- *κ maps S to a sub-sample of S of size at most d.*
- *ρ maps $\kappa(S)$ to a hypothesis $\rho(\kappa(S)) : \mathcal{X} \to \{\pm 1\}$; this is the output of the learning algorithm induced by the selection scheme.*

Following Littlestone and Warmuth, David et al. showed that every selection scheme does not overfit its data [11]: Let (κ, ρ) be a selection scheme of size d. Let S be a sample of m independent examples from an arbitrary distribution D that are labelled by some fixed concept c, and let $K(S) = \rho(\kappa(S))$ be the output of the selection scheme. For a hypothesis h, let $L_D(h) = \Pr_D[h \neq c]$ denote the true error of h and $L_S(h) = \frac{1}{m} \sum_{i=1}^m \mathbf{1}_{h(x_i) \neq c(x_i)}$ denote the empirical error of h.

Theorem 6 ([11]). *For every $\delta > 0$,*

$$\Pr_S \left[|L_D(K(S)) - L_S(K(S))| \geq \sqrt{\varepsilon \cdot L_S(K(S))} + \varepsilon \right] \leq \delta,$$

where

$$\varepsilon = 50 \frac{d \log(m/d) + \log(1/\delta)}{m}.$$

Proof (Theorem 5). Consider the following selection scheme of size $1/\varepsilon^2$ that agrees with the perceptron on samples with margin at least ε: If the input sample S has $\mathsf{marg}(S) \geq \varepsilon$, apply the perceptron (which gives a compression of size $1/\varepsilon^2$). Else, compress it to the emptyset and reconstruct it to some dummy hypothesis. The theorem now follows by applying Theorem 6 to this selection scheme and by the assumption that $\mathsf{marg}(S) \geq \varepsilon$ for $1 - \delta/2$ of the space (note that $L_S(K(S)) = 0$ when $\mathsf{marg}(S) \geq \varepsilon$). □

6   Robust Concepts

Here we follow the theme of robust concepts presented by Arriaga and Vempala [3]. Let $\mathcal{X} \subset \mathbb{R}^d$ be of size n so that $\max_{x \in \mathcal{X}} \|x\| = 1$. Think of $\mathcal{X}$ as representing a collection of high resolution images. As in many learning scenarios, some assumptions on the learning problem should be made in order to make it accessible. A typical assumption is that the unknown function to be learnt belongs to some specific class of functions. Here we focus on the class of all ε-separated partitions of $\mathcal{X}$; these are functions $Y : \mathcal{X} \to \{\pm 1\}$ that are linearly separable with margin at least ε. Such partitions are called robust concepts in [3] and correspond to "easy" classification problems.

Arriaga and Vempala demonstrated the difference between robust concepts and non-robust concept with the following analogy; it is much easier to distinguish between "Elephant" and "Dog" than between "African Elephant" and "Indian Elephant." They proved that random projections can help to perform efficient dimension reduction for ε-separated learning problems (and more general examples). They also described "neuronal" devices for performing it, and discussed their advantages. Similar dimension reductions were used in several other works in learning e.g. [4,6,15,16,21].

We observe that the perceptron's compression allows to deduce a *simultaneous* dimension reduction. Namely, the dimension reduction works simultaneously for the entire class of robust concepts. This follows from results in Ben-David et al. [5], who studied limitations of embedding learning problems in linearly separated classes.

We now explain this in more detail. The first step in the proof is the following theorem.

Theorem 7 ([5]). *The number of ε-separated partitions of $\mathcal{X}$ is at most $(2(n+1))^{1/\varepsilon^2}$.*

Proof. Given an ε-partition of the set $\mathcal{X}$, the perceptron algorithm finds a separating hyperplane after making at most $1/\varepsilon^2$ updates. It follows that every ε-partition can be represented by a multiset of $\mathcal{X}$ together with the corresponding signs. The total number of options is at most $(n+1)^{1/\varepsilon^2} \cdot 2^{1/\varepsilon^2}$. □

The theorem is sharp in the following sense.

Example 1. Let $e_1, \ldots, e_n \in \mathbb{R}^n$ be the n standard unit vectors. Every subset of the form $(e_i)_{i \in I}$ for $I \subset [n]$ of size k is $\Omega(1/\sqrt{k})$-separated, and there are $\binom{n}{k}$ such subsets.

The example also allows to lower bound the number of updates of any perceptron-like algorithm. If there is an algorithm that given $Y : \mathcal{X} \to \{\pm 1\}$ of margin ε is able to find w so that $Y(x) = \text{sign}(w \cdot x)$ for $x \in \mathcal{X}$ that can be described by at most K of the points in $\mathcal{X}$ then K should be at least $\Omega(1/\varepsilon^2)$.

The upper bound in the theorem allows to perform dimension reduction that simultaneously works well on the entire concept class. Let A be a $k \times d$ matrix with i.i.d. entries that are normally distributed $(N(0,1))^5$ with $k \geq C \log(n/\delta)/\varepsilon^4$ where $C > 0$ is an absolute constant. Given A, we can consider

$$A\mathcal{X} = \{Ax : x \in \mathcal{X}\} \subset \mathbb{R}^k$$

in a potentially smaller dimension space. The map $x \mapsto Ax$ is almost surely one-to-one on $\mathcal{X}$. So, every subset of $\mathcal{X}$ corresponds to a subset of $A\mathcal{X}$ and vice versa. The following theorem shows that it preserves all well-separated partitions.

Theorem 8 (implicit in [5]). *With probability of at least $1 - \delta$ over the choice of A, all ε-partitions of $\mathcal{X}$ are $\varepsilon/2$-partitions of $A\mathcal{X}$ and all $\varepsilon/2$-partitions of $A\mathcal{X}$ are $\varepsilon/4$-partitions of $\mathcal{X}$.*

The proof of the above theorem is a simple application of Theorem 7 together with the Johnson-Lindenstrauss lemma.

Lemma 2 ([19]). *Let $x_1, \ldots, x_N \in \mathbb{R}^d$ with $\|x_i\| \leq 1$ for all $i \in [N]$. Then, for every $\varepsilon > 0$ and $0 < \delta < 1/2$,*

$$\mathbb{P}\Big[\exists i, j \in [N] \ |(Ax_i \cdot Ax_j) - (x_i \cdot x_j)| > \varepsilon\Big] < \delta,$$

where $k = O(\log(N/\delta)/\varepsilon^2)$ and A is a $k \times d$ matrix with i.i.d. entries that are $N(0,1)$.

References

1. Andoni, A., Panigrahy, R., Valiant, G., Zhang, L.: Learning polynomials with neural networks. PMLR **32**(2), 1908–1916 (2014)
2. Anlauf, J.K., Biehl, M.: The AdaTron: an adaptive perceptron algorithm. EPL **10**, 687 (1989)
3. Arriaga, R.I., Vempala, S.: An algorithmic theory of learning: robust concepts and random projection. Mach. Learn. **63**(2), 161–182 (2006)
4. Balcan, N., Blum, A., Vempala, S.: On kernels, margins and low-dimensional mappings. In: ALT (2004)
5. Ben-David, S., Eiron, N., Simon, H.U.: Limitations of learning via embeddings in Euclidean half spaces. JMLR **3**, 441–461 (2002)
6. Blum, A., Kannan, R.: Learning an intersection of k halfspaces over a uniform distribution. In: FOCS (1993)
7. Boser, B.E., Guyon, I.M., Vapnik, V.N.: A training algorithm for optimal margin classifiers. In: COLT, pp. 144–152 (1992)

---

[5] Other distributions will work just as well.

8. Cesa-Bianchi, N., Conconi, A., Gentile, C.: On the generalization ability of on-line learning algorithms. IEEE Trans. Inf. Theory **50**(9), 2050–2057 (2004)
9. Collobert, R., Bengio, S.: Links between perceptrons, MLPs and SVMs. IDIAP (2004)
10. Crammer, K., Dekel, O., Keshet, J., Shalev-Shwartz, S., Singer, Y.: Online passive-aggressive algorithms. J. Mach. Learn. Res. **7**, 551–585 (2006)
11. David, O., Moran, S., Yehudayoff, A.: Supervised learning through the lens of compression. In: NIPS, pp. 2784–2792 (2016)
12. Farkas, G.: Über die Theorie der Einfachen Ungleichungen. Journal für die Reine und Angewandte Mathematik **124**(124), 1–27 (1902)
13. Freund, Y.: Boosting a weak learning algorithm by majority. Inf. Comput. **121**(2), 256–285 (1995)
14. Freund, Y., Schapire, R.E.: Large margin classification using the perceptron algorithm. Mach. Learn. **37**, 277–296 (1999). https://doi.org/10.1023/A:1007662407062
15. Garg, A., Har-Peled, S., Roth, D.: On generalization bounds, projection profile, and margin distribution. In: ICML, pp. 171–178 (2002)
16. Garg, A., Roth, D.: Margin distribution and learning. In: ICML, pp. 210–217 (2003)
17. Gentile, C.: A new approximate maximal margin classification algorithm. J. Mach. Learn. Res. **2**, 213–242 (2001)
18. Graepel, T., Herbrich, R., Shawe-Taylor, J.: PAC-Bayesian compression bounds on the prediction error of learning algorithms for classification. Mach. Learn. **59**, 55–76 (2005). https://doi.org/10.1007/s10994-005-0462-7
19. Johnson, W.B., Lindenstrauss, J.: Extensions of Lipschitz mappings into a Hilbert space. In: Conference in Modern Analysis and Probability (1982)
20. Khardon, R., Wachman, G.: Noise tolerant variants of the perceptron algorithm. J. Mach. Learn. Res. **8**, 227–248 (2007)
21. Klivans, A.R., Servedio, R.A.: Learning intersections of halfspaces with a margin. In: Shawe-Taylor, J., Singer, Y. (eds.) COLT 2004. LNCS (LNAI), vol. 3120, pp. 348–362. Springer, Heidelberg (2004). https://doi.org/10.1007/978-3-540-27819-1_24
22. Korzeń, M., Klęsk, P.: Maximal margin estimation with perceptron-like algorithm. In: Rutkowski, L., Tadeusiewicz, R., Zadeh, L.A., Zurada, J.M. (eds.) ICAISC 2008. LNCS (LNAI), vol. 5097, pp. 597–608. Springer, Heidelberg (2008). https://doi.org/10.1007/978-3-540-69731-2_58
23. Krauth, W., Mézard, M.: Learning algorithms with optimal stablilty in neural networks. J. Phys. A: Math. Gen. **20**, L745–L752 (1987)
24. LeCun, Y., Cortes, C.: The MNIST database of handwritten digits (1998)
25. Littlestone, N., Warmuth, M.: Relating data compression and learnability (1986, unpublished)
26. Matoušek, J.: On variants of the Johnson-Lindenstrauss lemma. Random Struct. Algorithms **33**(2), 142–156 (2008)
27. Mohri, M., Rostamizadeh, A.: Perceptron Mistake Bounds. arXiv:1305.0208
28. Moran, S., Yehudayoff, A.: Sample compression schemes for VC classes. JACM **63**(3), 1–21 (2016)
29. von Neumann, J.: Zur Theorie der Gesellschaftsspiele. Math. Ann. **100**, 295–320 (1928)
30. Novikoff, A.B.J.: On convergence proofs for perceptrons. In: Proceedings of the Symposium on the Mathematical Theory of Automata, vol. 12, pp. 615–622 (1962)
31. Rosenblatt, F.: The perceptron: a probabilistic model for information storage and organization in the brain. Psychol. Rev. **65**(6), 386–408 (1958)

32. Schapire, R.E., Freund, Y., Bartlett, P., Lee, W.S.: Boosting the margin: a new explanation for the effectiveness of voting methods. Ann. Stat. **26**(5), 1651–1686 (1998)
33. Shalev-Shwartz, S., Ben-David, S.: Understanding Machine Learning: From Theory to Algorithms. Cambridge University Press, Cambridge (2014)
34. Shalev-Shwartz, S., Singer, Y., Srebro, N., Cotter, A.: Pegasos: primal estimated sub-gradient solver for SVM. Math. Program. **127**(1), 3–30 (2011). https://doi.org/10.1007/s10107-010-0420-4
35. Simard, P.Y., Steinkraus, D., Platt, J.C.: Best practices for convolutional neural networks applied to visual document analysis. ICDAR **3**, 958–962 (2003)
36. Soudry, D., Hoffer, E., Srebro, N.: The implicit bias of gradient descent on separable data. arXiv:1710.10345 (2017)
37. Wendemuth, A.: Learning the unlearnable. J. Phys. A: Math. Gen. **28**, 5423 (1995)

#P-completeness of Counting Update Digraphs, Cacti, and Series-Parallel Decomposition Method

Kévin Perrot[(⊠)], Sylvain Sené, and Lucas Venturini

Université publique, Marseille, France
kevin.perrot@lis-lab.fr, sylvain.sene@lis-lab.fr,
lucas.venturini@ens-lyon.fr

Abstract. Automata networks are a very general model of interacting entities, with applications to biological phenomena such as gene regulation. In many contexts, the order in which entities update their state is unknown, and the dynamics may be very sensitive to changes in this schedule of updates. Since the works of Aracena et al., it is known that update digraphs are pertinent objects to study non-equivalent block-sequential update schedules. We prove that counting the number of equivalence classes, that is a tight upper bound on the synchronism sensitivity of a given network, is #P-complete. The problem is nevertheless computable in quasi-quadratic time for oriented cacti, and for oriented series-parallel graphs thanks to a decomposition method.

1 Introduction

Since their introduction by McCulloch and Pitts in the 1940s through the well known formal neural networks [20], automata networks (ANs) are a general model of interacting entities in finite state spaces. The field has important contributions to computer science, with Kleene's finite state automata [17], linear shift registers [14] and linear networks [12]. At the end of the 1960s, Kauffman and Thomas (independently) developed the use of ANs for the modeling of biological phenomena such as gene regulation [16, 28], providing a fruitful theoretical framework [26].

ANs can be considered as a collection of local functions (one per component), and influences among components may be represented as a so called *interaction digraph*. In many applications the order of components update is a priori unknown, and different schedules may greatly impact the dynamical properties of the system. It is known since the works of Aracena *et al.* in [4] that *update*

Mainly funded by our salaries as French State agents (affiliated to Laboratoire Cogitamus (CN), Aix-Marseille Univ., Univ. de Toulon, CNRS, LIS, Marseille, France (KP, SS and LV), Univ. Côte d'Azur, CNRS, I3S, Sophia Antipolis, France (KP), and École normale supérieure de Lyon, Computer Science department, Lyon, France (LV)), and secondarily by the projects ANR-18-CE40-0002 FANs, ECOS-CONICYT C16E01, STIC AmSud 19-STIC-03 (Campus France 43478PD).

The original version of this chapter was revised: The fictitious author was removed. The correction is available at https://doi.org/10.1007/978-3-030-51466-2_34

© Springer Nature Switzerland AG 2020, corrected publication 2021
M. Anselmo et al. (Eds.): CiE 2020, LNCS 12098, pp. 326–338, 2020.
https://doi.org/10.1007/978-3-030-51466-2_30

$$f_1(x) = 0 \qquad f_3(x) = x_1 \vee \neg x_4$$
$$f_2(x) = \neg x_4 \qquad f_4(x) = x_3$$

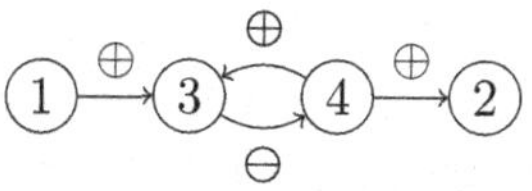

Fig. 1. Example of an AN f on the Boolean alphabet $[q] = \{1,2\}$ (conventionally renamed $\{0,1\}$), its interaction digraph, and a $\{\oplus, \ominus\}$-labeling $\mathrm{lab}_B = \mathrm{lab}_{B'}$ corresponding to the two equivalent update schedules $B = (\{1,2,3\},\{4\})$ and $B' = (\{1,3\},\{2,4\})$.

digraphs (consisting of labeling the arcs of the interaction digraphs with $\oplus$ and $\ominus$) capture the correct notion to consider a biologically meaningful family of update schedules called *block-sequential* in the literature. Since another work of Aracena *et al.* [3] a precise characterization of the valid labelings has been known, but their combinatorics remains puzzling. After formal definitions and known results in Sects. 2 and 3, we propose in Sect. 4 an explanation for this difficulty, through the lens of computational complexity theory: we prove that counting the number of update digraphs (valid $\{\oplus, \ominus\}$-labelings) is #P-complete. In Sect. 5 we consider the problem restricted to the family of oriented cactus graphs and give a $\mathcal{O}(n^2 \log n \log \log n)$ time algorithm, and finally in Sect. 6 we present a decomposition method leading to a $\mathcal{O}(n^2 \log^2 n \log \log n)$ algorithm for oriented series-parallel graphs.

2   Definitions

Given a finite alphabet $[q] = \{1, \ldots, q\}$, an *automata network* (AN) of size n is a function $f : [q]^n \to [q]^n$. We denote x_i the *component* $i \in [n]$ of some *configuration* $x \in [q]^n$. ANs are more conveniently seen as n *local functions* $f_i : [q]^n \to [q]$ describing the update of each component, *i.e.* with $f_i(x) = f(x)_i$. The *interaction digraph* captures the effective dependencies among components, and is defined as the digraph $G_f = ([n], A_f)$ with

$$(i,j) \in A_f \iff f_j(x) \neq f_j(y) \text{ for some } x, y \in [q]^n \text{ with } x_{i'} = y_{i'} \text{ for all } i' \neq i.$$

It is well known that the schedule of components update may have a great impact on the dynamics [5,13,21,23]. A *block-sequential update schedule* $B = (B_1, \ldots, B_t)$ is an ordered partition of $[n]$, defining the following dynamics

$$f^{(B)} = f^{(B_t)} \circ \cdots \circ f^{(B_2)} \circ f^{(B_1)} \quad \text{with} \quad f^{(B_i)}(x)_j = \begin{cases} f_j(x) & \text{if } j \in B_i \\ x_j & \text{if } j \notin B_i \end{cases}$$

i.e., parts are updated sequentially one after the other, and components within a part are updated in parallel. For the parallel update schedule $B^{\mathrm{par}} = ([n])$, we have $f^{(B^{\mathrm{par}})} = f$. Block-sequential update schedules are a classical family of update schedules considered in the literature, because they are perfectly fair: every local function is applied exactly once during each step. Equipped with an update schedule, $f^{(B)}$ is a discrete dynamical system on $[q]^n$. In the following we will shortly say *update schedule* to mean *block-sequential update schedule*.

It turns out quite intuitively that some update schedules will lead to the same dynamics, when the ordered partitions are very close and the difference relies on components far apart in the interaction digraph (see an example on Fig. 1). Aracena *et al.* introduced in [4] the notion of *update digraph* to capture this fact. To an update schedule one can associate its *update digraph*, which is a $\{\oplus, \ominus\}$-labeling of the arcs of the interaction digraph of the AN, such that (i, j) is negative $(\ominus)$ when i is updated strictly before j, and positive $(\oplus)$ otherwise. Formally, given an update schedule $B = (B_1, \ldots, B_t)$,

$$\forall (i,j) \in A_f : \mathrm{lab}_B((i,j)) = \begin{cases} \oplus & \text{if } i \in B_{t_i} \text{ and } j \in B_{t_j} \text{ with } t_i \geq t_j, \\ \ominus & \text{if } i \in B_{t_i} \text{ and } j \in B_{t_j} \text{ with } t_i < t_j. \end{cases}$$

Remark 1. Loops are always labeled $\oplus$, hence we consider our digraphs *loopless*.

The following result has been established: given two update schedules, if the relative order of updates among all adjacent components are identical, then the dynamics are identical. It leads naturally to an *equivalence relation* on update schedules, at the heart of the present work.

Theorem 1 ([4]). *Given an AN f and two update schedules B, B', if $\mathrm{lab}_B = \mathrm{lab}_{B'}$ then $f^{(B)} = f^{(B')}$. Hence we denote $B \equiv B'$ if and only if $\mathrm{lab}_B = \mathrm{lab}_{B'}$.*

It is very important to note that, though every update schedule corresponds to a $\{\oplus, \ominus\}$-labeling of G_f, the reciprocal of this fact is not true. For example, a cycle with all arcs labeled $\ominus$ would lead to a contradiction where components are updated strictly before themselves. Aracena *et al.* gave a precise characterization of *valid* update digraphs (*i.e.* the ones corresponding to at least one update schedule).

Theorem 2 ([3]). *A labeling function $\mathrm{lab} : A \to \{\oplus, \ominus\}$ is valid if and only if there is no cycle $(i_0, i_1, \ldots, i_k)$, with $i_0 = i_k$, of length $k > 0$ such that both:*

- $\forall\, 0 \leq j < k : \mathrm{lab}((i_j, i_{j+1})) = \oplus \vee \mathrm{lab}((i_{j+1}, i_j)) = \ominus,$
- $\exists\, 0 \leq j < k : \mathrm{lab}((i_{j+1}, i_j)) = \ominus.$

In words, the multidigraph where the orientation of negative arcs is reversed, does not contain a cycle with at least one negative arc (forbidden cycle). ◾

As a corollary, one can decide in polynomial time whether a labeling is valid (**Valid-UD Problem** is in P). We are interested in the following.

> **Update Digraphs Counting (#UD)**
> *Input:* a digraph $G = (V, A)$.
> *Output:* $\#\mathrm{UD}(G) = |\{\mathrm{lab} : A \to \{\ominus, \oplus\} \mid \mathrm{lab} \text{ is valid}\}|.$

The following definition is motivated by Theorem 2.

Definition 1. *Given a digraph $G = (V, A)$, let $\bar{G} = (V, \bar{A})$ denote the undirected multigraph underlying G, i.e. with an edge $\{i, j\} \in \bar{A}$ for each $(i, j) \in A$.*

Remark 2. We can restrict our study to connected digraphs (that is, such that $\bar{G}$ is connected), because according to Theorem 2 the only invalid labelings contain (forbidden) cycles. Given some G with $V_1, \ldots, V_k$ its connected components, and $G[V_i]$ the subdigraph induced by V_i, we straightforwardly have $\#\mathtt{UD}(G) = \prod_{i \in [k]} \#\mathtt{UD}(G[V_i])$, and this decomposition can be computed in linear time from folklore algorithms.

Theorem 3. *$\#\mathrm{UD}$ is in $\#\mathsf{P}$.*

Proof. The following non-deterministic algorithm runs in polynomial time (**Valid-UD Problem** is in P), and its number of accepting branches equals $\#\mathtt{UD}(G)$:

1. guess a labeling $\mathrm{lab} : A \to \{\oplus, \ominus\}$ (polynomial space),
2. accept if lab is valid, otherwise reject.

$\square$

3   Further Known Results

The consideration of update digraphs has been initiated by Aracena *et al.* in 2009 [4], with their characterization (Theorem 2) in [3]. In Sect. 4 we will present a problem closely related to **#UD** that has been proven to be NP-complete in [3], **UD Problem**, and bounds that we can deduce on **#UD** (Corollary 1, from [2]). In [1] the authors present an algorithm to enumerate update digraphs, and prove its correctness. They also consider a surprisingly complex question: given an AN f, knowing whether there exist two block-sequential update schedules B, B' such that $f^{(B)} \neq f^{(B')}$, is NP-complete. The value of $\#\mathtt{UD}(G)$ is known to be $3^n - 2^{n+1} + 2$ for bidirected cycles on n vertices [23], and to equal $n!$ if and only if the digraph is a tournament on n vertices [2].

4   Counting Update Digraphs Is #P-complete

The authors of [3] have exhibited an insightful relation between valid labelings and feedback arc sets of a digraph. We recall that a *feedback arc set* (FAS) of $G = (V, A)$ is a subset of arcs $F \subseteq A$ such that the digraph $(V, A \setminus F)$ is acyclic, and its size is $|F|$. This relation is developed inside the proof of NP-completeness of the following decision problem. We reproduce it as a Lemma.

Update Digraph Problem (UD Problem)
Input: a digraph $G = (V, A)$ and an integer k.
Question: does there exist a valid labeling of size at most k?

The *size* of a labeling is its number of $\oplus$ labels. It is clear that minimizing the number of $\oplus$ labels (or equivalently maximizing the number of $\ominus$ labels) is the difficult direction, the contrary being easy because $\mathrm{lab}(a) = \oplus$ for all $a \in A$ is always valid (and corresponds to the parallel update schedule $B^{\mathtt{par}}$).

Fig. 2. $F = \{(1,2),(2,3),(3,1)\}$ is a FAS, but the corresponding labeling is not valid: component 3 is updated prior to 2, 1 not prior to 2, and 3 not prior to 1, which is impossible.

Lemma 1 (appears in [3, Theorem 16]). *There exists a bijection between minimal valid labelings and minimal feedback arc sets of a digraph $G = (V, A)$.*

Proof (sketch). To get the bijection, we simply identify a labeling lab with its set of arcs labeled $\oplus$, denoted $F_{\text{lab}} = \{a \in A \mid \text{lab}(a) = \oplus\}$.   □

Any valid labeling corresponds to a FAS, and every minimal FAS corresponds to a valid labeling, hence the following bounds hold. The strict inequality for the lower bound comes from the fact that labeling all arcs $\oplus$ does not give a minimal FAS, as noted in [2] where the authors also consider the relation between update digraphs (valid labelings) and feedback arc sets, but from another perspective.

Corollary 1 ([3]). *For any digraph G, let #FAS(G) and #MFAS(G) be the number of FAS and minimal FAS of G, then #MFAS$(G) <$ #UD$(G) \leq$ #FAS(G).*

From Lemma 1 and results on the complexity of FAS counting problems presented in [22], we have the following corollary (minimum FAS are minimal, hence the identity is a parsimonious reduction from the same problems on FAS).

Corollary 2. *Counting the number of valid labelings of minimal size is #P-complete, and of minimum size is #·OptP[log n]-complete.*

However the correspondence given in Lemma 1 does not hold in general: there may exist some FAS F such that lab with $F_{\text{lab}} = F$ is not a valid labeling (see Fig. 2 for an example). As a consequence we do not directly get a counting reduction to **#UD**. It nevertheless holds that **#UD** is #P-hard, with the following reduction.

Theorem 4. *#UD is #P-hard.*

Proof. We present a (polynomial time) parsimonious reduction from the problem of counting the number of acyclic orientations of an undirected graph, proven to be #P-hard in [18].

Given an undirected graph $G = (V, E)$, let $\prec$ denote an arbitrary total order on V. Construct the digraph $G' = (V, A)$ with A the orientation of E according to $\prec$, i.e. $(u, v) \in A \iff \{u, v\} \in E$ and $u \prec v$. An example is given on Fig. 3 (left). A key property is that G' is acyclic, because A is constructed from an order $\prec$ on V (a cycle would have at least one arc (u, v) with $v \prec u$).

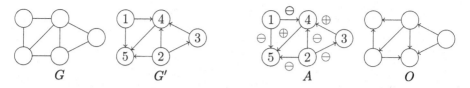

Fig. 3. Left: an undirected graph G (instance of acyclic orientation counting), and the obtained digraph G' (instance of update digraph counting). Right: a valid labeling A of G', and the corresponding orientation O of G.

We claim that there is a bijection between the valid labelings of G' and the acyclic orientations of G: to a valid labeling lab : $A \to \{\oplus, \ominus\}$ of G' we associate the orientation

$$O = \quad \{(u,v) \mid (u,v) \in A \text{ and } \mathrm{lab}((u,v)) = \oplus\}$$
$$\cup \{(v,u) \mid (u,v) \in A \text{ and } \mathrm{lab}((u,v)) = \ominus\}.$$

First remark that O is indeed an orientation of E: each edge of E is transformed into an arc of A, and each arc of A is transformed into an arc of O. An example is given on Fig. 3 (right). Now observe that O is exactly obtained from G' by reversing the orientation of arcs labeled $\ominus$ by lab. Furthermore, a cycle in O must contain at least one arc labeled $\ominus$ by lab, because G' is acyclic and $\oplus$ labels copy the orientation of G'. The claim therefore follows directly from the characterization of Theorem 2. $\qquad\qquad\square$

5   Quasi-quadratic Time Algorithm for Oriented Cacti

The difficulty of counting the number of update digraphs comes from the interplay between various possible cycles, as is assessed by the parsimonious reduction from acyclic orientations counting problem to #**UD**. Answering the problem for an oriented tree with m arcs is for example very simple: all of the 2^m labelings are valid. Cactus undirected graphs are defined in terms of very restricted entanglement of cycles, which we can exploit to compute the number of update digraphs for any orientation of its edges.

Definition 2. A cactus *is a connected undirected graph such that any vertex (or equivalently any edge) belongs to at most one simple cycle (cycle without vertex repetition).* An oriented cactus G *is a digraph such that $\bar{G}$ is a cactus.*

Cacti may intuitively be thought as trees with cycles. This is indeed the idea behind the *skeleton* of a cactus introduced in [9], via the following notions:

- a *c-vertex* is a vertex of degree two included in exactly one cycle,
- a *g-vertex* is a vertex not included in any cycle,
- remaining vertices are *h-vertices*,

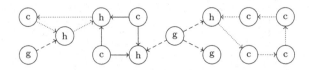

Fig. 4. An oriented cactus G, with $\{c, g, h\}$-vertex labels. Graft arcs are dashed, cycles forming directed cycles are dotted, and cycles not forming directed cycles are solid. Theorem 5 counts $\#\mathtt{UD}(G) = 2^1 2^3 (2^3 - 1)(2^5 - 1)(2^4 - 2) = 48\,608$.

and a *graft* is a maximal subtree of g- and h-vertices with no two h-vertices belonging to the same cycle. Then a cactus can be decomposed as grafts and cycles (two classes called *blocks*), connected at h-vertices according to a tree skeleton. These notions directly apply to oriented cacti (see an example on Fig. 4).

Theorem 5. $\#\mathbf{UD}$ *is computable in time* $\mathcal{O}(n^2 \log n \log \log n)$ *for oriented cacti.*

Proof. The result is obtained from the skeleton of an oriented cactus G, since potential forbidden cycles are limited to within blocks of the skeleton. From this independence, any union of valid labelings on blocks is valid, and we have the product

$$\#\mathtt{UD}(G) = \prod_{H \in \mathcal{G}} 2^{|H|} \prod_{H \in \vec{\mathcal{C}}} (2^{|H|} - 1) \prod_{H \in \mathcal{C}} (2^{|H|} - 2)$$

where $\mathcal{G}$ is the set of grafts of G, $\vec{\mathcal{C}}$ is the set of cycles forming directed cycles, $\mathcal{C}$ is the set of cycles not forming directed cycles, and $|H|$ is the number of arcs in block H. Indeed, grafts cannot create forbidden cycles hence any $\{\oplus, \ominus\}$-labeling will be valid, cycles forming a directed cycle can create exactly one forbidden cycle (with $\ominus$ labels on all arcs), and cycles not forming a directed cycle can create exactly two forbidden cycles (one for each possible direction of the cycle). In a first step the skeleton of a cactus can be computed in linear time [9]. Then, since the size n of the input is equal (up to a constant) to the number of arcs, the size of the output contains $\mathcal{O}(n)$ bits (upper bounded by the number of $\{\oplus, \ominus\}$-labelings), thus naively we have $\mathcal{O}(n)$ terms, each of $\mathcal{O}(n)$ bits, and the $\mathcal{O}(n \log n \log \log n)$ Schönhage–Strassen integer multiplication algorithm gives the result. □

Remark 3. Assuming multiplications to be done in constant time in the above result would be misleading, because we are multiplying integers having a number of digits in the magnitude of the input size. Also, the result may be slightly strengthened by considering the $\mathcal{O}(n \log n \, 2^{2 \log^* n})$ algorithm by Fürer in 2007.

6   Series-Parallel Decomposition Method

In this section we present a divide and conquer method in order to solve $\#\mathbf{UD}$, *i.e.* in order to count the number of valid labelings (update digraphs) of a given digraph. What will be essential in this decomposition method is not the orientation of arcs, but rather the topology of the underlying undirected (multi)graph $\bar{G}$. The (de)composition is based on defining two endpoints on our digraphs, and composing them at their endpoints. It turns out to be closely related to series-parallel graphs first formalized to model electric networks in 1892 [19]. In Subsect. 6.1 we present the operations of composition, and in Subsect. 6.2 we show how it applies to the family of oriented series-parallel graphs.

6.1   Sequential, Parallel, and Free Compositions

Let us first introduce some notations and terminology on the characterization of valid labelings provided by Theorem 2. Given lab : $A \to \{\oplus, \ominus\}$, we denote $\tilde{G}_{\text{lab}} = (V, \tilde{A})$ the multidigraph obtained by reversing the orientation of negative arcs:

$$(i, j) \in \tilde{A} \quad \Longleftrightarrow \quad \begin{array}{l} (i, j) \in A \text{ and } \text{lab}((i, j)) = \oplus, \\ \text{or } (j, i) \in A \text{ and } \text{lab}((j, i)) = \ominus. \end{array}$$

For simplicity we abuse the notation and still denote lab the labeling of the arcs of $\tilde{G}_{\text{lab}}$ (arcs keep their label from G to $\tilde{G}_{\text{lab}}$). From Theorem 2, lab is a valid labeling if and only if $\tilde{G}_{\text{lab}}$ does not contain any cycle with at least one arc labeled $\ominus$, called *forbidden cycle* (it may contain cycles with all arcs labeled $\oplus$). A path from i to j in $\tilde{G}_{\text{lab}}$ is called *negative* if it contains at least one arc labeled $\ominus$, and *positive* otherwise.

Definition 3. *A source-sink labeled graph (ss-graph) (G, α, β) is a multigraph G with two distinguished vertices $\alpha \neq \beta$. A triple (G, α, β) with G a digraph such that $(\bar{G}, \alpha, \beta)$ is a ss-graph, is called an* oriented ss-graph (oss-graph).

We can decompose the set of update digraphs (denoted $\text{UD}(G) = \{\text{lab} : A \to \{\oplus, \ominus\} \mid \text{lab is valid}\}$) into an oss-graph (G, α, β), based on the follow sets.

$$\text{UD}(G)^+_{\alpha \to \beta} = \{\text{lab} \in \text{UD}(G) \mid \text{there \textbf{exists} a path from } \alpha \text{ to } \beta \text{ in } \tilde{G}_{\text{lab}},$$
$$\text{and \textbf{all} paths from } \alpha \text{ to } \beta \text{ in } \tilde{G}_{\text{lab}} \text{ are \textbf{positive}}\}$$

$$\text{UD}(G)^-_{\alpha \to \beta} = \{\text{lab} \in \text{UD}(G) \mid \text{there \textbf{exists} a \textbf{negative} path from } \alpha \text{ to } \beta \text{ in } \tilde{G}_{\text{lab}}\}$$

$$\text{UD}(G)^{\varnothing}_{\alpha \to \beta} = \{\text{lab} \in \text{UD}(G) \mid \text{there exist \textbf{nopath} from } \alpha \text{ to } \beta \text{ in } \tilde{G}_{\text{lab}}\}$$

We define analogously $\text{UD}(G)^+_{\beta \to \alpha}$, $\text{UD}(G)^-_{\beta \to \alpha}$, $\text{UD}(G)^{\varnothing}_{\beta \to \alpha}$, and partition $\text{UD}(G)$ as:

1. $\text{UD}(G)^{+,+}_{\alpha,\beta} = \text{UD}(G)^+_{\alpha \to \beta} \cap \text{UD}(G)^+_{\beta \to \alpha}$ 4. $\text{UD}(G)^{\varnothing,+}_{\alpha,\beta} = \text{UD}(G)^{\varnothing}_{\alpha \to \beta} \cap \text{UD}(G)^+_{\beta \to \alpha}$
2. $\text{UD}(G)^{+,\varnothing}_{\alpha,\beta} = \text{UD}(G)^+_{\alpha \to \beta} \cap \text{UD}(G)^{\varnothing}_{\beta \to \alpha}$ 5. $\text{UD}(G)^{\varnothing,-}_{\alpha,\beta} = \text{UD}(G)^{\varnothing}_{\alpha \to \beta} \cap \text{UD}(G)^-_{\beta \to \alpha}$
3. $\text{UD}(G)^{-,\varnothing}_{\alpha,\beta} = \text{UD}(G)^-_{\alpha \to \beta} \cap \text{UD}(G)^{\varnothing}_{\beta \to \alpha}$ 6. $\text{UD}(G)^{\varnothing,\varnothing}_{\alpha,\beta} = \text{UD}(G)^{\varnothing}_{\alpha \to \beta} \cap \text{UD}(G)^{\varnothing}_{\beta \to \alpha}$

Notice that the three missing combinations, $\text{UD}(G)^{+,-}_{\alpha,\beta}$, $\text{UD}(G)^{-,+}_{\alpha,\beta}$ and $\text{UD}(G)^{-,-}_{\alpha,\beta}$, would always be empty because such labelings contain a forbidden cycle. For

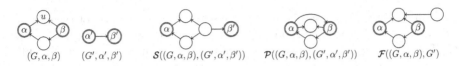

(G, α, β) (G', α', β') $\mathcal{S}((G, \alpha, \beta), (G', \alpha', \beta'))$ $\mathcal{P}((G, \alpha, \beta), (G', \alpha', \beta'))$ $\mathcal{F}((G, \alpha, \beta), G')$

Fig. 5. Example of series and parallel compositions, and a free composition at u, β'.

convenience let us denote $S = \{(+, +), (+, \varnothing), (-, \varnothing), (\varnothing, +), (\varnothing, -), (\varnothing, \varnothing)\}$. Given any oss-graph (G, α, β) we have

$$\#\mathrm{UD}(G) = \sum_{(s,t)\in S} \#\mathrm{UD}(G)^{s,t}_{\alpha,\beta} \tag{1}$$

where $\#\mathrm{UD}(G)^{s,t}_{\alpha,\beta} = |\mathrm{UD}(G)^{s,t}_{\alpha,\beta}|$. Oss-graphs may be thought as black boxes, we will compose them using the values of $\#\mathrm{UD}(G)^{s,t}_{\alpha,\beta}$, regardless of their inner topologies.

Definition 4. *We define three types of compositions (see Fig. 5).*

- *The* series composition *of two oss-graphs (G, α, β) and (G', α', β') with $V \cap V' = \emptyset$, is the oss-graph $\mathcal{S}((G, \alpha, \beta), (G', \alpha', \beta')) = (D, \alpha, \beta')$ with D the one-point join of G and G' identifying components β, α' as one single component.*
- *The* parallel composition *of two oss-graphs (G, α, β) and (G', α', β') with $V \cap V' = \emptyset$, is the oss-graph $\mathcal{P}((G, \alpha, \beta), (G', \alpha', \beta')) = (D, \alpha, \beta)$ with D the two-points join of G and G' identifying components α, α' and β, β' as two single components.*
- *The* free composition *at v, v' of an oss-graph $(G = (V, A), \alpha, \beta)$ and a digraph $G' = (V', A')$ with $V \cap V' = \emptyset$, $v \in V$, $v' \in V'$, is the oss-graph $\mathcal{F}((G, \alpha, \beta), G') = (D, \alpha, \beta)$ with D the one-point join of G and G' identifying v, v' as one single component.*

Remark that the three types of compositions from Definition 4 also apply to (undirected) ss-graph $(\bar{G}, \alpha, \beta)$. Series and free compositions differ on the endpoints of the obtained oss-graph, which has important consequences on counting the number of update digraphs, as stated in the following results. We will see in Theorem 6 from Subsect. 6.2 that both series and free compositions are needed in order to decompose the family of (general) *oriented series-parallel graphs* (to be defined).

Lemma 2. *For $(D, \alpha, \beta') = \mathcal{S}((G, \alpha, \beta), (G', \alpha', \beta'))$, the values of $\#\mathrm{UD}(D)^{s,t}_{\alpha,\beta'}$ for all $(s, t) \in S$ can be computed in time $\mathcal{O}(n \log n \log \log n)$ (with n the binary length of the values) from the values of $\#\mathrm{UD}(G)^{s,t}_{\alpha,\beta}$ and $\#\mathrm{UD}(G')^{s,t}_{\alpha',\beta'}$ for all $(s, t) \in S$.*

Proof (sketch). The result is obtained by considering the 36 couples of some $\mathrm{UD}(G)^{s,t}_{\alpha,\beta}$ and some $\mathrm{UD}(G')^{s',t'}_{\alpha',\beta'}$, each combination giving an element of $\mathrm{UD}(D)^{s'',t''}_{\alpha,\beta'}$ for some $(s'', t'') \in S$. □

Lemma 3. *For $(D, \alpha, \beta) = \mathcal{P}((G, \alpha, \beta), (G', \alpha', \beta'))$, the values of $\#\mathit{UD}(D)^{s,t}_{\alpha,\beta}$ for all $(s, t) \in S$ can be computed in time $\mathcal{O}(n \log n \log \log n)$ (with n the binary length of the values) from the values of $\#\mathit{UD}(G)^{s,t}_{\alpha,\beta}$ and $\#\mathit{UD}(G')^{s,t}_{\alpha',\beta'}$ for all $(s, t) \in S$.*

Proof (sketch). The proof is analogous to Lemma 2, except that some couples may create invalid labelings. □

Note that Remark 3 also applies to Lemmas 2 and 3. For the free composition the count is easier.

Lemma 4. *For $(D, \alpha, \beta) = \mathcal{F}((G, \alpha, \beta), G')$, we have $\#\mathit{UD}(D)^{s,t}_{\alpha,\beta} = \#\mathit{UD}(G)^{s,t}_{\alpha,\beta}$ $\#\mathit{UD}(G')$ for all $(s, t) \in S$.*

Proof. The endpoints of the oss-graph (D, α, β) are the endpoints of the oss-graph (G, α, β), and it is not possible to create a forbidden cycle in the union of a valid labeling on G and a valid labeling of G', therefore the union is always a valid labeling of D, each one belonging to the part (s, t) of (D, α, β) corresponding to the part $(s, t) \in S$ of (G, α, β). □

6.2   Application to Oriented Series-Parallel Graphs

The series and parallel compositions of Definition 4 correspond exactly to the class of two-terminal series-parallel graphs from [25, 29].

Definition 5. *A ss-graph (G, α, β) is two-terminal series-parallel (a ttsp-graph) if and only if one the following holds.*

- *(G, α, β) is a base ss-graph with two vertices α, β and one edge $\{\alpha, \beta\}$.*
- *(G, α, β) is obtained by a series or parallel composition[1] of two ttsp-graphs.*

In this case G alone is called a blind ttsp-graph.

Adding the free composition allows to go from two-terminal series-parallel graphs to (general) series-parallel graphs [11, 29]. More precisely, it allows exactly to add tree structures to ttsp-graphs, as we argue now (ttsp-graphs do not contain arbitrary trees, its only acyclic graphs being simple paths; *e.g.* one cannot build a *claw* from Definition 5).

Definition 6. *A multigraph G is series-parallel (sp-graph) if and only if all its 2-connected components are blind ttsp-graphs. A digraph G such that $\bar{G}$ is an sp-graph, is called an oriented sp-graph (osp-graph).*

The family of sp-graphs corresponds to the multigraphs obtained by series, parallel and free compositions from base ss-graphs.

Theorem 6. *G is an sp-graph if and only if (G, α, β) is obtained by series, parallel and free compositions from base ss-graphs, for some α, β.*

---

[1] With Definition 4 applied to (undirected) ss-graphs.

Proof (sketch). Free compositions allow to build all sp-graphs, because it offers the possibility to create the missing tree structures of ttsp-graphs: arbitrary 1-connected components linking 2-connected ttsp-graphs. Moreover free compositions do not go beyond sp-graphs, since the obtained multigraphs still have treewidth 2.   □

Theorem 7. #**UD** *is solvable in time* $\mathcal{O}(n^2 \log^2 n \log \log n)$ *on osp-graphs (without promise).*

Proof (sketch). This is a direct consequence of Lemmas 2, 3 and 4, because all values are in $\mathcal{O}(2^n)$ (the number of $\{\oplus, \ominus\}$-labelings) hence on $\mathcal{O}(n)$ bits, the values of $\#\mathtt{UD}(G)^{s,t}_{\alpha,\beta}$ are trivial for oriented base ss-graphs, and we perform $\mathcal{O}(n \log n)$ compositions (to reach Formula 1). The absence of promise comes from a linear time recognition algorithm in [29] for ttsp-graphs, which also provides the decomposition structure.   □

Again, Remark 3 applies to Theorem 7.

7   Conclusion

Our main result is the #P-completeness of #**UD**, *i.e.* of counting the number of non-equivalent block-sequential update schedules of a given AN f. We proved that this count can nevertheless be done in $\mathcal{O}(n^2 \log n \log \log n)$ time for oriented cacti, and in $\mathcal{O}(n^2 \log^2 n \log \log n)$ time for oriented series-parallel graphs. This last result has been obtained via a decomposition method providing a divide-and-conquer algorithm.

Remark that cliques or tournaments are intuitively difficult instances of #**UD**, because of the intertwined structure of potential forbidden cycles. It turns out that K_4 is the smallest clique that cannot be build with series, parallel and free decompositions, and that series-parallel graphs (Definition 6) correspond exactly to the family of K_4-minor-free graphs [11] (it is indeed closed by minor [27]). In further works we would like to extend this characterization and the decomposition method to (di)graphs with multiple endpoints.

The complexity analysis of the algorithms presented in Theorems 5 and 7 may be improved, and adapted to the parallel setting using the algorithms presented in [7, 15]. One may also ask for which other classes of digraphs is $\#\mathtt{UD}(G)$ computable efficiently (in polynomial time)? Since we found such an algorithm for graphs of treewidth 2, could it be that the problem is fixed parameter tractable on bounded treewidth digraphs? Rephrased more directly, could a general tree decomposition (which, according to the proof of Theorem 6, is closely related to the series-parallel decomposition for treewidth 2) be exploited to compute the solution to #**UD**? Alternatively, what other types of decompositions one can consider in order to ease the computation of $\#\mathtt{UD}(G)$?

Finally, from the multiplication obtained for one-point join of two graphs (Lemma 4 on free composition), we may ask whether $\#\mathtt{UD}(G)$ is an evaluation of the Tutte polynomial? From its universality [8], it remains to know whether there is a deletion-contradiction reduction. However defining a Tutte polynomial for directed graphs is still an active area of research [6, 10, 24, 30].

References

1. Aracena, J., Demongeot, J., Fanchon, É., Montalva, M.: On the number of different dynamics in Boolean networks with deterministic update schedules. Math. Biosci. **242**, 188–194 (2013)
2. Aracena, J., Demongeot, J., Fanchon, É., Montalva, M.: On the number of update digraphs and its relation with the feedback arc sets and tournaments. Discrete Appl. Math. **161**, 1345–1355 (2013)
3. Aracena, J., Fanchon, É., Montalva, M., Noual, M.: Combinatorics on update digraphs in Boolean networks. Discrete Appl. Math. **159**, 401–409 (2011)
4. Aracena, J., Goles, E., Moreira, A., Salinas, L.: On the robustness of update schedules in Boolean networks. Biosystems **97**, 1–8 (2009)
5. Aracena, J., Gómez, L., Salinas, L.: Limit cycles and update digraphs in Boolean networks. Discrete Appl. Math. **161**, 1–12 (2013)
6. Awan, J., Bernardi, O.: Tutte polynomials for directed graphs. J. Comb. Theory Ser. B **140**, 192–247 (2020)
7. Bodlaender, H.L., de Fluiter, B.: Parallel algorithms for series parallel graphs. In: Diaz, J., Serna, M. (eds.) ESA 1996. LNCS, vol. 1136, pp. 277–289. Springer, Heidelberg (1996)
8. Brylawski, T.H.: A decomposition for combinatorial geometries. Trans. Am. Math. Soc. **171**, 235–282 (1972)
9. Burkard, R.E., Krarup, J.: A linear algorithm for the pos/neg-weighted 1-median problem on a cactus. Computing **60**, 193–215 (1998)
10. Chan, S.H.: Abelian sandpile model and Biggs-Merino polynomial for directed graphs. J. Comb. Theory Ser. A **154**, 145–171 (2018)
11. Duffin, R.J.: Topology of series-parallel networks. J. Math. Anal. Appl. **10**, 303–318 (1965)
12. Elspas, B.: The theory of autonomous linear sequential networks. IRE Trans. Circuit Theory **6**, 45–60 (1959)
13. Fatès, N.: A guided tour of asynchronous cellular automata. J. Cell. Automata **9**, 387–416 (2014)
14. Huffman, D.A.: Canonical forms for information-lossless finite-state logical machines. IRE Trans. Inf. Theory **5**, 41–59 (1959)
15. Já Já, J.: An Introduction to Parallel Algorithms. Addison-Wesley, Boston (1992)
16. Kauffman, S.A.: Metabolic stability and epigenesis in randomly constructed genetic nets. J. Theor. Biol. **22**, 437–467 (1969)
17. Kleene, S.C.: Representation of events in nerve nets and finite automata. Project RAND RM-704, US Air Force (1951)
18. Linial, N.: Hard enumeration problems in geometry and combinatorics. SIAM J. Algebraic Discrete Methods **7**, 331–335 (1986)
19. MacMahon, P.A.: The combinations of resistances. The Electrician **28**, 601–602 (1892)
20. McCulloch, W.S., Pitts, W.: A logical calculus of the ideas immanent in nervous activity. J. Math. Biophys. **5**, 115–133 (1943)
21. Noual, M., Sené, S.: Synchronism versus asynchronism in monotonic Boolean automata networks. Nat. Comput. **17**, 393–402 (2017)
22. Perrot, K.: On the complexity of counting feedback arc sets. arXiv:1909.03339 (2019)
23. Perrot, K., Montalva-Medel, M., de Oliveira, P.P.B., Ruivo, E.L.P.: Maximum sensitivity to update schedule of elementary cellular automata over periodic configurations. Nat. Comput. **19**, 51–90 (2020)

24. Perrot, K., Pham, V.T.: Chip-firing game and partial Tutte polynomial for Eulerian digraphs. Electron. J. Comb. **23**, P1.57 (2016)
25. Riordan, J., Shannon, C.E.: The number of two-terminal series-parallel networks. J. Math. Phys. **21**, 83–93 (1942)
26. Robert, F.: Blocs-H-matrices et convergence des méthodes itératives classiques par blocs. Linear Algebra Appl. **2**, 223–265 (1969)
27. Robertson, N., Seymour, P.D.: Graph minors. XX. Wagner's conjecture. J. Comb.Theory Ser. B **92**, 325–357 (2004)
28. Thomas, R.: Boolean formalization of genetic control circuits. J. Theor. Biol. **42**, 563–585 (1973)
29. Valdes, J., Tarjan, R.E., Lawler, E.L.: The recognition of series parallel digraphs. In: Proceedings of STOC 1979, pp. 1–12 (1979)
30. Yow, K.S.: Tutte-Whitney polynomials for directed graphs and maps. Ph.D. thesis, Monash University (2019)

Faster Online Computation of the Succinct Longest Previous Factor Array

Nicola Prezza[1,2] and Giovanna Rosone[1(✉)]

[1] University of Pisa, Pisa, Italy
nicola.prezza@di.unipi.it, giovanna.rosone@unipi.it
[2] LUISS Guido Carli, Rome, Italy
nprezza@luiss.it

Abstract. We consider the problem of computing online the Longest Previous Factor array $LPF[1, n]$ of a text T of length n. For each $1 \leq i \leq n$, $LPF[i]$ stores the length of the longest factor of T with at least two occurrences, one ending at i and the other at a previous position $j < i$. We present an improvement over the previous solution by Okanohara and Sadakane (ESA 2008): our solution uses less space (compressed instead of succinct) and runs in $O(n \log^2 n)$ time, thus being faster by a logarithmic factor. As a by-product, we also obtain the first online algorithm computing the Longest Common Suffix (LCS) array (that is, the LCP array of the reversed text) in $O(n \log^2 n)$ time and compressed space. We also observe that the LPF array can be represented succinctly in $2n$ bits. Our online algorithm computes directly the succinct LPF and LCS arrays.

Keywords: Longest Previous Factor · Online · Compressed data structures

1 Introduction

This paper focuses on the problem of computing the *Longest Previous Factor* (LPF) array which stores, for each position i in a string S, the length of the longest factor (substring) of S that ends both at i and to the left of i in S. While the notion of Longest Previous Factor has been introduced in [10], an array with the same definition already appeared in McCreight's suffix tree construction algorithm [18] (the *head* array) and recently in [12] (the π array).

The concept of LPF array is close to that of *Longest Common Prefix* (*LCP*) and *Permuted Longest Common Prefix* (PLCP) arrays, structures that are usually associated with the suffix array (SA) data structure to speed up particular queries on strings (for example, pattern matching).

Supported by the project MIUR-SIR CMACBioSeq ("Combinatorial methods for analysis and compression of biological sequences") grant n. RBSI146R5L.

© Springer Nature Switzerland AG 2020
M. Anselmo et al. (Eds.): CiE 2020, LNCS 12098, pp. 339–352, 2020.
https://doi.org/10.1007/978-3-030-51466-2_31

The problem of searching for the longest previous factor is fundamental in many applications [10], including data compression and pattern analysis. For example, the LPF array can be used to derive the Ziv-Lempel factorization [25], a very powerful text compression tool based on longest previous factors [9,10].

Methods to compute the LPF array [6,9,10,21] can be broadly classified into two categories: batch (offline) and online algorithms. For instance, in [10] the authors give two offline linear-time algorithms for computing the Longest Previous Factor (LPF) array. The idea of the first algorithm is that, given SA, for any position i, they only need to consider the suffixes starting to the left of i in S which are closest to the suffix starting at position i in SA. In the second algorithm (see also [7] for a similar in spirit but independent work), the authors use a similar idea, but they take advantage of the fact that this variant processes the suffix array in one pass and requires less memory space.

In [11], the authors show how an algorithm similar to the one of [9,10] can compute the LPF array in linear running time by reading SA left-to-right (that is, online on SA) using a stack that reduces the memory space to $O(\sqrt{n})$ for a string of length n in addition to the SA, LCP and LPF arrays. This algorithm requires less than $2\sqrt{2n} + O(1)$ integer cells in addition to its input and output.

Unlike batch algorithms, an online algorithm for the problem should report the longest match just after reading each character. The online version of the problem can be defined as follows: given a history $T[1, i-1]$, and the next character $c = T[i]$, the goal is to find the longest substring that matches the current suffix: $T[j, \ldots, j+l-1] = T[i-l+1, \ldots, i]$, and report the position and the length of the matched substring. This process must be performed for all $i = 1, \ldots, n$.

Okanohara and Sadakane in [21] propose an online algorithm that relies on the incremental construction of Enhanced Suffix Arrays (ESA) [1] in a similar way to Weiner's suffix tree construction algorithm [24]. They employ compressed full-text indexing methods [20] to represent ESA dynamically in succinct space. Their algorithm requires $n \log \sigma + o(n \log \sigma) + O(n) + \sigma \log n$ bits of working space[1], $O(n \log^3 n)$ total time, and $O(\log^3 n)$ delay per character, where n is the input size and σ is the alphabet size.

Another online construction of the LCP array, in this case of a *string collection*, appears in [8]. In this work, the authors show how to update the LCP of a string collection when all strings are extended by one character.

Our work is a direct improvement over Okanohara and Sadakane's [21] algorithm. The bottleneck in their strategy is the use of a dynamic Range Minimum Query (RMQ) data structure over the (dynamic) LCP array. In this paper, we observe that the RMQ is not needed at all since we can update our structures by computing, with direct character comparisons, just irreducible LCP values. Since it is well-known that the sum of such values amounts to $O(n \log n)$, this yields a logarithmic improvement over the algorithm described in [21]. On the other hand, our strategy offers a worse delay of $O(n \log n)$ per input character.

---

[1] In their analysis they do not report the term $\sigma \log n$, which however should be included since they use a prefix sum structure over the alphabet's symbols.

2   Definitions

A *string* $S = s_1 s_2 \ldots s_n$ is a sequence of $n = |S|$ symbols from alphabet $\Sigma = [1, \sigma]$, with $\sigma \leq n$. A *text* T is a string beginning with special symbol $\# = 1$, not appearing elsewhere in T. A *factor* (or *substring*) of a string S is written as $S[i, j] = s_i \cdots s_j$ with $1 \leq i \leq j \leq n$. When defining an array A, we use the same notation $A[1, k]$ to indicate that A has k entries enumerated from 1 to k.

In this work we use text indices based on the principle of co-lexicographically sorting the prefixes of a text, rather than lexicographically sorting its suffixes. This is the same approach adopted in [21] and is required by the online left-to-right nature of the problem we consider. Given a string $S \in \Sigma^n$, we denote by $<$ the standard co-lexicographic ordering among the prefixes of S.

The *Prefix* array $PA[1, n]$ of a string $S[1, n]$ [23] is an array containing the permutation of the integers $1, 2, \ldots, n$ that arranges the ending positions of the prefixes of S into co-lexicographical order, i.e., for all $1 \leq i < j \leq n$, $S[1, PA[i]] < S[1, PA[j]]$. The *Inverse Prefix* array $IPA[1, n]$ is the inverse permutation of PA, i.e., $IPA[i] = j$ if and only if $PA[j] = i$.

The *C-array* of a string S is an array $C[1, \sigma]$ such that $C[i]$ contains the number of characters lexicographically smaller than i in S, plus one (S will be clear from the context). It is well-known that this array can be kept within $\sigma \log n + o(\sigma \log n)$ bits of space on a dynamic string by using succinct searchable partial sums [3,14], which support all operations in $O(\log n)$ time.

The *co-lexicographic Burrows-Wheeler Transform* $BWT[1, n]$ of a text T is a reversible transformation that permutes its symbols as $BWT[i] = T[PA[i] + 1]$ if $PA[i] < n$, and $BWT[i] = \#$ otherwise [5].

The *Longest Common Suffix* array $LCS[1, n]$ of a string S [17] is an array storing in $LCS[i]$ the length of the longest common suffix shared by the $(i-1)$-th and i-th co-lexicographically smallest text prefixes if $i > 1$, and $LCS[1] = 0$ otherwise. Function $LCS(i, j)$ generalizes this array: given a string $S[1, n]$, $LCS(i, j)$ denotes the longest common suffix between $S[1, PA[i]]$ and $S[1, PA[j]]$. The *Permuted Longest Common Suffix* array $PLCS[1, n]$ of a string S stores LCS values in string order, rather than co-lexicographic order: $PLCS[i] = LCS[IPA[i]]$.

Function $S.rank_c(i)$, where $c \in \Sigma$, returns the number of characters equal to c in $S[1, i-1]$. Similarly, function $S.select_c(i)$ returns the position of S containing the i-th occurrence of c.

Given $BWT[1, n]$ of a text $T[1, n]$, the *LF mapping* is a function $BWT.LF(i)$ that, given the BWT position containing character $T[j]$ (with $j = PA[i] + 1$), returns the BWT position i' of character $T[j + 1]$. This function can be implemented with a *rank* operation on BWT and one access to the C array. Similarly, the *FL mapping* is the reverse of LF: this is the function $BWT.FL(i)$ that, given the BWT position containing character $T[j]$ (with $j = PA[i] + 1$), returns the BWT position i' of character $T[j - 1]$ (assume for simplicity that $j > 1$; otherwise, $BWT[i] = \#$). This function can be implemented with a *select* operation on BWT and a search on the C array.

3   Succinct PLCS and LPF Arrays

We start by formally introducing the definition of *LPF array*.

Definition 1 (Longest Previous Factor array). *The* Longest Previous Factor array $LPF[1, n]$ *of a string* $S[1, n]$ *is the array containing, at each location* $LPF[i]$, *the largest integer* k *such that there exists* $j < i$ *for which the longest common suffix between* $S[1, j]$ *and* $S[1, i]$ *has length* k.

Kasai et al. [16] observe that the Permuted Longest Common Prefix array is almost increasing: $PLCP[i + 1] \geq PLCP[i] - 1$. Of course, this still holds true for the Permuted Longest Common Suffix array that we consider in our work. Specifically, the symmetric relation $PLCS[i + 1] \leq PLCS[i] + 1$ holds. In the next lemma we observe that the same property is true also for the LPF array.

Lemma 1. *For any* $i < n$, *it holds* $LPF[i + 1] \leq LPF[i] + 1$.

Proof. Let $LPF[i] = k$. Then, k is the largest integer such that the substring $T[i - k + 1, i]$ starts at another position $j < i - k + 1$. Assume, for contradiction, that $LPF[i + 1] = k' > k + 1$. Then, this means that $s = T[(i + 1) - k' + 1, i + 1]$ occurs at another position $j' < (i + 1) - k' + 1$. But then, also the prefix $T[(i + 1) - k' + 1, i]$ of s occurs at j'. This is a contradiction, since the length of $T[(i + 1) - k' + 1, i]$ is $k' - 1 > k = LPF[i]$. □

Note that $PLCS[1] = LPF[1] = 0$, thus the two arrays can be encoded and updated succinctly with the same technique, described in Lemma 2.

Lemma 2. *Let* $A[1, n]$ *be a non-negative integer array satisfying properties (a)* $A[1] = 0$ *and (b)* $A[i + 1] \leq A[i] + 1$ *for* $i < n$. *Then, there is a data structure of* $2n + o(n)$ *bits supporting the following operations in* $O(\log n / \log \log n)$ *time:*

(1) access any $A[i]$,
(2) append a new element $A[n + 1]$ *at the end of* A, *and*
(3) update: $A[i] \leftarrow A[i] + \Delta$,

provided that operations (2) and (3) do not violate properties (a) and (b). Running time of operation (2) is amortized $(O(n \log n / \log \log n)$ *in the worst case).*

Proof. We encode A as a bitvector A' of length *at most* $2n$ bits, defined as follows. We start with $A' = 01$ and, for $i = 2, \ldots, n$ we append the bit sequence $0^{A[i-1]+1-A[i]}1$ to the end of A'. The intuition is that every bit set increases the previous value $A[i - 1]$ by 1, and every bit equal to 0 decreases it by 1. Then, the value $A[i]$ can be retrieved simply as $2i - A'.select_1(i)$. Clearly, A' contains n bits equal to 1 since for each $1 \leq i \leq n$ we insert a bit equal to 1 in A'. Since the total number of 1s is n and A is non-negative, A' contains at most n bits equal to 0 as well. It follows that A' contains at most $2n$ bits in total. In order to support updates, we encode the bitvector with the dynamic string data structure of Munro and Nekrich [19], which takes at most $2n + o(n)$ bits of space and supports queries and updates in $O(\log n / \log \log n)$ worst-case time. We already showed how operation (1) reduces to *select* on A'. Let $\Delta = A[n]+1-A[n+1]$. To support operation (2), we need to append the bit sequence $0^\Delta 1$ at the end of A'. In the worst case, this operation takes $O(\Delta \log n/\log \log n) = O(n \log n)$ time. However, the sum of all Δ is equal to the total number of 0s in the bitvector; this implies that, over a sequence of n insertions, this operation takes $O(\log n/\log \log n)$ amortized time. Finally, operation $A[i] \leftarrow A[i] + \Delta$ can be implemented by moving the bit at position $A'.select_1(i)$ by Δ positions to the left, which requires just one *delete* and one *insert* operation on A' ($O(\log n/\log \log n)$ time). Note that this is always possible, provided that the update operation does not violate properties (a) and (b) on the underlying array A.   □

4   Online Algorithm

We first give a sketch of our idea, and then proceed with the details. Similarly to Okanohara and Sadakane [21], we build online the BWT and the compressed LCS array of the text, and use the latter component to output online array LPF. This is possible by means of a simple observation: after reading character $T[i]$, entry $LPF[i]$ is equal to the maximum between $LCS[IPA[i]]$ and $LCS[IPA[i] + 1]$. As in [21], array LCS is represented in compressed form by storing PLCS (in $2n + o(n)$ bits, Lemma 2) and a sampling of the prefix array PA which, together with BWT, allows computing any $PA[i]$ in $O(\log^2 n)$ time. Then, we can retrieve any LCS value in $O(\log^2 n)$ time as $LCS[i] = PLCS[PA[i]]$.

The bottleneck of Okanohara and Sadakane's strategy is the update of LCS. This operation requires being able to compute the longest common suffix between two arbitrary text's prefixes $T[1, PA[i]]$ and $T[1, PA[j]]$ (see [21] for all the details). By a well-known relation, this value is equal to $\min(LCS[i,j])$ (assume $i < j$ w.l.o.g.). In Okanohara and Sadakane's work, this is achieved using a dynamic Range Minimum Query (RMQ) data structure on top of LCS. The RMQ is a balanced tree whose leaves cover $\Theta(\log n)$ LCS values each and therefore requires accessing $O(\log n)$ LCS values in order to compute $\min(LCS[i,j])$, for a total running time of $O(\log^3 n)$. We note that this running time cannot be improved by simply replacing the dynamic RMQ structure of [21] with more recent structures. Brodal et al. in [4] describe a dynamic RMQ structure supporting queries and updates in $O(\log n/\log \log n)$ time, but the required space is $O(n)$ words. Heliou et al. in [13] reduce this space to $O(n)$ bits, but they require, as in [21], $O(\log n)$ accesses to the underlying array.

Our improvement over Okanohara and Sadakane's algorithm stems from the observation that the RMQ structure is not needed at all, as we actually need to compute by direct symbol comparisons just *irreducible* LCS values:

Definition 2. $LCS[i]$ *is said to be* irreducible *if and only if either* $i = 0$ *or* $BWT[i] \neq BWT[i-1]$ *hold.*

Irreducible LCS values enjoy the following property:

Lemma 3 ([15], **Thm. 1**). *The sum of all irreducible LCS values is at most* $2n \log n$.

As a result, we will spend overall just $O(n \log^2 n)$ time to compute all irreducible LCS values. This is less than the time $O(n \log^3 n)$ needed in [21] to compute $O(n)$ minima on the LCS.

4.1   Data Structures

Dynamic BWT. Let $T[1, i]$ be the text prefix seen so far. As in [21], we keep a dynamic BWT data structure to store the BWT of $T[1, i]$. In our case, this structure is represented using Munro and Nekrich's dynamic string [19] and takes $nH_k + o(n \log \sigma) + \sigma \log n + o(\sigma \log n)$ bits of space, for any $k \in o(\log_\sigma n)$. The latter two space components are needed for the C array encoded with succinct searchable partial sums [3,14]. The structure supports *rank*, *select*, and *access* in $O(\log n / \log \log n)$ time, while *appending* a character at the end of T and computing the LF and FL mappings are supported in $O(\log n)$ time (the bottleneck are succinct searchable partial sums, which cannot support all operations simultaneously in $O(\log n / \log \log n)$ time by current implementations [3,14]).

Dynamic Sparse Prefix Array. As in Okanohara and Sadakane's solution, we also keep a dynamic Prefix Array sampling. Let $D = \lceil \log n \rceil$ be the sample rate. We store in a dynamic sequence PA' all integers $x_j = j/D$ such that $j \bmod D = 0$, for $j \le i$ (i.e. we sample one out of D text positions and re-enumerate them starting from 1). Letting $j_1 < \cdots < j_k$ be the co-lexicographic order of the sampled text positions seen so far, the corresponding integers are stored in PA' in the order $x_{j_1}, \ldots, x_{j_k}$. In the next paragraph we describe the structure used to represent PA' (as well as its inverse), which will support queries and updates in $O(\log n)$ time. We use again Munro and Nekrich's dynamic string [19] to keep a dynamic bitvector B_{PA} of length i (i being the length of the current text prefix) that marks with a bit set sampled entries of PA. Since we sample one out of $D = \lceil \log n \rceil$ text's positions and the bitvector is entropy-compressed, its size is $o(n)$ bits. At this point, any $PA[j]$ can be retrieved by returning $D \cdot PA'[B_{PA}.rank_1(j) + 1]$ if $B_{PA}[j] = 1$ or performing at most D LF mapping steps otherwise, for a total running time of $O(\log^2 n)$. Note that, by the way we re-enumerate sampled text positions, the sequence PA' is a permutation.

Dynamic Sparse Inverse Prefix Array. The first difference with Okanohara and Sadakane's solution is that we keep the inverse of PA' as well, that is, a (dynamic) sparse inverse prefix array: we denote this array by IPA' and define it as $IPA'[PA'[j]] = j$, for all $1 \le j \le |PA'|$. First, note that we insert integers in PA' in increasing order: $x = 1, 2, 3, \ldots$. Inserting a new integer x at some position t in PA' has the following effect in IPA': first, all elements $IPA'[k] \ge t$ are increased by 1. Then, value t is appended at the end of IPA'.

Example 1. Let PA' and $IPA' = (PA')^{-1}$ be the following permutations: $PA' = \langle 3, 1, 2, 4 \rangle$ and $IPA' = \langle 2, 3, 1, 4 \rangle$. Suppose we insert integer 5 at position 2 in PA'. The updated permutations are: $PA' = \langle 3, 5, 1, 2, 4 \rangle$ and $IPA' = \langle 3, 4, 1, 5, 2 \rangle$.

Policriti and Prezza in [22] show how to represent a permutation of size k and its inverse upon insertions and access queries in $O(\log k)$ time per operation and $O(k)$ words of space. The idea is to store PA$'$ in a self-balancing tree, sorting its elements by the inverse permutation IPA$'$. Then, IPA$'$ is represented simply as a vector of pointers to the nodes of the tree. By enhancing the tree's nodes with the corresponding sub-tree sizes, the tree can be navigated top-down (to access PA$'$) and bottom-up (to access IPA$'$) in logarithmic time. Since we sample one out of $D = \lceil \log n \rceil$ positions, the structure takes $O(n)$ bits of space.

To compute any $IPA[j]$, we proceed similarly as for PA. We compute the sampled position $j - \delta$ (with $\delta \geq 0$) preceding j in the text, we find the corresponding position t on PA as $t = B_{PA}.select(IPA'[(j - \delta)/D])$, and finally perform $\delta \leq D$ steps of LF mapping to obtain $IPA[j]$. Note that, without loss of generality, we can consider position 1 to be always sampled since $IPA[1] = 1$ is constant. To sum up, computing any $IPA[j]$ requires $O(\log^2 n)$ time, while updating PA$'$ and IPA$'$ takes $O(\log n)$ time.

Dynamic PLCS Vector. We also keep the dynamic PLCS vector, stored using the structure of Lemma 2. When extending the current text prefix $T[1, i]$ by character $T[i + 1]$, LCS changes in two locations: first, a new value is inserted at position $IPA[i + 1]$. Then, the value $LCS[IPA[i + 1] + 1]$ (if this cell exists) can possibly increase, due to the insertion of a new text prefix before it in co-lexicographic order. As a consequence, PLCS changes in two places as well: (i) a new value $PLCS[i+1] = LCS[IPA[i+1]]$ is appended at the end, and (ii) value $PLCS[PA[IPA[i + 1] + 1]]$ (possibly) increases. Both operations are supported in $O(\log n)$ (amortized) time by Lemma 2.

The way these new PLCS values are calculated is where our algorithm differs from Okanohara and Sadakane's [21], and is described in the next section.

4.2   Updating the LCS Array

In this section we show how to update the LCS array (stored in compressed format as described in the previous sections).

Algorithm. We show how to compute the new LCS value to be inserted at position $IPA[i + 1]$ (after extending $T[1, i]$ with $T[i + 1]$). The other update, to $LCS[IPA[i + 1] + 1]$, is completely symmetric so we just sketch it. Finally, we analyze the amortized complexity of our algorithm.

Let $a = T[i + 1]$ be the new text symbol, and let k be the position such that $BWT[k] = \#$. We recall that the BWT extension algorithm works by replacing $BWT[k]$ with the new character $a = T[i + 1]$, and by inserting $\#$ at position $C[a] + BWT.rank_a(k) = IPA[i + 1]$. We also recall that the location of the first occurrence of a symbol a preceding/following $BWT[k] = \#$ can be easily found with one *rank* and one *select* operations on BWT.

Now, consider the BWT of $T[1, i]$. We distinguish three main cases (in [21], all these cases were treated with a single Range Minimum Query):

(a) $BWT[1, k]$ does not contain occurrences of character a. Then, $T[1, i + 1]$ is the co-lexicographically smallest prefix ending with a, therefore the new LCS value to be inserted at position $IPA[i + 1]$ is 0.

(b) $BWT[k-1] = a$. Then, prefix $T[1, PA[k-1]+1]$ (ending with $BWT[k-1] = a$) immediately precedes $T[1, i+1]$ in co-lexicographic order. It follows that the LCS between these two prefixes is equal to 1 plus the LCS between $T[1, PA[k-1]]$ and $T[1, i]$, i.e. $1 + LCS[IPA[i]]$. This is the new LCS value to be inserted at position $IPA[i+1]$.

(c) The previous letter equal to a in $BWT[1, k]$ occurs at position $j < k-1$. The goal here is to compute the LCS $\ell = LCS(j, k)$ between prefixes $T[1, PA[k]]$ and $T[1, PA[j]]$. Integer $\ell + 1$ is the new LCS value to be inserted at position $IPA[i + 1]$. We distinguish two further sub-cases.

  (c.1) String $BWT[k+1, i]$ does not contain occurrences of character a. Then, we compare the two prefixes $T[1, PA[k]]$ and $T[1, PA[j]]$ right-to-left simply by repeatedly applying function FL from BWT positions j and k. The number of performed symbol comparisons is $LCS(j, k)$.

  (c.2) There is an occurrence of a after position k. Let $q > k$ be the smallest position such that $BWT[q] = a$. Table 1 reports an example of this case. Then, $T[1, PA[j]+1]$ and $T[1, PA[q]+1]$ are adjacent in co-lexicographic order, thus we can compute $\ell' = LCS(j, q)$ as follows. Letting $j' = BWT.LF(j)$ and $q' = BWT.LF(q) = j' + 1$ (that is, the co-lexicographic ranks of $T[1, PA[j]+1]$ and $T[1, PA[q]+1]$, respectively), we have $\ell' = LCS[q']-1$. Since $j < k < q$, we have $LCS(j, k) = \ell \geq \ell'$. In order to compute $\ell = LCS(j, k)$, the idea is to skip the first ℓ' comparisons (which we know will return a positive result), and only compare the remaining $\ell - \ell'$ characters in the two prefixes, that is, compare the two prefixes $T[1, PA[k]-\ell']$ and $T[1, PA[j] - \ell']$. This can be achieved by finding the co-lexicographic ranks of these two prefixes, that is $IPA[PA[k] - \ell']$ and $IPA[PA[j] - \ell']$ respectively, and applying the FL function from these positions to extract the $\ell - \ell'$ remaining matching characters in the prefixes. The number of performed symbol comparisons is $\ell - \ell' = LCS(j, k) - LCS(j, q)$.

As noted above, the other update to be performed at position $LCS[IPA[i + 1] + 1]$ is completely symmetric so we just sketch it here. The cases where $BWT[k, i]$ does not contain occurrences of a or where $BWT[k + 1] = a$ correspond to cases (a) and (b). If the first occurrence of a following position k appears at position $q > k + 1$, on the other hand, we distinguish two further cases. The case where $BWT[1, k]$ does not contain occurrences of a is handled as case (c.1) above (by direct character comparisons between two text prefixes). Otherwise, we find the first occurrence of $a = BWT[j]$ before position k and proceed as in case (c.2), by finding the LCS ℓ' between the suffixes $T[1, PA[q]]$ and $T[1, PA[j]]$, and comparing prefixes $T[1, PA[k] - \ell']$ and $T[1, PA[q] - \ell']$.

Amortized Analysis. In the following, by *symbol comparisons* we indicate the comparisons performed in case (c) to compute LCS values (by means of iterating the FL mapping). For simplicity, we count only comparisons resulting in a match between the two compared characters: every time we encounter a mismatch, the comparison is interrupted; this can happen at most $2n$ times (as we update at most two LCS values per iteration), therefore it adds at most $O(n \log n)$ to our final running time (as every FL step takes $O(\log n)$ time).

We now show that the number of symbol comparisons performed in case (c) is always upper-bounded by the sum of irreducible LCS values.

Definition 3. *A BWT position $k > 1$ is said to be a* relevant run break *if and only if:*

(i) $BWT[k-1] \neq BWT[k]$,
(ii) there exists $j < k - 1$ such that $BWT[j] = BWT[k]$, and
(iii) if $BWT[k-1] = \#$, then $k > 2$ and $BWT[k-2] \neq BWT[k]$.

Condition (i) requires k to be on the border of an equal-letter BWT run. Condition (ii) requires that there is a character equal to $BWT[k]$ before position $k - 1$, and condition (iii) states that $\#$ does not contribute in forming relevant run breaks (e.g. in string $a\#a$, the second occurrence of a is not a relevant run break; however, in $ac\#a$ the second occurrence of a is). Intuitively, condition (iii) is required since extending the text by one character might result in two runs of the same letter separated by just $\#$ to be merged (e.g. $aaa\#a$ becomes $aaaaa$ after replacing $\#$ with a). Without condition (iii), after such a merge we could have characters inside a run that are charged with symbol comparisons.

In Lemma 4 we prove that our algorithm maintains the following invariant:

Invariant 1. *Consider the structures BWT and LCS for $T[1, i]$ at step i. Moreover, let k be a relevant run break, and let $j < k - 1$ be the largest position such that $BWT[k] = BWT[j]$. Then:*

1. Position k is charged with $c_k = LCS(j, k)$ symbol comparisons, and
2. Only relevant run breaks are charged with symbol comparisons.

Lemma 4. *Invariant 1 is true after every step $i = 1, \ldots, n$ of our algorithm.*

Proof. After step $i = 1$, we have processed just $T[1]$ and the property is trivially true as there are no relevant run breaks. Assume by inductive hypothesis that the property holds at step i, i.e. after building all structures (BWT, LCS) for $T[1, i]$. We show that the application of cases (a-c) maintains the invariant true.

Case (a) does not perform symbol comparisons. Moreover, it does not destroy any relevant run break. The only critical case is $BWT[k+1] = a$, since replacing $BWT[k] = \#$ with a destroys the run break at position $k + 1$. However, note that $k + 1$ cannot be a *relevant* run break, since $BWT[1, k]$ does not contain occurrences of a. It follows that case (a) maintains the invariant.

Also case (b) does not perform symbol comparisons and does not destroy any relevant run break. The only critical case is $BWT[k + 1] = a$, since replacing $BWT[k] = \#$ with a destroys the run break at position $k+1$. However, note that $k + 1$ cannot be a *relevant* run break, since $BWT[k - 1] = a$ and $BWT[k+1] =$

a are separated by $BWT[k] = \#$, which by definition does not contribute in forming relevant run breaks. It follows that case (b) maintains the invariant.

(c.1) Consider the BWT of $T[1, i]$, and let k be the terminator position: $BWT[k] = \#$. Note that, by Definition 3, k is not a relevant run break since no other position contains the terminator, and thus by Invariant 1 it is not charged yet with any symbol comparison. Case (c.1) compares the k-th and j-th co-lexicographically smallest text prefixes, where $j < k-1$ is the previous occurrence of a in the BWT. Clearly, the number of comparisons performed is exactly $c_k = LCS(j, k)$: we charge this quantity to BWT position k. Then, we update the BWT by (i) replacing $BWT[k] = \#$ with a, which makes k a valid relevant run break since $BWT[k-1] \neq a$, $BWT[j] = BWT[k]$, and $j < k-1$ and (ii) inserting $\#$ in some BWT position, which (possibly) shifts position k to $k' \in \{k, k + 1\}$ (depending whether $\#$ is inserted before or after k) but does not alter the value of $c_k = LCS(j, k')$, so k' is a relevant run break and is charged correctly as of Invariant 1. Finally, note that (1) the new BWT position containing $\#$ is not charged with any symbol comparison (since we just inserted it), (2) that, if two runs get merged after replacing $\#$ with $T[i + 1]$ then, thanks to Condition (iii) of Definition 3 and Invariant 1 at step i, no position inside a equal-letter run is charged with symbol comparisons, and (3) if the new $\#$ is inserted inside a equal-letter run a^t, thus breaking it as $a^{t_1}\#a^{t_2}$ with $t = t_1 + t_2$ and $t_1 > 0$, then the position following $\#$ is not charged with any symbol comparison. (1–3) imply that we still charge only relevant run breaks with symbol comparisons: Invariant 1 is therefore true at step $i + 1$.

(c.2) Consider the BWT of $T[1, i]$, and let k, j, q, with $j < k - 1 < k < q$, be the terminator position ($BWT[k] = \#$) and the immediately preceding and following positions containing $a = BWT[j] = BWT[q]$. Note that q is a relevant run-break, charged with $c_q = LCS(j, q)$ symbol comparisons by Invariant 1. Assume that $LCS(j, k) \geq LCS(k, q)$: the other case is symmetric and we discuss it below. Then, $LCS(k, q) = LCS(j, q) = c_q$. First, we "lift" the $c_q = LCS(j, q)$ symbol comparisons from position q and re-assign them to position k. By definition, case (c.2) of our algorithm performs $LCS(j, k) - LCS(j, q)$ symbol comparisons; we charge also these symbol comparisons to position k. After replacing $BWT[k]$ with letter a, position k becomes a relevant run break, and is charged with $c_q + (LCS(j, k) - LCS(j, q)) = LCS(j, k) = c_k$ symbol comparisons. Position q, on the other hand, is now charged with 0 symbol comparisons; note that this is required if $q = k + 1$ (as in the example of Table 1), since in that case q is no longer a relevant run break (as we replaced $\#$ with a). Finally, we insert $\#$ in some BWT position which, as observed above, does not break Invariant 1.

The other case is $LCS(j, k) < LCS(k, q)$. Then $LCS(j, k) = LCS(j, q)$, and therefore case (c.2) does not perform additional symbol comparisons to compute $LCS[IPA[i + 1]]$. On the other hand, the symmetric of case (c.2) (i.e. the case where we update $LCS[IPA[i + 1] + 1]$) performs $LCS(k, q) - LCS(j, q)$ symbol comparisons if $q > k + 1$ (none otherwise); these are all charged to position q and, added to the $LCS(j, q)$ comparisons already charged to q, sum up to $LCS(k, q)$ comparisons. This is correct, since in that case q remains a relevant run break. If, on the other hand, $q = k+1$, then no additional comparisons need to be made to update $LCS[IPA[i + 1] + 1]$, and we simply lift the $LCS(j, q)$ comparisons from q (which is no longer a relevant run break) and charge them to k (which becomes a relevant run break). This is correct, since k is now charged with $LCS(j, q) = LCS(j, k)$ symbol comparisons. □

Table 1. The example illustrates case (c.2). Column L is the BWT. The other columns contain the sorted text prefixes. Left: structures for $T[1, i] = \#abaaabbaababa$. We are about to extend the text with letter a. Positions k, j, q contain $\#$ (to be replaced with a) and the immediately preceding and succeeding BWT positions containing a. To find $LCS(j, q)$, apply LF to j, q, obtaining positions j' and q'. Then, $LCS(j, q) = LCS[q'] - 1 = 2$, emphasized in italic. At this point, $LCS(j, k)$ is computed by comparing the j-th and k-th smallest prefixes outside the italic zone (found using IPA and PA). In the example, we find 1 additional match (underlined). It follows that the new LCS to be inserted between positions j' and q' is $1 + LCS(j, k) = 1 + (LCS(j, q) + 1) = 4$. Right: structures updated after appending a to the text. In bold on column LCS: the new LCS value inserted $(LCS[IPA[i + 1]] = 4)$ and the one updated by the symmetric of case (c.2) $(LCS[q'] = 3$; in this example, the value doesn't change). In bold on column F: last letters of the j'-th and q'-th smallest text's prefixes, interleaved with $T[1, i+1]$.

F	L	LCS	
b a a a b b a a b a b a #	a	0	
a a a b b a a b a b a # a	b	0	
b a a b a b a # a b a a a	b	1	
b b a a b a b a # a b a a	a	2	j'
a b a # a b a a a b b a a	b	3	q'
a b b a a b a b a # a b a	a	1	j
a # a b a a a b b a a b a	b	3	
a b a a a b b a a b a b a	#	3	k
b a b a # a b a a a b b a	a	2	q
a a b b a a b a b a # a b	a	0	
a a b a b a # a b a a a b	b	2	
b a # a b a a a b b a a b	a	3	
# a b a a a b b a a b a b	a	2	
a b a b a # a b a a a b b	a	1	

F	L	LCS	
b a a a b b a a b a b a a #	a	0	
a a a b b a a b a b a a # a	b	0	
b a a b a b a a # a b a a a	b	1	
b b a a b a b a a # a b a a	a	2	j'
a b a a a b b a a b a b a a	#	**4**	
a b a a # a b a a a b b a a	b	**3**	q'
a b b a a b a b a a # a b a	a	1	j
a a # a b a a a b b a a b a	b	3	
# a b a a a b b a a b a b a	a	3	k
b a b a a # a b a a a b b a	a	2	q
a a b b a a b a b a a # a b	a	0	
a a b a b a a # a b a a a b	b	2	
b a a # a b a a a b b a a b	a	3	
a # a b a a a b b a a b a b	a	2	
a b a b a a # a b a a a b b	a	1	

Lemma 5. *At any step* $i = 1, \ldots, n$, *let* $k_1, \ldots, k_r$ *be the relevant run breaks and* $c_{k_1}, \ldots, c_{k_r}$ *be the symbol comparisons charged to them, respectively. Then,* $\sum_{t=1}^{r} c_{k_t} \leq 2i \log i$.

Proof. By definition of c_{k_t}, we have $c_{k_t} = LCS(j, k_t) \leq LCS(k_t - 1, k_t) = LCS[k_t]$, where $j < k_t - 1$ is the largest position to the left of $k_t - 1$ containing symbol $BWT[k_t]$. Moreover, note that $\{k_1, \ldots, k_r\}$ is a subset of the BWT run breaks $\{k : BWT[k - 1] \neq BWT[k]\}$, therefore each $LCS[k_t]$ is irreducible. Let S be the sum of irreducible LCS values. By applying Lemma 3, we obtain:

$$\sum_{j=1}^{r} c_{k_t} \leq \sum_{j=1}^{r} LCS[k_t] \leq S \leq 2i \log i$$

We obtain our main result:

Theorem 2 (Online succinct LPF and LCS arrays). *The succinct LPF and LCS arrays of a text $T \in [1, \sigma]^n$ can be computed online in $O(n \log^2 n)$ time and $O(n \log n)$ delay per character using $nH_k + o(n \log \sigma) + O(n) + \sigma \log n + o(\sigma \log n)$ bits of working space (including the output), for any $k \in o(\log_\sigma n)$.*

Proof. After LCS and IPA have been updated at step i, we can compute $LPF[i]$ simply as $LPF[i] = \max\{LCS[IPA[i]], LCS[IPA[i] + 1]\}$ in $O(\log^2 n)$ time. This value can be appended at the end of the succinct representation of LPF (Lemma 2) in $O(\log n)$ amortized time (which in the worst case becomes $O(n \log n)$). Updating BWT, PA', and IPA' takes $O(\log n)$ time per character. The most expensive part is updating the structures representing LCS: at each step we need to perform a constant number of accesses to arrays PA, IPA, and LCS, which alone takes $O(\log^2 n)$ time per character. Updating PLCS takes $O(\log n)$ amortized time per character (which in the worst case becomes $O(n \log n)$) by Lemma 2. By Lemma 5 we perform overall $O(n \log n)$ symbol comparisons, each requiring two FL steps and two BWT accesses, for a total of $O(n \log^2 n)$ time. Note that a single comparison between two text prefixes cannot extend for more than n characters, therefore in the worst case a single step takes $O(n \log n)$ time. This is our delay per character. To conclude, in Sect. 4.1 we showed that our data structures take at most $nH_k + o(n \log \sigma) + O(n) + \sigma \log n + o(\sigma \log n)$ bits of space.   □

Finally we note that, at each step i, we can output also the *location* of the longest previous factor: this requires just one access to the prefix array PA.

5   Conclusions

We improved the state-of-the-art algorithm, from Okanohara and Sadakane [21], computing online the (succinct) LPF and LCS arrays of the text. Our improvement stems from the observation that a dynamic RMQ structure over the LCS array is not needed, as the LCS can be updated by performing a number of character comparisons that is upper-bounded by the sum of irreducible LCS values. Future extensions of this work will include reducing the delay of our algorithm (currently $O(n \log n)$). We observe that it is rather simple to obtain $O(\log^2 n)$ delay at the cost of randomizing the algorithm by employing an online Karp-Rabin fingerprinting structure such as the one described in [2]: once fast fingerprinting is available, one can quickly find the LCS between any two text prefixes by binary search. It would also be interesting to reduce the overall running time of our algorithm. This, however, does not seem straightforward to achieve, as it would require finding a faster implementation of a dynamic compressed prefix array (and its inverse) and finding a faster way of updating LCS values (possibly, with a faster dynamic succinct RMQ structure).

References

1. Abouelhoda, M.I., Kurtz, S., Ohlebusch, E.: Replacing suffix trees with enhanced suffix arrays. J. Discrete Algorithms **2**(1), 53–86 (2004). https://doi.org/10.1016/S1570-8667(03)00065-0

2. Alzamel, M., et al.: Online algorithms on antipowers and antiperiods. In: Brisaboa, N.R., Puglisi, S.J. (eds.) SPIRE 2019. LNCS, vol. 11811, pp. 175–188. Springer, Cham (2019). https://doi.org/10.1007/978-3-030-32686-9_13

3. Bille, P., Christiansen, A.R., Prezza, N., Skjoldjensen, F.R.: Succinct partial sums and fenwick trees. In: Fici, G., Sciortino, M., Venturini, R. (eds.) SPIRE 2017. LNCS, vol. 10508, pp. 91–96. Springer, Cham (2017). https://doi.org/10.1007/978-3-319-67428-5_8

4. Brodal, G.S., Davoodi, P., Srinivasa Rao, S.: Path minima queries in dynamic weighted trees. In: Dehne, F., Iacono, J., Sack, J.-R. (eds.) WADS 2011. LNCS, vol. 6844, pp. 290–301. Springer, Heidelberg (2011). https://doi.org/10.1007/978-3-642-22300-6_25

5. Burrows, M., Wheeler, D.: A block sorting data compression algorithm. Technical report, DEC Systems Research Center (1994)

6. Chairungsee, S., Charuphanthuset, T.: An approach for LPF table computation. In: Anderst-Kotsis, G., et al. (eds.) DEXA 2019. CCIS, vol. 1062, pp. 3–7. Springer, Cham (2019). https://doi.org/10.1007/978-3-030-27684-3_1

7. Chen, G., Puglisi, S.J., Smyth, W.F.: Fast and practical algorithms for computing all the runs in a string. In: Ma, B., Zhang, K. (eds.) CPM 2007. LNCS, vol. 4580, pp. 307–315. Springer, Heidelberg (2007). https://doi.org/10.1007/978-3-540-73437-6_31

8. Cox, A.J., Garofalo, F., Rosone, G., Sciortino, M.: Lightweight LCP construction for very large collections of strings. J. Discrete Algorithms **37**, 17–33 (2016). https://doi.org/10.1016/j.jda.2016.03.003

9. Crochemore, M., Ilie, L., Smyth, W.F.: A simple algorithm for computing the Lempel Ziv factorization. In: Data Compression Conference (DCC 2008), pp. 482–488 (2008). https://doi.org/10.1109/DCC.2008.36

10. Crochemore, M., Ilie, L.: Computing longest previous factor in linear time and applications. Inf. Process. Lett. **106**(2), 75–80 (2008). https://doi.org/10.1016/j.ipl.2007.10.006

11. Crochemore, M., Ilie, L., Iliopoulos, C.S., Kubica, M., Rytter, W., Waleń, T.: Computing the longest previous factor. Eur. J. Comb. **34**(1), 15–26 (2013). https://doi.org/10.1016/j.ejc.2012.07.011

12. Franěk, F., Holub, J., Smyth, W.F., Xiao, X.: Computing quasi suffix arrays. J. Autom. Lang. Comb. **8**(4), 593–606 (2003)

13. Heliou, A., Léonard, M., Mouchard, L., Salson, M.: Efficient dynamic range minimum query. Theor. Comput. Sci. **656**(PB), 108–117 (2016). https://doi.org/10.1016/j.tcs.2016.07.002

14. Hon, W.K., Sadakane, K., Sung, W.K.: Succinct data structures for searchable partial sums with optimal worst-case performance. Theor. Comput. Sci. **412**(39), 5176–5186 (2011)

15. Kärkkäinen, J., Manzini, G., Puglisi, S.J.: Permuted longest-common-prefix array. In: Kucherov, G., Ukkonen, E. (eds.) CPM 2009. LNCS, vol. 5577, pp. 181–192. Springer, Heidelberg (2009). https://doi.org/10.1007/978-3-642-02441-2_17

16. Kasai, T., Lee, G., Arimura, H., Arikawa, S., Park, K.: Linear-time longest-common-prefix computation in suffix arrays and its applications. In: Amir, A. (ed.) CPM 2001. LNCS, vol. 2089, pp. 181–192. Springer, Heidelberg (2001). https://doi.org/10.1007/3-540-48194-X_17

17. Manber, U., Myers, G.: Suffix arrays: a new method for on-line string searches. SIAM J. Comput. **22**(5), 935–948 (1993). https://doi.org/10.1137/0222058

18. McCreight, E.M.: A space-economical suffix tree construction algorithm. J. ACM **23**(2), 262–272 (1976). https://doi.org/10.1145/321941.321946

19. Munro, J.I., Nekrich, Y.: Compressed data structures for dynamic sequences. In: Bansal, N., Finocchi, I. (eds.) ESA 2015. LNCS, vol. 9294, pp. 891–902. Springer, Heidelberg (2015). https://doi.org/10.1007/978-3-662-48350-3_74

20. Navarro, G., Mäkinen, V.: Compressed full-text indexes. ACM Comput. Surv. textbf39(1), 2-es (2007). https://doi.org/10.1145/1216370.1216372

21. Okanohara, D., Sadakane, K.: An online algorithm for finding the longest previous factors. In: Halperin, D., Mehlhorn, K. (eds.) ESA 2008. LNCS, vol. 5193, pp. 696–707. Springer, Heidelberg (2008). https://doi.org/10.1007/978-3-540-87744-8_58

22. Policriti, A., Prezza, N.: From LZ77 to the run-length encoded Burrows-Wheeler transform, and back. In: 28th Annual Symposium on Combinatorial Pattern Matching. Schloß Dagstuhl (2017)

23. Puglisi, S.J., Turpin, A.: Space-time tradeoffs for longest-common-prefix array computation. In: Hong, S.-H., Nagamochi, H., Fukunaga, T. (eds.) ISAAC 2008. LNCS, vol. 5369, pp. 124–135. Springer, Heidelberg (2008). https://doi.org/10.1007/978-3-540-92182-0_14

24. Weiner, P.: Linear pattern matching algorithms. In: 14th Annual Symposium on Switching and Automata Theory (SWAT), pp. 1–11 (1973). https://doi.org/10.1109/SWAT.1973.13

25. Ziv, J., Lempel, A.: A universal algorithm for sequential data compression. IEEE Trans. Inf. Theory **23**(3), 337–343 (1977). https://doi.org/10.1109/TIT.1977.1055714

Recent Advances in Text-to-Pattern Distance Algorithms

Przemysław Uznański$^{(\boxtimes)}$ ⓘ

Institute of Computer Science, University of Wrocław, Wrocław, Poland
puznanski@cs.uni.wroc.pl

Abstract. Computing text-to-pattern distances is a fundamental problem in pattern matching. Given a text of length n and a pattern of length m, we are asked to output the distance between the pattern and every n-substring of the text. A basic variant of this problem is computation of Hamming distances, that is counting the number of mismatches (different characters aligned), for each alignment. Other popular variants include ℓ_1 distance (Manhattan distance), ℓ_2 distance (Euclidean distance) and general ℓ_p distance. While each of those problems trivially generalizes classical pattern-matching, the efficient algorithms for them require a broader set of tools, usually involving both algebraic and combinatorial insights. We briefly survey the history of the problems, and then focus on the progress made in the past few years in many specific settings: fine-grained complexity and lower-bounds, $(1 + \varepsilon)$ multiplicative approximations, k-bounded relaxations, streaming algorithms, purely combinatorial algorithms, and other recently proposed variants.

1 Hamming Distance

A most fundamental problem in stringology is that of pattern matching: given pattern P and text T, find all occurrences of P in T where by occurrence we mean a substring (a consecutive fragment) of T that is identical to P. A huge efforts have been put into advancement of understanding of pattern matching by the community. One particular variant to consider is finding occurrences or almost-occurrences of P in T. For this, we need to specify almost-occurrences: e.g. introduce some form of measure of distance between words, and then look for substrings of T which are close to P. We are interested in measures that are position-based, that is they are defined over strings of equal length, and are based upon distances between letters on corresponding positions (thus e.g. edit distance is out of scope of this survey). Consider for example

Definition 1 (Hamming distance). *For strings A, B of equal length, their Hamming distance is defined as*

$$Ham(A, B) = |\{i : A[i] \neq B[i]\}|.$$

Supported by Polish National Science Centre grant 2019/33/B/ST6/00298.

M. Anselmo et al. (Eds.): CiE 2020, LNCS 12098, pp. 353–365, 2020.
https://doi.org/10.1007/978-3-030-51466-2_32

Hence, the Hamming distance counts the number of *mismatches* between two words. This leads us to the core problem considered in this survey.

Definition 2 (Text-to-pattern Hamming distance). *For a text $T[1, n]$ and a pattern $P[1, m]$, the text-to-pattern Hamming distance asks for an output array $O[1, n - m + 1]$ such that*

$$O[i] = Ham(P, T[i, i + m - 1]).$$

Observe that this problem generalizes the detection of almost-occurrences – one can scan the output array and output positions with small distance to the pattern.

1.1 Convolution in Text-to-Pattern Distance

Convolution of two vectors (arrays) is defined as follow

Definition 3 (Convolution). *For 0-based vectors A and B we define their convolution $(A \circ B)$ as a vector:*

$$(A \circ B)[k] = \sum_{i+j=k} A[i] \cdot B[j].$$

Such a definition has a natural interpretation e.g. in terms of polynomial product: if we interpret a vector as coefficients of a polynomial that is $A(x) = \sum_i A[i] \cdot x^i$ and $B(x) = \sum_i B[i] \cdot x^i$, then $(A \circ B)$ are coefficients of $A(x) \cdot B(x)$.

Convolution over integers is computed by Fast Fourier transform (FFT) in time $\mathcal{O}(n \log n)$. This requires actual embedding of integers into field, e.g. $\mathbb{F}_p$ or $\mathbb{C}$. This comes at a cost, if e.g. we were to consider text-to-pattern distance over (non-integer) alphabets that admit only field operations, e.g. matrices or geometric points. Convolution can be computed using a "simpler" set of operations, that is just with ring operations in e.g. $\mathbb{Z}_p$ using Toom-Cook multiplication [35], which is a generalization of famous divide-and-conquer Karatsuba algorithm [20]. However, not using FFT makes the algorithm slower, with Toom-Cook algorithm taking time $\mathcal{O}(n2^{\sqrt{2 \log n}} \log n)$, and increases the complexity of the algorithm.

Fischer and Paterson in [16] observed that convolution can be used to compute text-to-pattern Hamming distance for small alphabets. Consider the following observation: for binary P and T, denote by P' the *reversed* P. Then we have the following property:

$$Ham(P, T[i, i + m - 1]) = \sum_j \left(T[i + j](1 - P[j]) + (1 - T[i + j])P[j] \right)$$

$$= \sum_{j+k=m+1+i} \left(T[j]\overline{P'[k]} + \overline{T[j]}P'[k] \right)$$

$$= (T \circ \overline{P'})[m + 1 + i] + (\overline{T} \circ P')[m + 1 + i]$$

where e.g. $\overline{T}$ denotes *negating* every entry of T. Thus the whole algorithm is done by computing two convolutions in time $\mathcal{O}(n \log m)$.[1] This approach in fact generalizes to arbitrary size alphabets by following observation: "contribution" of single $c \in \Sigma$ to number of mismatches for all positions can be computed with single convolution. This results in $\mathcal{O}(|\Sigma|n \log m)$ time algorithm.

The natural question is whether faster (than naive quadratic-time) algorithms for large alphabets exist. The answer is affirmative, by (almost simultaneous) results of Abrahamson [1] and Kosaraju [24]. The insight is that for any letter $c \in |\Sigma|$, we can compute its "contribution" twofold:

- by FFT in time $\mathcal{O}(n \log m)$,
- or in time $\mathcal{O}(t)$ per each of n alignments, where t is the number of occurrences of c in lets say pattern.

The insight is that we apply the former for letters that appear often ("dense" case) and latter for sparse letters. Since there can be at most m/T letters that appear at least T times in pattern each, the total running time is $\mathcal{O}(n \log m \cdot \frac{m}{T} + Tn)$ which is minimized when $T = \sqrt{m \log m}$ with run-time $\mathcal{O}(n\sqrt{m \log m})$.

This form of mixing combinatorial and algebraical insights is typical for the type of problems considered in this paper, and we will see more of it in the following sections. As a side-note, the complexity of $\mathcal{O}(n\sqrt{m \log m})$ remains state-of-the-art.

1.2   Relaxation: k-Bounded Distances

The lack of progress in Hamming text-to-pattern distance complexity sparked interest in searching for relaxations of the problem, in hope of reaching linear (or almost linear) run-time. For example if we consider reporting only the values not exceeding a certain threshold value k, then we have the so-called k-approximated distance. The motivation comes from the fact that if we are looking for almost-occurrences, then if the distance is larger than a certain threshold value, the text fragment is too dissimilar to pattern and we are safe to discard it.

The very first solution to this problem was shown by Landau and Vishkin [26] working in time $\mathcal{O}(nk)$, using an essentially combinatorial approach of taking $\mathcal{O}(1)$ time per mismatch per alignment using LCP queries (Longest Common Prefix queries), where $\mathrm{LCP}(i,j)$ returns maximal k such that $T[i, i+k] = P[j, j+k]$. This solution requires preprocessing of T and P with e.g. suffix tree, which is a standard tool-set of stringology. This solution still is slower than naive algorithm for $k = m^{1/2+\delta}$, but has the nice property of using actually $\mathcal{O}(n)$ time for constant k. This technique is also known as *kangaroo jumps*.

This initiated a series of improvements to the complexity, with algorithms of complexity $\mathcal{O}(n\sqrt{k \log k})$ and $\mathcal{O}((k^3 \log k + m) \cdot n/m)$ by Amir et al. [4]. First algorithm is an adaptation of general algorithm of Abrahamson with balancing

---

[1] Its $\log m$ not $\log n$ by standard trick of reducing the problem to $\lceil n/m \rceil$ instances with pattern P of length m and text of length $2m$.

of "sparse" vs. "dense" case done w.r.t. k instead of m (some further combinatorial insights are required to make the cases work with proper run-time). Such trade-off has this nice property that for $k = m$ the complexity matches that of Abrahamson's algorithm. Second algorithm is more interesting, since it shows that for non-trivial values of k (in this case, $k = \mathcal{O}(m^{1/3})$) near-linear time algorithms are possible.

The later complexity was then improved to $\mathcal{O}((k^2 \log k + m \operatorname{poly} \log m) \cdot n/m)$ by Clifford et al. [13]. We now discuss the techniques of this algorithm, starting with *kernelization* technique.

Definition 4 ([13]). *An integer $\pi > 0$ is an x-period of a string $S[1, m]$, if $Ham(S[\pi + 1, m], S[1, m - \pi]) \leq x$ and π is minimal such integer.*

Such definition should be compared with regular definition of a period, where π is a period of string S if $S[\pi + 1, m] = S[1, m - \pi]$.

We then observe the following:

Lemma 1 ([13]). *If ℓ is a $2x$-period of the pattern, then any two occurrences of the pattern in the text with at most x mismatches are at offset distance at least ℓ.*

The first step of the algorithm is to determine some small $\mathcal{O}(k)$-period of the pattern. This actually does not require any specialized machinery and can be done with a 2-approximate algorithm for text-to-pattern Hamming distance (multiplicative approximations are a topic of the following section). We then distinguish two cases, where small means $\mathcal{O}(k)$.

No small k-period. This is an "easy" case, where a filtering step allows us to keep only $\mathcal{O}(n/k)$ alignments that are candidates for $\leq k$-distance matches. A "kangaroo jumps" technique of Landau and Vishkin allows us to verify each one of them in $\mathcal{O}(k)$ time, resulting in linear time spent in this case.

Small $2k$-period. This is a case where we can deduce some regularity properties. Denote the $2k$-period as ℓ. First, P can be decomposed into ℓ words from its arithmetic progressions of positions, with step ℓ and every possible offset. From the definition of ℓ being $2k$-period, we know that the total number of runs in those words is small. The more interesting property is that even though the text T can be arbitrary, if T is not regular enough it can be discarded (and this actually concerns any part of the text that is not regular). More precisely, there is a substring of text that is regular enough and contains all the alignments of P that are at Hamming distance at most k (assuming $n = 2m$, which we can always guarantee).

What remains is to observe that finding ℓ, compressing of P into arithmetic progressions and finding compressible region T' of T all can be done in $\widetilde{\mathcal{O}}(n)$ time, and that all of alignments of text to pattern correspond to alignments of those arithmetic progressions, and can be solved in $\mathcal{O}(k^2)$ time.

Final step in the sequence of improvements to this problem was done by Gawrychowski and Uznański [17]. They observe that the algorithm from [13] can be interpreted in terms of reduction: instance of k-bounded text-to-pattern Hamming distance with T and P is reduced to new T' and P', where T' and P' are possibly of the same length, but have total number of runs in their Run-length encoding (RLE) representation bounded as $\mathcal{O}(k)$. The algorithm from [13] then falls back to brute force $\mathcal{O}(k^2)$ time computation. While $\mathcal{O}(k^{2-\delta})$ algorithm for RLE-compressed pattern matching would falsify 3-SUM conjecture (c.f. [10]), some structural properties of the instances can be leveraged based on the fact that they are RLE-compressed from inputs of length m. A balancing argument (in style of one from [4] or [1]) follows, allowing to solve this sub-problem in time $\mathcal{O}(k\sqrt{m\log m})$. The final complexity for the whole algorithm becomes then $\widetilde{\mathcal{O}}((m + k\sqrt{m}) \cdot n/m)$.

1.3   Relaxation: $1 \pm \varepsilon$ Approximation

Another way to relax to text-to-pattern distance is to consider multiplicative approximation when reporting number of mismatches. The very elegant argument made by Karloff [21] states the following.

Observation 1. *Consider a randomly chosen projection* $\varphi : \Sigma \rightarrow \{0,1\}$ *(each letters mapping is chosen independently and uniformly at random) and words* A, B. *Then*

$$\mathbb{E}[Ham(\varphi(A), \varphi(B))] = \frac{1}{2} Ham(A, B),$$

where $\varphi(A)$ *denotes applying* φ *to each letter of* A *separately.*

Thus the algorithm consists of: (i) choosing independently at random K random projections; (ii) for each projection, computing text-to-pattern Hamming distance over projected input; (iii) averaging answers. A concentration argument then follows, giving standard $K = \mathcal{O}(\frac{\log n}{\varepsilon^2})$ independent repetitions guaranteeing that average recovers actual Hamming distance with $(1 \pm \varepsilon)$ multiplicative guarantee, with high probability. This gives total run-time $\widetilde{\mathcal{O}}(n/\varepsilon^2)$.

The $\frac{1}{\varepsilon^2}$ dependency was believed to be inherent, as is the case for e.g. space complexity of sketching of Hamming distance, cf. [8,19,38]. However, for approximate pattern matching that was refuted in Kopelowitz and Porat [22,23], where randomized algorithms were provided with complexity $\mathcal{O}(\frac{n}{\varepsilon} \log n \log m \log \frac{1}{\varepsilon} \log |\Sigma|)$ and $\mathcal{O}(\frac{n}{\varepsilon} \log n \log m)$ respectively. The second mentioned algorithm is actually surprisingly simple: instead of projecting onto binary alphabet, random projections $\Sigma \rightarrow [u]$ are used, where $u = \mathcal{O}(\varepsilon^{-1})$. Such projections collapse in expectation only an ε-fraction of mismatches, introducing systematic $1 + \varepsilon$ multiplicative error. A simple Markov bound argument follows, that since *expected* error is within desired bound, taking few (lets say $\log n$) repetitions and taking median guarantees recovery of good approximate answer with high probability. What remains to observe is that exact counting of text-to-pattern distance over projected alphabet takes u repetitions of convolution, so the total runtime is $\widetilde{\mathcal{O}}(n/\varepsilon)$. An alternative exposition to this result was provided in [34].

2   Other Norms

A natural extension to counting mismatches is to consider other norms (e.g. ℓ_1, ℓ_2, general ℓ_p norm or ℓ_∞ norm), or to move beyond norms (so called *threshold pattern matching* c.f. Atallah and Duket [6] or *dominance pattern matching* c.f. Amir and Farach [3]).

Definition 5 (ℓ_p **distance**). *For two strings of equal length over integer alphabet and constant $p > 0$, their ℓ_p distance is defined as*

$$\|A - B\|_p = \left(\sum_i \Big| A[i] - B[i] \Big|^p \right)^{1/p}.$$

Definition 6 (ℓ_∞ **distance**). *For two strings of equal length over integer alphabet, their ℓ_∞ distance is defined as*

$$\|A - B\|_\infty = \max_i \Big| A[i] - B[i] \Big|.$$

2.1   Exact Algorithms

To see that the link between convolution and text-to-pattern distance is relevant when considering other norms, consider the case of computing ℓ_2 distances. We are computing output array $O[]$ such that $O[i] = \sum_j (T[i+j] - P[j])^2$. However, this is equivalent to computing, for every i simultaneously, value $\sum_j T[i+j]^2 + \sum_j P[j]^2 - 2\sum_j T[i+j]P[j]$. While the terms $\sum_j T[i+j]^2$ and $\sum_j P[j]^2$ can be easily precomputed in $\mathcal{O}(n)$ time, we observe (following [29]) that $\sum_j T[i+j]P[j]$ is essentially convolution. Indeed, consider P' such that $P'[j] = P[m+1-j]$, and then what follows.

We now consider ℓ_1 distance. Using techniques similar to Hamming distance, the $\mathcal{O}(n\sqrt{n \log n})$ complexity algorithms were developed independently in 2005 by Clifford et al. [11] and Amir et al. [5] for reporting all ℓ_1 distances. The algorithms use a balancing argument, starting with observation that alphabet can be partitioned into buckets, where each bucket is a consecutive interval of alphabet. The contribution of characters from the same interval is counted in one phase, and contribution of characters from distinct intervals is counted in second phase.

Interestingly, no known algorithm for *exact* computation of text-to-pattern ℓ_p distance for arbitrary value of p is known. By the folklore observation, for any even p we can reduce it to convolution and have $\mathcal{O}(n \log m)$ time algorithm (c.f. Lipsky and Porat [29], with $\mathcal{O}$ hiding p^2 dependency). By the results of Labib et al. [25] any odd-value integer p admits $\widetilde{\mathcal{O}}(n\sqrt{m \log m})$ time algorithm (the algorithm is given implicitly, by providing a reduction from ℓ_p to Hamming distance, with $\widetilde{\mathcal{O}}$ hiding $(\log m)^{\mathcal{O}(p)}$ dependency).

2.2   Approximate and k-Bounded Algorithms

Once again, the topic spurs interest in approximation algorithm for distance functions. In [29] a deterministic algorithm with a run time of $\mathcal{O}(\frac{n}{\varepsilon^2} \log m \log U)$ was given, while later in [17] the complexity has been improved to a (randomized) $\mathcal{O}(\frac{n}{\varepsilon} \log^2 n \log m \log U)$, where U is the maximal integer value on the input. Later [34] it was shown that such complexity is in fact achievable (up to poly-log factors) with a deterministic solution. All those solutions follow similar framework of *linearity-preserving* reductions, which has actually broader applications. The framework is as follow: imagine we want to approximate some distance function $d : \Sigma \times \Sigma \to \mathbb{R}$. We build small number of pairs of projections, $L_1, \ldots, L_t, R_1, \ldots, R_t$, with the following property: $d(x, y) \approx \sum_i L_i(x) \cdot R_i(x)$.[2] Given such formulation, by linearity, text-to-pattern of A and B using distance function d is approximated by a linear combination of convolutions of $L_i(A)$ and $R_i(B)$. The complexity of the solutions follows from the number of different projections that need to be used.

For ℓ_∞ distances, in [29] a $\widetilde{\mathcal{O}}(n/\varepsilon)$ time approximate solution was given, while in Lipsky and Porat [27] a k-bounded ℓ_∞ distance algorithm with time $\mathcal{O}(nk \log m)$ was given. For k-bounded ℓ_1 distances, [5] a $\mathcal{O}(n\sqrt{k \log k})$ run-time algorithm was given, while in [17] an algorithm with run-time $\widetilde{\mathcal{O}}((m + k\sqrt{m}) \cdot n/m)$ was given. The fact that those run-times are (up to poly-logs) identical to corresponding run-times of k-bounded Hamming distances is not a coincidence, as [17] have shown that k-bounded ℓ_1 is at least as easy as k-bounded Hamming distance reporting.

A folklore result (c.f. [29]) states that the randomized algorithm with a run time of $\widetilde{\mathcal{O}}(\frac{n}{\varepsilon^2})$ is in fact possible for any ℓ_p distance, $0 < p \leq 2$, with use of p-stable distributions and convolution. Such distributions exist only when $p \leq 2$, which puts a limit on this approach. See [30] for wider discussion on p-stable distributions. Porat and Efremenko [32] has shown how to approximate general distance functions between pattern and text in time $\widetilde{\mathcal{O}}(\frac{n}{\varepsilon^2})$. Their solution does not immediately translates to ℓ_p distances, since it allows only for score functions of form $\sum_j d(t_{i+j}, p_j)$ where d is arbitrary metric over Σ. Authors state that their techniques generalize to computation of ℓ_2 distances, and in fact those generalize further to ℓ_p distances as well, but the ε^{-2} dependency in their approach is unavoidable. Finally, for any $p > 0$ there is ℓ_p distance $(1 \pm \varepsilon)$-approximate algorithm running in time $\widetilde{\mathcal{O}}(n/\varepsilon)$ by results shown in [34]. Final result follows the framework of linearity-preserving reductions.

3   Lower Bounds

It is a major open problem whether near-linear time algorithm, or even $\mathcal{O}(n^{3/2-\delta})$ time algorithms, are possible for such problems. A conditional lower bound was shown in [12], via a reduction from matrix multiplication. This means

---

[2] Here we used $\approx$ since its in the context of approximate algorithms. The same framework applies to exact algorithms, then we replace $\approx$ with $=$.

that existence of combinatorial algorithm with run-time $\mathcal{O}(n^{3/2-\delta})$ solving the problem for Hamming distances implies combinatorial algorithms for Boolean matrix multiplication with $\mathcal{O}(n^{3-\delta})$ run-time, which existence is unlikely. Looking for unconditional bounds, we can state this as a lower-bound of $\Omega(n^{\omega/2})$ for Hamming distances pattern matching, where $2 \leq \omega < 2.373$ is a matrix multiplication exponent. In fact those techniques can be generalized to take into account k-bounded version of this problem:

Theorem 2 ([17]). *For any positive $\varepsilon, \alpha, \kappa$ such that $\frac{1}{2}\alpha \leq \kappa \leq \alpha \leq 1$ there is no combinatorial algorithm solving pattern matching with $k = \Theta(n^\kappa)$ mismatches in time $\mathcal{O}((k\sqrt{m})^{1-\varepsilon}) \cdot n/m)$ for a text of length n and a pattern of length $m = \Theta(n^\alpha)$, unless the combinatorial matrix multiplication conjecture fails.*

Complexity of pattern matching under Hamming distance and under ℓ_1 distance was proven to be identical (up to poly-logarithmic terms) in [25]. This equivalence in fact applies to a wider range of distance functions and in general other score functions. The result shows that a wide class of functions are equivalent under linearity-preserving reductions to computation of Hamming distances. The class includes e.g. dominance score, ℓ_1 distance, threshold score, ℓ_{2p+1} distance, any of above with wildcards, and in fact a wider class called piece-wise polynomial functions.

Definition 7. *For integers A, B, C and polynomial $P(x, y)$ we say that the function $P(x, y) \cdot \mathbb{1}[Ax + By + C > 0]$ is half-plane polynomial. We call a sum of half-plane polynomial functions a piece-wise polynomial. We say that a function is axis-orthogonal piece-wise polynomial, if it is piece-wise polynomial and for every i, $A_i = 0$ or $B_i = 0$.*

Observe that $\mathrm{Ham}(x, y) = \mathbb{1}[x > y] + \mathbb{1}[x < y]$, $\max(x, y) = x \cdot \mathbb{1}[x \geq y] + y \cdot \mathbb{1}[x < y]$, $|x - y|^{2p+1} = (x - y)^{2p+1} \cdot \mathbb{1}[x > y] + (y - x)^{2p+1} \cdot \mathbb{1}[x < y]$, and e.g. threshold function can be defined as $\mathrm{thr}_\delta(x, y) \overset{\text{def}}{=} \mathbb{1}[|x - y| \geq \delta] = \mathbb{1}[x \leq y - \delta] + \mathbb{1}[x \geq y + \delta]$.

Theorem 3. *Let $\diamond$ be a piece-wise polynomial of constant degree and $\mathrm{poly}\log n$ number of summands.*

- *If $\diamond$ is axis orthogonal, then $\diamond$ is "easy": $(+, \diamond)$ convolution takes $\widetilde{O}(n)$ time, $(+, \diamond)$ matrix multiplication takes $\widetilde{O}(n^\omega)$ time.*
- *Otherwise, $\diamond$ is Hamming distance complete: under one-to-polylog reductions, on inputs bounded in absolute value by $\mathrm{poly}(n)$, $(+, \diamond)$ product is equivalent to Hamming distance, $(+, \diamond)$ convolution is equivalent to text-to-pattern Hamming distance and $(+, \diamond)$ matrix product is equivalent to Hamming-distance matrix product.*

Some of those reduction (for specific problems) were presented in literature, c.f. [28, 37, 39], but never as a generic class-of-problems equivalence.

This means that the encountered barrier for all of the induced text-to-pattern distance problems is in fact the same barrier, and we should not expect algorithms with dependency $n^{3/2-\delta}$ without some major breakthrough. Unfortunately such reductions do not preserve properties of k-bounded instances or $1 \pm \varepsilon$-approximate ones, so this result tells us nothing about relative complexity of relaxed problems, and it is a major open problem to do so.

4 Streaming Algorithms

In streaming algorithms, the goal is to process text in a streaming fashion, and answer in a real-time about the distance between last m characters of text and a pattern. The primary measure of efficiency is the memory complexity of the algorithm, that is we assume that the whole input (or even the whole pattern) is too large to fit into the memory and some for of small-space representation is required. The time to process each character is the secondary measure of efficiency, since it usually is linked to memory efficiency. By folklore result, exact reporting of e.g. Hamming distances is impossible in $o(m)$ memory, so the focus of the research has been on relaxed problems, that is k-bounded and $(1 \pm \varepsilon)$-approximate reporting.

For k-bounded reporting of Hamming distances, in Porat and Porat [31] a $\mathcal{O}(k^3)$ space and $\mathcal{O}(k^2)$ time per character streaming algorithm was presented. It was later improved in [13] to $\widetilde{\mathcal{O}}(k^2)$ space and $\widetilde{\mathcal{O}}(\sqrt{k})$ time per character, and then in Clifford et al. [14] to $\widetilde{\mathcal{O}}(k)$ space keeping $\widetilde{\mathcal{O}}(\sqrt{k})$ time per character. Many interesting techniques were developed for this problem. As an example, k-mismatch problem can be reduced to (k^2 many instances of) 1-mismatch problem (c.f. [13]), which in fact reduces to exact pattern matching in streaming model (c.f. [31]). Other approach is to construct efficient rolling sketches for k-mismatch problem, based on Reed-Solomon error correcting codes (c.f. [14]).

For $1 \pm \varepsilon$, two interesting approaches are possible. First approach was presented by Clifford and Starikovskaya [15] and later refined in Svagerka et al. [33]. This approach consists of using rolling sketches of text started every $\sim \sqrt{m}$ positions, and additionally $\sim \sqrt{m}$ sketches of substrings of length $m - \sqrt{m}$ of pattern are maintained (guaranteeing that at least one sketch in text is aligned to one sketch of long pattern fragment). One way of building rolling sketches for approximate Hamming distance is to use random projections to binary alphabet and reduce the problem to one for binary alphabet, where binary alphabet uses Johnson-Lindenstrauss type of constructions. This approach results in $\widetilde{\mathcal{O}}(\sqrt{m}/\varepsilon^2)$ memory and $\widetilde{\mathcal{O}}(1/\varepsilon^2)$ time per character.

Alternative approach was proposed in recent work of Chan et al. [9]. They start with observation that the Hamming distance can be estimated by checking mismatches at a random subset of positions. Their algorithm uses a random subset as follow: the algorithm picks a random prime p (of an appropriately chosen size) and a random offset b, and considers a subset of positions $\{b, b+p, b+2p, \ldots\}$. The structured nature of the subset enables more efficient computation. It turns out that even better efficiency is achieved by using multiple (but still relatively few) offsets. When approximating the Hamming distance of the pattern at subsequent text locations, the set of sampled positions in the text changes, and so a straightforward implementation seems too costly. To overcome this challenge, a key idea is to shift the sample a few times in the pattern and a few times in the text (namely, for a trade-off parameter z, our algorithm considers z shifts in the pattern and p/z shifts in the text). Interestingly, the proposed solution is even more efficient when considering a $(1\pm\varepsilon)$-approximate k-bounded reporting of Hamming distances.

Theorem 4 ([9]). *There is an algorithm that reports $(1 \pm \varepsilon)$-approximate k-bounded Hamming distances in a streaming setting that uses $\widetilde{\mathcal{O}}(\min(\sqrt{k}/\varepsilon^2, \sqrt{m}/\varepsilon^{1.5}))$ space and takes $\mathcal{O}(1/\varepsilon^3)$ time per character.*

Focusing on other norms, we note that in [33] a sublinear space algorithms for ℓ_p norms for $0 < p \leq 2$ was presented. The specific details of construction vary between different values of p, and the techniques include: using p-stable distributions (c.f. [18]), range-summable hash functions (c.f. [7]) and Johnson-Lindenstrauss projections (c.f. [2]).

Theorem 5 ([33]). *Let $\sigma = n^{\mathcal{O}(1)}$ denote size of alphabet. There is a streaming algorithm that computes a $(1 \pm \varepsilon)$-approximation of the ℓ_p distances. The parameters of the algorithm are*

1. *in $\widetilde{\mathcal{O}}(\varepsilon^{-2}\sqrt{n} + \log\sigma)$ space, and $\widetilde{\mathcal{O}}(\varepsilon^{-2})$ time per arrival when $p = 0$ (Hamming distance);*
2. *in $\widetilde{\mathcal{O}}(\varepsilon^{-2}\sqrt{n} + \log^2\sigma)$ space and $\widetilde{\mathcal{O}}(\sqrt{n}\log\sigma)$ time per arrival when $p = 1$;*
3. *in $\widetilde{\mathcal{O}}(\varepsilon^{-2}\sqrt{n}+\log^2\sigma)$ space and $\widetilde{\mathcal{O}}(\varepsilon^{-2}\sqrt{n})$ time per arrival when $0 < p < 1/2$;*
4. *in $\widetilde{\mathcal{O}}(\varepsilon^{-2}\sqrt{n} + \log^2\sigma)$ space and $\widetilde{\mathcal{O}}(\varepsilon^{-3}\sqrt{n})$ time per arrival when $p = 1/2$;*
5. *in $\widetilde{\mathcal{O}}(\varepsilon^{-2}\sqrt{n}+\log^2\sigma)$ space and $\widetilde{\mathcal{O}}(\sigma^{\frac{2p-1}{1-p}}\sqrt{n}/\varepsilon^{2+3\cdot\frac{2p-1}{1-p}})$ time per arrival when $1/2 < p < 1$;*
6. *in $\widetilde{\mathcal{O}}(\varepsilon^{-2-p/2}\sqrt{n}\log^2\sigma)$ space and $\mathcal{O}(\varepsilon^{-p/2}\sqrt{n} + \varepsilon^{-2}\log\sigma)$ time per arrival for $1 < p \leq 2$.*

5   Open Problems

Below we list several open problems of the area, which we believe are the most promising research directions and/or pressing questions.

1. Show deterministic algorithm for $(1 \pm \varepsilon)$-approximate ℓ_p reporting for $0 < p < 1$, preferably in time $\widetilde{\mathcal{O}}(n/\varepsilon)$.
2. What is the time complexity of exact ℓ_p reporting for non-integer p?
3. Show conditional lower bound for exact Hamming distance reporting from stronger hypotheses, like 3SUM-HARDNESS.
4. Lower bounds for $1 \pm \varepsilon$ approximations (conditional between problems, or from external problems), for any of the discussed problems.
5. What is the true space complexity dependency in streaming $(1 \pm \varepsilon)$ approximate Hamming distance reporting? Is $\sqrt{m}\varepsilon^{-1.5}$ complexity optimal?

6. Can we close the gap between streaming complexity of approximate ℓ_p algorithms and streaming complexity of approximate Hamming distance?
7. Can we design effective "combinatorial" algorithms for all mentioned problems (e.g. not relying on convolution)? For Hamming, ℓ_1 and ℓ_2 distances answer is at least partially yes (c.f. [9] and [36]).

References

1. Abrahamson, K.R.: Generalized string matching. SIAM J. Comput. **16**(6), 1039–1051 (1987)
2. Achlioptas, D.: Database-friendly random projections: Johnson-Lindenstrauss with binary coins. J. Comput. Syst. Sci. **66**(4), 671–687 (2003). https://doi.org/10.1016/S0022-0000(03)00025-4
3. Amir, A., Farach, M.: Efficient matching of nonrectangular shapes. Ann. Math. Artif. Intell. **4**(3), 211–224 (1991). https://doi.org/10.1007/BF01531057
4. Amir, A., Lewenstein, M., Porat, E.: Faster algorithms for string matching with k mismatches. J. Algorithms **50**(2), 257–275 (2004). https://doi.org/10.1016/S0196-6774(03)00097-X
5. Amir, A., Lipsky, O., Porat, E., Umanski, J.: Approximate Matching in the L_1 Metric. In: Apostolico, A., Crochemore, M., Park, K. (eds.) CPM 2005. LNCS, vol. 3537, pp. 91–103. Springer, Heidelberg (2005). https://doi.org/10.1007/11496656_9
6. Atallah, M.J., Duket, T.W.: Pattern matching in the hamming distance with thresholds. Inf. Process. Lett. **111**(14), 674–677 (2011). https://doi.org/10.1016/j.ipl.2011.04.004
7. Calderbank, A.R., Gilbert, A.C., Levchenko, K., Muthukrishnan, S., Strauss, M.: Improved range-summable random variable construction algorithms. In: SODA, pp. 840–849 (2005)
8. Chakrabarti, A., Regev, O.: An optimal lower bound on the communication complexity of gap-hamming-distance. SIAM J. Comput. **41**(5), 1299–1317 (2012). https://doi.org/10.1137/120861072
9. Chan, T.M., Golan, S., Kociumaka, T., Kopelowitz, T., Porat, E.: Approximating text-to-pattern hamming distances. In: STOC 2020 (2020)
10. Chen, K.-Y., Hsu, P.-H., Chao, K.-M.: Approximate matching for run-length encoded strings is 3SUM-hard. In: Kucherov, G., Ukkonen, E. (eds.) CPM 2009. LNCS, vol. 5577, pp. 168–179. Springer, Heidelberg (2009). https://doi.org/10.1007/978-3-642-02441-2_15
11. Clifford, P., Clifford, R., Iliopoulos, C.: Faster algorithms for δ, γ-matching and related problems. In: Apostolico, A., Crochemore, M., Park, K. (eds.) CPM 2005. LNCS, vol. 3537, pp. 68–78. Springer, Heidelberg (2005). https://doi.org/10.1007/11496656_7
12. Clifford, R.: Matrix multiplication and pattern matching under Hamming norm. http://www.cs.bris.ac.uk/Research/Algorithms/events/BAD09/BAD09/Talks/BAD09-Hammingnotes.pdf. Accessed Mar 2017
13. Clifford, R., Fontaine, A., Porat, E., Sach, B., Starikovskaya, T.: The k-mismatch problem revisited. In: SODA, pp. 2039–2052 (2016). https://doi.org/10.1137/1.9781611974331.ch142
14. Clifford, R., Kociumaka, T., Porat, E.: The streaming k-mismatch problem. In: SODA, pp. 1106–1125 (2019). https://doi.org/10.1137/1.9781611975482.68

15. Clifford, R., Starikovskaya, T.: Approximate hamming distance in a stream. In: ICALP, pp. 20:1–20:14 (2016). https://doi.org/10.4230/LIPIcs.ICALP.2016.20
16. Fischer, M.J., Paterson, M.S.: String-matching and other products. Technical report (1974)
17. Gawrychowski, P., Uznański, P.: Towards unified approximate pattern matching for hamming and L_1 distance. In: ICALP, pp. 62:1–62:13 (2018). https://doi.org/10.4230/LIPIcs.ICALP.2018.62
18. Indyk, P.: Stable distributions, pseudorandom generators, embeddings, and data stream computation. J. ACM **53**(3), 307–323 (2006). https://doi.org/10.1145/1147954.1147955
19. Jayram, T.S., Kumar, R., Sivakumar, D.: The one-way communication complexity of hamming distance. Theory Comput. **4**(1), 129–135 (2008). https://doi.org/10.4086/toc.2008.v004a006
20. Karatsuba, A.: Multiplication of multidigit numbers on automata. Soviet physics doklady **7**, 595–596 (1963)
21. Karloff, H.J.: Fast algorithms for approximately counting mismatches. Inf. Process. Lett. **48**(2), 53–60 (1993). https://doi.org/10.1016/0020-0190(93)90177-B
22. Kopelowitz, T., Porat, E.: Breaking the variance: approximating the hamming distance in $1/\epsilon$ time per alignment. In: FOCS, pp. 601–613 (2015). https://doi.org/10.1109/FOCS.2015.43
23. Kopelowitz, T., Porat, E.: A simple algorithm for approximating the text-to-pattern hamming distance. In: SOSA@SODA, pp. 10:1–10:5 (2018). https://doi.org/10.4230/OASIcs.SOSA.2018.10
24. Kosaraju, S.R.: Efficient string matching (1987). Manuscript
25. Labib, K., Uznański, P., Wolleb-Graf, D.: Hamming distance completeness. In: CPM, pp. 14:1–14:17 (2019). https://doi.org/10.4230/LIPIcs.CPM.2019.14
26. Landau, G.M., Vishkin, U.: Efficient string matching with k mismatches. Theor. Comput. Sci. **43**, 239–249 (1986). https://doi.org/10.1016/0304-3975(86)90178-7
27. Lipsky, O., Porat, E.: Approximate matching in the L_∞ metric. Inf. Process. Lett. **105**(4), 138–140 (2008). https://doi.org/10.1016/j.ipl.2007.08.012
28. Lipsky, O., Porat, E.: L_1 pattern matching lower bound. Inf. Process. Lett. **105**(4), 141–143 (2008). https://doi.org/10.1016/j.ipl.2007.08.011
29. Lipsky, O., Porat, E.: Approximate pattern matching with the L_1, L_2 and L_∞ metrics. Algorithmica **60**(2), 335–348 (2011). https://doi.org/10.1007/s00453-009-9345-9
30. Nolan, J.: Stable Distributions: Models for Heavy-Tailed Data. Birkhauser, New York (2003)
31. Porat, B., Porat, E.: Exact and approximate pattern matching in the streaming model. In: FOCS, pp. 315–323 (2009). https://doi.org/10.1109/FOCS.2009.11
32. Porat, E., Efremenko, K.: Approximating general metric distances between a pattern and a text. In: SODA, pp. 419–427 (2008). http://dl.acm.org/citation.cfm?id=1347082.1347128
33. Starikovskaya, T., Svagerka, M., Uznański, P.: L_p pattern matching in a stream. CoRR abs/1907.04405 (2019)
34. Studený, J., Uznański, P.: Approximating approximate pattern matching. In: CPM, vol. 128, pp. 15:1–15:13 (2019). https://doi.org/10.4230/LIPIcs.CPM.2019.15
35. Toom, A.: The complexity of a scheme of functional elements simulating the multiplication of integers. In: Doklady Akademii Nauk, vol. 150, pp. 496–498. Russian Academy of Sciences (1963)

36. Uznański, P.: Approximating text-to-pattern distance via dimensionality reduction. CoRR abs/2002.03459 (2020)
37. Vassilevska, V.: Efficient algorithms for path problems in weighted graphs. Ph.D. thesis, Carnegie Mellon University (2008)
38. Woodruff, D.P.: Optimal space lower bounds for all frequency moments. In: SODA, pp. 167–175 (2004). http://dl.acm.org/citation.cfm?id=982792.982817
39. Zhang, P., Atallah, M.J.: On approximate pattern matching with thresholds. Inf. Process. Lett. **123**, 21–26 (2017). https://doi.org/10.1016/j.ipl.2017.03.001

Insertion-Deletion Systems
with Substitutions I

Martin Vu[1] and Henning Fernau[2]([⊠]) [iD]

[1] FB 3 - Informatik, Universität Bremen, Bremen, Germany
martin.vu@uni-bremen.de
[2] Universität Trier, Fachber. 4 – Abteilung Informatikwissenschaften,
54286 Trier, Germany
fernau@uni-trier.de

Abstract. With good biological motivation, we add substitutions as a further type of operations to (in particular, context-free) insertion-deletion systems. This way, we obtain new characterizations of and normal forms for context-sensitive and recursively enumerable languages.

Keywords: Computational completeness · Context-sensitive · Insertions · Deletions · Substitutions

1 Introduction

Insertion-deletion systems, or ins-del systems for short, are well established as computational devices and as a research topic within Formal Languages throughout the past nearly 30 years, starting off with the PhD thesis of Kari [3].

However, from its very beginning, papers highlighting the potential use of such systems in modelling DNA computing also discussed the replacement of single letters (possibly within some context) by other letters, an operation called *substitution* in [2,4]. Interestingly, all theoretical studies on grammatical mechanisms involving insertions and deletions omitted including the substitution operation in their studies. With this paper, we are stepping into this gap by studying ins-del systems with substitutions, or ins-del-sub systems for short.

We put special emphasis on extending context-free ins-del systems with substitutions. We observe quite diverse effects, depending on whether the substitutions are context-free, one-sided or two-sided. We can characterize the context-sensitive languages by extending context-free insertion systems with substitutions, which can be seen as a new normal form for monotone Chomsky grammars. For omitted proofs and further results, see [13].

2 Basic Definitions and Observations

We assume the reader to be familiar with the basics of formal language theory.

© Springer Nature Switzerland AG 2020
M. Anselmo et al. (Eds.): CiE 2020, LNCS 12098, pp. 366–378, 2020.
https://doi.org/10.1007/978-3-030-51466-2_33

An *ins-del system* is a 5-tuple $ID = (V, T, A, I, D)$, consisting of two alphabets V and T with $T \subseteq V$, a finite language A over V, a set of *insertion* rules I and a set of *deletion* rules D. Both sets of rules are formally defined as sets of triples of the form (u, a, v) with $a, u, v \in V^*$ and $a \neq \lambda$. We call elements occurring in T *terminal* symbols, while referring to elements of $V \backslash T$ as *nonterminals*. Elements of A are called *axioms*.

Let $w_1 u v w_2$, with $w_1, u, v, w_2 \in V^*$, be a string. Applying the insertion rule $(u, a, v) \in I$ inserts the string $a \in V^*$ between u and v, which results in the string $w_1 u a v w_2$.

The application of a deletion rule $(u, a, v) \in D$ results in the removal of an substring a from the context (u, v). More formally let $w_1 u a v w_2 \in V^*$ be a string. Then, applying $(u, a, v) \in D$ results in the string $w_1 u v w_2$.

We define the relation $\Longrightarrow$ as follows: Let $x, y \in V^*$. Then we write $x \Longrightarrow_{\text{ins}} y$ if y can be obtained by applying an insertion or deletion rule to x. We also write $(u, a, v)_{\text{ins}}$ or $(u, a, v)_{\text{del}}$ to specify whether the applied rule has been an insertion or a deletion rule. Consider $(u, a, v)_{\text{ins}}$ or $(u, a, v)_{\text{del}}$. Then we refer to u as the *left context* and to v as the *right context* of $(u, a, v)_{\text{ins}}/(u, a, v)_{\text{del}}$.

Let $ID = (V, T, A, I, D)$ be an ins-del system. The language generated by ID is defined by $L(ID) = \{w \in T^* \mid \alpha \Longrightarrow^* w, \alpha \in A\}$.

The *size* of ID describes its descriptional complexity and is defined by a tuple $(n, m, m'; p, q, q')$, where

$$n = \max\{|a| \mid (u, a, v) \in I\}, \quad p = \max\{|a| \mid (u, a, v) \in D\}$$
$$m = \max\{|u| \mid (u, a, v) \in I\}, \quad q = \max\{|u| \mid (u, a, v) \in D\}$$
$$m' = \max\{|v| \mid (u, a, v) \in I\}, \quad q' = \max\{|v| \mid (u, a, v) \in D\}.$$

By $\text{INS}_n^{m,m'} \text{DEL}_p^{q,q'}$ we denote the family of all ins-del systems of size $(n, m, m'; p, q, q')$ [1,12]. Depending on the context, we also denote the family of languages characterized by ins-del systems of size $(n, m, m'; p, q, q')$ by $\text{INS}_n^{m,m'} \text{DEL}_p^{q,q'}$. We call a family $\text{INS}_n^{0,0} \text{DEL}_p^{0,0}$ a family of *context-free* ins-del systems, while we call a family $\text{INS}_n^{m,m'} \text{DEL}_p^{q,q'}$ with $m + m' > 0 \wedge mm' = 0$ or $(q + q' > 0 \wedge qq' = 0)$ a family of *one-sided* ins-del systems. According to [1], an ins-del system

$$ID' = (V \cup \{\$\}, T, A', I', D' \cup \{(\lambda, \$, \lambda)\})$$

of size $(n, m, m'; p, q, q')$ is said to be *in normal form* if

- for any $(u, a, v) \in I'$, it holds that $|a| = n$, $|u| = m$ and $|v| = m'$, and
- for any $(u, a, v) \in D'$, it holds that $|a| = p$, $|u| = q$ and $|v| = q'$.

Alhazov *et al.* [1,12] have shown the following auxiliary result:

Theorem 1. *For every ins-del system ID, one can construct an insertion-deletion system ID' in normal form of the same size with $L(ID') = L(ID)$.*

In the following sections, we use a modified normal form for ins-del systems of size $(1, 1, 1; 1, 1, 1)$. Given an arbitrary ins-del system of size $(1, 1, 1; 1, 1, 1)$, the construction of this modified normal form is as follows:

Construction 1. *Let $ID = (V, T, A, I, D)$ be an ins-del system of size $(1, 1, 1; 1, 1, 1)$. Without loss of generality, we assume $\{\$, X\} \cap V = \emptyset$ and $\$ \neq X$. We construct $ID'' = (V \cup \{\$, X\}, T, A'', I'', D'')$ as follows:*

$$A'' = \{X\$a\$X \mid \alpha \in A\}$$
$$I'' = \{(z_1, s, z_2) \mid (r, s, t) \in I, \ z_1 = r \ if \ r \neq \lambda \ and \ z_1 = \$ \ otherwise,$$
$$z_2 = t \ if \ t \neq \lambda \ and \ z_2 = \$ \ otherwise\}$$
$$\cup \{(z_1, \$, z_2) \mid z_1, z_2 \in (\{\$\} \cup V)\}$$
$$D'' = \{(z_1, a, z_2) \mid (u, a, v) \in D, \ z_1 = u \ if \ u \neq \lambda \ and \ z_1 = \$ \ otherwise,$$
$$z_2 = v \ if \ v \neq \lambda \ and \ z_2 = \$ \ otherwise\}$$
$$\cup \{(z_1, \$, z_2) \mid z_1, z_2 \in (\{\$\} \cup \{X\} \cup V)\} \cup \{(\lambda, X, \lambda)\}$$

The basic idea of Construction 1 is the same as in the usual normal form constructions (see Theorem 1): the symbol $\$$ is used as a padding symbol to ensure that the left and right contexts of all rules are of the required size. We can show that Construction 1 is equivalent to the usual normal form construction.

Theorem 2. *Let $ID' = (V \cup \{\$\}, T, A', I', D' \cup \{(\lambda, \$, \lambda)\})$ be an ins-del system of size $(1, 1, 1; 1, 1, 1)$ in normal form and $ID'' = (V \cup \{\$, X\}, T, A'', I'', D'')$ be defined according to Construction 1. Then, $L(ID') = L(ID'')$.*

Unlike the usual normal form construction, context-free deletions can only occur at the beginning and the end of a sentential form in the case of Construction 1. This fact will prove useful below.

Ins-del systems have been extensively studied regarding the question if they can describe all of the recursively enumerable languages. Let us summarize these results first by listing the classes of languages known to be equal to RE: $INS_1^{1,1}DEL_1^{1,1}$ [10], $INS_3^{0,0}DEL_2^{0,0}$ and $INS_2^{0,0}DEL_3^{0,0}$ [7], $INS_1^{1,1}DEL_1^{0,0}$ [9, Theorem 6.3], $INS_2^{0,0}DEL_1^{1,1}$ [5], $INS_2^{0,1}DEL_2^{0,0}$ and $INS_1^{1,2}DEL_1^{1,0}$ [8], $INS_1^{1,0}DEL_1^{1,2}$ [5]. By way of contrast, the following language families are known not to be equal to RE, the first one is even a subset of CF: $INS_2^{0,0}DEL_2^{0,0}$ [12], $INS_1^{1,1}DEL_1^{1,0}$ [8], $INS_1^{1,0}DEL_1^{1,1}$ [5], $INS_1^{1,0}DEL_2^{0,0}$ and $INS_2^{0,0}DEL_1^{1,0}$ [6].

We define *substitution rules* to be of the form $(u, a \to b, v)$; $u, v \in V^*$; $a, b \in V$. Let w_1uavw_2; $w_1, w_2 \in V^*$ be a string over V. Then applying the substitution rule $(u, a \to b, v)$ allows us to substitute a single letter a with another letter b in the context of u and v, resulting in the string w_1ubvw_2. Formally, we define an ins-del-sub system to be a 6-tuple $ID_r = (V, T, A, I, D, S)$, where V, T, A, I and D are defined as in the case of usual ins-del systems and S is a set of substitution rules. Substitution rules define a relation $\Longrightarrow_{sub}$: Let $x = w_1uavw_2$ and $y = w_1ubvw_2$ be strings over V. We write $x \Longrightarrow_{sub} y$ iff there is a substitution rule $(u, a \to b, v)$. In the context of ins-del-sub systems, we write $\Longrightarrow$ to denote any of the relations $\Longrightarrow_{ins}$, $\Longrightarrow_{del}$ or $\Longrightarrow_{sub}$. We define the closures $\Longrightarrow^*$ and $\Longrightarrow^+$ as usual. The language generated by an ins-del-sub system ID_r is defined as $L(ID_r) = \{w \in T^* \mid \alpha \Longrightarrow^* w, \ \alpha \in A\}$.

As with usual ins-del system, we measure the complexity of an ins-del-sub system $ID_r = (V, T, A, I, D, S)$ via its *size*, which is defined as a tuple

$(n, m, m'; p, q, q'; r, r')$, where n, m, m', p, q and q' are defined as in the case of usual ins-del systems, while r and r' limit the maximal length of the left and right context of a substitution rule, respectively, i.e., $r = \max\{|u| \mid (u, a \to b, v) \in S\}$, $r' = \max\{|v| \mid (u, a \to b, v) \in S\}$. $\mathrm{INS}_n^{m,m'} \mathrm{DEL}_p^{q,q'} \mathrm{SUB}^{r,r'}$ denotes the family of all ins-del-sub systems of size $(n, m, m'; p, q, q'; r, r')$. Note that, as only one letter is replaced by any substitution rule, there is no subscript below SUB. Depending on the context, we also refer to the family of languages generated by ins-del-sub systems of size $(n, m, m'; p, q, q'; r, r')$ by $\mathrm{INS}_n^{m,m'} \mathrm{DEL}_p^{q,q'} \mathrm{SUB}^{r,r'}$. Expanding our previous terminology, we call substitution rules of the form $(\lambda, a \to b, \lambda)$ *context-free*, while substitution rules of the form $(u, a \to b, \lambda)$ or $(\lambda, a \to b, v)$ with $u \neq \lambda \neq v$ are called *one-sided*. Substitution rules of the form $(u, a \to b, v)$ with $u \neq \lambda \neq v$ are referred to as *two-sided*.

Let ^{R} be the the reversal (mirror) operator. For a language L and its mirror L^R the following lemma holds.

Lemma 1. $L \in \mathrm{INS}_n^{m,m'} \mathrm{DEL}_p^{q,q'} \mathrm{SUB}^{r,r'}$ *iff* $L^R \in \mathrm{INS}_n^{m',m} \mathrm{DEL}_p^{q',q} \mathrm{SUB}^{r',r}$.

We will now define the term *resolve*. Let $ID_r = (V, T, A, I, D, S)$ be an ins-del-sub system. We say that a nonterminal X of ID_r is resolved if X is either deleted or substituted. It is easy to see that in any terminal derivation of ID_r all nonterminals must be resolved at some point of the derivation. We remark that a nonterminal X may be resolved by being substituted with a nonterminal Y, which in turn must be resolved.

As in the case of ins-del systems without substitution rules, we define a *normal form* for ins-del-sub systems. An ins-del-sub system

$$ID_r = (V \cup \{\$\}, T, A, I, D \cup \{(\lambda, \$, \lambda)\}, S)$$

of size $(n, m, m'; p, q, q'; r, r')$ is said to be in normal form if

- for any $(u, a, v) \in I$, it holds that $|a| = n$, $|u| = m$ and $|v| = m'$;
- for any $(u, a, v) \in D$, it holds that $|a| = p$, $|u| = q$ and $|v| = q'$;
- for any $(u, a \to b, v) \in S$, it holds that $|u| = r$ and $|v| = r'$.

Theorem 3. *For every ins-del-sub system ID_r of size $(n, m, m'; p, q, q'; r, r')$, one can construct an ins-del-sub system ID'_r of the same size in normal form, with $L(ID'_r) = L(ID_r)$.*

Proof. Let $ID_r = (V, T, A, I, D, S)$ be an ins-del-sub system of size $(n, m, m'; p, q, q'; r, r')$. The basic idea is similar to the normal form construction for ins-del systems in [1,12]. In fact, the sets of insertion and deletion rules of $ID'_r = (V \cup \{\$\}, T, A', I', D' \cup \{(\lambda, \$, \lambda)\}, S')$ are constructed as in the ins-del system normal form construction. S' and A' are defined as follows:

$$S' = \{z_1, a \to b, z_2 \mid (u, a \to b, v) \in S, \ z_1 \in u \sqcup \$^*, \ |z_1| = r, \ z_2 \in v \sqcup \$^*, \ |z_2| = r'\},$$
$$A' = \{\$^i \alpha \$^t \$^j \mid \alpha \in A, \ i = \max\{m, q, r\}, \ j = \max\{m', q', r'\}, \ t = \max\{p - |w|, 0\}\}.$$

As in Theorem 1, \$ is a new symbol, that is, $\$ \notin V$, which is introduced to be the padding symbol.

Let $h : V \cup \{\$\} \to V$ be a homomorphism with $h(x) = x$ if $x \in V$ and $h(\$) = \lambda$. $\alpha \underset{ID_r}{\overset{\frown}{\Longrightarrow}}{}^* w$, $\alpha \in A$ if and only if $\$^i \alpha \$^t \$^j \underset{ID'_r}{\overset{\frown}{\Longrightarrow}}{}^* w'$ with $h(w') = w$ can be shown by induction. While the only-if part follows easily, consider the following for the if part. We can assume that in any derivation of ID'_r the first i and the last j letters of the axiom are not deleted until the very end of derivation. Hence, insertion rules of the form $(z_1, \$^n, z_2)$ with $z_1 \in (\$ \cup V)^m$, $z_2 \in (\$ \cup V)^{m'}$ are applicable until the very end. It is clear that due to insertion rules of the form $(z_1, \$^n, z_2)$ and the deletion rule $(\lambda, \$, \lambda)$ it is possible to generate an arbitrary number of \$ at an arbitrary position of a sentential form of ID'_r. $\qquad \square$

The following result will be useful in subsequent proofs; compare to Lemma 1.

Lemma 2. *Let $\mathcal{L}$ be a family of languages that is closed under reversal. Then:*

1. $\mathcal{L} \subseteq INS_n^{m,m'} DEL_p^{q,q'} SUB^{r,r'}$ *iff* $\mathcal{L} \subseteq INS_n^{m',m} DEL_p^{q',q} SUB^{r',r}$.
2. $INS_n^{m,m'} DEL_p^{q,q'} SUB^{r,r'} \subseteq \mathcal{L}$ *iff* $INS_n^{m',m} DEL_p^{q',q} SUB^{r',r} \subseteq \mathcal{L}$.

Due to the definition of ins-del-sub systems, the following result is clear.

Lemma 3. $INS_n^{m,m'} DEL_p^{q,q'} \subseteq INS_n^{m,m'} DEL_p^{q,q'} SUB^{r,r'}$.

Whether this inclusion is proper, is the question, that will be addressed in the following sections. We will see that while in some cases an arbitrary system of size $(n, m, m'; p, q, q', r, r')$ can be simulated by a system of size $(n, m, m'; p, q, q')$, this is not the general case. Furthermore, we will see that families $INS_n^{m,m'} DEL_p^{q,q'}$, which are not computationally complete, may reach computational completeness via an extension with substitution rules. Additionally, we will see below that families of ins-del systems which are equally powerful may no longer be after being extended with the same class of substitution rules, i.e., we have $INS_{n_1}^{m_1,m'_1} DEL_{p_1}^{q_1,q'_1} = INS_{n_2}^{m_2,m'_2} DEL_{p_2}^{q_2,q'_2}$, but possibly $INS_{n_1}^{m_1,m'_1} DEL_{p_1}^{q_1,q'_1} SUB^{r,r'} \subset INS_{n_2}^{m_2,m'_2} DEL_{p_2}^{q_2,q'_2} SUB^{r,r'}$. The reverse case might occur, as well.

Because the application of an insertion rule (u, x, v) corresponds to the application of the monotone rewriting rule $uv \to uav$ and the application of a substitution rule $(u, a \to b, v)$ corresponds to the application of the monotone rewriting rule $uav \to ubv$, a monotone grammar can simulate derivations of an insertion-substitution system. (More technically speaking, we have to do the replacements on the level of pseudo-terminals N_a for each terminal a and also add rules $N_a \to a$, but these are minor details.) Hence, we can conclude:

Theorem 4. *For any integers $m, m', n, r, r' \geq 0$, $INS_n^{m,m'} DEL_0^{0,0} SUB^{r,r'} \subseteq CS$.*

3   Main Results

We will focus on context-free ins-del systems, which are extended with substitution rules. More precisely, we will analyze the computational power of the family of systems $INS_n^{0,0} DEL_p^{0,0} SUB^{r,r'}$.

We are going to analyze substitution rules of the form $(\lambda, a \to b, \lambda)$, which means that letters may be substituted regardless of any context. We will show that extending context-free ins-del systems with context-free substitution rules does not result in a more powerful system. In fact, a context-free ins-del-sub system of size $(n, 0, 0; p, 0, 0; 0, 0)$ can be simulated by an ins-del system of size $(n, 0, 0; p, 0, 0)$.

Theorem 5. *Let* $ID_r \in INS_n^{0,0} DEL_p^{0,0} SUB^{0,0}$, *then there exists an ins-del system* $ID \in INS_n^{0,0} DEL_p^{0,0}$ *such that* $L(ID_r) = L(ID)$.

Proof (Sketch). Let $ID_r = (V, T, A, I, D, S) \in INS_n^{0,0} DEL_p^{0,0} SUB^{0,0}$. It is clear that any letter a, that is to be replaced by a substitution rule $(\lambda, a \to b, \lambda)$, has been introduced by either an insertion rule $(\lambda, w_1 a w_2, \lambda)$ or as part of an axiom $w_1' a w_2'$ at some point before executing the substitution. As a serves no purpose other than to be replaced (i.e., it is not used as a context), the basic idea is to skip introducing a altogether and introduce b instead. More formally: instead of applying an insertion rule $(\lambda, w_1 a w_2, \lambda)$/an axiom $w_1' a w_2'$ and replacing a via $(\lambda, a \to b, \lambda)$ at a later point, we introduce a new insertion rule $(\lambda, w_1 b w_2, \lambda)$/a new axiom $w_1' b w_2'$, which we apply instead of $(\lambda, w_1 a w_2, \lambda)/w_1' a w_2'$. This idea can be cast into an algorithm to produce an ins-del system $ID = (V, T, A', I', D)$ with $L(ID_r) = L(ID)$. As only context-free insertion rules of size maximum $(n, 0, 0)$ are added to I', it is clear that $ID \in INS_n^{0,0} DEL_p^{0,0}$ holds. $\square$

Considering the question about the generative power of context-free ins-del systems with context-free substitution rules compared to usual context-free ins-del systems, Theorem 5 and Lemma 3 together yield:

Corollary 1. $INS_n^{0,0} DEL_p^{0,0} SUB^{0,0} = INS_n^{0,0} DEL_p^{0,0}$.

Example 1. Consider the ins-del-sub system

$$ID_r = (\{a, b, c\}, \{a, b, c\}, \{\lambda\}, \{(\lambda, aaa, \lambda)\}, \emptyset, S)$$

with $S = \{(\lambda, a \to b, \lambda), (\lambda, b \to c, \lambda)\}$. The language generated by ID_r is $L(ID_r) = \{w \mid w \in \{a, b, c\}^*, |w| = 3n, n \in \mathbb{N}\}$. Using the construction introduced in Theorem 5 yields the ins-del system $ID = (\{a, b, c\}, \{a, b, c\}, \{\lambda\}, I, \emptyset)$, with $I = \{(\lambda, x_1 x_2 x_3, \lambda) \mid x_1, x_2, x_3 \in \{a, b, c\}\}$. While it is clear that $L(ID) = L(ID_r)$, we remark that this example shows that the construction method of Theorem 5 may yield an ins-del system whose number of rules is exponentially greater than the number of rules of the system with substitutions.

3.1   Extension with One-Sided Substitution

Now, we will analyze the effect of one-sided substitution rules if used to extend a context-free ins-del system. We will show that using one-sided substitution rules can greatly increase the computational power of context-free insertion and deletion rules. In some cases, we even get computationally completeness results.

We will now construct an ins-del-sub system ID'_r of size $(1,0,0;1,0,0;1,0)$ which simulates ID_r of size $(1,1,0;1,1,0;1,0)$. The system ID'_r is constructed in the following manner:

Construction 2. *We assume the system* $ID_r = (V, T, A, I, D, S)$ *to be in normal form and any rule of* ID_r *to be labelled in a one-to-one manner, i.e., there is a bijection between a set of labels and the rule set. Let* $\$$ *be the padding symbol used in the construction of the normal form of* ID_r*. The system* $ID'_r = (V', T, A, I', D', S \cup S')$ *is constructed as follows. For each rule of* ID_r*, we introduce a new nonterminal* X_i *and define*

$$V' = V \cup \{X_i \mid i \text{ is the label of a rule of } ID_r\}.$$

The set I' *of insertion rules of* ID'_r *contains all* (λ, X_i, λ)*, where* i *is the label of an insertion rule of* ID_r*, while the set* D' *of deletion rules as contains* $(\lambda, \$, \lambda)$ *and all* (λ, X_i, λ)*, where* i *is the label of a deletion rule of* ID_r*. Furthermore, we define the set of substitution rules* $S' = S_1 \cup S_2$*, with*

$$S_1 = \{(u, X_i \to a, \lambda) \mid i \text{ is the label of an insertion rule } (u, a, \lambda) \text{ of } ID_r\} \text{ and}$$

$$S_2 = \{(u, a \to X_i, \lambda) \mid i \text{ is the label of a deletion rule } (u, a, \lambda) \text{ of } ID_r, u \neq \lambda\}.$$

Each deletion rule $(u, a, \lambda) \in D$ of ID_r, where i is the label of (u, a, λ), corresponds to a deletion rule $(\lambda, X_i, \lambda) \in D'$ and a substitution rule $(u, a \to X_i, \lambda) \in S_2$ of ID'_r. The basic idea of the construction is to simulate a deletion rule $(u, a, \lambda) \in D$ by substituting the letter a with left context u via $(u, a \to X_i, v) \in S_2$. The introduced nonterminal X_i is then deleted at some point by the deletion rule $(\lambda, X_i, \lambda) \in D'$. It is clear that a derivation of the form

$$w_1 u a w_2 \overset{\wedge}{\Longrightarrow} w_1 u X_i w_2 \overset{\wedge}{\Longrightarrow} w_1 u w_2,$$

in which the application of $(\lambda, X_i, \lambda) \in D'$ succeeds an application of $(u, a \to X_i, \lambda) \in S_2$ immediately, is equivalent to the application of a deletion rule $(u, a, v) \in D$. It needs much more care to prove the following converse:

Proposition 1. *Let* $\alpha \in A$*. Consider a derivation* $\alpha \overset{\wedge}{\Longrightarrow}^* w \in T^*$ *of* ID'_r *. Then, there is an alternative derivation of* ID'_r*, leading from* α *to* w*, in which all nonterminals* $X_i \in V' \backslash V$ *are resolved immediately after being introduced.*

This allows us to state:

Theorem 6. $INS_1^{0,0} DEL_1^{0,0} SUB^{1,0} = INS_1^{1,0} DEL_1^{1,0} SUB^{1,0}$.

Consider ins-del systems of size $(1,0,0;1,0,0)$ extended with one-sided substitution rules; the increase in computational power is quite significant:

$$INS_1^{0,0} DEL_1^{0,0} \subset INS_1^{1,0} DEL_1^{1,0} \subset INS_1^{1,0} DEL_1^{1,0} SUB^{1,0} = INS_1^{0,0} DEL_1^{0,0} SUB^{1,0}$$

Both inclusions are proper. First, observe that $ba^+ \in INS_1^{1,0} DEL_0^{0,0} \backslash INS_1^{0,0} DEL_1^{0,0}$. The system of size $(1,1,0;0,0,0;1,0)$ presented in Example 2 generates $(ba)^+$. Verlan [12, Theorem 5.3] has shown that even ins-del systems of size $(1,1,0;1,1,1)$ cannot generate the language $(ba)^+$.

Example 2. Consider the following ins-del-sub system: $ID_s = (V, T, I, \emptyset, S)$, with $V = \{X_1, X_2, X_3, a, b\}$, $T = \{a, b\}$ and $A = \{ba, bX_1X_3\}$. The set of insertion rules is defined as $I = \{(X_1, X_2, \lambda), (X_2, X_1, \lambda)\}$, while the set of substitution rules is $S = \{(b, X_1 \rightarrow a, \lambda), (a, X_2 \rightarrow b, \lambda), (b, X_3 \rightarrow a, \lambda)\}$. The generated language is $L(ID_s) = (ba)^+$, as we can easily see that any generated word begins with a letter b and ends with a letter a. Furthermore, any word generated by ID_s is not of the form w_1bbw_2, as the only way to introduce the terminal symbol b (except for the b introduced via the axiom) is by substituting a nonterminal X_2 with b. However, this substitution requires a left context a, which means that at some point the letter to the left of any b has been a. There are no insertion rules which can insert an additional b or X_2 between a and b. Furthermore, there are no deletion rules at all, which means that no a can be deleted. Therefore, the letter to the left of any b cannot be another b. Using the same argumentation, we can see, that any word generated by ID_s is not of the form w_1aaw_2, either.

It is easy to see that a result identical to Theorem 6 can be shown analogously for the mirrors of $\mathrm{INS}_1^{0,0}\mathrm{DEL}_1^{0,0}\mathrm{SUB}^{1,0}$ and $\mathrm{INS}_1^{1,0}\mathrm{DEL}_1^{1,0}\mathrm{SUB}^{1,0}$. Therefore:

Corollary 2. $INS_1^{0,0} DEL_1^{0,0} SUB^{0,1} = INS_1^{0,1} DEL_1^{0,1} SUB^{0,1}$.

We now analyze the computational power of ins-del-sub systems of size $(2, 0, 0; 2, 0, 0; 0, 1)$. While the family of ins-del systems of size $(2, 0, 0; 2, 0, 0)$ is known to be a proper subset of CF, see [11,12], we will show that an extension with substitution rules of the form $(\lambda, A \rightarrow B, C)$ results in a significant increase in computational power. More precisely, by simulating an ins-del systems of size $(2, 0, 1; 2, 0, 0)$, we will show that $\mathrm{INS}_2^{0,0}\mathrm{DEL}_2^{0,0}\mathrm{SUB}^{0,1} = \mathrm{RE}$ holds.

Construction 3. *Let* $ID = (V, T, A, I, D)$ *be a system of size* $(2, 0, 1; 2, 0, 0)$ *in normal form and all insertion rules of ID be labelled in a one-to-one manner. We construct the system* $ID_r = (V', T, A, I', D, S')$, *which simulates ID, as follows:*

$$V' = V \cup \{N_{i,2}, N_{i,1} \mid i \text{ is the label of an insertion rule } (\lambda, ab, c) \in I\}$$
$$I' = \{(\lambda, N_{i,2}N_{i,1}, \lambda) \mid i \text{ is the label of an insertion rule } (\lambda, ab, c) \in I\}$$
$$S' = \{(\lambda, N_{i,1} \rightarrow b, c), (\lambda, N_{i,2} \rightarrow a, b) \mid i \text{ is the label of a rule } (\lambda, ab, c) \in I\}$$

The basic idea of Construction 3 is essentially the same as in Construction 2: as context-free insertion rules cannot scan for contexts (by definition), this task is handled by the corresponding substitution rules. Consider an insertion rule $(\lambda, N_{i,2}N_{i,1}, \lambda)$ of ID_r where i is the label of an insertion rule $(\lambda, ab, c) \in I$. Then the substitution rules, corresponding to this rule, are $(\lambda, N_{i,1} \rightarrow b, c)$ and $(\lambda, N_{i,2} \rightarrow a, b)$. This idea leads us to:

Theorem 7. $INS_2^{0,1} DEL_2^{0,0} \subseteq INS_2^{0,0} DEL_2^{0,0} SUB^{0,1}$.

As $\mathrm{INS}_2^{0,1}\mathrm{DEL}_2^{0,0} = \mathrm{RE}$ holds according to [8, Theorem 5], we conclude:

Corollary 3. $INS_2^{0,0} DEL_2^{0,0} SUB^{0,1} = RE$.

This is an interesting result as the families of ins-del systems of size $(2, 0, 0; 2, 0, 0)$ and of size $(2, 0, 0; 0, 0, 0)$ are known to be equal [12, Theorem 4.7], yet both classes extended with the same class of substitution rules differ in computational power. As RE and CS are closed under reversal, the next corollary follows with Lemma 2 and Theorem 4.

Corollary 4. $INS_2^{0,0} DEL_0^{0,0} SUB^{1,0} \subseteq CS$ and $INS_2^{0,0} DEL_2^{0,0} SUB^{1,0} = RE$.

3.2   Extension with Two-Sided Substitution

After analyzing the effect of context-free and one-sided substitution rules on context-free ins-del systems, we will now proceed to two-sided substitution rules, i.e., substitution rules with left and right context. Somehow surprisingly, this lifts the computational power of even the 'weakest' ins-del systems, that is, systems of size $(1, 0, 0; 1, 0, 0)$, up to the level of RE. Let $ID \in INS_1^{1,1} DEL_1^{1,1}$. We will show that there is a system $ID_r \in INS_1^{0,0} DEL_1^{0,0} SUB^{1,1}$ capable of simulating ID. The basic idea is that the context checks, necessary for simulating rules with left and right context, are performed by the substitution rules. The system ID_r is constructed in the following manner:

Construction 4. *Let* $ID = (V, T, A, I, D) \in INS_1^{1,1} DEL_1^{1,1}$ *be in normal form according to Construction 1. For each rule of* ID, *we have a unique label, say,* i, *and we introduce a new nonterminal* X_i. *Define* $ID_r = (V', T, A, I', D', S)$ *with*

$$V' = V \cup \{X_i \mid i \text{ is the label of a rule in } I \text{ or } D\},$$
$$I' = \{(\lambda, X_i, \lambda) \mid i \text{ is the label of an insertion rule } (u, a, v)\},$$
$$D' = \{(\lambda, X_i, \lambda) \mid i \text{ labels a deletion rule } (u, a, v) \neq (\lambda, X, \lambda)\} \cup \{(\lambda, X, \lambda)\},$$

where X *is defined as in Construction 1. Finally,* $S = S_1 \cup S_2$ *with*

$$S_1 = \{(u, X_i \rightarrow a, v) \mid i \text{ is the label of an insertion rule } (u, a, v)\} \text{ and}$$
$$S_2 = \{(u, a \rightarrow X_i, v) \mid i \text{ is the label of a deletion rule } (u, a, v) \neq (\lambda, X, \lambda)\}.$$

The basic idea is similar to Construction 2. Each deletion rule $(u, a, v) \in D$ of ID, where i is the label of (u, a, v), corresponds to a deletion rule $(\lambda, X_i, \lambda) \in D'$ and a substitution rule $(u, a \rightarrow X_i, v) \in S_2$. We leave the context checks to the substitution rules. The same idea is applied to the insertion rules. With some technical effort, we can prove the following result.

Proposition 2. *Let* $\alpha \in A$. *Consider a derivation* $\alpha \Longrightarrow^* w \in T^*$. *Then, there is an alternative derivation, leading from* α *to* w, *in which all nonterminals* $X_i \in V' \backslash V$ *are resolved immediately after being introduced.*

This property is the key to show that for a system ID_r of size $(1, 0, 0; 1, 0, 0; 1, 1)$ constructed from a given ins-del system ID of size $(1, 1, 1; 1, 1, 1)$ in normal form according to Construction 4, we find $L(ID) = L(ID_r)$. As such ins-del systems are known to be computational complete, we conclude:

Corollary 5. $INS_1^{1,1}DEL_1^{1,1} = INS_1^{0,0}DEL_1^{0,0}SUB^{1,1} = RE.$

We now analyze the power of ins-del-sub systems of size $(1,0,0;0,0,0;1,1)$. By definition, it is clear that $INS_1^{0,0}DEL_0^{0,0}SUB^{1,1} \subseteq INS_1^{0,0}DEL_1^{0,0}SUB^{1,1}$ holds. In the following, we will show that this inclusion is proper. To be more precise, we will show that ins-del-sub systems of size $(1,0,0;0,0,0;1,1)$ characterize the context-sensitive languages. By Theorem 4, we are left to prove $CS \subseteq INS_1^{0,0}DEL_0^{0,0}SUB^{1,1}$.

For every context-sensitive language L, there is a linear bounded automaton (LBA) $LB = (Q,T,\Gamma,q_0,\delta,\square,F)$ accepting L. We are going to construct an ins-del-sub system of size $(1,0,0;0,0,0;1,1)$ to simulate LB. We first give a brief sketch of the basic idea behind this simulation in the following paragraph. The simulation evolves around strings of the form

$$(u_1,\$v_1)(u_2,v_2)\ldots(u_{i-1},v_{i-1})(u_i,q_jv_i)(u_{i+1},v_{i+1})\ldots(u_{n-1},v_{n-1})(u_n,v_n\#)$$

with $u_1,\ldots,u_n \in T$; $q_j \in Q$ and $v_1,\ldots,v_n \in \Gamma$. The concatenation of the first component of each tuple, that is, $u_1\ldots u_n$, is the input word of the linear bounded automaton LB, while the concatenation of the second component of each tuple, that is, $\$v_1v_2\ldots v_{i-1}q_jv_iv_{i+1}\ldots v_{n-1}v_n\#$, represents a configuration of LB running on the input word $u_1\ldots u_n$. The simulation of LB runs entirely on the second components of the tuples. If $\$v_1v_2\ldots v_{i-1}q_jv_iv_{i+1}\ldots v_{n-1}v_n\#$ is an accepting configuration, i.e., $q_i \in F$, we substitute all tuples with their respective first component. For instance (u_k,v_k) is substituted with u_k. In short, this means that if (the simulation of) LB running on $u_1\ldots u_n$ halts in an accepting configuration, we generate the word $u_1\ldots u_n$. More details follow:

Construction 5. *Consider an arbitrary LBA $LB = (Q,T,\Gamma,q_0,\delta,\square,F)$ accepting $L \subseteq T^*$. Let $\$$ be the left and $\#$ be the right endmarker of LB. We define $\underline{L} := \{\lambda,\$\}$ and $\underline{R} := \{\lambda,\#\}$. Then, the ins-del-sub system $ID_r = (V \cup T, T, A, I, \emptyset, S)$ with $V = V_1 \cup V_2 \cup V_3$, where*

$$V_1 = \{X_0\} \cup \{X_a \mid a \in T\}$$
$$V_2 = \{(a,q_ib),(a,\$q_ib),(a,q_i\$b),(a,q_ib\#),(a,bq_i\#) \mid a \in T, b \in \Gamma, q_i \in Q\}$$
$$\qquad \cup \{(a,b),(a,\$b),(a,b\#) \mid a \in T, b \in \Gamma\}$$
$$V_3 = \{X_{(a,b\underline{r});q_i;L} \mid a \in T, b \in \Gamma, q_i \in Q, \underline{r} \in \underline{R}\}$$
$$\qquad \cup \{X_{(a,\underline{l}b);q_i;R} \mid a \in T, b \in \Gamma, q_i \in Q, \underline{l} \in \underline{L}\}$$
$$\qquad \cup \{X_{(a,q_ib\underline{r}),r}, X_{(a,\underline{l}q_ib),l} \mid a \in T, b \in \Gamma, q_i \in Q, \underline{r} \in \underline{R}, \underline{l} \in \underline{L}\}$$

generates the language L by simulating LB. Strings of the form

$$(u_1,\$v_1)(u_2,v_2)\ldots(u_{i-1},v_{i-1})(u_i,q_jv_i)(u_{i+1},v_{i+1})\ldots(u_{n-1},v_{n-1})(u_n,v_n\#)$$

consist of symbols in V_2, while the symbols in V_1 are auxiliary symbols used to generate such strings. The symbols in V_3 are used to simulate LB's transitions. We define $A = \{X_0(a,a\#) \mid a \in T\} \cup \{a \mid a \in L \cap T\}$ and $I = \{(\lambda,X_a,\lambda) \mid a \in$

T}. If $\lambda \in L$, we add λ to the axiom. The set of substitution rules is defined as $S = S_{init} \cup S_N \cup S_R \cup S_L \cup S_{endmarker,L} \cup S_{endmarker,R} \cup S_{final}$. In

$$S_{init} = \{(X_0, X_a \to (a, a), \lambda) \mid a \in T\} \cup \{(\lambda, X_0 \to (a, \$q_0 a), \lambda) \mid a \in T\},$$

we collect substitution rules used to initialize the simulation.

The substitution rules in the set S_N are used to simulate the application of a transition $\delta(q_i, b) \ni (q_j, c, N)$. S_N consists of substitution rules of the form

$$(\lambda, (a, q_i b) \to (a, q_j c), \lambda), \ (\lambda, (a, \$q_i b) \to (a, \$q_j c), \lambda), \ (\lambda, (a, q_i b \#) \to (a, q_j c \#), \lambda),$$

with $a \in T$, $q_i, q_j \in Q$, $b, c \in \Gamma$ and $\delta(q_i, b) \ni (q_j, c, N)$.

The substitution rules in the set S_L are used to simulate left moves. For each transition $\delta(q_i, b) \ni (q_j, c, L)$ of LB, we add substitution rules

$$(\lambda, (a, q_i b\underline{r}) \to X_{(a, c\underline{r}); q_j; L}, \lambda), \ (\lambda, (d, \underline{l}e) \to X_{(d, \underline{l}q_j e), l}, X_{(a, c\underline{r}); q_j; L}),$$

$$(X_{(d, \underline{l}q_j e), l}, X_{(a, c\underline{r}); q_j; L} \to (a, c\underline{r}), \lambda), \ (\lambda, X_{(d, \underline{l}q_j e), l} \to (d, \underline{l}q_j e), (a, c\underline{r}))$$

to S_L with $a, d \in T$, $b, c, e \in \Gamma$, $q_i, q_j \in Q$, $\underline{l} \in \underline{L}$, $\underline{r} \in \underline{R}$. Similarly, the substitution rules in S_R can simulate right moves $\delta(q_i, b) \ni (q_j, c, R)$ with:

$$(\lambda, (a, \underline{l}q_i b) \to X_{(a, \underline{l}c); q_j; R}, \lambda), \ (X_{(a, \underline{l}c); q_j; R}, (d, e\underline{r}) \to X_{(d, q_j e\underline{r}), r}, \lambda)$$

$$(\lambda, X_{(a, \underline{l}c); q_j; R} \to (a, \underline{l}c), X_{(d, q_j e\underline{r}), r}), \ ((a, \underline{l}c), X_{(d, q_j e\underline{r}), r} \to (d, q_j e\underline{r}), \lambda).$$

The set $S_{endmarker,L}$ consists of substitution rules of the form

$$(\lambda, (a, \$q_i b) \to (a, q_j \$c), \lambda), \ (\lambda, (a, q_i \$b) \to (a, \$q_j b), \lambda)$$

with $a \in T$, $b, c \in \Gamma$, $q_i, q_j \in Q$, $\delta(q_i, b) \ni (q_j, c, L)$ and $\delta(q_i, \$) \ni (q_j, \$, R)$.
The set $S_{endmarker,R}$ consists of substitution rules of the form

$$(\lambda, (a, q_i b \#) \to (a, cq_j \#), \lambda), \ (\lambda, (a, bq_i \#) \to (a, q_j b \#), \lambda)$$

with $a \in T$, $b, c \in \Gamma$, $q_i, q_j \in Q$, $\delta(q_i, b) \ni (q_j, c, R)$ and $\delta(q_i, \#) \ni (q_j, \#, L)$. Both sets are used for the simulation of $\delta(q_i, b) \ni (q_j, c, L)$ and $\delta(q_i, b) \ni (q_j, c, R)$ as well, in the case the read/write head moves to/from an endmarker. The set $S_{final} = S_{f_1} \cup S_{f_2}$ is used to generate a word $w \in T^*$ if w has been accepted by the simulated linear bounded automaton LB. S_{f_1} consists of the substitutions

$$(\lambda, (a, q_f b) \to a, \lambda), (\lambda, (a, \$q_f b) \to a, \lambda), (\lambda, (a, q_f \$b) \to a, \lambda),$$
$$(\lambda, (a, q_f b \#) \to a, \lambda), (\lambda, (a, bq_f \#) \to a, \lambda)$$

and S_{f_2} consists of

$$(\lambda, (a, b) \to a, c), (c, (a, b) \to a, \lambda), (\lambda, (a, \$b) \to a, c), (c, (a, b \#) \to a, \lambda)$$

with $a, c \in T$, $b \in \Gamma$, $q_f \in F$.

Working out the correctness of this construction, we can show:

Theorem 8. $INS_1^{0,0}DEL_0^{0,0}SUB^{1,1} = CS.$

As a consequence of Theorem 8, we can formulate the following Penttonen-style normal form theorem for context-sensitive languages. We believe that this could be useful in particular when dealing with variations of insertion systems.

Theorem 9. *For every context-sensitive language L, $\lambda \notin L$, there is a context-sensitive grammar $G = (N, T, P, S)$, such that $L = L(G)$, with rules of the forms*

$$A \to AB$$
$$AB \to AC, AB \to CB$$
$$A \to a, \ \text{with } a \in T, \ A, B, C \in N.$$

Allowing erasing productions on top, we also arrive at a new characterization of the family of recursively enumerable languages. By different methods, we could even prove that either of the two non-context-free forms suffices to achieve RE.

4 Summary and Main Open Questions

We have shown that the addition of substitution rules to ins-del systems yields new characterizations of RE and CS. In particular we have shown the following equalities: $INS_2^{0,0}DEL_2^{0,0}SUB^{1,0} = RE$, $INS_1^{0,0}DEL_1^{0,0}SUB^{1,1} = RE$ and $INS_1^{0,0}DEL_1^{0,0}SUB^{1,1} = CS$. Additionally we have shown $INS_1^{1,0}DEL_1^{1,0}SUB^{1,0} = INS_1^{0,0}DEL_1^{0,0}SUB^{1,0}$. While in the above cases an extension with (non-context-free) substitution rules leads to an increase in computational power, we have also shown that the addition of context-free substitution rules to context-free ins-del systems does not affect the computational power.

The main open question is if $INS_1^{0,0}DEL_1^{0,0}SUB^{1,0}$ is computationally complete. We conjecture this not to be the case, as with only left context rules, information can only be propagated in one direction. Yet, should $INS_1^{0,0}DEL_1^{0,0}SUB^{1,0}$ equal RE, this would provide an interesting new normal form. A minor open question is the strictness of the inclusion $INS_2^{0,0}DEL_0^{0,0}SUB^{0,1} \subseteq CS$.

References

1. Alhazov, A., Krassovitskiy, A., Rogozhin, Y., Verlan, S.: Small size insertion and deletion systems. In: Martin-Vide, C. (ed.) Applications of Language Methods, pp. 459–515. Imperial College Press (2010)
2. Beaver, D.: Computing with DNA. J. Comput. Biol. **2**(1), 1–7 (1995)
3. Kari, L.: On insertions and deletions in formal languages. Ph.D. thesis, University of Turku, Finland (1991)
4. Karl, L.: DNA computing: arrival of biological mathematics. Math. Intell. **19**(2), 9–22 (1997). https://doi.org/10.1007/BF03024425
5. Krassovitskiy, A., Rogozhin, Y., Verlan, S.: Further results on insertion-deletion systems with one-sided contexts. In: Martín-Vide, C., Otto, F., Fernau, H. (eds.) LATA 2008. LNCS, vol. 5196, pp. 333–344. Springer, Heidelberg (2008). https://doi.org/10.1007/978-3-540-88282-4_31

6. Krassovitskiy, A., Rogozhin, Y., Verlan, S.: Computational power of insertion-deletion (P) systems with rules of size two. Nat. Comput. **10**, 835–852 (2011). https://doi.org/10.1007/s11047-010-9208-y

7. Margenstern, M., Păun, G., Rogozhin, Y., Verlan, S.: Context-free insertion-deletion systems. Theor. Comput. Sci. **330**(2), 339–348 (2005)

8. Matveevici, A., Rogozhin, Y., Verlan, S.: Insertion-deletion systems with one-sided contexts. In: Durand-Lose, J., Margenstern, M. (eds.) MCU 2007. LNCS, vol. 4664, pp. 205–217. Springer, Heidelberg (2007). https://doi.org/10.1007/978-3-540-74593-8_18

9. Păun, G., Rozenberg, G., Salomaa, A.: DNA Computing: New Computing Paradigms. Springer, Heidelberg (1998). https://doi.org/10.1007/978-3-662-03563-4

10. Takahara, A., Yokomori, T.: On the computational power of insertion-deletion systems. Nat. Comput. **2**(4), 321–336 (2003). https://doi.org/10.1023/B:NACO.0000006769.27984.23

11. Verlan, S.: On minimal context-free insertion-deletion systems. J. Autom. Lang. Comb. **12**(1–2), 317–328 (2007)

12. Verlan, S.: Recent developments on insertion-deletion systems. Comput. Sci. J. Moldova **18**(2), 210–245 (2010)

13. Vu, M.: On insertion-deletion systems with substitution rules. Master's thesis, Informatikwissenschaften, Universität Trier, Germany (2019)

Correction to: #P-completeness of Counting Update Digraphs, Cacti, and Series-Parallel Decomposition Method

Kévin Perrot, Sylvain Sené, and Lucas Venturini

Correction to:
Chapter "#P-completeness of Counting Update Digraphs,
Cacti, and Series-Parallel Decomposition Method"
in: M. Anselmo et al. (Eds.): *Beyond the Horizon*
***of Computability*, LNCS 12098,**
https://doi.org/10.1007/978-3-030-51466-2_30

The author originally listed as Camille Noûs on this conference paper [1] is fictitious (http://www.cogitamus.fr/indexen.html) and as such does not fulfill the requirements for authorship. The correct authorship list is: Kévin Perrot, Sylvain Sené and Lucas Venturini. This has been corrected.

[1] Perrot, K., Sené, S., Venturini, L.: #P-completeness of Counting Update Digraphs, Cacti, and Series-Parallel Decomposition Method. In: Anselmo, M., Della Vedova, G., Manea F., Pauly, A. (eds.) CiE 2020. LNCS, vol. 12098, pp. 326-338. Springer, Cham (2020). https://doi.org/10.1007/978-3-030-51466-2_30

The updated version of this chapter can be found at
https://doi.org/10.1007/978-3-030-51466-2_30

© Springer Nature Switzerland AG 2021
M. Anselmo et al. (Eds.): CiE 2020, LNCS 12098, p. C1, 2021.
https://doi.org/10.1007/978-3-030-51466-2_34

Author Index

Printed in the United States
by Baker & Taylor Publisher Services